Mathematical Methods
in Physics
and Engineering

Mathematical Methods in Physics and Engineering

John W. Dettman

PROFESSOR OF MATHEMATICS
OAKLAND UNIVERSITY

Dover Publications, Inc.

New York

Published in Canada by General Publishing Company, Ltd., 30 Lesmill Road, Don Mills, Toronto, Ontario.
Published in the United Kingdom by Constable and Company, Ltd.

This Dover edition, first published in 1988, is an unabridged, corrected republication of the second edition, 1969, of the work first published by McGraw-Hill Book Company, New York in 1962, as part of the International Series in Pure and Applied Mathematics. All the chapter-end bibliographies have been brought up to date.

Manufactured in the United States of America
Dover Publications, Inc.
31 East 2nd Street
Mineola, N.Y. 11501

Library of Congress Cataloging-in-Publication Data

Dettman, John W. (John Warren)
    Mathematical methods in physics and engineering / John W. Dettman.
        p.   cm.
    Originally published: New York : McGraw-Hill, 1969.
    Includes bibliographical references and index.
    ISBN 0-486-65649-7 (pbk.)
    1. Mathematics—1961- 2. Engineering—Mathematics. I. Title.
QA37.2.D47   1988
510—dc19   87-33229
                                                            CIP

Preface

In the seven years since the first edition was published the trend toward introducing mathematical topics earlier in the undergraduate curriculum has continued. For example, the Committee on the Undergraduate Program in Mathematics (CUPM) of the Mathematical Association of America has recommended that all students of mathematics and the physical sciences be given a course in linear algebra as early as the sophomore year. This probably obviates the devotion of a chapter of this book to linear algebra. However, since it may be several years before it is common for advanced undergraduate students of engineering and physics to be sufficiently well prepared in linear algebra, I have decided to retain the linear algebra in the first chapter. For those students who have an adequate background, this chapter can serve as a review and also can serve to define the algebraic prerequisites for the rest of the book. The sections on vibration problems and linear programming illustrate some of the applications of the linear algebra in physics and engineering. On the other hand, with students better prepared in algebra it was felt that a more abstract treatment of applied mathematics is now appropriate. Consequently, I have developed a chapter on Hilbert spaces which carries the subject through the spectral theory for self-adjoint completely continuous operators. This makes it possible to tighten up the later treatment of boundary-value problems and integral equations.

There are three other significant changes from the first edition. A section on generalized functions has been added to Chapter 5 to help explain the use of the Dirac delta function in connection with Green's functions. Another important change is that the section on series solutions of ordinary differential equations has been completely rewritten so that it is now independent of complex variable theory. This means that the first six chapters can be studied by a student without prior knowledge of complex variables. However, since Chapter 8 depends heavily on analytic functions of a complex

vii

variable, a new Chapter 7 on analytic function theory has been written.   This has been done so that the prerequisites could be held to a minimum.   I now feel that the entire book can be comprehended by an undergraduate student with a good course in calculus which includes infinite series and uniform convergence.

I have specifically not tried to make the various chapters self-contained, realizing full well that this might limit the book somewhat in its use as a reference.   It seems to me that the interrelations between the various parts of the theory should be emphasized, so that the reader does not get the impression that applied mathematics is a collection of disjointed topics.   There are many recurring themes which thread their way through the whole development.   One of these is the vector space concept, which gives a framework for linear problems in general.   Another is the eigenvalue problem, which is first met in an algebraic context but later reappears in the study of ordinary and partial differential equations and integral equations.   The calculus of variations presents a unified method of handling many problems.   The same can be said of transform techniques.   The method of constructing Green's functions is applicable to many types of boundary-value problems, and it allows one to formulate problems in terms of integral equations, thus opening up new possibilities for solution.

The exercises which follow every section are generally of three types: (1) those which test the understanding of the material just covered, (2) those which extend the results to new situations, and (3) those which lay the groundwork for new concepts to be introduced later.   Those which fall into the third category will be marked with an asterisk, not necessarily because they are more difficult, but as a guide to the instructor.   These exercises should be assigned for the maximum continuity.

One can never adequately acknowledge his indebtedness for material for a book as broad as this.   It would be impossible to develop an adequate bibliography.   However, for the sake of the reader, a short list of references is included at the end of each chapter.

I am deeply grateful for the comments and constructive criticisms of many readers of the first edition.   I can only hope that they find the second edition an up-to-date improvement.

John W. Dettman

Contents

Chapter 1. Linear Algebra

1.1 Linear Equations. Summation Convention

The problem of solving systems of linear algebraic equations frequently arises in analysis and in its applications to physical problems. Such a system can be written as

$$a_{11}x_1 + a_{12}x_2 + \cdots + a_{1n}x_n = c_1$$
$$a_{21}x_1 + a_{22}x_2 + \cdots + a_{2n}x_n = c_2$$
$$\cdots\cdots\cdots\cdots\cdots\cdots\cdots\cdots\cdots\cdots$$
$$a_{m1}x_1 + a_{m2}x_2 + \cdots + a_{mn}x_n = c_m$$

The problem is: Given constants $a_{11}, a_{12}, \ldots, a_{mn}$ and $c_1, c_2, \ldots, c_m$, under what conditions will there exist a set $x_1, x_2, \ldots, x_n$ which simultaneously satisfies the system of equations, and how may this set of unknowns be determined? Before we turn to the solution of this problem, let us first consider some alternative ways of writing the system of linear equations.

By use of the Σ notation for summation, the system can be written as follows:

$$\sum_{j=1}^{n} a_{1j}x_j = c_1$$
$$\sum_{j=1}^{n} a_{2j}x_j = c_2$$
$$\cdots\cdots\cdots\cdots\cdots$$
$$\sum_{j=1}^{n} a_{mj}x_j = c_m$$

Realizing that the first subscript on the a's and the subscript on the c's agree in each equation and that in the system this subscript takes on the values 1 to m, successively, we can write

$$\sum_{j=1}^{n} a_{ij}x_j = c_i \qquad i = 1, 2, 3, \ldots, m$$

1

Furthermore, we note that the Σ is really not necessary, since in each equation the summation takes place over the repeated subscript j; that is, in the left-hand side of each equation j takes on the values 1 through n, and the terms are then summed.   For this reason we now drop the Σ and adopt the following convention, known as the **summation convention.** *If in a given term in an expression a Latin[1] subscript is repeated, a summation over that subscript is implied.   The term is the sum of all possible such terms with the repeated subscript taking on the successive values 1 through n.* The value of n must be stated, unless it is understood ahead of time and no ambiguity can arise.   By use of the summation convention, the system of linear equations can be written simply as

$$a_{ij}x_j = c_i \qquad \begin{matrix} i = 1, 2, \ldots, m \\ j = 1, 2, \ldots, n \end{matrix}$$

The summation convention can be a very useful tool and can produce a great saving in writing.   Skillful use of it will shorten many proofs and clarify one's thinking involving lengthy expressions, where one might get bogged down in details if the expressions are written out in full.   However, until thoroughly familiar with its use, one must be careful to learn exactly what operations involving the summation convention are legitimate.   The following discussion and examples may help to clarify some of these points.

The repeated subscript is often called a **dummy subscript,** because the letter designating it can be changed without changing the expression, provided it is understood that the new letter takes on the same values in the summation as the old.   For example, the system of linear equations can just as well be written as

$$a_{ik}x_k = c_i \qquad \begin{matrix} i = 1, 2, \ldots, m \\ k = 1, 2, \ldots, n \end{matrix}$$

that is, the dummy j has been changed to k without affecting the system of equations, since the same summation takes place because of the repeated k. Or as another example,

$$a_i b_i = a_j b_j = a_1 b_1 + a_2 b_2 + \cdots + a_n b_n$$

A subscript which is not repeated in the same term of an equation is called a **free subscript.**   For example, in $a_{ij}x_j = c_i$, i is a free subscript.   The range 1 to m of i indicates the number of different equations in the system, while the range 1 to n of j indicates the number of terms in the summation which occurs in each equation.   When i takes on the value 10, it refers to the tenth equation

---

[1] Here we are anticipating certain cases where we may need a repeated subscript without summation.   In that case, we shall use Greek subscripts, having made the summation convention apply specifically to Latin subscripts.

in the system. In the equation

$$a_i + b_i = c_i$$

i is a free subscript. Although it is repeated on the left-hand side of the equation, it is not repeated in a single term. However, in an expression like

$$c_i(a_i + b_i) = c_1(a_1 + b_1) + c_2(a_2 + b_2) + \cdots + c_n(a_n + b_n)$$

it is a dummy, and a summation is implied.

Unless it is otherwise stated, "numbers" like the a's, c's, and x's which appear in expressions like $a_{ij}x_j = c_i$ come from a **field** where the following rules of algebra hold:[1]

1. **Associative law of addition:** $(a + b) + c = a + (b + c)$
2. **Commutative law of addition:** $a + b = b + a$
3. **Associative law of multiplication:** $(ab)c = a(bc)$
4. **Commutative law of multiplication:** $ab = ba$
5. **Distributive law:** $a(b + c) = ab + ac$

Assuming "numbers" from a field, it is not difficult to prove the following statements involving the summation convention:

$$a_i b_i = b_i a_i$$
$$a_i(b_i + c_i) = a_i b_i + a_i c_i$$
$$(a_i b_i)(c_j d_j) = (a_i c_j)(b_i d_j)$$

The proof of the third in the case where i and j take on the value $1, 2, 3$ would proceed as follows:

$$\begin{aligned}
(a_i b_i)(c_j d_j) &= (a_1 b_1 + a_2 b_2 + a_3 b_3)(c_1 d_1 + c_2 d_2 + c_3 d_3) \\
&= a_1 b_1 c_1 d_1 + a_1 b_1 c_2 d_2 + a_1 b_1 c_3 d_3 \\
&\quad + a_2 b_2 c_1 d_1 + a_2 b_2 c_2 d_2 + a_2 b_2 c_3 d_3 \\
&\quad + a_3 b_3 c_1 d_1 + a_3 b_3 c_2 d_2 + a_3 b_3 c_3 d_3 \\
&= a_1 c_1 b_1 d_1 + a_1 c_2 b_1 d_2 + a_1 c_3 b_1 d_3 \\
&\quad + a_2 c_1 b_2 d_1 + a_2 c_2 b_2 d_2 + a_2 c_3 b_2 d_3 \\
&\quad + a_3 c_1 b_3 d_1 + a_3 c_2 b_3 d_2 + a_3 c_3 b_3 d_3 \\
&= (a_i c_j)(b_i d_j)
\end{aligned}$$

In this expression there is a summation over i and j, so that both are dummies. A subscript should never be repeated more than once in the same term, or

---

[1] For the other postulates of a field consult a book on algebra such as G. Birkhoff and S. MacLane, "A Survey of Modern Algebra," 3d ed., The Macmillan Company, New York, 1965. Some examples of a field are the real numbers and the complex numbers. The field of complex numbers includes the field of the real numbers and is called an extension of the real-number field. In this book we shall be dealing exclusively with the real or complex numbers.

confusion may arise. For example, one might be tempted to write

$$(a_i b_i)(c_j d_j) = (a_i b_i)(c_i d_i)$$
$$= (a_i c_i)(b_i d_i)$$
$$= (a_i c_i)(b_j d_j)$$

This, of course, is incorrect and can be avoided by careful observation of the rule that a *dummy may be repeated only once in a single term*. Upon substitution from one expression into another it may be necessary to change the "name" of a dummy in order to adhere to this rule. For example, if $y_i = a_{ij} x_j$ and $z_i = b_{ij} x_j$, then $y_i z_i = a_{ij} x_j b_{ik} x_k = a_{ij} b_{ik} x_j x_k$. Here, the dummy j is changed to k in one expression in order to avoid the second repetition of j in the final expression. We cannot express something like $a_1 b_1 c_1 + a_2 b_2 c_2 + a_3 b_3 c_3$ using the summation convention, but this does not turn out to be a serious penalty to pay.

The identities illustrated above are true for finite sums but would not necessarily hold for infinite sums. For this reason, we shall not apply the summation convention to infinite sums but shall go back to the notation $\sum\limits_{i=1}^{\infty}$ when we wish to write an infinite sum. *Otherwise the summation convention will be implied, unless the contrary is stated.*

The summation convention can also be used to some advantage in expressions involving partial derivatives. For example, if $w(x_1, x_2, x_3, \ldots, x_n)$ is a function of $x_1, x_2, \ldots, x_n$, which are in turn functions of a parameter t, then

$$\frac{dw}{dt} = \frac{\partial w}{\partial x_i} \frac{dx_i}{dt}$$

Here the summation in the usual "chain rule" is implied by the repetition of the subscript i. As another example, consider a vector $\mathbf{V}$ with components v_1, v_2, v_3 which are functions of three space variables x_1, x_2, x_3. Then the divergence of $\mathbf{V}$, written $\nabla \cdot \mathbf{V}$, is

$$\nabla \cdot \mathbf{V} = \frac{\partial v_1}{\partial x_1} + \frac{\partial v_2}{\partial x_2} + \frac{\partial v_3}{\partial x_3} = \frac{\partial v_i}{\partial x_i}$$

Exercises 1.1[1]

1. Write out and hence verify the identities $a_i b_i = b_i a_i$ and

$$a_i (b_i + c_i) = a_i b_i + a_i c_i$$

*2. If $y_i = b_{ij} x_j$ and $z_i = a_{ij} y_j$, show that $z_i = c_{ij} x_j$ where $c_{ij} = a_{ik} b_{kj}$.
*3. If $a_{ij} = a_{ji}$ and $b_{ij} = -b_{ji}$, show that $a_{ij} b_{ij} = 0$.

---

[1] The starred exercises contain results which will be used in later sections of the book.

*4. We define the **Kronecker delta** δ_{ij} as follows: $\delta_{ij} = 1$ if $i = j$; and $\delta_{ij} = 0$ if $i \neq j$. Show that $a_{ij}\delta_{jk} = a_{ik} = \delta_{ij}a_{jk}$, where i, j, and k take on the values 1, 2, ... , n.

5. If $x_1, x_2, \ldots, x_n$ are a set of independent variables which are in turn functions of a set of independent variables $y_1, y_2, \ldots, y_n$, that is, $x_i = x_i(y_1, y_2, \ldots, y_n)$, and the jacobian of the transformation is not equal to zero, show that

$$\delta_{ij} = \frac{\partial x_i}{\partial x_j} = \frac{\partial x_i}{\partial y_k}\frac{\partial y_k}{\partial x_j}$$

*6. We define e_{ijk} as follows: $e_{ijk} = 0$ if $i = j$, $j = k$, or $i = k$; $e_{ijk} = 1$ if i, j, k are an even permutation of the numbers 1, 2, 3; and $e_{ijk} = -1$ if i, j, k are an odd permutation of 1, 2, 3. Show that $e_{ijk}a_{1i}a_{2j}a_{3k} =$

$$\begin{vmatrix} a_{11} & a_{12} & a_{13} \\ a_{21} & a_{22} & a_{23} \\ a_{31} & a_{32} & a_{33} \end{vmatrix}$$

7. If **U** and **V** are vectors with components u_1, u_2, u_3 and v_1, v_2, v_3, and

$$\mathbf{U} \cdot \mathbf{V} = u_i v_i$$

and the components of **U** x **V** are $e_{ijk}u_j v_k$, show the following, using subscript notation and summation convention:

$$\mathbf{U} \cdot \mathbf{V} = \mathbf{V} \cdot \mathbf{U}$$

$$\mathbf{U} \times \mathbf{V} = -\mathbf{V} \times \mathbf{U}$$

$$\mathbf{U} \cdot (\mathbf{U} \times \mathbf{V}) = 0$$

$$\mathbf{V} \cdot (\mathbf{U} \times \mathbf{V}) = 0$$

8. If **U** is a vector function of the three space variables x_1, x_2, x_3, show that the components of the curl of **U**, written $\nabla \times \mathbf{U}$, can be expressed as $e_{ijk}(\partial u_k / \partial x_j)$, where u_1, u_2, u_3 are the components of **U**.

9. Using subscript notation, prove the following vector identities:

$$\mathbf{U} \cdot (\mathbf{V} \times \mathbf{W}) = \begin{vmatrix} u_1 & u_2 & u_3 \\ v_1 & v_2 & v_3 \\ w_1 & w_2 & w_3 \end{vmatrix}$$

$$\mathbf{U} \cdot (\mathbf{V} \times \mathbf{W}) = -\mathbf{V} \cdot (\mathbf{U} \times \mathbf{W}) = -\mathbf{W} \cdot (\mathbf{V} \times \mathbf{U})$$

$$\nabla \cdot (\nabla \times \mathbf{U}) = 0$$

1.2   Matrices

Another way of writing a system of linear algebraic equations is in terms of matrices. A **matrix** is an $m \times n$ array of **elements**[1] from a field arranged

---

[1] For our purposes it will be sufficient to assume that the elements are from the field of complex numbers, which of course includes real numbers.

in rows and columns as follows:

$$A = \begin{Vmatrix} a_{11} & a_{12} & \cdots & a_{1n} \\ a_{21} & a_{22} & \cdots & a_{2n} \\ \cdots\cdots\cdots\cdots\cdots\cdots \\ a_{m1} & a_{m2} & \cdots & a_{mn} \end{Vmatrix}$$

and which obeys certain laws of equality, addition, subtraction, and multiplication, which we shall state presently. If we want to specify the elements of A, we write a_{ij}, $i = 1, 2, \ldots, m$, $j = 1, 2, \ldots, n$. The first subscript designates the row and the second the column from which the element is taken. If $m = n$, then we have a **square matrix of order n.**

Two matrices are **equal** if and only if their corresponding elements are equal. We may write this as follows:

$$A = B \text{ if and only if } a_{ij} = b_{ij} \qquad \begin{aligned} i &= 1, 2, \ldots, m \\ j &= 1, 2, \ldots, n \end{aligned}$$

The **sum** of two matrices is the matrix formed by adding corresponding elements, or

$$A + B = C \text{ if and only if } a_{ij} + b_{ij} = c_{ij} \qquad \begin{aligned} i &= 1, 2, \ldots, m \\ j &= 1, 2, \ldots, n \end{aligned}$$

The **difference** of two matrices is the matrix formed by subtracting corresponding elements, or

$$A - B = C \text{ if and only if } a_{ij} - b_{ij} = c_{ij} \qquad \begin{aligned} i &= 1, 2, \ldots, m \\ j &= 1, 2, \ldots. n \end{aligned}$$

The **product** of an $m \times n$ and an $n \times p$ matrix is an $m \times p$ matrix, formed as follows:

$$AB = C \text{ if and only if } c_{ij} = a_{ik}b_{kj} \qquad \begin{aligned} i &= 1, 2, \ldots, m \\ j &= 1, 2, \ldots, p \\ k &= 1, 2, \ldots, n \end{aligned}$$

Clearly we can add and subtract only matrices with the same number of rows and columns, and we can multiply matrices only where the matrix on the left has the same number of columns as the matrix on the right has rows.

If we define an $m \times n$ matrix A, with elements a_{ij}, an $n \times 1$ matrix X, with elements $x_1, x_2, \ldots, x_n$, and an $m \times 1$ matrix C, with elements $c_1, c_2, \ldots, c_m$, and use the definitions for matrix multiplication and equality, we can write the system of equations $a_{ij}x_j = c_i$ as

$$AX = C$$

In terms of linear algebraic equations, we see that the definitions of addition, subtraction, and multiplication do make sense. For example, if $a_{ij}x_j = c_i$ and

$b_{ij}x_j = d_i$, then $c_i + d_i = a_{ij}x_j + b_{ij}x_j = (a_{ij} + b_{ij})x_j$ and

$$c_i - d_i = a_{ij}x_j - b_{ij}x_j = (a_{ij} - b_{ij})x_j$$

In matrix notation these statements can be written

$AX = C$, $BX = D$ implies $(A + B)X = C + D$ and $(A - B)X = C - D$

Also, if $y_i = a_{ij}x_j$ and $z_i = b_{ij}y_j$, then $z_i = b_{ik}a_{kj}x_j = c_{ij}x_j$, where $c_{ij} = b_{ik}a_{kj}$.†
In matrix notation this can be written

$Y = AX$, $Z = BY$ implies $Z = B(AX) = (BA)X = CX$, where $BA = C$

By working with elements and their algebraic properties, one can easily show that the matrix operations satisfy the following laws:
1. **Associative law of addition:** $(A + B) + C = A + (B + C)$
2. **Commutative law of addition:** $A + B = B + A$
3. **Associative law of multiplication:** $(AB)C = A(BC)$
4. **Distributive laws:** $A(B + C) = AB + AC$; $(B + C)A = BA + CA$

Multiplication is not, in general, commutative, since $a_{ij}b_{jk} \neq b_{ij}a_{jk}$, even if the multiplication makes sense both ways. This is the reason why it was necessary to state two distributive laws; that is, $A(B + C) \neq (B + C)A$.

A **zero matrix** O is a matrix every one of whose elements is 0. It has the following obvious properties:

$$A - A = O$$
$$A + O = A$$
$$OA = AO = O$$

which hold for all matrices A for which these operations make sense.

The **identity matrix** I is the $n \times n$ matrix with elements δ_{ij} (Kronecker delta, defined in Exercises 1.1). It has the property

$$AI = IA = A$$

for all $n \times n$ matrices A (see Exercises 1.1).

The **transpose** A' of a matrix A is the matrix formed by interchanging the rows and columns of A. If A is $m \times n$, then A' is $n \times m$. We can express the elements of A' as follows:

$$a'_{ij} = a_{ji}$$

The following theorem is easily proved. *The transpose of the product of two matrices is the product of their transposes in the opposite order, or* $(AB)' = B'A'$. Let $C = AB$; then $c_{ik} = a_{ij}b_{jk}$; $c'_{ik} = c_{ki} = a_{kj}b_{ji} = b_{ji}a_{kj} = b'_{ij}a'_{jk}$. Therefore, $(AB)' = C' = B'A'$. The multiplications make sense, since, if A is $m \times n$ and B is $n \times p$, then A' is $n \times m$ and B' is $p \times n$, and B' and A' can be multiplied in that order.

† See exercise 2, Sec. 1.1.

Multiplication of a matrix A by a scalar k from the field which contains the elements of A results in a matrix whose elements are the elements of A multiplied by k; that is, $B = kA$ if and only if $b_{ij} = ka_{ij}$ for $i = 1, 2, \ldots, m$ and $j = 1, 2, \ldots, n$. The negative of a matrix, $-A$, is that matrix formed by multiplying each element by -1, or equivalently, by changing the sign of each element; that is, $-A$ has elements $-a_{ij}$. We now see that we could have defined subtraction in terms of addition of the negative.

$$A - B = A + (-B)$$

An $n \times n$ matrix A is **symmetric** if and only if it is equal to its transpose; that is, $A = A'$, or $a_{ij} = a_{ji}$.

An $n \times n$ matrix A is **skew-symmetric** if and only if it is equal to the negative of its transpose; that is, $A = -A'$, or $a_{ij} = -a_{ji}$.

If a matrix A has complex elements, then we define its **conjugate** $A*$ as the matrix formed from A by taking the complex conjugate of each element; that is, $A*$ has elements a_{ij}^*. Matrices with complex elements are important in quantum mechanics. Of particular importance are those called hermitian matrices. A **hermitian matrix** is a matrix which is equal to the transpose of its conjugate; that is, $A = (A*)'$, or $a_{ij} = a_{ji}^*$. If all the elements happen to be real, then the hermitian property is the same as the symmetry property.

Exercises 1.2

1. Verify the associative and commutative laws for matrix addition, the associative law for matrix multiplication, and the two distributive laws.

2. Construct an example to show that matrix multiplication is not, in general, commutative.

*3. If A and B are $n \times n$ matrices and $AB = O$, does this imply that $A = O$ or $B = O$? Explain.

*4. Show that every square matrix with complex elements can be written as the sum of a symmetric matrix and a skew-symmetric matrix. HINT: Write A as $\frac{1}{2}(A + A') + \frac{1}{2}(A - A')$.

*5. If X is a column matrix (a matrix with one column) with elements x_1, x_2, x_3 being the three rectangular cartesian coordinates of a point in three-dimensional space, A is a 3×3 matrix of real constants, B is a 1×3 row matrix of real constants, and k is a real constant, show that the general equation of a quadric surface is

$$X'AX + BX + k = 0$$

Also show that only the symmetric part of A enters the equation (see exercise 4).

*6. A **diagonal matrix** is an $n \times n$ matrix with zero elements except along the principal diagonal (upper left to lower right). Show that the elements of a diagonal matrix can be written as $a_{\alpha\beta} = d_\alpha \delta_{\alpha\beta}$, where $d_1, d_2, \ldots, d_n$ are the diagonal elements. Also show that every diagonal matrix is symmetric and that multiplication of two diagonal matrices is commutative.

7. Prove that if A and B are symmetric and $AB = D$, D diagonal, then $AB = BA$.

8. If A and B are hermitian, show that $-i(AB - BA)$ is hermitian.

*9. Suppose a matrix is made up of blocks of smaller matrices as follows:

$$M = \left\| \begin{array}{cc} A & B \\ C & D \end{array} \right\|$$

If M is $m \times n$, A is $p \times q$, and D is $r \times s$, what must be the dimensions of B and C? Prove the following:

$$\left\| \begin{array}{cc} A & B \\ C & D \end{array} \right\| \left\| \begin{array}{cc} E & F \\ G & H \end{array} \right\| = \left\| \begin{array}{cc} AE + BG & AF + BH \\ CE + DG & CF + DH \end{array} \right\|$$

1.3 Determinants

Before we return to the problem of solving systems of linear algebraic equations, we must first discuss determinants.

Every $n \times n$ matrix has associated with it a determinant which is a number from the field which contains the elements of the matrix. The **determinant** of the matrix A, written

$$\begin{vmatrix} a_{11} & a_{12} & \cdots & a_{1n} \\ a_{21} & a_{22} & \cdots & a_{2n} \\ \hdotsfor{4} \\ a_{n1} & a_{n2} & \cdots & a_{nn} \end{vmatrix}$$

or more simply $|A|$, is a number given by the following expression:

$$|A| = e_{i_1 i_2 i_3 \cdots i_n} a_{1i_1} a_{2i_2} a_{3i_3} \cdots a_{ni_n}$$

$e_{i_1 i_2 i_3 \cdots i_n} = 0$ if any pair of subscripts are equal

$\qquad\qquad = 1$ if $i_1, i_2, \ldots, i_n$ is an even permutation[1] of $1, 2, 3, \ldots, n$

$\qquad\qquad = -1$ if $i_1, i_2, \ldots, i_n$ is an odd permutation of $1, 2, 3, \ldots, n$

Note that the expression for $|A|$ is a single number because of the summations over the repeated subscripts $i_1, i_2, \ldots, i_n$. The definition can be stated as follows: Form all possible products of n elements from A, selecting first an element from the first row, second an element from the second row, and so on until one from each row has been selected for a given product, but being careful that no two elements come from the same column. Then attach a plus or

---

[1] A **permutation** of n different integers is an arrangement of the integers. There are, for example, six different permutations of the integers 1, 2, 3, that is, 1, 2, 3; 2, 3, 1; 3, 1, 2; 1, 3, 2; 2, 1, 3; and 3, 2, 1. An **inversion** of the order of a pair of adjacent integers changes the permutation. If it takes an even number of inversions to change a given permutation to the **normal order** 1, 2, 3, $\ldots$, n, then the permutation is said to be **even.** If an odd number of inversions is required to restore to normal order, then it is **odd.** Hence, 1, 2, 3; 2, 3, 1; and 3, 1, 2 are even, while 1, 3, 2; 2, 1, 3; and 3, 2, 1 are odd. It can be shown that evenness and oddness are independent of the specific set of inversions used to change the permutation to normal order.

minus sign to each product according to whether the column subscripts of the elements chosen form an even or an odd permutation of the integers 1 to n. Finally, add all such products with their attached signs. The resulting sum is the value of the determinant. Since for each product there are n ways of selecting a factor from the first row, $n - 1$ ways of selecting a factor from the second row, $n - 2$ ways of selecting a factor from the third row, etc., the number of terms in the sum which gives the value of the determinant is $n!$.

It is not hard to show that the above definition can be changed by replacing the word "row" with "column" and the word "column" with "row" throughout. In other words, starting from the definition, we can show that

$$|A| = e_{i_1 i_2 i_3 \cdots i_n} a_{i_1 1} a_{i_2 2} a_{i_3 3} \cdots a_{i_n n}$$

or simply $|A| = |A'|$. Since $e_{i_1 i_2 i_3 \cdots i_n}$ is zero if any pair of subscripts are equal, for the terms which appear in either of the above expansions the set of numbers $i_1, i_2, \ldots, i_n$ must include all the integers from 1 to n. Consider a particular term $e_{i_1 i_2 \cdots i_n} a_{i_1 1} a_{i_2 2} \cdots a_{i_n n}$ (no summations) in the expansion. We do not change the value of this product if we rearrange the factors so that the row subscripts are in normal order, that is,

$$a_{i_1 1} a_{i_2 2} a_{i_3 3} \cdots a_{i_n n} = a_{1 j_1} a_{2 j_2} a_{3 j_3} \cdots a_{n j_n}$$

This rearrangement of factors induces the permutation of the column subscripts $j_1, j_2, \ldots, j_n$. The term on the left appears once, and only once, in one expansion, while the term on the right appears once, and only once, in the other expansion. Therefore, the expansions will be equal, provided the corresponding terms appear with the same sign, or if $e_{i_1 i_2 \cdots i_n} = e_{j_1 j_2 \cdots j_n}$ for corresponding terms. This will be the case, because $i_1, i_2, \ldots, i_n$ is restored to normal order by an even or an odd number of inversions, the same number of which are applied to the normal order to produce $j_1, j_2, \ldots, j_n$.

This result is extremely important, for it indicates that any property of a determinant which depends on a property of a row or an operation on rows can be stated equally well for columns.

The following properties of determinants are easily proved from the definition:

1. If every element in a given row (or column) of a square matrix is zero, its determinant is zero.

2. If every element in a given row (or column) of a square matrix is multiplied by the same number k, the determinant is multiplied by k.

3. If any pair of rows (or columns) of a square matrix are interchanged, the sign of its determinant is changed.

4. If two rows (or columns) of a square matrix are proportional, its determinant is zero.

5. If each element of a given row (or column) of a square matrix can be written as the sum of two terms, then its determinant can be written as the

sum of two determinants, each of which contains one of the terms in the corresponding row (or column).

6. If to each element of a given row (or column) of a square matrix is added k times the corresponding element of another row (or column), the value of its determinant is unchanged.

By way of illustration, we shall prove property 3. We begin by showing that if two adjacent rows are interchanged, the sign of the determinant is changed.

$$|A| = e_{i_1 i_2 \cdots i_k i_{k+1} \cdots i_n} a_{1 i_1} a_{2 i_2} \cdots a_{k i_k} a_{k+1 i_{k+1}} \cdots a_{n i_n}$$
$$= e_{i_1 i_2 \cdots i_k i_{k+1} \cdots i_n} a_{1 i_1} a_{2 i_2} \cdots a_{k+1 i_{k+1}} a_{k i_k} \cdots a_{n i_n}$$

This follows because the order of multiplying the a's in each term is unimportant. Now i_k and i_{k+1} are dummy subscripts, so we replace i_k by i_{k+1} and i_{k+1} by i_k.

$$|A| = e_{i_1 i_2 \cdots i_{k+1} i_k \cdots i_n} a_{1 i_1} a_{2 i_2} \cdots a_{k+1 i_k} a_{k i_{k+1}} \cdots a_{n i_n}$$
$$= -e_{i_1 i_2 \cdots i_k i_{k+1} \cdots i_n} a_{1 i_1} a_{2 i_2} \cdots a_{k+1 i_k} a_{k i_{k+1}} \cdots a_{n i_n}$$

The minus sign in the last line is necessary because the inversion of i_k and i_{k+1} in $e_{i_1 i_2 \cdots i_{k+1} i_k \cdots i_n}$ causes every even permutation of subscripts to become odd and every odd permutation of subscripts to become even. The last line, except for the minus sign, is the expansion of the determinant of the matrix formed from A by interchanging the kth and the $(k + 1)$st rows. Property 3 now follows by observation that any pair of rows can be interchanged by a succession of an odd number of interchanges of adjacent rows, each of which changes the sign of the preceding determinant. For example, if we wish to interchange the jth and the kth rows, $k > j$, we can accomplish this by interchanging the jth row successively with the $k - j - 1$ intervening rows, by interchanging the jth and the kth rows (now adjacent), and finally by interchanging the kth row with the $k - j - 1$ rows between its present position and the original position of the jth row. This requires a total of $2(k - j - 1) + 1$ interchanges of adjacent rows, which is always an odd number.

An interesting outcome of defining matrix multiplication as we did is that the determinant of the product of two square matrices is the product of their determinants; that is,

$$|AB| = |A| \, |B|$$

To prove this, we begin by defining

$$e_{i_1 i_2 \cdots i_n} a_{j_1 i_1} a_{j_2 i_2} \cdots a_{j_n i_n}$$

If $j_1, j_2, \ldots, j_n = 1, 2, \ldots, n$ in normal order, then this expression gives the value of the determinant of A by definition. If any pair of the j's have the same value, then the value of the expression is zero, since it will then represent a determinant with two rows equal. If $j_1, j_2, \ldots, j_n$ is a permutation of $1, 2, \ldots, n$, then the expression gives either plus or minus the determinant of A, depending on whether it takes an even or an odd number of interchanges of rows to arrive at the expression for $|A|$. This can be summarized as follows:

$$e_{i_1 i_2 \cdots i_n} a_{j_1 i_1} a_{j_2 i_2} \cdots a_{j_n i_n} = |A| \, e_{j_1 j_2 \cdots j_n}$$

Similarly,

$$e_{i_1 i_2 \cdots i_n} a_{i_1 j_1} a_{i_2 j_2} \cdots a_{i_n j_n} = |A| \, e_{j_1 j_2 \cdots j_n}$$

The desired result now follows:

$$\begin{aligned} |A|\,|B| &= |A| \, e_{i_1 i_2 \cdots i_n} b_{i_1 1} b_{i_2 2} \cdots b_{i_n n} \\ &= e_{j_1 j_2 \cdots j_n} a_{j_1 i_1} a_{j_2 i_2} \cdots a_{j_n i_n} b_{i_1 1} b_{i_2 2} \cdots b_{i_n n} \\ &= e_{j_1 j_2 \cdots j_n} (a_{j_1 i_1} b_{i_1 1})(a_{j_2 i_2} b_{i_2 2}) \cdots (a_{j_n i_n} b_{i_n n}) \\ &= |AB| \end{aligned}$$

Another way of expanding the determinant of a square matrix is the so-called **expansion by cofactors.** Starting from the definition, we can write

$$\begin{aligned} |A| &= e_{i_1 i_2 \cdots i_n} a_{1 i_1} a_{2 i_2} \cdots a_{n i_n} \\ &= a_{1 i_1} e_{i_1 i_2 \cdots i_n} a_{2 i_2} \cdots a_{n i_n} \\ &= a_{1 i_1} A_{1 i_1} \end{aligned}$$

where

$$A_{1 i_1} e_{i_1 i_2 \cdots i_n} a_{2 i_2} \cdots a_{n i_n}$$

are the **cofactors** of the elements in the first row. In general,

$$A_{j i_j} = e_{i_1 \cdots i_{j-1} i_j i_{j+1} \cdots i_n} a_{1 i_1} \cdots a_{j-1 i_{j-1}} a_{j+1 i_{j+1}} \cdots a_{n i_n}$$

is the cofactor of the element in the jth row, i_jth column, and the expansion by cofactors of the jth row is given by

$$|A| = a_{j i_j} A_{j i_j} \qquad \text{(no summation on } j\text{)}$$

We also have an expansion by cofactors of column elements, as follows:

$$|A| = a_{i_j j} A_{i_j j} \qquad \text{(no summation on } j\text{)}$$

The cofactors themselves are, except possibly for a sign change, determinants of $(n-1) \times (n-1)$ square matrices formed from A by striking out one row and one column. Beginning from the definition of the cofactor, we have the following:

$$A_{j i_j} = (-1)^{j-1} e_{i_j i_1 i_2 \cdots i_{j-1} i_{j+1} \cdots i_n} a_{1 i_1} a_{2 i_2} \cdots a_{j-1 i_{j-1}} a_{j+1 i_{j+1}} \cdots a_{n i_n}$$

Here we have moved the subscript i_j ahead of the others by $j-1$ inversions, hence the factor $(-1)^{j-1}$. If it takes p inversions to put $i_1, i_2, \ldots, i_{j-1}$, $i_{j+1}, \ldots, i_n$ in normal order except for the missing integer i_j, then it takes $p + i_j - 1$ inversions to put $i_j, i_1, i_2, \ldots, i_{j-1}, i_{j+1}, \ldots, i_n$ in normal order. Hence, we can write

$$e_{i_j i_1 i_2 \cdots i_{j-1} i_{j+1} \cdots i_n} = (-1)^{i_j - 1} e_{k_1 k_2 \cdots k_{n-1}}$$

where $k_1, k_2, \ldots, k_{n-1}$ is the set of $n-1$ integers obtained from $i_1, i_2, \ldots,$ $i_{j-1}, i_{j+1}, \ldots, i_n$ by repeating an integer if it is less than i_j, and reducing by 1 any integer which is greater than i_j. It now follows that

$$A_{ji_j} = (-1)^{j+i_j} e_{k_1 k_2 \cdots k_{n-1}} b_{1k_1} b_{2k_2} \cdots b_{n-1 k_{n-1}}$$

The b's are the elements of an $(n-1) \times (n-1)$ matrix formed from A by striking out the jth row and the i_jth column. The final result is that the cofactor A_{ji_j} is $(-1)^{j+i_j}$ times the determinant of the matrix formed from A by omitting the jth row and i_jth column.[1]

The next question is what happens if we write down a cofactor expansion using the elements of a given row but the cofactors of a different row. In other words, what is the value of

$$a_{ji_k} A_{ki_k}$$

where $j \neq k$? Using the expression for the cofactor, we have

$$a_{ji_k} A_{ki_k} = a_{ji_k} e_{i_1 i_2 \cdots i_n} a_{1i_1} \cdots a_{ji_j} \cdots a_{k-1 i_{k-1}} a_{k+1 i_{k+1}} \cdots a_{ni_n}$$

$$= e_{i_1 i_2 \cdots i_n} a_{1i_1} \cdots a_{ji_j} \cdots a_{ji_k} \cdots a_{ni_n}$$

$$= 0$$

because the resulting expression is the expansion of a determinant with two rows equal. This result and the cofactor expansion of $|A|$ can be combined together in the single statement

$$a_{ij} A_{kj} = |A|\, \delta_{ik}$$

If $|A| \neq 0$, then we can write

$$a_{ij} \frac{A_{kj}}{|A|} = \delta_{ik}$$

or

$$a_{ij} \frac{A'_{jk}}{|A|} = \delta_{ik}$$

In terms of matrices, what we have just shown is that for every square matrix A for which $|A| \neq 0$ there exists a **right inverse** matrix A^{-1}, with elements

$$a_{ij}^{-1} = \frac{A_{ji}}{|A|}$$

such that

$$A A^{-1} = I$$

Starting from the cofactor expansion by column elements, it is not difficult to show that

$$\frac{A'_{ij}}{|A|}\, a_{jk} = \delta_{ik}$$

[1] This determinant is called the **minor** of a_{ji_j}.

or, in other words,

$$A^{-1}A = I$$

Therefore, this right inverse is also a left inverse, and we may refer to it simply as the inverse of A. The inverse is unique, for, if another inverse B existed, then $AB = I$ and

$$B = IB = A^{-1}AB = A^{-1}I = A^{-1}$$

If a matrix A has an inverse, then

$$AA^{-1} = I$$

and

$$|A|\,|A^{-1}| = |AA^{-1}| = |I| = 1$$

Therefore, $|A| \neq 0$, and we have the following theorem: *A square matrix A has an inverse if and only if $|A| \neq 0$.* Such a matrix is called **nonsingular.**

We are now partially able to solve systems of linear algebraic equations. Suppose that

$$AX = C$$

is a system of n linear algebraic equations in n unknowns and that $|A| \neq 0$. Since A is nonsingular, it has an inverse, A^{-1}. If the system of equations has a solution, it can be found by multiplication on the left by A^{-1}; that is,

$$A^{-1}AX = IX = X = A^{-1}C$$

In terms of elements,

$$x_i = \frac{A'_{ij}c_j}{|A|}$$

The numerators in this expression are determinants of $n \times n$ matrices formed from A by replacing the ith column with the column of c's. We have shown so far that, *if* there is a solution, it is given by $A^{-1}C$. To show that the solution actually satisfies the equations, we substitute as follows:

$$a_{ki}\frac{A'_{ij}}{|A|}c_j = \delta_{kj}c_j = c_k$$

We have arrived at **Cramer's rule:** *If a system of n linear algebraic equations in n unknowns,*

$$a_{ij}\,x_j = c_i$$

has a nonsingular coefficient matrix A, then the system has a unique solution given by

$$x_i = \frac{A'_{ij}c_j}{|A|}$$

We still have to resolve the question in the case where the coefficient matrix is singular and in the case the number of unknowns and the number of equations are different. These cases will be taken up in the next section.

We conclude this section with the definition of two special nonsingular matrices, which are defined in terms of their inverses.   The first of these is the **orthogonal matrix,** which is a square matrix whose transpose is its inverse; that is,

$$A' = A^{-1}$$

When we are dealing with matrices with complex elements, the counterpart of an orthogonal matrix is a **unitary matrix,** which is a square matrix the transpose of whose conjugate is its inverse; that is,

$$(A^*)' = A^{-1}$$

Exercises 1.3

1. Prove properties 1, 2, and 5 of determinants, starting from the definition.

2. Assuming properties 3 and 5, prove properties 4 and 6 of determinants.

3. Expand the following determinant, using (a) the definition, (b) cofactor expansions, and (c) properties 1 to 6 to reduce the problem to simpler determinants:

$$\begin{vmatrix} 1 & 3 & -1 & 2 \\ 2 & 1 & 3 & 1 \\ -1 & 2 & -1 & 3 \\ -2 & 1 & 2 & -3 \end{vmatrix}$$

*4. If A is orthogonal, prove that $|A| = \pm 1$.

*5. If A and B are nonsingular $n \times n$ matrices, show that $(AB)^{-1} = B^{-1}A^{-1}$.

6. The **adjoint** of a square matrix A is the transpose of the matrix of the cofactors of A.   Show that if A is nonsingular, the determinant of its adjoint is equal to $|A|^{n-1}$.

*7. We define as **elementary row operations** on a matrix the following: (a) interchange of two rows; (b) multiplication of a row by a number $k \neq 0$; (c) addition of two rows.

Show that each of these operations can be performed on an $n \times n$ matrix, A, by multiplying on the left by an **elementary matrix,** E, which is obtained from the identity matrix by the same operation.   Show that if A is nonsingular, A can be reduced to the identity matrix by a finite number of elementary row operations and that A^{-1} can be computed by performing the same operations in the same order on the identity matrix.

*8. Rephrase the results of exercise 7 in terms of elementary column operations.

1.4 Systems of Linear Algebraic Equations. Rank of a Matrix

We now return to the problem of solving a system of linear algebraic equations.   Recall that we were able to write this system in the matrix form

$$AX = C$$

where A is the given $m \times n$ **coefficient matrix,** C is a column matrix of given constants, and X is a column matrix representing the solution (if it exists). We define the **augmented matrix** B as the $m \times (n + 1)$ matrix formed from A by adding the column of c's as the last column.

Every matrix contains certain square matrices formed by deleting rows or columns or both.   We define the **rank of a matrix** as the order of the highest-order square matrix with a nonvanishing determinant contained in the matrix. It is obvious that the rank of the coefficient matrix is less than or equal to the rank of the augmented matrix, because every square matrix contained in the former is also contained in the latter.

We are now ready to prove the following theorem: *The system of linear algebraic equations has a solution if and only if the rank of the augmented matrix is equal to the rank of the coefficient matrix.*

We begin the proof of this theorem by considering **elementary row operations** and **elementary matrices.**   We define an elementary row operation as one of the following operations on a matrix: (1) the interchange of a pair of rows, (2) the multiplication of a row by a nonzero number, or (3) the addition of two rows.   It is not difficult to show that any of these operations can be performed on an $m \times n$ matrix by multiplying on the left by an elementary $m \times m$ matrix, obtained from the identity matrix by the corresponding operation.[1]   If we multiply both sides of the system of algebraic equations by an elementary matrix E; that is,

$$\bar{A}X = EAX = EC = \bar{C}$$

we retain the equality and the general form of the equations.   We have a new coefficient matrix $\bar{A}$ and a new augmented matrix $\bar{B}$, with

$$\bar{A} = EA$$
$$\bar{B} = EB$$

Multiplication on the left by elementary matrices has the following effects on the system of equations, corresponding to the three row operations listed above: (1) interchanges a pair of equations, (2) multiplies an equation through by a nonzero number, and (3) adds two equations together.   These are the three operations required for the **Gauss-Jordan reduction,** which we shall now outline.   First, arrange the equations so that $a_{11} \neq 0$.   Then multiply the first equation by $1/a_{11}$.   Next, add to the second equation $(-a_{21})$ times the first.   This makes the coefficient of x_1 in the second equation zero.   Repeat the process until the coefficients of x_1 in all equations but the first are zero. The next step is to rearrange the equations so that the coefficient of x_2 in the second equation is nonzero, divide through the second equation by this coefficient, and proceed as before to make the coefficient of x_2 in all equations but the second zero.   After a finite number of steps, the system of equations

---

[1] See Exercises 1.3.

reduces to

$$x_1 + \bar{a}_{1r+1}x_{r+1} + \cdots + \bar{a}_{1n}x_n = \bar{c}_1$$
$$x_2 + \bar{a}_{2r+1}x_{r+1} + \cdots + \bar{a}_{2n}x_n = \bar{c}_2$$
$$\cdots\cdots\cdots\cdots\cdots\cdots\cdots\cdots\cdots$$
$$x_r + \bar{a}_{rr+1}x_{r+1} + \cdots + \bar{a}_{rn}x_n = \bar{c}_r$$
$$0 = \bar{c}_{r+1}$$
$$0 = \bar{c}_{r+2}$$
$$\cdots\cdots\cdots$$
$$0 = \bar{c}_m$$

From this reduction, it is clear that if one or more of the $\bar{c}_{r+1}, \ldots, \bar{c}_m$ are not zero, there will be no solution, and we say that the system of equations is **inconsistent.** If $\bar{c}_{r+1} = \bar{c}_{r+2} = \cdots = \bar{c}_m = 0$, then we can write the solution

$$x_1 = \bar{c}_1 - \bar{a}_{1r+1}x_{r+1} - \cdots - \bar{a}_{1n}x_n$$
$$x_2 = \bar{c}_2 - \bar{a}_{2r+1}x_{r+1} - \cdots - \bar{a}_{2n}x_n$$
$$\cdots\cdots\cdots\cdots\cdots\cdots\cdots\cdots\cdots$$
$$x_r = \bar{c}_r - \bar{a}_{rr+1}x_{r+1} - \cdots - \bar{a}_{rn}x_n$$

In this case we say that the equations are **consistent.** If $r = n$, there is a unique solution. If $r < n, x_1, x_2, \ldots, x_r$ can be written in terms of $x_{r+1}, \ldots, x_n$, which can be assigned arbitrarily. In this case there is an $(n - r)$-fold infinity of solutions.

If $r = m = n$, the solution is unique and is given by Cramer's rule. Thus

$$AX = C$$
$$IAX = IC$$
$$E_p \cdots E_2E_1IAX = E_p \cdots E_2E_1IC$$
$$A^{-1}AX = A^{-1}C$$
$$X = A^{-1}C$$

Here $E_1, E_2, \ldots, E_p$ are the elementary row matrices which are involved in the reduction, and[1]

$$E_p \cdots E_2E_1I = A^{-1}$$

Finally, we state these results in terms of the ranks of the coefficient and augmented matrices. The reduction was performed by a finite number of row operations, none of which changes the rank of a matrix. Suppose M is a matrix of rank r and $\bar{M}$ is a matrix obtained from M by one of the three row operations. If $\bar{D}$ is an $(r + 1)$st-order determinant from $\bar{M}$, then:

[1] See Exercises 1.3.

Case 1. $\bar{D} = D$ or $\bar{D} = -D$, where in each cash D is an $(r + 1)$st-order determinant from M.   But $D = 0$, since M is of rank r.

Case 2. $\bar{D} = D$ or $\bar{D} = kD$, where $k \neq 0$ and D is an $(r + 1)$st-order determinant from M.   But $D = 0$, since M is of rank r.

Case 3. $\bar{D} = D$ or $\bar{D} = D + \tilde{D}$, where D and $\tilde{D}$ are $(r + 1)$st-order determinants from M.   But $D = \tilde{D} = 0$, since M is of rank r.

Therefore, a row operation cannot increase the rank of a matrix.   Likewise, a row operation cannot decrease the rank of a matrix, for in this case M can be obtained from $\bar{M}$ by a row operation which cannot increase the rank.

If we write the reduced form of the system of equations as

$$\bar{A}X = \bar{C}$$

then

$$\bar{A} = \begin{Vmatrix} 1 & 0 & 0 & \cdots & 0 & \bar{a}_{1r+1} & \bar{a}_{1r+2} & \cdots & \bar{a}_{1n} \\ 0 & 1 & 0 & \cdots & 0 & \bar{a}_{2r+1} & \bar{a}_{2r+2} & \cdots & \bar{a}_{2n} \\ \hdotsfor{9} \\ 0 & 0 & 0 & \cdots & 1 & \bar{a}_{rr+1} & \bar{a}_{rr+2} & \cdots & \bar{a}_{rn} \\ 0 & 0 & 0 & \cdots & 0 & 0 & 0 & \cdots & 0 \\ \hdotsfor{9} \\ 0 & 0 & 0 & \cdots & 0 & 0 & 0 & \cdots & 0 \end{Vmatrix}$$

and

$$\bar{B} = \begin{Vmatrix} 1 & 0 & 0 & \cdots & 0 & \bar{a}_{1r+1} & \bar{a}_{1r+2} & \cdots & \bar{a}_{1n} & \bar{c}_1 \\ 0 & 1 & 0 & \cdots & 0 & \bar{a}_{2r+1} & \bar{a}_{2r+2} & \cdots & \bar{a}_{2n} & \bar{c}_2 \\ \hdotsfor{10} \\ 0 & 0 & 0 & \cdots & 1 & \bar{a}_{rr+1} & \bar{a}_{rr+2} & \cdots & \bar{a}_{rn} & \bar{c}_r \\ 0 & 0 & 0 & \cdots & 0 & 0 & 0 & \cdots & 0 & \bar{c}_{r+1} \\ \hdotsfor{10} \\ 0 & 0 & 0 & \cdots & 0 & 0 & 0 & \cdots & 0 & \bar{c}_m \end{Vmatrix}$$

Obviously $\bar{A}$ is of rank r.   If $\bar{c}_{r+1} = \bar{c}_{r+2} = \cdots = \bar{c}_m = 0$, then the system of equations has a solution and the rank of $\bar{B}$ is also r.   If one or more of the $\bar{c}_{r+1}$ to $\bar{c}_m$ are not zero, then the rank of $\bar{B}$ is greater than r and the system of equations has no solution.   This completes the proof of the theorem.

A case of particular importance is the case where $c_1 = c_2 = \cdots = c_n = 0$. In this case, we say we have a system of **homogeneous linear equations.** Adding a column of zeros to the coefficient matrix cannot affect its rank; therefore the augmented matrix always has the same rank as the coefficient matrix, and such a system always has a solution.   This is not surprising since $x_1 = x_2 = \cdots = x_n = 0$ is obviously a solution.   This is called the **trivial**

solution. We are often concerned about whether the system has a **non-trivial solution,** where at least one of the x's is not zero. From the above discussion follow two important corollaries, the proofs of which will be left for the reader.

A system of m homogeneous linear algebraic equations in n unknowns always has a nontrivial solution if $m < n$.

A system of n homogeneous linear algebraic equations in n unknowns has a nontrivial solution if and only if the determinant of the coefficient matrix is zero.

Exercises 1.4

1. Prove the two corollaries at the end of this section.

2. Prove that multiplication of an $m \times n$ matrix A on the right by an $n \times n$ elementary matrix obtained from the identity by an elementary column operation cannot change the rank of A.

3. Determine if the following system has a solution:

$$2x + y - z + u = -2$$
$$x - y - z + u = 1$$
$$x - 4y - 2z + 2u = 6$$
$$4x + y - 3z + 3u = -1$$

*4. Determine the values of λ for which the following system has a nontrivial solution:

$$9x - 3y = \lambda x$$
$$-3x + 12y - 3z = \lambda y$$
$$-3y + 9z = \lambda z$$

Find a nontrivial solution in each case.

*5. If $X = X_0$ is a solution of $AX = O$, show that $X = kX_0$ is also a solution for any k.

*6. If X_0 is a solution of $AX = O$ and X_1 is a solution of $AX = C$, show that $kX_0 + X_1$ is also a solution of $AX = C$. State a criterion for uniqueness of the solution of $AX = C$.

7. Find a nontrivial solution of the following system:

$$x + y - z + u = 0$$
$$-x + 2y + z - u = 0$$
$$2x - y + 3z - 2u = 0$$

8. For what values of λ will the following equations have real nontrivial solutions

$$x + y = \lambda x$$
$$-x + y = \lambda y$$

1.5 Vector Spaces

It is assumed that the reader is familiar with three-dimensional vector analysis. The collection of vectors encountered there follows certain laws of combination which make it an example of an algebraic system called a vector space. We shall want to define and use the concept of vector space, but first let us recall some of the familiar notions about vectors from three-dimensional vector analysis.

One of the easiest ways to denote a three-dimensional vector is in terms of its three components. Let U be a vector with real components (u_1,u_2,u_3) and V be a vector with real components (v_1,v_2,v_3);† then the vector sum $U + V$ has components $(u_1 + v_1,\ u_2 + v_2,\ u_3 + v_3)$. We note at once that the sum is a vector and has the properties

$$U + V = V + U$$

$$(U + V) + W = U + (V + W)$$

We define the zero vector 0 as the vector with all three components 0. Then obviously,

$$U + 0 = U$$

for all U. The negative vector of the vector U, denoted by $-U$, is the vector with components $(-u_1,-u_2,-u_3)$. Clearly,

$$U + (-U) = 0$$

Multiplication by a real scalar a is defined as follows:

$$aU = (au_1,au_2,au_3)$$

This operation leads to a vector and has the following properties:

$$a(U + V) = aU + aV$$

$$(a + b)U = aU + bU$$

$$(ab)U = a(bU)$$

$$1U = U$$

These familiar properties suggest the postulates of a **vector space** which we now define. *A vector space over a field is a set of vectors together with two operations, addition and multiplication by a scalar from the field, satisfying the following postulates:*

1. **Closure under addition:** For every pair of vectors U and V there is a unique sum denoted by $U + V$, which is a vector.
2. **Addition is associative:** $(U + V) + W = U + (V + W)$.
3. **Addition is commutative:** $U + V = V + U$.
4. A **zero vector** 0 exists, such that $U + 0 = U$ for all U.
5. A **negative vector** $-U$ exists for all U, such that $U + (-U) = 0$.

---

† In terms of the familiar i, j, k notation,

$$U = u_1i + u_2j + u_3k \text{ and } V = v_1i + v_2j + v_3k$$

6. **Closure under multiplication by a scalar:** For every scalar a from the field and every vector $\mathbf{U}$ there is a unique vector $a\mathbf{U}$.

7. $a(\mathbf{U} + \mathbf{V}) = a\mathbf{U} + a\mathbf{V}$.

8. $(a + b)\mathbf{U} = a\mathbf{U} + b\mathbf{U}$.

9. $(ab)\mathbf{U} = a(b\mathbf{U})$.

10. $1\mathbf{U} = \mathbf{U}$.

From these ten postulates many theorems can be proved. For example, the connection between the zero scalar and the zero vector is

$$0\mathbf{U} = \mathbf{0}$$

for all $\mathbf{U}$. This can be proved as follows:

$$(1 + 0)\mathbf{U} = 1\mathbf{U} + 0\mathbf{U}$$
$$1\mathbf{U} = \mathbf{U} + 0\mathbf{U}$$
$$\mathbf{U} = \mathbf{U} + 0\mathbf{U}$$
$$\mathbf{U} + (-\mathbf{U}) = \mathbf{U} + (-\mathbf{U}) + 0\mathbf{U}$$
$$\mathbf{0} = \mathbf{0} + 0\mathbf{U}$$
$$\mathbf{0} = 0\mathbf{U}$$

The negative of the field and the negative vector are connected by the relation

$$(-1)\mathbf{U} = -\mathbf{U}$$

for all $\mathbf{U}$. This is easy to demonstrate.

$$(1 - 1)\mathbf{U} = 1\mathbf{U} + (-1)\mathbf{U}$$
$$0\mathbf{U} = \mathbf{U} + (-1)\mathbf{U}$$
$$\mathbf{0} = \mathbf{U} + (-1)\mathbf{U}$$
$$-\mathbf{U} + \mathbf{0} = -\mathbf{U} + \mathbf{U} + (-1)\mathbf{U}$$
$$-\mathbf{U} = (-1)\mathbf{U}$$

The proofs of some other useful properties will be left for the exercises.

We have already seen that the vectors of three-dimensional vector analysis form a vector space. There are many other examples. Consider, for example, the set of n-tuples of complex numbers. They form a vector space over the field of complex numbers, if we define addition and multiplication by a scalar as follows. If $\mathbf{U} = (u_1, u_2, \ldots, u_n)$ and $\mathbf{V} = (v_1, v_2, \ldots, v_n)$, then

$$\mathbf{U} + \mathbf{V} = (u_1 + v_1, u_2 + v_2, \ldots, u_n + v_n)$$

If a is a complex scalar, then

$$a\mathbf{U} = (au_1, au_2, \ldots, au_n)$$

Obviously, we have closure under addition and multiplication by a scalar.

Also, from the properties of complex numbers it is easy to verify postulates 2, 3, 7, 8, 9, and 10. The zero vector is defined as $\mathbf{0} = (0,0, \ldots, 0)$, and it clearly satisfies postulate 4. The negative of $\mathbf{U}$ is defined as

$$-\mathbf{U} = (-u_1, -u_2, \ldots, -u_n)$$

and it satisfies postulate 5.

As another example, consider the set of continuous real-valued functions of the real variable x defined on the interval $0 \le x \le 1$. This is a vector space over the field of real numbers, if we define sum as $\mathbf{f} + \mathbf{g} = f(x) + g(x)$, and multiplication by a scalar as $a\mathbf{f} = af(x)$. The zero vector is the function which is identically zero in the interval, and the negative is defined by $-\mathbf{f} = -f(x)$. It is not difficult to show that the ten postulates of a vector space are satisfied.

A set of vectors $\mathbf{X}_1, \mathbf{X}_2, \ldots, \mathbf{X}_n$ is said to be **linearly dependent** if there exists a set of scalars $c_1, c_2, \ldots, c_n$, not all zero, such that

$$c_i\mathbf{X}_i = \mathbf{0}$$

If a set of vectors is not linearly dependent, then it is said to be **linearly independent.** In other words, if $\mathbf{X}_1, \mathbf{X}_2, \ldots, \mathbf{X}_n$ are independent, then $c_i\mathbf{X}_i = \mathbf{0}$ implies that $c_i = 0$ for all i.

A set of vectors $\mathbf{X}_1, \mathbf{X}_2, \ldots, \mathbf{X}_n$ in a vector space is said to span the space if every vector in the space can be written as a linear combination of the set; that is, for every $\mathbf{U}$ in the space there exists a set of scalars $c_1, c_2, \ldots, c_n$, such that

$$\mathbf{U} = c_i\mathbf{X}_i$$

A set $\mathbf{X}_1, \mathbf{X}_2, \ldots, \mathbf{X}_n$ may span a vector space and still not be linearly independent. However, if it is dependent we can select from the set a subset which is linearly independent and also spans the the space.[1] Suppose $\mathbf{X}_1, \mathbf{X}_2, \ldots, \mathbf{X}_n$ are dependent; then there exists a set of scalars $\gamma_1, \gamma_2, \ldots, \gamma_n$, not all zero, such that

$$\gamma_i\mathbf{X}_i = \mathbf{0}$$

Without loss of generality, we can assume that $\gamma_n \ne 0$, and we can solve for $\mathbf{X}_n$ in terms of $\mathbf{X}_1, \mathbf{X}_2, \ldots, \mathbf{X}_{n-1}$.

$$\mathbf{X}_n = -\frac{\gamma_1}{\gamma_n}\mathbf{X}_1 - \frac{\gamma_2}{\gamma_n}\mathbf{X}_2 - \cdots - \frac{\gamma_{n-1}}{\gamma_n}\mathbf{X}_{n-1}$$

Hence, for any $\mathbf{U}$ in the vector space

$$\mathbf{U} = c_1\mathbf{X}_1 + c_2\mathbf{X}_2 + \cdots + c_{n-1}\mathbf{X}_{n-1}$$
$$+ c_n\left(-\frac{\gamma_1}{\gamma_n}\mathbf{X}_1 - \frac{\gamma_2}{\gamma_n}\mathbf{X}_2 - \cdots - \frac{\gamma_{n-1}}{\gamma_n}\mathbf{X}_{n-1}\right)$$

[1] Here we must exclude the so-called **zero space,** consisting of just the zero vector. The zero space is spanned by the zero vector which is dependent.

Therefore, the subset $X_1, X_2, \ldots, X_{n-1}$ spans the space. If this subset is not independent, we can repeat the process again. Eventually we shall arrive at a subset $X_1, X_2, \ldots, X_m$ with $m < n$, which spans the space and is linearly independent.

We define the **dimension** of a vector space as the minimum number of nonzero vectors which span the space.[1] For example, in the vector space of n-tuples of complex numbers we may choose the vectors $X_1 = (1,0,0,\ldots,0)$, $X_2 = (0,1,0,\ldots,0), \ldots, X_n = (0,0,0,\ldots,1)$. They span the space, for any vector $U = (u_1, u_2, u_3, \ldots, u_n)$ can be written as a linear combination of the X's; that is,

$$U = u_1 X_1 + u_2 X_2 + u_3 X_3 + \cdots + u_n X_n$$

The X's are linearly independent, for

$$c_1 X_1 + c_2 X_2 + c_3 X_3 + \cdots + c_n X_n = (c_1, c_2, c_3, \ldots, c_n) = 0$$

implies that $c_1 = c_2 = c_3 = \cdots = c_n = 0$. As we shall show, this implies that n is the minimum number of vectors which span the space, and therefore the dimension of the space is n.

It may not be possible to find a finite number of vectors which span the space. For example, the vector space of continuous functions on the interval $0 \leq x \leq 1$ is not spanned by a finite number of vectors.[2] We say that a vector space is **finite-dimensional** if and only if it is spanned by a finite number of vectors. Otherwise it is **infinite-dimensional**. At first, we shall confine out attention to finite-dimensional vector spaces. Some things have to be modified in the case of infinite-dimensional vector spaces, and we shall return to this case in a later chapter.

A **basis** of a vector space is a set of linearly independent vectors which spans the space. We have already shown that from any finite set of vectors which spans the space we can select a linearly independent subset which spans the space. Such a subset forms a basis for the space. There may be many bases for the same vector space. For example, we have already given a basis for the vector space of n-tuples of complex numbers, but another one would consist of the vectors

$$Y_1 = (1,0,0,\ldots,0) \qquad Y_2 = (1,1,0,\ldots,0)$$
$$Y_3 = (1,1,1,0,\ldots,0), \ldots \qquad Y_n = (1,1,1,\ldots,1)$$

However, it is easily shown that *every basis must contain the same number of vectors*.

Suppose that we have two bases $X_1, X_2, \ldots, X_n$ and $Y_1, Y_2, \ldots, Y_m$. Since the X's span the space, Y_1 can be expressed as a linear combination of

---

[1] According to this definition the zero space has dimension zero.
[2] The functions $1, x, x^2, \ldots$ are all continuous and independent.

the X's; that is,

$$Y_1 = c_1 X_1 + c_2 X_2 + \cdots + c_n X_n$$

where at least one of the c's is not zero.   Assume $c_n \neq 0$ (otherwise renumber the X's so that it is).   Then

$$X_n = \frac{1}{c_n} (Y_1 - c_1 X_1 - c_2 X_2 - \cdots - c_{n-1} X_{n-1})$$

Therefore, the set of vectors $Y_1, X_1, X_2, \ldots, X_{n-1}$ spans the vector space, and Y_2 can be expressed as a linear combination of this set; that is,

$$Y_2 = b_1 Y_1 + \gamma_1 X_1 + \gamma_2 X_2 + \cdots + \gamma_{n-1} X_{n-1}$$

Here at least one of the γ's is not zero.   Otherwise Y_1 and Y_2 would not be independent.   Solve for one of the X's whose coefficient is not zero in terms of Y_1, Y_2, and the $n - 2$ other X's which span the vector space.   After repeating this process m times, we arrive at a set of vectors $Y_1, Y_2, \ldots, Y_m$ and $n - m$ of the X's which span the space.   From this we conclude that $n \geq m$.   The same argument can be applied with the roles of the X's and Y's interchanged, leading to the conclusion that $m \geq n$.   The final result that $n = m$ follows.

To find the dimension of a finite-dimensional vector space it is sufficient to demonstrate any basis and count the number of vectors in the set.   This verifies our conclusion that the space of n-tuples of complex numbers is an n-dimensional space.

The fact that a basis spans a vector space means that every vector in the space has a representation as a linear combination of basis vectors.   The fact that the basis is a set of linearly independent vectors implies that *the representation in terms of a given basis is unique*.   Suppose that U has two representations in terms of the basis $X_1, X_2, \ldots, X_n$,

$$U = a_i X_i = b_i X_i$$

Then $$0 = U - U = (a_i - b_i) X_i$$

But the X's are independent, so that $a_i = b_i$ for all i.

Suppose that we have an n-dimensional vector space over the field of complex numbers, with a basis $X_1, X_2, \ldots, X_n$.   Relative to this basis every vector in the space has a unique representation in terms of n scalars.   There is, therefore, a one-to-one correspondence between the vector space and the vector space of n-tuples of complex numbers.   Furthermore, if $U = a_i X_i$ and $V = b_i X_i$, then $U + V = (a_i + b_i) X_i$.   If c is a complex scalar, $cU = (ca_i) X_i$. Therefore, addition of vectors and multiplication of a vector by a scalar correspond to addition and multiplication by a scalar of the corresponding vectors in the space of n-tuples.   Such a relation between two vector spaces is called an **isomorphism**.

A **subspace** of a vector space is the space consisting of all possible linear combinations of a subset X_i of vectors from the vector space. That every subspace is also a vector space is derived from the fact that every vector in the subspace can be expressed as a linear combination of vectors. For example, if $U = a_i X_i$, $V = b_i X_i$, and $W = c_i X_i$, then

$$U + V = (a_i + b_i)X_i = (b_i + a_i)X_i = V + U$$

$$(U + V) + W = [(a_i + b_i) + c_i]X_i = [a_i + (b_i + c_i)]X_i = U + (V + W)$$

The subspace always contains the zero vector

$$0 = 0X_1 + 0X_2 + \cdots + 0X_m$$

and a negative for every vector in the subspace; that is, if $U = a_i X_i$, then

$$-U = (-a_i)X_i$$

Multiplication of a vector U by a scalar c is defined by

$$cU = (ca_i)X_i$$

and the remaining postulates follow immediately.

If the subset X_i is a basis of the vector space, then the subspace is the whole space. If the subset X_i is linearly independent and not a basis, then the subspace is a **proper subspace**; i.e., it is not the whole space. An example of a subspace would be the space spanned by the vectors $(1,-1,0)$ and $(0,1,-1)$ in three-dimensional euclidean space. This two-dimensional subspace consists of the plane containing the two given vectors and the origin.

Exercises 1.5

1. Establish the following statements:

a. The collection of complex numbers is a two-dimensional vector space over the field of real scalars.

b. The collection of $n \times 1$ column matrices with complex elements in an n-dimensional vector space over the field of complex scalars.

c. The collection of all polynomials of degree three or less is a four-dimensional subspace of the vector space of continuous real-valued functions of the real variable x defined on the interval $0 \le x \le 1$.

d. The collection of all $m \times n$ matrices with complex elements is an mn-dimensional vector space over the field of complex scalars.

2. Prove the following theorems pertaining to vector spaces:

a. The zero vector is unique.

b. The negative of a vector is unique.

c. $a0 = 0$ for any scalar a.

d. If $aU = 0$, then either $a = 0$, or $U = 0$, or both.

3. Are the vectors $(0,-1,0)$, $(0,1,-1)$, $(1,-2,1)$ linearly independent? Can $(-2,1,-3)$ be expressed as a linear combination of these three? Express your result both geometrically and in terms of solutions of linear algebraic equations.

4. Prove that two finite-dimensional vector spaces which are isomorphic have the same dimension.

*5. An n-dimensional vector space over the field of complex scalars is isomorphic to the vector space of n-tuples of complex numbers. If m vectors are given with $m \leq n$, state a test for linear dependence in terms of the rank of an $n \times m$ matrix formed by using the m n-tuples as columns. What happens in the case $m > n$?

*6. Prove that in an n-dimensional vector space any set of $n + 1$ vectors is linearly dependent.

*7. Consider an $n \times m$ matrix A. Define the rank of A in terms of the dimension of the subspace spanned by the m vectors which form the columns of A. Prove that your definition is equivalent to the one given in Sec. 1.4.

8. Prove that a subset of a vector space is a subspace if and only if it is closed under addition and multiplication by a scalar.

1.6   Scalar Product

In three-dimensional vector analysis we define the scalar product of two vectors as

$$(\mathbf{U},\mathbf{V}) = u_1 v_1 + u_2 v_2 + u_3 v_3$$

This is directly related to the magnitudes of the vectors and the cosine of the angle between the two vectors. It has the following properties, which are easily verified.

$$(\mathbf{U},\mathbf{V}) = (\mathbf{V},\mathbf{U})$$
$$(\mathbf{U},\mathbf{V} + \mathbf{W}) = (\mathbf{U},\mathbf{V}) + (\mathbf{U},\mathbf{W})$$
$$(a\mathbf{U},\mathbf{V}) = a(\mathbf{U},\mathbf{V})$$
$$(\mathbf{U},\mathbf{U}) \geq 0$$
$$(\mathbf{U},\mathbf{U}) = 0 \text{ if and only if } \mathbf{U} = \mathbf{0}$$

The concept of scalar product can be generalized to vector spaces other than the familiar one mentioned above. We shall retain essentially the same properties only slightly modified to include the possibility of the scalar product being a complex number. In the applications, we are usually concerned with vector spaces over the field of complex numbers, and in this case we define a **scalar product**[1] $(\mathbf{U},\mathbf{V})$ which has a complex value and satisfies the following postulates:

1. $(\mathbf{U},\mathbf{V}) = (\mathbf{V},\mathbf{U})^*$
2. $(\mathbf{U},\mathbf{V} + \mathbf{W}) = (\mathbf{U},\mathbf{V}) + (\mathbf{U},\mathbf{W})$
3. $(a\mathbf{U},\mathbf{V}) = a^*(\mathbf{U},\mathbf{V})$
4. $(\mathbf{U},\mathbf{U}) \geq 0$
5. $(\mathbf{U},\mathbf{U}) = 0$ if and only if $\mathbf{U} = \mathbf{0}$

Properties 4 and 5 are equivalent to the statement that the *scalar product of a vector with itself is positive-definite*. Recall that, in three-dimensional vector analysis, the scalar product of a vector with itself is the square of the length of the vector. Although "length of a vector" does not have the usual

---

[1] Sometimes referred to as an inner product.

meaning in a generalized vector space, we still define length, or more commonly **norm,** of a vector as

$$\|\mathbf{U}\| = \sqrt{(\mathbf{U},\mathbf{U})}$$

and it turns out that this has most of the properties we usually associate with length or distance.

From the postulates for scalar product, we can prove some additional useful properties. For example,

$$(\mathbf{U},a\mathbf{V}) = a(\mathbf{U},\mathbf{V})$$

This is proved as follows:

$$(\mathbf{U},a\mathbf{V}) = (a\mathbf{V},\mathbf{U})^* = [a^*(\mathbf{V},\mathbf{U})]^* = a(\mathbf{V},\mathbf{U})^* = a(\mathbf{U},\mathbf{V})$$

The proofs of the following two properties will be left for the exercises.

$$(\mathbf{U} + \mathbf{V}, \mathbf{W}) = (\mathbf{U},\mathbf{W}) + (\mathbf{V},\mathbf{W})$$

$$\|a\mathbf{U}\| = |a| \, \|\mathbf{U}\|$$

A very important property is **Schwarz's inequality**

$$|(\mathbf{U},\mathbf{V})| \leq \|\mathbf{U}\| \, \|\mathbf{V}\|$$

This can be proved as follows. For any scalar a

$$
\begin{aligned}
0 \leq \|\mathbf{U} + a\mathbf{V}\|^2 &= (\mathbf{U} + a\mathbf{V}, \mathbf{U} + a\mathbf{V}) \\
&= (\mathbf{U},\mathbf{U}) + (a\mathbf{V},\mathbf{U}) + (\mathbf{U},a\mathbf{V}) + (a\mathbf{V},a\mathbf{V}) \\
&= \|\mathbf{U}\|^2 + a^*(\mathbf{V},\mathbf{U}) + a(\mathbf{U},\mathbf{V}) + |a|^2 \, \|\mathbf{V}\|^2
\end{aligned}
$$

Now let $a = \lambda(\mathbf{U},\mathbf{V})^*/|(\mathbf{U},\mathbf{V})|$, with λ real. This is possible if $(\mathbf{U},\mathbf{V}) \neq 0$, but if $(\mathbf{U},\mathbf{V}) = 0$, then Schwarz's inequality is trivial. Making this substitution in the above, we have

$$0 \leq \|\mathbf{U}\|^2 + 2\lambda \, |(\mathbf{U},\mathbf{V})| + \lambda^2 \, \|\mathbf{V}\|^2$$

This is a quadratic expression in the real variable λ with real coefficients. Therefore, the discriminant must be less than or equal to zero. Hence,

$$4 \, |(\mathbf{U},\mathbf{V})|^2 - 4 \, \|\mathbf{U}\|^2 \, \|\mathbf{V}\|^2 \leq 0$$

or

$$|(\mathbf{U},\mathbf{V})| \leq \|\mathbf{U}\| \, \|\mathbf{V}\|$$

From Schwarz's inequality follows another important inequality, known as the **triangle inequality,**

$$\|\mathbf{U} + \mathbf{V}\| \leq \|\mathbf{U}\| + \|\mathbf{V}\|$$

For any pair of vectors

$$
\begin{aligned}
\|\mathbf{U} + \mathbf{V}\|^2 &= \|\mathbf{U}\|^2 + \|\mathbf{V}\|^2 + (\mathbf{U},\mathbf{V}) + (\mathbf{V},\mathbf{U}) \\
&\leq \|\mathbf{U}\|^2 + \|\mathbf{V}\|^2 + 2 \, |(\mathbf{U},\mathbf{V})| \\
&\leq \|\mathbf{U}\|^2 + \|\mathbf{V}\|^2 + 2 \, \|\mathbf{U}\| \, \|\mathbf{V}\| \\
&\leq (\|\mathbf{U}\| + \|\mathbf{V}\|)^2
\end{aligned}
$$

from which it follows that

$$\|\mathbf{U} + \mathbf{V}\| \leq \|\mathbf{U}\| + \|\mathbf{V}\|$$

Let us now consider some examples of scalar products.   Consider the vector space of n-tuples of complex numbers.   Here we can define a scalar product as

$$(\mathbf{U},\mathbf{V}) = u_1^* v_1 + u_2^* v_2 + \cdots + u_n^* v_n = u_i^* v_i$$

Then

$$(\mathbf{V},\mathbf{U}) = v_i^* u_i = u_i v_i^* = (u_i^* v_i)^* = (\mathbf{U},\mathbf{V})^*$$

$$(\mathbf{U},\mathbf{V} + \mathbf{W}) = u_i^*(v_i + w_i) = u_i^* v_i + u_i^* w_i$$

$$= (\mathbf{U},\mathbf{V}) + (\mathbf{U},\mathbf{W})$$

$$(a\mathbf{U},\mathbf{V}) = (au_i)^* v_i = a^* u_i^* v_i = a^*(\mathbf{U},\mathbf{V})$$

$$(\mathbf{U},\mathbf{U}) = u_i^* u_i = |u_1|^2 + |u_2|^2 + \cdots + |u_n|^2$$

Since $(\mathbf{U},\mathbf{U})$ is a sum of squares of real numbers, it will be greater than or equal to zero, and it will be zero if and only if $u_1 = u_2 = \cdots = u_n = 0$.   Hence, we have verified that this definition of scalar product satisfies all the postulates and therefore the other properties follow without further proof.   For example, Schwarz's inequality takes the form

$$|u_i^* v_i| \leq \sqrt{(u_i^* u_i)(v_j^* v_j)}$$

and the triangle inequality becomes

$$\sqrt{(u_i^* + v_i^*)(u_i + v_i)} \leq \sqrt{u_i^* u_i} + \sqrt{v_i^* v_i}$$

As another example, consider the set of complex-valued functions of the real variable x defined on the interval $0 \leq x \leq 1$ and which have at most a finite number of discontinuities.   Let us assume that they are Riemann integrable and square integrable; that is, the integrals

$$\int_0^1 f(x)\, dx \qquad \text{and} \qquad \int_0^1 |f(x)|^2\, dx$$

exist, where these may be improper integrals.   We define addition in the space by

$$\mathbf{f} + \mathbf{g} = f(x) + g(x) \qquad 0 \leq x \leq 1$$

Obviously $\mathbf{f} + \mathbf{g}$ has only a finite number of discontinuities and is integrable. If f and g are both bounded, then f^*g and g^*f are bounded integrable functions, and hence

$$|f + g|^2 = (f^* + g^*)(f + g)$$

$$= |f|^2 + |g|^2 + f^*g + g^*f$$

is integrable.   If either f or g are unbounded, then we are dealing with improper integrals and to show the integrability of $|f + g|^2$ we have to make use of the inequality

$$|f + g|^2 = |f|^2 + |g|^2 + f^*g + g^*f$$
$$\leq |f|^2 + |g|^2 + 2|f^*g|$$
$$\leq |f|^2 + |g|^2 + 2|f|\,|g|$$
$$\leq 2|f|^2 + 2|g|^2$$

which holds at each point of the interval. Therefore

$$\int_0^1 |f + g|^2\,dx \leq 2\int_0^1 |f|^2\,dx + 2\int_0^1 |g|^2\,dx$$

which shows the square integrability of $f + g$. This proves the closure property for addition.

Multiplication by a complex scalar is defined as

$$a\mathbf{f} = af(x) \qquad 0 \leq x \leq 1$$

Closure under multiplication by a scalar is easily demonstrated.

$$\int_0^1 |af|^2\,dx = \int_0^1 |a|^2\,|f|^2\,dx = |a|^2 \int_0^1 |f|^2\,dx < \infty$$

The other postulates of a vector space are easily verified.

For scalar product in this space we define

$$(\mathbf{f},\mathbf{g}) = \int_0^1 f^*g\,dx$$

This quantity exists for every pair of square integrable functions, for

$$|f^*g| = |f|\,|g| \leq \tfrac{1}{2}(|f|^2 + |g|^2)$$

and

$$\int_0^1 |f^*g|\,dx \leq \frac{1}{2}\int_0^1 (|f|^2 + |g|^2)\,dx < \infty$$

and a function which is absolutely integrable is integrable.

This scalar product satisfies the postulates since

$$(\mathbf{g},\mathbf{f})^* = \left(\int_0^1 g^*f\,dx\right)^* = \int_0^1 (g^*f)^*\,dx$$

$$= \int_0^1 f^*g\,dx = (\mathbf{f},\mathbf{g})$$

$$(\mathbf{f},\mathbf{g} + \mathbf{h}) = \int_0^1 f^*(g + h)\,dx$$

$$= \int_0^1 f^*g\,dx + \int_0^1 f^*h\,dx$$

$$= (\mathbf{f},\mathbf{g}) + (\mathbf{f},\mathbf{h})$$

$$(\mathbf{f},\mathbf{f}) = \int_0^1 f^*f\,dx = \int_0^1 |f|^2\,dx \geq 0$$

If $(\mathbf{f},\mathbf{f}) = 0$, then $f(x) = 0$ *almost everywhere*.   By this we mean that $f(x) = 0$ except on a finite number of points, a set which does not contribute to the integral.   In this case the zero vector of our space is any function which is zero almost everywhere.   A more natural setting for this discussion would be the theory of Lebesgue integration.   One could then talk about the set of Lebesgue square integrable functions $L_2(0,1)$ and define the scalar product in terms of the Lebesgue integral.   All the above properties would hold where *almost everywhere* now refers to *except on a set of measure zero*.   The finite sets we have used are of measure zero but there are also infinite sets of measure zero.   The interested reader should refer to a book on integration theory like A. E. Taylor, "General Theory of Functions and Integration," Blaisdell Publishing Company (division of Ginn and Company), Waltham, Mass., 1965.

Schwarz's inequality for this function space takes the form

$$\left| \int_0^1 f^* g \, dx \right| \leq \sqrt{\int_0^1 |f|^2 \, dx} \sqrt{\int_0^1 |g|^2 \, dx}$$

and the triangle inequality is

$$\left(\int_0^1 |f + g|^2 \, dx \right)^{\frac{1}{2}} \leq \left(\int_0^1 |f|^2 \, dx \right)^{\frac{1}{2}} + \left(\int_0^1 |g|^2 \, dx \right)^{\frac{1}{2}}$$

In a vector space we sometimes refer to a vector as a **point**.   We then define the **distance between two points** as the norm of the difference of the two vectors; that is,

$$d(\mathbf{U},\mathbf{V}) = \|\mathbf{U} - \mathbf{V}\|$$

We see that this definition of distance satisfies all the usual properties which we associate with a distance; that is,

1. The distance is positive unless the two points coincide.

$$\|\mathbf{U} - \mathbf{V}\| \geq 0$$
$$\|\mathbf{U} - \mathbf{V}\| = 0 \text{ if and only if } \mathbf{U} = \mathbf{V}$$

2. The distance is symmetric.

$$\|\mathbf{U} - \mathbf{V}\| = \|\mathbf{V} - \mathbf{U}\|$$

3. The triangle inequality is satisfied.

$$\|\mathbf{U} - \mathbf{V}\| \leq \|\mathbf{U} - \mathbf{W}\| + \|\mathbf{W} - \mathbf{V}\|$$

A vector space in which distance between points is defined with these three properties is called a **metric space**.[1]   A vector space may be a metric space without having a scalar product (see exercises 5 and 6).

---

[1] $d(\mathbf{U},\mathbf{V})$ is called the **metric** for the space.

In the case of the vector space of n-tuples of complex numbers, the distance formula is

$$d(\mathbf{U},\mathbf{V}) = \sqrt{(u_i - v_i)^*(u_i - v_i)} = \|\mathbf{U} - \mathbf{V}\|$$

In the case of the vector space of square integrable functions,

$$d(\mathbf{f},\mathbf{g}) = \sqrt{\int_0^1 |f - g|^2 \, dx} = \|\mathbf{f} - \mathbf{g}\|$$

In this case $d = 0$ implies $f = g$ almost everywhere.   Here two functions are considered equal even though they may differ on a set of points which does not contribute to the integral.

Exercises 1.6

1. Prove the following properties of scalar product from the postulates:

a. $(\mathbf{U} + \mathbf{V}, \mathbf{W}) = (\mathbf{U},\mathbf{W}) + (\mathbf{V},\mathbf{W})$

b. $\|a\mathbf{U}\| = |a| \|\mathbf{U}\|$

2. Prove the following statements for a vector space with a scalar product:

a. The parallelogram rule

$$\|\mathbf{U} + \mathbf{V}\|^2 + \|\mathbf{U} - \mathbf{V}\|^2 = 2 \|\mathbf{U}\|^2 + 2 \|\mathbf{V}\|^2$$

b. The pythagorean theorem

$$\|\mathbf{U} + \mathbf{V}\|^2 = \|\mathbf{U}\|^2 + \|\mathbf{V}\|^2 \text{ if } (\mathbf{U},\mathbf{V}) = 0$$

c. $\|\mathbf{U} - \mathbf{V}\| \geq | \|\mathbf{U}\| - \|\mathbf{V}\| |$

3. Show that Schwarz's inequality is an equality if and only if the two vectors are proportional.

*4. Show that the triangle inequality is an equality if and only if the two vectors are proportional and the constant of proportionality is a nonnegative real number.

*5. Suppose that in a vector space over the field of real numbers a positive-definite norm is defined for each vector which satisfies the triangle inequality and $\|a\mathbf{U}\| = |a| \|\mathbf{U}\|$. Show that a real-valued scalar product can be defined as follows:

$$(\mathbf{U},\mathbf{V}) = \tfrac{1}{2}\{\|\mathbf{U} + \mathbf{V}\|^2 - \|\mathbf{U}\|^2 - \|\mathbf{V}\|^2\}$$

which satisfies the postulates, if the following identity is satisfied by the norms:

$$\|\mathbf{U} + \mathbf{V}\|^2 + \|\mathbf{U} - \mathbf{V}\|^2 = 2 \|\mathbf{U}\|^2 + 2 \|\mathbf{V}\|^2$$

*6. Consider the vector space of n-tuples of real numbers with norm defined by $\|\mathbf{U}\| = |u_1| + |u_2| + \cdots + |u_n|$. Show that this has the desired properties of a norm, that is, positive-definiteness, $\|a\mathbf{U}\| = |a| \|\mathbf{U}\|$, and triangle inequality, but that one cannot define from it a scalar product as in exercise 5.

1.7 Orthonormal Basis. Linear Transformations

We have already seen that in an n-dimensional vector space there are many bases, all of which are sets of n linearly independent vectors which span the space.   If in the space we can define a scalar product, we can select certain bases with special properties.   The most important of these is known as an **orthonormal basis.**

Two vectors are **orthogonal** if and only if their scalar product is zero.   A vector is said to be **normalized** if its norm is one.   A set of vectors $\mathbf{X}_i$ is **orthonormal** if

$$(\mathbf{X}_i, \mathbf{X}_j) = \delta_{ij}$$

Suppose in an n-dimensional vector space we have a set of n orthonormal vectors which span the space; then they must be linearly independent, for

$$a_i \mathbf{X}_i = \mathbf{0}$$

implies that $a_i = 0$ for all i, since

$$0 = (\mathbf{X}_j, \mathbf{0}) = (\mathbf{X}_j, a_i \mathbf{X}_i) = a_i \delta_{ji} = a_j$$

Therefore, the set of vectors $\mathbf{X}_i$ is a basis.   Having such a basis is very convenient, because the representation of any vector in the space is very easy to find in terms of an orthonormal basis.

Suppose $\mathbf{U}$ is a vector with a representation

$$\mathbf{U} = u_i \mathbf{X}_i$$

in terms of the orthonormal basis $\mathbf{X}_i$.   Then

$$\begin{aligned}(\mathbf{X}_j, \mathbf{U}) &= u_i (\mathbf{X}_j, \mathbf{X}_i) \\ &= u_i \delta_{ji} \\ &= u_j \end{aligned}$$

We call the u_i the **components** of $\mathbf{U}$ relative to the orthonormal basis $\mathbf{X}_i$.   If we have another vector $\mathbf{V}$ with representation

$$\mathbf{V} = v_i \mathbf{X}_i$$

then

$$\begin{aligned}(\mathbf{U}, \mathbf{V}) &= (u_i \mathbf{X}_i, v_j \mathbf{X}_j) \\ &= u_i^* v_j (\mathbf{X}_i, \mathbf{X}_j) \\ &= u_i^* v_j \delta_{ij} \\ &= u_i^* v_i \end{aligned}$$

Recall that we have an isomorphism between the n-dimensional vector space and the space of n-tuples of complex numbers and that the scalar product in the space of n-tuples for a pair of vectors $(u_1, u_2, \ldots, u_n)$ and $(v_1, v_2, \ldots, v_n)$ is $u_i^* v_i$.   We now see that under this isomorphism scalar product is preserved; that is,

$$(\mathbf{U}, \mathbf{V}) = u_i^* v_i$$

We now show that from any linearly independent set of vectors which span the vector space we can construct an orthonormal basis. This process is known as the **Schmidt process**. Let Y_i be a set of linearly independent vectors not orthonormal. Then compute

$$X_1 = \frac{Y_1}{\|Y_1\|}$$

X_1 is then normalized, for

$$\|X_1\|^2 = \frac{(Y_1, Y_1)}{\|Y_1\|^2} = 1$$

Next compute

$$X_2 = \frac{Y_2 - c_1 X_1}{\|Y_2 - c_1 X_1\|}$$

where c_1 is determined so that $(X_1, X_2) = 0$; that is,

$$(X_1, X_2) = \frac{(X_1, Y_2) - c_1}{\|Y_2 - c_1 X_1\|} = 0$$

$$c_1 = (X_1, Y_2)$$

Next compute

$$X_3 = \frac{Y_3 - c_2 X_1 - c_3 X_2}{\|Y_3 - c_2 X_1 - c_3 X_2\|}$$

with

$$c_2 = (X_1, Y_3)$$

$$c_3 = (X_2, Y_3)$$

This process is continued until all the Y's are used up and as many X's are computed as there are Y's. None of the X's can be the zero vector, because that would imply that a linear combination of the Y's gives the zero vector, contradicting the linear independence of the Y's.

Let us now consider the possibility of changing the representation of a vector space by a change of basis. Let X_i and Y_i be two bases for the same vector space; then any vector U has a representation in terms of each basis.

$$U = u_i X_i$$

$$U = \bar{u}_i Y_i$$

Also the X's have representations in terms of the Y's.

$$X_i = a_{ji} Y_j$$

Substituting, we have

$$U = \bar{u}_i Y_i = u_j X_j = u_j a_{ij} Y_i = a_{ij} u_j Y_i$$

Since the representation of a vector in terms of a given basis is unique, we have

$$\bar{u}_i = a_{ij} u_j$$

Let U be a column matrix with elements u_i, $\bar{U}$ be a column matrix with elements $\bar{u}_i$, and A be a square matrix with elements a_{ij}; then the change of representation can be written in terms of matrices as

$$\bar{U} = AU$$

If α is any scalar, then

$$\alpha\mathbf{U} = \alpha u_i\mathbf{X}_i = \alpha\bar{u}_i\mathbf{Y}_i$$

and the change of representation can be written as

$$A(\alpha U) = \alpha(AU) = \alpha\bar{U}$$

If we have two vectors $\mathbf{U}$ and $\mathbf{V}$, then

$$\mathbf{U} + \mathbf{V} = (u_i + v_i)\mathbf{X}_i = (\bar{u}_i + \bar{v}_i)\mathbf{Y}_i$$

and

$$A(U + V) = AU + AV = \bar{U} + \bar{V}$$

Putting the last two statements together, we have

$$A(\alpha U + \beta V) = \alpha AU + \beta AV = \alpha\bar{U} + \beta\bar{V}$$

In summary, we can say that the change of representation, which can be characterized by the square matrix A, is a **linear transformation** of the vector space.

Operators which have the property

$$\mathcal{O}(\alpha p + \beta q) = \alpha\mathcal{O}(p) + \beta\mathcal{O}(q)$$

are called **linear operators.** Besides the one considered above, some familiar examples are the derivative and the integral; that is,

$$\frac{d}{dx}[\alpha f(x) + \beta q(x)] = \alpha\frac{d}{dx}f(x) + \beta\frac{d}{dx}q(x)$$

$$\int[\alpha f(x) + \beta q(x)]\,dx = \alpha\int f(x)\,dx + \beta\int q(x)\,dx$$

Now let us specialize to the case where we change from one orthonormal basis $\mathbf{X}_i$ to another orthonormal basis $\mathbf{Y}_i$. We have the relation

$$\mathbf{X}_i = a_{ji}\mathbf{Y}_j$$

and

$$\begin{aligned}
\delta_{ik} = (\mathbf{X}_i, \mathbf{X}_k) &= (a_{ji}\mathbf{Y}_j, a_{mk}\mathbf{Y}_m) \\
&= a_{ji}^* a_{mk}(\mathbf{Y}_j, \mathbf{Y}_m) \\
&= a_{ji}^* a_{mk}\delta_{jm} \\
&= a_{ji}^* a_{jk} \\
&= (a_{ij}^*)' a_{jk}
\end{aligned}$$

or

$$(A^*)'A = I$$

This says that A is a **unitary matrix.**   If we are working with a vector space over the field of real numbers, the matrix A will be real and A will be **orthogonal.**   A transformation of a vector space

$$\bar{U} = AU$$

where A is unitary is called a **unitary transformation.**

One of the important properties of a unitary transformation is that it leaves the scalar product invariant; that is,

$$\begin{aligned}
\bar{u}_i^* \bar{v}_i = (\bar{U}^*)' \bar{V} &= (A^* U^*)' AV \\
&= (U^*)'(A^*)' AV \\
&= (U^*)' IV \\
&= (U^*)' V
\end{aligned}$$

Since the length or norm of a vector can be expressed in terms of the scalar product, we sometimes say that a *unitary transformation is a norm preserving transformation.*

Exercises 1.7

1. Test the following set of vectors for linear independence and construct from it an orthonormal basis: $(1,0,1,0)$, $(1,-1,0,1)$, $(0,1,-1,1)$, $(1,-1,1,-1)$.

*2. Consider two sets of rectangular cartesian-coordinate systems for euclidean three space, with a common origin.   Unit vectors in the direction of the three coordinate axes for the two coordinate systems serve as two different orthonormal bases.   In terms of the cosines of the angles between the two sets of axes, find the matrix A relating the two coordinate systems

$$\bar{X} = AX$$

and show that A is orthogonal.

3. Given two bases $\mathbf{X}_i$ and $\mathbf{Y}_i$ such that $\mathbf{X}_i = a_{ji}\mathbf{Y}_j$.   Prove that the matrix A with elements a_{ij} is nonsingular.   Hence, find the inverse transformation.

4. Consider the set of all linear transformations of a finite-dimensional vector space, corresponding to change of bases.   Define the product of two transformations as follows.   If $\bar{U} = AU$ and $\bar{\bar{U}} = B\bar{U}$, then $\bar{\bar{U}} = BAU = CU$, where $C = BA$.   Prove the following:

  a. The product of three transformations is associative.
  b. There exists an identity transformation such that $AI = IA = A$ for every A.
  c. For every transformation A there exists an inverse A^{-1}, such that

$$A^{-1}A = AA^{-1} = I$$

Hence, the set of transformations is a **group.**

*5. A set of functions $f_n(x)$ is said to be orthonormal on the interval $a \leq x \leq b$, if

$$\int_a^b f_n^*(x) f_m(x)\, dx = \delta_{nm}$$

Show that $f_n(x) = (1/\sqrt{\pi}) \sin nx$ is orthonormal on the interval $0 \leq x \leq 2\pi$.

*6. The set of functions $1, x, x^2, \ldots, x^n, \ldots$ is linearly independent. It is not, however, orthonormal. Construct the first four of a set of polynomials from these which are orthonormal on the interval $-1 \leq x \leq 1$. These polynomials are proportional to the **Legendre polynomials** (see Sec. 4.3).

7. Given an m-dimensional subspace in an n-dimensional vector space, show that any vector in the vector space can be uniquely represented as the sum of a vector in the subspace and a vector orthogonal to every vector in the subspace.

*8. Find c_i, $i = 1, 2, \ldots, m$, which minimize $\|\mathbf{U} - c_i\mathbf{X}_i\|^2$, where $\mathbf{U}$ is a given vector in n-dimensional vector space, and $\mathbf{X}_i$ is a set of m orthonormal vectors. Also prove **Bessel's inequality**

$$\|\mathbf{U}\|^2 \geq \sum_{i=1}^{m} |(\mathbf{X}_i, \mathbf{U})|^2$$

When does Bessel's inequality become an equality?

*9. Prove **Parseval's equation** for a finite-dimensional vector space with scalar product

$$(\mathbf{U}, \mathbf{V}) = (\mathbf{X}_i, \mathbf{U})^*(\mathbf{X}_i, \mathbf{V})$$

where $\mathbf{X}_i$ is an orthonormal basis.

*10. A matrix A represents a **self-adjoint** linear transformation if, for every vector pair $\mathbf{X}$ and $\mathbf{Y}$ on which it may act, $(Y, AX) = (AY, X)$. Prove that A is hermitian if and only if it is self-adjoint in the whole space.

11. Let f be a function which assigns to each vector of an n-dimensional vector space S_n a unique vector of an m-dimensional vector space S_m, where f is linear; that is, $f(\alpha\mathbf{U} + \beta\mathbf{V}) = \alpha f(\mathbf{U}) + \beta f(\mathbf{V})$. Let S_n have a basis $\mathbf{X}_i$ and S_m have a basis $\mathbf{Y}_j$. Show that f can be characterized by the $m \times n$ matrix A with elements a_{ji}, where $f(\mathbf{X}_i) = a_{ji}\mathbf{Y}_j$.

12. Consider the vector space of n-tuples of complex numbers. In terms of a given basis $\mathbf{X}_i$, a linear transformation of the space into itself is given by $f(\mathbf{X}_i) = a_{ji}\mathbf{X}_j$, $f(\mathbf{U}) = f(u_i\mathbf{X}_i) = a_{ji}u_i\mathbf{X}_j = \mathbf{V} = v_j\mathbf{X}_j$; that is, $V = AU$. Now suppose we change the basis as follows, $\mathbf{X}_i = b_{ji}\mathbf{Y}_j$. Show that in terms of the new basis $\bar{V} = (B^{-1}AB)\bar{U}$.

1.8 Quadratic Forms. Hermitian Forms

In terms of three-dimensional rectangular cartesian coordinates x_1, x_2, x_3, we can write the equation of the most general quadric surface as[1]

$$X'AX + BX + k = 0$$

where

$$X = \begin{Vmatrix} x_1 \\ x_2 \\ x_3 \end{Vmatrix}$$

is a column matrix representing the vector from the origin to a point with coordinates x_1, x_2, x_3 on the surface. A is a 3×3 symmetric matrix of real constants, B is a 1×3 row matrix of real constants, and k is a real constant.

---

[1] See exercise 5, Sec. 1.2.

The first term in this equation,

$$Q = X'AX = a_{11}x_1^2 + a_{22}x_2^2 + a_{33}x_3^2 + 2a_{12}x_1x_2 + 2a_{13}x_1x_3 + 2a_{23}x_2x_3$$

is a quadratic form in the three variables. A standard problem in analytic geometry is to determine an orthogonal transformation of coordinates (rotation of axes)

$$X = T\bar{X}$$

which will reduce Q to the diagonal form

$$Q = \bar{X}'T'AT\bar{X} = \bar{X}'D\bar{X} = \lambda_1\bar{x}_1^2 + \lambda_2\bar{x}_2^2 + \lambda_3\bar{x}_3^2$$

This problem occurs frequently in other applications, so we shall study it in its more general form in n-dimensional vector space.

Let

$$X = \begin{Vmatrix} x_1 \\ x_2 \\ \cdot \\ \cdot \\ \cdot \\ x_n \end{Vmatrix}$$

be a column matrix representing the n-dimensional vector $(x_1, x_2, \ldots, x_n)$ with real components.

$$Q = X'AX$$

is a quadratic form in $x_1, x_2, \ldots, x_n$, where A is an $n \times n$ real matrix. Without loss of generality, we can assume that A is symmetric, for

$$Q = X'[\tfrac{1}{2}(A + A') + \tfrac{1}{2}(A - A')]X$$
$$= X'[\tfrac{1}{2}(A + A')]X$$

and $\tfrac{1}{2}(A + A')$ is symmetric. The problem is to find an orthogonal transformation

$$X = T\bar{X}$$

which will reduce the quadratic form Q to the diagonal form

$$Q = X'AX = \bar{X}'T'AT\bar{X} = \bar{X}'D\bar{X} = \lambda_1\bar{x}_1^2 + \lambda_2\bar{x}_2^2 + \cdots + \lambda_n\bar{x}_n^2$$

In other words, the problem reduces to finding a matrix T such that $T'AT = D$, where D is diagonal. This can always be done for a real symmetric matrix A. The proof follows.

Since we are assuming that T is orthogonal, we have[1]

$$TT'AT = AT = TD$$
$$a_{ij}t_{j\alpha} = t_{ij}d_{j\alpha} = \lambda_\alpha t_{i\alpha}$$

[1] Recall that the summation convention applies only to Latin subscripts. Therefore, there is no summation on α.

Let the αth column of T be Y_α, a column matrix; then the last equation becomes

$$A Y_\alpha = \lambda_\alpha Y_\alpha$$

Dropping the subscript α, we are seeking vectors Y such that

$$A Y = \lambda Y$$

This equation can also be written

$$(A - \lambda I) Y = 0$$

This equation will have a nontrivial solution if and only if

$$|A - \lambda I| = 0$$

This is called the **characteristic equation.** It is an nth-degree polynomial equation in λ, which has n roots, $\lambda_1, \lambda_2, \ldots, \lambda_n$, not always distinct. These are called the **characteristic roots** or **eigenvalues** of the matrix A.

We first show that all eigenvalues are real. For each eigenvalue λ there is a nontrivial **eigenvector** Y such that

$$A Y = \lambda Y$$

Taking the conjugate of both sides of the equation, we have

$$A Y^* = \lambda^* Y^*$$

Taking the transpose,

$$(Y^*)' A = \lambda^* (Y^*)'$$

Then

$$(Y^*)' A Y = \lambda^* (Y^*)' Y = \lambda^* \, \|\mathbf{Y}\|^2$$

Also

$$(Y^*)' A Y = \lambda (Y^*)' Y = \lambda \, \|\mathbf{Y}\|^2$$

Subtracting,

$$(\lambda^* - \lambda) \, \|\mathbf{Y}\|^2 = 0$$

from which we conclude that λ is real.

Next we show that for two different eigenvalues the corresponding eigenvectors are orthogonal. Assuming $\lambda_\alpha \neq \lambda_\beta$, then

$$A Y_\alpha = \lambda_\alpha Y_\alpha$$
$$A Y_\beta = \lambda_\beta Y_\beta$$
$$Y'_\alpha A = \lambda_\alpha Y'_\alpha$$
$$Y'_\alpha A Y_\beta = \lambda_\alpha Y'_\alpha Y_\beta = \lambda_\alpha (\mathbf{Y}_\alpha, \mathbf{Y}_\beta)$$
$$Y'_\alpha A Y_\beta = \lambda_\beta Y'_\alpha Y_\beta = \lambda_\beta (\mathbf{Y}_\alpha, \mathbf{Y}_\beta)$$

Subtracting,

$$(\lambda_\alpha - \lambda_\beta)(\mathbf{Y}_\alpha, \mathbf{Y}_\beta) = 0$$

Therefore,

$$(\mathbf{Y}_\alpha, \mathbf{Y}_\beta) = 0$$

If all eigenvalues are different, then the columns of T, treated as n-dimensional vectors, are orthogonal to one another. They can also be normalized, since if

$$AY = \lambda Y$$

then

$$A\frac{Y}{\|\mathbf{Y}\|} = \lambda \frac{Y}{\|\mathbf{Y}\|}$$

Therefore,

$$T'T = I$$

and in this case our problem is solved. The diagonal elements of D are the n different eigenvalues of A.

If some of the eigenvalues are equal, then we have to proceed differently. In this case, we can select a subset of eigenvalues $\lambda_1, \lambda_2, \ldots, \lambda_m$, with $1 \leq m \leq n$, which are all different. The corresponding eigenvectors $\mathbf{Y}_1, \mathbf{Y}_2, \ldots, \mathbf{Y}_m$ are an orthonormal set. Now we define an orthogonal matrix

$$S = \|Y_1 Y_2 \cdots Y_m Z_1 Z_2 \cdots Z_{n-m}\|$$

where the Z's are column matrices obtained by solving the equations

$$(\mathbf{Y}_i, \mathbf{Z}_j) = 0 \qquad i = 1, 2, \ldots, m$$
$$(\mathbf{Z}_k, \mathbf{Z}_j) = 0 \qquad k = 1, 2, \ldots, j - 1$$

These equations always have a nontrivial solution, since they are a set of $m + k$ homogeneous linear equations in n unknowns with $m + k < n$. We also assume that the $\mathbf{Z}$'s have been normalized. If we perform the multiplication $S'AS$, we have

$$S'AS = \begin{Vmatrix} \lambda_1 & 0 & 0 & \cdots & 0 & 0 & 0 & \cdots & 0 \\ 0 & \lambda_2 & 0 & \cdots & 0 & 0 & 0 & \cdots & 0 \\ 0 & 0 & \lambda_3 & \cdots & 0 & 0 & 0 & \cdots & 0 \\ \cdots & \cdots & \cdots & \cdots & \cdots & \cdots & \cdots & \cdots & \cdots \\ 0 & 0 & 0 & \cdots & \lambda_m & 0 & 0 & \cdots & 0 \\ 0 & 0 & 0 & \cdots & 0 & c_{11} & c_{12} & \cdots & c_{1\,n-m} \\ 0 & 0 & 0 & \cdots & 0 & c_{21} & c_{22} & \cdots & c_{2\,n-m} \\ \cdots & \cdots & \cdots & \cdots & \cdots & \cdots & \cdots & \cdots & \cdots \\ 0 & 0 & 0 & \cdots & 0 & c_{n-m\,1} & c_{n-m\,2} & \cdots & c_{n-m\,n-m} \end{Vmatrix}$$

The matrix C is a real symmetric $(n - m) \times (n - m)$ matrix. It is not diagonal, because the $\mathbf{Z}$'s do not satisfy the eigenvector equation. Under the orthogonal transformation $X = S\bar{X}$ the quadratic form becomes

$$Q = \bar{X}'S'AS\bar{X} = \lambda_1\bar{x}_1^2 + \lambda_2\bar{x}_2^2 + \cdots + \lambda_m\bar{x}_m^2 + \bar{Q}$$

where

$$\bar{Q} = U'CU \qquad U = \left\| \begin{matrix} \bar{x}_{m+1} \\ \bar{x}_{m+2} \\ \cdot \\ \cdot \\ \cdot \\ \bar{x}_n \end{matrix} \right\|$$

The eigenvalues of C are also eigenvalues of A, since

$$|S'AS - \lambda I| = |S'AS - \lambda S'IS|$$
$$= |S'| \, |S| \, |A - \lambda I|$$
$$= (\lambda_1 - \lambda)(\lambda_2 - \lambda) \cdots (\lambda_m - \lambda) \, |C - \lambda I|$$

Therefore, the eigenvalues of C are the repeated eigenvalues of A.

The next step is to reduce $\bar{Q}$ to diagonal form by finding an orthogonal matrix R such that $R'CR = \bar{D}$, a diagonal matrix. This is just the problem we had before, only in an $(n - m)$-dimensional space. For simplicity, let us assume that the eigenvalues of C are distinct. If not we can repeat the above process again. The eigenvalues of C and corresponding eigenvectors satisfy

$$CV_\alpha = \lambda_\alpha V_\alpha \qquad \alpha = m + 1, m + 2, \ldots, n$$

Consider the matrix

$$\bar{R} = \left\| \begin{matrix} 1 & 0 & 0 & \cdots & 0 & 0 & 0 & \cdots & 0 \\ 0 & 1 & 0 & \cdots & 0 & 0 & 0 & \cdots & 0 \\ 0 & 0 & 1 & \cdots & 0 & 0 & 0 & \cdots & 0 \\ \multicolumn{9}{c}{\cdots\cdots\cdots\cdots\cdots\cdots\cdots\cdots} \\ 0 & 0 & 0 & \cdots & 1 & 0 & 0 & \cdots & 0 \\ 0 & 0 & 0 & \cdots & 0 & r_{11} & r_{12} & \cdots & r_{1n-m} \\ 0 & 0 & 0 & \cdots & 0 & r_{21} & r_{22} & \cdots & r_{2n-m} \\ \multicolumn{9}{c}{\cdots\cdots\cdots\cdots\cdots\cdots\cdots\cdots} \\ 0 & 0 & 0 & \cdots & 0 & r_{n-m1} & r_{n-m2} & \cdots & r_{n-m\ n-m} \end{matrix} \right\|$$

Since R is orthogonal, it is easily shown that $\bar{R}$ is orthogonal. If the orthogonal transformation $\bar{R}'$ is applied to $\bar{X}$, it does not affect the part of the quadratic form already diagonalized, but it does diagonalize $\bar{Q}$. Therefore, we have

$$X = S\bar{X} = S\bar{R}\bar{\bar{X}} = T\bar{\bar{X}}$$

a transformation which reduces Q to diagonal form. T is orthogonal, since

$$T^{-1} = (S\bar{R})^{-1} = \bar{R}^{-1}S^{-1} = \bar{R}'S' = (S\bar{R})' = T'$$

Therefore, T is the desired orthogonal transformation. The columns of T are the complete set of eigenvectors. Since orthonormal vectors are linearly independent, the eigenvalue problem leads to precisely n linearly independent vectors, whether there are repeated eigenvalues or not. If a certain eigenvalue is a multiple root of the characteristic equation, we say that the eigenvalue is **degenerate**. If it occurs as a root m times, we say that it is m-**fold degenerate**. Nevertheless, there are still m linearly independent eigenvectors, each satisfying the eigenvector equation for the same value of λ, and these vectors span an m-dimensional subspace.

Since the eigenvectors associated with a given real symmetric $n \times n$ matrix are a set of n linearly independent vectors, they form a basis for the n-dimensional space. Therefore, any vector in the space can be expanded as a linear combination of the eigenvectors. This fact allows us to solve certain systems of equations in terms of the eigenvectors of a matrix appearing in the equations. Consider, for example, the nonhomogeneous system of equations

$$AX - \lambda X = C$$

where X is an $n \times 1$ column matrix representing the n unknowns, A is a given $n \times n$ real symmetric matrix, λ is a given constant, and C is a given $n \times 1$ column matrix of real constants. Since X is an n-dimensional vector, it can be written as a linear combination of the eigenvectors Y_α of A; that is,

$$X = \sum_{\alpha=1}^{n} \gamma_\alpha Y_\alpha$$

Then the system of equations becomes

$$A \sum_{\alpha=1}^{n} \gamma_\alpha Y_\alpha - \lambda \sum_{\alpha=1}^{n} \gamma_\alpha Y_\alpha = C$$

$$\sum_{\alpha=1}^{n} \lambda_\alpha \gamma_\alpha Y_\alpha - \lambda \sum_{\alpha=1}^{n} \gamma_\alpha Y_\alpha = C$$

$$\sum_{\alpha=1}^{n} \lambda_\alpha \gamma_\alpha (\mathbf{Y}_\alpha, \mathbf{Y}_\beta) - \lambda \sum_{\alpha=1}^{n} \gamma_\alpha (\mathbf{Y}_\alpha, \mathbf{Y}_\beta) = (\mathbf{C}, \mathbf{Y}_\beta)$$

$$\lambda_\beta \gamma_\beta - \lambda \gamma_\beta = (\mathbf{C}, \mathbf{Y}_\beta)$$

$$\gamma_\beta = \frac{(\mathbf{C}, \mathbf{Y}_\beta)}{\lambda_\beta - \lambda}$$

Thus the problem is solved, provided $\lambda \neq \lambda_\beta$ for any β. If $\lambda = \lambda_\alpha$ for some β, then there is no solution to the problem unless $(\mathbf{Y}_\beta, \mathbf{C}) = 0$. In this case, γ_β is arbitrary, and there are infinitely many solutions.

A real quadratic form is said to be **positive-definite** if $Q = X'AX > 0$ unless $X = 0$. Obviously, if Q is in the diagonal form

$$Q = \lambda_1 x_1^2 + \lambda_2 x_2^2 + \cdots + \lambda_n x_n^2$$

and all the λ's are positive, then $Q = 0$ implies $x_1 = x_2 = \cdots = x_n = 0$, or $X = O$. Otherwise $Q > 0$, and therefore Q is positive-definite. If Q is not in diagonal form, it can nevertheless be reduced to diagonal form by an orthogonal transformation $X = T\bar{X}$. Under this transformation

$$Q = X'AX = \bar{X}'D\bar{X} = \lambda_1\bar{x}_1^2 + \lambda_2\bar{x}_2^2 + \cdots + \lambda_n\bar{x}_n^2$$

where $\lambda_1, \lambda_2, \ldots, \lambda_n$ are the eigenvalues of A. If all the λ's are positive, then $Q > 0$ unless $\bar{X} = O$. But under the transformation, $X = TO = O$, so that Q is positive-definite if all the eigenvalues of A are positive. On the other hand, if any one eigenvalue $\lambda_k \leq 0$, then there exists a nonzero vector $\bar{X}$ which will make $Q \leq 0$. The corresponding $X = T\bar{X}$ is also nonzero, and therefore Q is not positive-definite. The conclusion is that *a quadratic form $Q = X'AX$ is positive-definite if and only if all the eigenvalues of A are positive.*

A pair of quadratic forms $Q_1 = X'AX$ and $Q_2 = X'BX$ can be simultaneously reduced to diagonal form if one of them is positive-definite. Let Q_1 be positive-definite; then there exists an orthogonal transformation $X = T\bar{X}$ which reduces it to the diagonal form

$$Q_1 = \lambda_1\bar{x}_1^2 + \lambda_2\bar{x}_2^2 + \cdots + \lambda_n\bar{x}_n^2$$

with all the λ's positive. The transformation $\bar{X} = SY$, with

$$S = \begin{Vmatrix} \lambda_1^{-\frac{1}{2}} & 0 & 0 & \cdots & \cdots & 0 \\ 0 & \lambda_2^{-\frac{1}{2}} & 0 & \cdots & \cdots & 0 \\ 0 & 0 & \lambda_3^{-\frac{1}{2}} & \cdots & \cdots & 0 \\ \cdots & \cdots & \cdots & \cdots & \cdots & \cdots \\ \cdots & \cdots & \cdots & \cdots & \cdots & 0 \\ 0 & 0 & 0 & \cdots & \cdots & \lambda_n^{-\frac{1}{2}} \end{Vmatrix}$$

reduces Q_1 to

$$Q_1 = y_1^2 + y_2^2 + \cdots + y_n^2 = Y'Y$$

The same transformation $X = TSY$ changes Q_2 to

$$Q_2 = Y'S'T'BTSY = Y'CY$$

C is real and symmetric since

$$C' = S'T'B'T''S'' = S'T'BTS = C$$

so that Q_2 is a quadratic form in Y which can be reduced to diagonal form by an orthogonal transformation $Y = R\bar{Y}$, giving

$$Q_2 = \mu_1\bar{y}_1^2 + \mu_2\bar{y}_2^2 + \cdots + \mu_n\bar{y}_n^2$$

The transformation $Y = R\bar{Y}$ does not affect the form of Q_1, since

$$Q_1 = Y'Y = \bar{Y}'R'R\bar{Y} = \bar{Y}'\bar{Y}$$

Hence, the total transformation $X = TSR\bar{Y}$ reduces Q_1 and Q_2 simultaneously to diagonal form. The transformation is not, in general, orthogonal.

The problem corresponding to reduction of quadratic forms in the space of vectors with complex components is the problem of reducing hermitian forms to diagonal form. A **hermitian form** is an expression of the form

$$H = (X^*)'AX$$

where A is hermitian; that is, $(A^*)' = A$. The value of H is always real, for

$$H^* = (H^*)' = (X'A^*X^*)' = (X^*)'(A^*)'X = (X^*)'AX = H$$

The main theorem dealing with hermitian forms is that *a hermitian form can always be reduced to diagonal form by a unitary transformation.* The proof follows.

We seek a unitary transformation $X = T\bar{X}$, such that $(T^*)' = T^{-1}$, which will reduce H to diagonal form; that is,

$$H = (X^*)'AX = (\bar{X}^*)'(T^*)'AT\bar{X} = (\bar{X}^*)'D\bar{X}$$
$$= \lambda_1 |\bar{x}_1|^2 + \lambda_2 |\bar{x}_2|^2 + \cdots + \lambda_n |\bar{x}_n|^2$$

In other words, we wish to find a unitary matrix, such that $(T^*)'AT = D$ or $AT = TD$. In subscript notation this is $a_{ij}t_{j\alpha} = \lambda_\alpha t_{i\alpha}$. Let Y_α be the αth column of T; then

$$AY_\alpha = \lambda_\alpha Y_\alpha$$

Thus we are led to the eigenvalue problem $AY = \lambda Y$ and the characteristic equation $|A - \lambda I| = 0$. The characteristic equation has n solutions $\lambda_1, \lambda_2, \ldots, \lambda_n$, not necessarily distinct.

Since A has complex elements, the eigenvectors Y_α have complex components. However, because A is hermitian, the eigenvalues are real.

$$A^*Y^* = \lambda^*Y^*$$
$$(Y^*)'(A^*)' = (Y^*)'A = \lambda^*(Y^*)'$$
$$(Y^*)'AY = \lambda^*(Y^*)'Y = \lambda^* \|\mathbf{Y}\|^2$$
$$(Y^*)'AY = \lambda(Y^*)'Y = \lambda \|\mathbf{Y}\|^2$$

Subtracting, $(\lambda^* - \lambda) \|\mathbf{Y}\|^2 = 0$, and since $\|\mathbf{Y}\|^2 > 0$, λ is real. Also, for two different eigenvalues the eigenvectors are orthogonal. If $\lambda_\alpha \neq \lambda_\beta$,

$$AY_\alpha = \lambda_\alpha Y_\alpha$$
$$AY_\beta = \lambda_\beta Y_\beta$$
$$(Y_\alpha^*)'(A^*)' = (Y_\alpha^*)'A = \lambda_\alpha(Y_\alpha^*)'$$
$$(Y_\alpha^*)'AY_\beta = \lambda_\alpha(Y_\alpha^*)'Y_\beta = \lambda_\alpha(\mathbf{Y}_\alpha,\mathbf{Y}_\beta)$$
$$(Y_\alpha^*)'AY_\beta = \lambda_\beta(Y_\alpha^*)'Y_\beta = \lambda_\beta(\mathbf{Y}_\alpha,\mathbf{Y}_\beta)$$

Subtracting, $(\lambda_\alpha - \lambda_\beta)(\mathbf{Y}_\alpha,\mathbf{Y}_\beta) = 0$, and therefore $(\mathbf{Y}_\alpha,\mathbf{Y}_\beta) = 0$.

If all the eigenvalues are distinct, we have a set of n orthogonal eigenvectors. The eigenvectors can also be normalized, and therefore they constitute an orthonormal set of vectors. The matrix

$$T = \| Y_1 Y_2 \cdots Y_n \|$$

is therefore unitary, and this is the matrix which reduces H to diagonal form. If some of the eigenvalues are equal, we have the degenerate case, and we proceed in a manner quite similar to that used in the case of the quadratic form. The details will not be given.

Exercises 1.8

1. Show that the general equation of a quadric surface in three-dimensional euclidean space,

$$X'AX + BX + k = 0$$

can be reduced to the form $Y'AY = c$ if A is nonsingular.

2. On the basis of the eigenvalues of A, classify the quadric surfaces

$$X'AX + BX + k = 0$$

into ellipsoids, hyperboloids, paraboloids, and cylinders.

3. Find an orthogonal transformation which will reduce

$$Q = 2x_1^2 + 2x_2^2 + 2x_3^2 - x_1x_2 - x_2x_3$$

to diagonal form.

4. Find a transformation which will simultaneously reduce

$$Q_1 = 3x_1^2 - 2x_1x_2 + 3x_2^2$$

and $Q_2 = x_1x_2$ to diagonal form.

5. Show that a pair of hermitian forms can be simultaneously reduced to diagonal form if one of them is positive-definite.

6. The $n \times n$ matrix A is said to be **equivalent** to the $n \times n$ matrix B, written $A \sim B$, if there exist nonsingular matrices S and T such that $SAT = B$. Show that this equivalence relation satisfies the three required properties:

a. Reflexive: for all A, $A \sim A$.

b. Symmetric: if $A \sim B$, then $B \sim A$.

c. Transitive: if $A \sim B$ and $B \sim C$, then $A \sim C$.

7. If T is nonsingular, $T^{-1}AT$ is called a **similarity transformation.** Show that if A and B are equivalent to diagonal matrices under the same similarity transformation, then A and B commute.

8. If $\lambda_1, \lambda_2, \ldots, \lambda_n$ are the eigenvalues of a real symmetric matrix A with corresponding normalized eigenvectors $Y_1, Y_2, \ldots, Y_n$, show that the following equations are satisfied:

a. $(A - \lambda_1 I)a_1 Y_1 = O$

b. $(A - \lambda_2 I)(A - \lambda_1 I)(a_1 Y_1 + a_2 Y_2) = O$

c. $(A - \lambda_n I)(A - \lambda_{n-1} I) \cdots (A - \lambda_1 I)(a_1 Y_1 + a_2 Y_2 + \cdots + a_n Y_n) = O$

for arbitrary $a_1, a_2, \ldots, a_n$. Hence show that $F(A) = O$, where

$$F(\lambda) = (\lambda - \lambda_n)(\lambda - \lambda_{n-1}) \cdots (\lambda - \lambda_1) = 0$$

is the characteristic equation of A.   This is a special case of the **Cayley-Hamilton theorem.**

*9. Let A be an $m \times n$ matrix with elements $a_{ij}(t)$ which are differentiable functions of t.   If we define the derivative of A by

$$\frac{dA}{dt} = \lim_{\Delta t \to 0} \frac{A(t + \Delta t) - A(t)}{\Delta t}$$

show that dA/dt has elements $(d/dt)a_{ij}$.

*10. Let $H(X) = (X^*)'AX$ be a positive-definite hermitian form.   Define the **associated bilinear form** $H(X,Y) = (X^*)'AY$.   Show that $H(X,Y)$ satisfies the postulates for a scalar product.

11. A is **unitarily similar** to B if there exists a unitary matrix T such that $T^{-1}AT = B$.   Prove that A is unitarily similar to a diagonal matrix with real elements if and only if it is hermitian.

12. Prove that an $n \times n$ matrix is similar to a diagonal matrix if and only if it has n independent eigenvectors.

13. Prove that an $n \times n$ matrix is similar to a diagonal matrix if it has n distinct eigenvalues.   HINT: Assume that some subset of the eigenvectors is linearly dependent and obtain a contradiction.

1.9   Systems of Ordinary Differential Equations. Vibration Problems

As an application of some of the ideas developed in the previous sections, we now consider the problem of solving certain systems of ordinary differential equations.   The system of differential equations

$$\alpha_m \frac{d^m x_1}{dt^m} + \alpha_{m-1} \frac{d^{m-1} x_1}{dt^{m-1}} + \cdots + \alpha_1 \frac{dx_1}{dt} = a_{11}x_1 + a_{12}x_2 + \cdots + a_{1n}x_n$$

$$\alpha_m \frac{d^m x_2}{dt^m} + \alpha_{m-1} \frac{d^{m-1} x_2}{dt^{m-1}} + \cdots + \alpha_1 \frac{dx_2}{dt} = a_{21}x_1 + a_{22}x_2 + \cdots + a_{2n}x_n$$

$$\cdots\cdots\cdots\cdots\cdots\cdots\cdots\cdots\cdots\cdots\cdots\cdots\cdots\cdots\cdots\cdots\cdots\cdots$$

$$\alpha_m \frac{d^m x_n}{dt^m} + \alpha_{m-1} \frac{d^{m-1} x_n}{dt^{m-1}} + \cdots + \alpha_1 \frac{dx_n}{dt} = a_{n1}x_1 + a_{n2}x_2 + \cdots + a_{nn}x_n$$

where $\alpha_1, \alpha_2, \ldots, \alpha_m$ and a_{ij} are real constants, is a system of n linear ordinary differential equations in n unknowns.   This system can be written in matrix notation as

$$\mathscr{D}X = AX$$

where $$\mathscr{D} = \alpha_m \frac{d^m}{dt^m} + \alpha_{m-1} \frac{d^{m-1}}{dt^{m-1}} + \cdots + \alpha_1 \frac{d}{dt}$$

is an mth-order linear differential operator.   If A is symmetric, we know that there exists an orthogonal matrix T such that $T'AT = D$, where D is diagonal.   This fact leads to a solution of the system.

Let $X = T\bar{X}$; then $\mathscr{D}X = T\mathscr{D}\bar{X} = AT\bar{X}$.   Multiplying on the left by T gives $\mathscr{D}\bar{X} = T'AT\bar{X} = D\bar{X}$.   In terms of the components of $\bar{X}$, $\mathscr{D}\bar{x}_\beta = \lambda_\beta \bar{x}_\beta$, where the λ's are the eigenvalues of A.   In terms of $\bar{X}$ the variables are separated, so the problem reduces to solving n mth-order ordinary differential equations.   Each $\bar{x}_\beta$ contains m arbitrary constants, so that the general solution of the system has mn arbitrary constants.

Problems of this type occur frequently in the theory of small vibrations. The general theory of small vibrations will be treated in Chap. 3, but we can consider some examples here.   In the case of undamped vibrations, the operator $\mathscr{D}$ is just the second derivative, and the system of differential equations can be written as

$$\ddot{X} = \frac{d^2X}{dt^2} = AX$$

Under the orthogonal transformation $X = T\bar{X}$, the system becomes

$$\frac{d^2\bar{x}_\beta}{dt^2} = \lambda_\beta \bar{x}_\beta$$

which has the solution

$$\bar{x}_\beta = c_\beta \sin{(\sqrt{-\lambda_\beta}\, t + \phi_\beta)}$$

where c_β and ϕ_β are arbitrary constants.   The solution of the original system is $X = T\bar{X}$, and the $2n$ constants c_β and ϕ_β are evaluated in terms of the initial conditions on the x's.

If the initial conditions are properly chosen, it is possible to make $c_\beta = 1$ and $c_i = 0$ for $i \neq \beta$.   For such a choice of initial conditions we have the particular solution

$$x_{i\beta} = t_{ij}\bar{x}_{j\beta} = t_{i\beta} \sin{(\sqrt{-\lambda_\beta}\, t + \phi_\beta)}$$

$$= y_{i\beta} \sin{(\sqrt{-\lambda_\beta}\, t + \phi_\beta)}$$

where the $y_{i\beta}$ is the ith component of the βth eigenvector of A, which satisfies the eigenvalue equation $AY_\beta = \lambda_\beta Y_\beta$.   We notice that this particular solution is one in which every component of the solution has the same circular frequency $\omega_\beta = \sqrt{-\lambda_\beta}$.   Such a solution is called a **normal mode** of the system, with a **natural frequency** ω_β.   The normal modes are directly related to the eigenvalues and eigenvectors of A by the equation

$$\mathbf{X}_\beta = \mathbf{Y}_\beta \sin{(\omega_\beta t + \phi_\beta)}$$

When we substitute $\mathbf{X}_\beta$ into the system of differential equations, we have

$$\frac{d^2}{dt^2}X_\beta = -\omega_\beta^2 Y_\beta \sin{(\omega_\beta t + \phi_\beta)} = AY_\beta \sin{(\omega_\beta t + \phi_\beta)}$$

or

$$AY_\beta = \lambda_\beta Y_\beta$$

the eigenvalue problem. Anticipating this result, we could have started out by seeking those particular solutions all of whose components have the same frequency. This would have led to the eigenvalue problem $AY = \lambda Y$ and characteristic equation $|A - \lambda I| = 0$, the roots of which give the natural frequencies. The general solution is then a linear combination of the normal modes $\mathbf{X} = c_k \mathbf{X}_k$. The normal modes are orthogonal; that is, for $\alpha \neq \beta$

$$(\mathbf{X}_\alpha, \mathbf{X}_\beta) = (\mathbf{Y}_\alpha, \mathbf{Y}_\beta) \sin(\omega_\alpha t + \phi_\alpha) \sin(\omega_\beta t + \phi_\beta) = 0$$

because the eigenvectors are orthogonal. This makes it easy to evaluate the unknown constants in terms of the initial conditions:[1]

$$\mathbf{X}(0) = \sum_{\alpha=1}^{n} c_\alpha \mathbf{Y}_\alpha \sin \phi_\alpha$$

$$[\mathbf{X}(0), \mathbf{Y}_\beta] = c_\beta \sin \phi_\beta$$

$$\dot{\mathbf{X}}(0) = \left(\frac{dX}{dt}\right)_{t=0} = \sum_{\alpha=1}^{n} \omega_\alpha c_\alpha \mathbf{Y}_\alpha \cos \phi_\alpha$$

$$[\dot{\mathbf{X}}(0), \mathbf{Y}_\beta] = \omega_\beta c_\beta \cos \phi_\beta$$

$$c_\beta = \left\{[\mathbf{X}(0), \mathbf{Y}_\beta]^2 - \frac{[\dot{\mathbf{X}}(0), \mathbf{Y}_\beta]^2}{\lambda_\beta}\right\}^{\frac{1}{2}}$$

$$\phi_\beta = \tan^{-1} \frac{\omega_\beta [\mathbf{X}(0), \mathbf{Y}_\beta]}{[\dot{\mathbf{X}}(0), \mathbf{Y}_\beta]}$$

As an example of the foregoing discussion, let us consider the small vibrations of the following system of masses and springs:

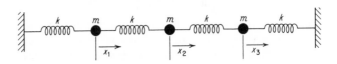

FIGURE 1

The differential equations governing the motion are

$$m\ddot{x}_1 = -kx_1 + k(x_2 - x_1)$$
$$m\ddot{x}_2 = -k(x_2 - x_1) + k(x_3 - x_2)$$
$$m\ddot{x}_3 = k(x_2 - x_3) - kx_3$$

In matrix form they are $\ddot{X} = AX$, where

$$A = \left\| \begin{array}{ccc} -2\alpha & \alpha & 0 \\ \alpha & -2\alpha & \alpha \\ 0 & \alpha & -2\alpha \end{array} \right\|$$

[1] The value of $\tan^{-1}$ is chosen to give $\cos \phi_\beta$ and $\sin \phi_\beta$ the correct sign.

$\alpha = k/m$.   The characteristic equation is

$$-(\lambda + 2\alpha)^3 + 2\alpha^2(\lambda + 2\alpha) = 0$$

and it has roots $\lambda_1 = -2\alpha$, $\lambda_2 = (-2 + \sqrt{2})\alpha$, $\lambda_3 = (-2 - \sqrt{2})\alpha$, giving the natural frequencies $\omega_1 = \sqrt{2k/m}$, $\omega_2 = \sqrt{(2 - \sqrt{2})k/m}$, $\omega_3 = \sqrt{(2 + \sqrt{2})k/m}$. The corresponding normal modes are

$$\mathbf{X}_1 = \left\| \begin{array}{c} \dfrac{1}{\sqrt{2}} \\[2mm] 0 \\[2mm] -\dfrac{1}{\sqrt{2}} \end{array} \right\| \sin(\omega_1 t + \phi_1)$$

$$\mathbf{X}_2 = \left\| \begin{array}{c} \tfrac{1}{2} \\[2mm] \dfrac{1}{\sqrt{2}} \\[2mm] \tfrac{1}{2} \end{array} \right\| \sin(\omega_2 t + \phi_2)$$

$$\mathbf{X}_3 = \left\| \begin{array}{c} \tfrac{1}{2} \\[2mm] -\dfrac{1}{\sqrt{2}} \\[2mm] \tfrac{1}{2} \end{array} \right\| \sin(\omega_3 t + \phi_3)$$

The first mode represents a motion in which the center mass remains fixed and the masses on the sides move in opposite directions with equal amplitude.   In the second mode all masses move in the same direction, the side masses moving with equal amplitude, and the center mass moving with an amplitude $\sqrt{2}$ times the amplitude of the others.   The third mode is like the second, except that the center mass moves in the opposite direction from the side masses. Any other possible motion of the system is a linear combination of these normal modes.

The case of forced vibrations with periodic forcing functions can also be handled very readily.   Suppose in the above example each mass has an external periodic force acting on it; that is,

$$f_1 = b_1 \sin(\mu_1 t + \theta_1)$$
$$f_2 = b_2 \sin(\mu_2 t + \theta_2)$$
$$f_3 = b_3 \sin(\mu_3 t + \theta_3)$$

are the forces acting on masses with displacements x_1, x_2, x_3, respectively.

The differential equations can now be written

$$\dot{X} = AX + F_1 + F_2 + F_3$$

where
$$F_1 = \begin{Vmatrix} \dfrac{b_1}{m} \\ 0 \\ 0 \end{Vmatrix} \sin(\mu_1 t + \theta_1) \qquad F_2 = \begin{Vmatrix} 0 \\ \dfrac{b_2}{m} \\ 0 \end{Vmatrix} \sin(\mu_2 t + \theta_2)$$

and
$$F_3 = \begin{Vmatrix} 0 \\ 0 \\ \dfrac{b_3}{m} \end{Vmatrix} \sin(\mu_3 t + \theta_3)$$

We have the system of nonhomogeneous differential equations

$$\dot{X} - AX = F_1 + F_2 + F_3$$

There is a complementary solution X_0 which satisfies the homogeneous differential equations $\dot{X}_0 - AX_0 = 0$ and contains six arbitrary constants. Furthermore, there are three particular solutions Z_1, Z_2, Z_3 which satisfy the equations $\dot{Z}_i - AZ_i = F_i$, $i = 1, 2, 3$. Since the equations are linear,

$$X = X_0 + Z_1 + Z_2 + Z_3$$

satisfies the differential equations $\dot{X} - AX = F_1 + F_2 + F_3$ and has the required number of arbitrary constants, and it is, therefore, the general solution. Our problem is thus one of solving for a particular solution

$$\dot{Z} - AZ = F = C \sin(\mu t + \theta)$$

Let $Z = B \sin(\mu t + \theta)$; then $\dot{Z} = -\mu^2 B \sin(\mu t + \theta)$, and $-\mu^2 B - AB = C$.† B can be written as a linear combination of the eigenvectors of A; that is,

$$B = \sum_{\alpha=1}^{n} \gamma_\alpha Y_\alpha$$

Substituting for B,

$$-\mu^2 \sum_{\alpha=1}^{n} \gamma_\alpha Y_\alpha - A \sum_{\alpha=1}^{n} \gamma_\alpha Y_\alpha = C$$

$$-\mu^2 \sum_{\alpha=1}^{n} \gamma_\alpha Y_\alpha - \sum_{\alpha=1}^{n} \lambda_\alpha \gamma_\alpha Y_\alpha = C$$

$$-\mu^2 \sum_{\alpha=1}^{n} \gamma_\alpha (\mathbf{Y}_\alpha, \mathbf{Y}_\beta) - \sum_{\alpha=1}^{n} \lambda_\alpha \gamma_\alpha (\mathbf{Y}_\alpha, \mathbf{Y}_\beta) = (\mathbf{C}, \mathbf{Y}_\beta)$$

$$-\mu^2 \gamma_\beta - \lambda_\beta \gamma_\beta = (\mathbf{C}, \mathbf{Y}_\beta)$$

$$\gamma_\beta = \frac{(\mathbf{C}, \mathbf{Y}_\beta)}{-\mu^2 - \lambda_\beta}$$

† Compare with the nonhomogeneous linear equations in Sec. 1.8.

provided $\lambda_\beta \neq -\mu^2$. If $-\mu^2 = \lambda_\beta$ for some β, then we have the case of **resonance** where the amplitude of vibration builds up with time, and this must be handled in a different way.

In the case of resonance, $-\mu^2 = \lambda_\beta$ for some β, the problem can still be solved by diagonalizing the matrix A, assuming A is real and symmetric. There exists an orthogonal transformation $Z = T\bar{Z}$ which reduces the equations to

$$\ddot{\bar{z}}_\alpha - \lambda_\alpha \bar{z}_\alpha = \bar{c}_\alpha \sin{(\mu t + \theta)}$$

Particular solutions for these equations are

$$\bar{z}_\alpha = \frac{\bar{c}_\alpha}{-\mu^2 - \lambda_\alpha} \sin{(\mu t + \theta)}$$

except for $\alpha = \beta$, where the method of undetermined coefficients gives the particular solution

$$\bar{z}_\beta = \frac{\bar{c}_\beta}{-2\mu} t \cos{(\mu t + \theta)}$$

The particular solution $Z = T\bar{Z}$ is a linear combination of the particular solutions one of which builds up in amplitude with the factor t. This is typical of resonance without damping.

If the applied forces are not sinusoidal but are periodic, then they can be expanded in Fourier series. Then the response to each component can be calculated by the methods just described. Since the problem is linear, the total response can be calculated by adding the responses of all the components, provided the resulting Fourier series can be shown to converge.

Exercises 1.9

1. Three equal masses are attached to a string at equal distances from one another and from the ends, which are fastened to supports so that there is a tension T in the string. Compute the normal modes and natural frequencies of small transverse vibrations about equilibrium. Describe the normal modes. HINT: Assume the tension constant, neglect gravity, and assume the string weightless.

2. In the example of the three masses mounted on springs given in the text, assume that the system is started from rest in the equilibrium position by a force $mg \sin 2\sqrt{k/m}\, t$ acting on the center mass. Find the complete solution for the motion.

3. Assuming damping forces $-\beta^2 \dot{x}_1$, $-\beta^2 \dot{x}_2$, and $-\beta^2 \dot{x}_3$ acting on the three masses in exercise 2, find the transient solution for arbitrary initial conditions.

4. Do exercise 2 with damping forces as in exercise 3 and with force $mg \sin \sqrt{2k/m}\, t$ acting on the center mass.

5. In the example of the three masses mounted on four springs of the text, remove the end springs and the walls (assume the masses are resting on a frictionless table). Investigate small longitudinal vibrations of the system. What is the meaning of the eigenvalue zero?

6. Repeat exercise 5 with the end spring and wall removed on one side only.

7. Two identical pendulums are hung side by side with a weightless spring connecting the two pendulum bobs. Investigate small vibrations of the system in the plane of the equilibrium configuration.

1.10   Linear Programming

As another example of the use of linear algebra, we consider the solution of a certain type of optimization problem which has become known as **linear programming.** The most common applications of this technique have been in business and economics. However, linear programming is considered introductory to the related fields of nonlinear programming, dynamic programming, etc., which have been used in engineering analysis.

To illustrate the kind of problem encountered in linear programming, consider the following: a manufacturer can produce two products in quantities x and y out of three kinds of raw materials, which we designate A, B, and C. The first product requires 2 units of A, 2 units of B, and 1 unit of C, while the second requires 2 units of A, 1 unit of B, and 2 units of C. The totals of raw materials available are 900 units of A, 700 units of B, and 800 units of C. In order not to exceed the available supplies, we require that

$$2x + 2y \leq 900$$
$$2x + y \leq 700$$
$$x + 2y \leq 800$$

The manufacturer finds that his unit profit on the first product is two-thirds of his unit profit on the second. He wishes to maximize his profit, and so he wishes to maximize the quantity $z = 2x + 3y$. In addition to the above inequalities, we require that x and y be nonnegative because it does not make any sense in this problem to produce a negative number of a certain product. The problem is then to maximize[1] the linear expression $2x + 3y$ subject to the linear constraints

$$2x + 2y \leq 900$$
$$2x + y \leq 700$$
$$x + 2y \leq 800$$
$$x \geq 0$$
$$y \geq 0$$

We have purposely picked an example with a small number of variables so that we can solve the problem geometrically, thus giving the reader an easy, intuitive picture of the problem and its solution. The constraints $x \geq 0$ and $y \geq 0$ require that the solution lie in the first quadrant. The other constraints require that the solution lie in the common part of three

---

[1] In some problems the minimum of a linear expression may be required.

half planes (see Fig. 2). To satisfy all the constraints, the solution must lie
in the shaded region. This is a convex region because it is the intersection
of a finite number of half planes (see exercise 2, Sec. 1.10). A convex region
is one such that whenever it contains the end points of a line segment it
contains the entire line segment. We can now obtain the solution to the
problem geometrically as follows. If we think of z as a parameter, the
expression $z = 2x + 3y$ represents a family of lines all with slope $-2/3$ and
with x and y intercepts $z/2$ and $z/3$, respectively. Clearly we can maximize
z by picking from this family of lines the one farthest from the origin which
intersects the convex region where the constraints are satisfied. This line is
indicated in the figure by the dashed line. It passes through the "corner"
at the point (100,350) where the maximum z of 1,250 is achieved.

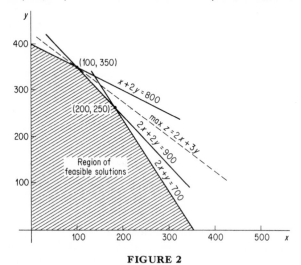

FIGURE 2

This example is typical of linear programming problems. There is a
nonempty convex region of points satisfying the constraints which consists
of the intersection of a finite number of half spaces; that is, the constraints
are all linear inequalities. It can be proved[1] that if there is a finite solution
to the problem it occurs at a vertex of a convex polyhedron whose faces are
in the boundaries of half spaces determined by the constraints. Since there
are but a finite number of these vertices, a search through all of them will
produce the required solution. We shall present in the remainder of this
section a systematic method for performing this search, called the **simplex
method.** However, first we must comment on a couple of special cases of
which we must be aware.

[1] See G. Hadley, "Linear Programming," Addison-Wesley Publishing Company, Inc.,
Reading, Mass., 1962.

In the above example, if $z = x + y$, then these lines would be parallel to the line $2x + 2y = 900$ and we would achieve maximum z all along the line segment lying between $(100,350)$ and $(200,250)$. Of course we still find the maximum z at two vertices, although the solution is no longer unique. Another case is more troublesome. Consider the problem to maximize $z = x + y$ subject to $x \geq 0$, $y \geq 0$, and $x \leq 5$. Clearly there is no finite maximum z because y can be made arbitrarily large with the other constraints still satisfied. It turns out that the simplex method will determine when this phenomenon occurs. If a practical linear programming problem is properly formulated, it should not lead to max $z = \infty$. Nevertheless, it is nice to have a method which will determine when a problem is not well formulated.

In the simplex method we shall be concerned with the problem of maximizing a linear **objective function**

$$z = c_1 x_1 + c_2 x_2 + \cdots + c_r x_r$$

subject to the conditions that the variables be nonnegative and that they satisfy a set of m linear inequalities[1]

$$a_{11} x_1 + a_{12} x_2 + \cdots + a_{1r} x_r \leq b_1$$
$$a_{21} x_1 + a_{22} x_2 + \cdots + a_{2r} x_r \leq b_2$$
$$\cdots\cdots\cdots\cdots\cdots\cdots\cdots\cdots\cdots\cdots\cdots$$
$$a_{m1} x_1 + a_{m2} x_2 + \cdots + a_{mr} x_r \leq b_m$$

We first turn these inequalities into equations by introducing **slack variables** $x_{r+1}, x_{r+2}, \ldots, x_{r+m}$, which are the quantities which must be added to the left-hand sides to produce equality. The slack variables are hence nonnegative and they satisfy

$$a_{11} x_1 + a_{12} x_2 + \cdots + a_{1r} x_r + x_{r+1} = b_1$$
$$a_{21} x_1 + a_{22} x_2 + \cdots + a_{2r} x_r + x_{r+2} = b_2$$
$$\cdots\cdots\cdots\cdots\cdots\cdots\cdots\cdots\cdots\cdots\cdots\cdots$$
$$a_{m1} x_1 + a_{m2} x_2 + \cdots + a_{mr} x_r + x_{r+m} = b_m$$

In matrix notation our problem is now to maximize

$$z = C'X$$

where C is an $n = (r + m) \times 1$ matrix with zeros in all the rows after the mth and X is an $n \times 1$ matrix of the variables including the slacks, subject to the conditions that the variables be nonnegative and

$$AX = B$$

[1] If the constraints have the inequalities reversed, we may multiply by -1 and put them in this form.

where A is $m \times n$ and B is $m \times 1$. We shall assume that these equations are consistent, that is, that the ranks of A and of the augmented matrix are the same. We shall also assume that the rank of A is m. This assures us that there are at least as many variables as equations and that none of the equations are redundant. The case where $m = n$ is of little interest, because then there would be a unique solution which would optimize z if the $x_i \geq 0$, $i = 1, 2, \ldots, n$, and otherwise would not solve the problem.

We say that a solution to $AX = B$ is **feasible** if all the $x_i \geq 0$. Out of A let us select an $m \times m$ nonsingular matrix by eliminating all but m columns. The variables corresponding to the m remaining columns we shall call a **basis**. A feasible solution shall be called a **basic feasible solution** if all of the variables outside the basis have the value zero. We shall first prove that *if $AX = B$ has a feasible solution it has a basic feasible solution.* We have a feasible solution, so we can assume (by renumbering if necessary) that $x_1, x_2, \ldots, x_p > 0$ and that $x_{p+1} = x_{p+2} = \cdots = x_n = 0$. Let $\mathbf{A}_1, \mathbf{A}_2, \ldots, \mathbf{A}_p$ be the columns of A corresponding to $x_1, x_2, \ldots, x_p$, respectively. Then

$$x_i \mathbf{A}_i = \mathbf{B} \qquad i = 1, 2, \ldots, p$$

where $\mathbf{B}$ is the m-dimensional vector of constants on the right-hand sides of the equations. If $\mathbf{A}_1, \mathbf{A}_2, \ldots, \mathbf{A}_p$ are independent, then $p \leq m$ because m is the rank of A. In this case we have a basic feasible solution. This is obvious if $p = m$. If $p < m$ we can add $m - p$ columns each multiplied by zero so that $x_1, x_2, \ldots, x_p > 0$ and $x_{p+1} = x_{p+2} = \cdots = x_m = 0$ is a basic feasible solution. If $p > m$, then $\mathbf{A}_1, \mathbf{A}_2, \ldots, \mathbf{A}_p$ are dependent. Then there exists a $\gamma_k > 0$ such that

$$\gamma_k \mathbf{A}_k = - \sum_{\substack{i=1 \\ i \neq k}}^{p} \gamma_i \mathbf{A}_i$$

and we have

$$\sum_{\substack{i=1 \\ i \neq k}}^{p} x_i \mathbf{A}_i + x_k \mathbf{A}_k = \mathbf{B}$$

$$\sum_{\substack{i=1 \\ i \neq k}}^{p} \left(x_i - x_k \frac{\gamma_i}{\gamma_k} \right) \mathbf{A}_i = \mathbf{B}$$

Hence, we have a new solution

$$\bar{x}_i = x_i - x_k \frac{\gamma_i}{\gamma_k} \qquad \text{(no summation)}$$

with one fewer nonzero variable. This solution will be feasible if $\bar{x}_i \geq 0$. We show that we can always choose k so that it is feasible. Since $x_k > 0$ and $\gamma_k > 0$, $\bar{x}_i > 0$ if $\gamma_i \leq 0$. There will always be some $\gamma_i > 0$. We examine all the ratios x_i/γ_i such that $\gamma_i > 0$. We then pick k so that

$$\frac{x_k}{\gamma_k} = \min_{\gamma_i > 0} \frac{x_i}{\gamma_i}$$

and hence

$$\frac{\bar{x}_i}{\gamma_i} = \frac{x_i}{\gamma_i} - \frac{x_k}{\gamma_k} \geq 0$$

showing that $\bar{x}_i \geq 0$. On the assumption that $p > m$ we have obtained a new feasible solution with at least one less positive variable. After a finite number of such steps we shall arrive at a basic feasible solution. This completes the proof.

We may now complete the discussion of the simplex method. We find a basic feasible solution. This corresponds to a vertex of a convex polyhedron. If this does not maximize the objective, we change the basis by moving one variable out of the basis and another variable in so as to increase the objective. This corresponds to moving from one vertex of the polyhedron to an adjacent vertex. We continue this process until either the objective is maximized or an indication is given that the objective can be made as large as we please for various feasible solutions.

Let us assume that the first m rows of A are linearly independent, so that $x_1, x_2, x_3, \ldots, x_m$ form a basis. It may be necessary to renumber the variables to achieve this. We arrange the equations with the objective as follows:

$$a_{11}x_1 + a_{12}x_2 + \cdots + a_{1m}x_m + \cdots + a_{1n}x_n = b_1$$
$$a_{21}x_1 + a_{22}x_2 + \cdots + a_{2m}x_m + \cdots + a_{2n}x_n = b_2$$
$$\cdots\cdots\cdots\cdots\cdots\cdots\cdots\cdots\cdots\cdots\cdots\cdots\cdots\cdots$$
$$a_{m1}x_1 + a_{m2}x_2 + \cdots + a_{mm}x_m + \cdots + a_{mn}x_n = b_m$$
$$c_1x_1 + c_2x_2 + \cdots + c_mx_m + \cdots + c_nx_n = z$$

It is clear that we can obtain a Gauss-Jordan reduction which will transform the equations to the form

$$x_1 + \bar{a}_{1m+1}x_{m+1} + \cdots + \bar{a}_{1n}x_n = \bar{b}_1$$
$$x_2 + \bar{a}_{2m+1}x_{m+1} + \cdots + \bar{a}_{2n}x_n = \bar{b}_2$$
$$\cdots\cdots\cdots\cdots\cdots\cdots\cdots\cdots\cdots\cdots\cdots\cdots$$
$$x_m + \bar{a}_{mm+1}x_{m+1} + \cdots + \bar{a}_{mn}x_n = \bar{b}_m$$
$$\bar{c}_{m+1}x_{m+1} + \cdots + \bar{c}_nx_n = z - N_1$$

If $\bar{b}_1, \bar{b}_2, \ldots, \bar{b}_m > 0$, then putting $x_{m+1} = x_{m+2} = \cdots = x_n = 0$ will give us a basic feasible solution, $x_1 = \bar{b}_1$, $x_2 = \bar{b}_2$, $\ldots, x_m = \bar{b}_m$. We shall assume this to be the case. For this solution $z = N_1$ and so in the process we have evaluated the objective.

Let us assume that $\bar{c}_k > 0$ for some $m + 1 \leq k \leq n$. If we increase x_k away from zero while keeping $x_{m+1} = x_{m+2} = \cdots = x_{k-1} = x_{k+1} = \cdots = x_n = 0$, then we increase z to the value $N_1 + \bar{c}_k x_k$. In the process, $x_i = \bar{b}_i - \bar{a}_{ik} x_k$, $i = 1, 2, 3, \ldots, m$, and our solution remains feasible provided $\bar{b}_i - \bar{a}_{ik} x_k \geq 0$, $i = 1, 2, 3, \ldots, m$. If $\bar{a}_{ik} \leq 0$ for $i = 1, 2, 3, \ldots, m$, then we can increase x_k without bound and always remain feasible. In this case $\max z = \infty$. If $\bar{a}_{ik} > 0$ for some values of i, $1 \leq i \leq m$, then there is a limitation on how large we can make x_k and still remain feasible. In fact, if j is a subscript such that

$$\theta = \frac{\bar{b}_j}{\bar{a}_{jk}} = \min_{\bar{a}_{ik} > 0} \frac{\bar{b}_i}{\bar{a}_{ik}}$$

then for $\bar{a}_{ik} > 0$

$$\frac{x_i}{\bar{a}_{ik}} = \frac{\bar{b}_i}{\bar{a}_{ik}} - \theta = \frac{\bar{b}_i}{\bar{a}_{ik}} - \frac{\bar{b}_j}{\bar{a}_{jk}} \geq 0$$

This shows that the solution remains feasible as long as $x_k \leq \theta$ or that θ is the maximum value we can assign to x_k in this process and still have a feasible solution. In so doing we have increased z to the value $N_1 + \bar{c}_k \theta$. Notice that our new feasible solution is a basic feasible solution because in the process we have made the variable $x_j = \bar{b}_j - \bar{a}_{jk} \theta = 0$ and $\bar{a}_{jk} \neq 0$. In effect we have taken x_j out of the basis and brought x_k into the basis.

Let us summarize what we have said so far. If there is a nondegenerate basic feasible solution,[1] then by a Gauss-Jordan reduction we can arrange the equations plus the objective as above and will be presented with three possibilities:

1. All of the $\bar{c}_{m+1}, \bar{c}_{m+2}, \ldots, \bar{c}_n \leq 0$, and hence bringing another variable into the basis cannot increase the objective. In this case z is already maximized.

2. For some $\bar{c}_k > 0$, all of the $\bar{a}_{ik} \leq 0$, $i = 1, 2, 3, \ldots, m$, and hence x_k can be increased without bound and $\max z = \infty$.

3. For some $\bar{c}_k > 0$, some of the $\bar{a}_{ik} > 0$, $1 \leq i \leq m$, and in this case x_k can be brought into the basis so that z is increased.

In cases 1 and 2, we have gone as far as we need go. In case 3, we can bring x_k into the basis replacing x_j as determined above and obtain a new reduced form of the problem by a Gauss-Jordan reduction. We are again faced with the same three possibilities. If $\max z \neq \infty$ and we never run into a degenerate basic feasible solution, after a finite number of steps we must find the maximum value of the objective.

Before illustrating the simplex method with an example, we must comment about the case of a degenerate basic feasible solution. Suppose in the above

---

[1] A degenerate basic feasible solution is one in which one of the variables in the basis is zero. This would correspond in the above to the case where one of the $\bar{b}_1, \bar{b}_2, \ldots, \bar{b}_m$ is zero.

$b_p = 0$, $1 \leq p \leq m$. Then $x_p = 0$ in the basis, and we have a degenerate basic feasible solution. Therefore, for $x_k > 0$, $x_p = -\bar{a}_{pk}x_k$ will be feasible only if $\bar{a}_{pk} \leq 0$. In this case we may still increase z and remain feasible if $\bar{c}_k > 0$. On the other hand, if $\bar{a}_{pk} > 0$, then $\theta = 0$ and the process of bringing x_k into the basis and taking x_p out will lead to another degenerate basic feasible solution and will not change the objective. If $\bar{c}_s > 0$ for $k \neq s$ and $\bar{a}_{ps} \leq 0$, we may move out of this stalemate by choosing a different variable to bring into the basis, so there are ways of avoiding this difficulty. On the other hand, there is nothing in the above discussion which rules out the possibility of moving into a degenerate basic feasible solution when the basis is changed. In fact this will be precisely the case if

$$\theta = \min_{\bar{a}_{ik} > 0} \frac{b_i}{\bar{a}_{ik}}$$

occurs for more than one value of i. Although the case of degeneracy is certainly a theoretical possibility, according to Hadley[1] it does not cause practical problems to cycle, that is, iterate indefinitely without reaching a solution.

We conclude this section by resolving our original example by the simplex method. We arrange the problem as follows with x, u, and w in the basis:

$$
\begin{aligned}
2x + u \quad\;\; + 2y \quad\;\; &= 900 \\
2x \qquad\quad + y + v &= 700 \\
x \quad\; + w + 2y \quad\;\; &= 800 \\
2x \qquad\quad + 3y \quad\;\; &= z
\end{aligned}
$$

Our first reduction leads to

$$
\begin{aligned}
x \qquad\qquad + \tfrac{1}{2}y + \tfrac{1}{2}v &= 350 \\
u \quad\; + y - v &= 200 \\
w + \tfrac{3}{2}y - \tfrac{1}{2}v &= 450 \\
2y - v &= z - 700
\end{aligned}
$$

This gives us a basic feasible solution $x = 350$, $u = 200$, $w = 450$, $y = 0$, $v = 0$, with $z = 700$. Using the above notation, $\bar{c}_4 = 2$, $\theta = \bar{b}_2/\bar{a}_{24} = 200$. Therefore, we put y into the basis and take u out. We make the new reduction and we have

$$
\begin{aligned}
x \qquad\qquad - \tfrac{1}{2}u + v &= 250 \\
y \quad\; + u - v &= 200 \\
w - \tfrac{3}{2}u + v &= 150 \\
-2u + v &= z - 700 - 400
\end{aligned}
$$

[1] Hadley, *op. cit.*, p. 113.

This gives us a basic feasible solution $x = 250$, $y = 200$, $w = 150$, $u = 0$, $v = 0$, with $z = 1{,}100$. This time $\bar{c}_5 = 1$, $\theta = \bar{b}_3/\bar{a}_{35} = 150$. Therefore, we put v into the basis and take w out. We make the new reduction and we have

$$
\begin{aligned}
x \quad\quad\quad + u - w &= 100 \\
y \quad - \tfrac{1}{2}u + w &= 350 \\
v - \tfrac{3}{2}u + w &= 150 \\
- \tfrac{1}{2}u - w &= z - 700 - 400 - 150
\end{aligned}
$$

This time $\bar{c}_4 = -\tfrac{1}{2}$ and $\bar{c}_5 = -1$ are both negative and so z is maximized at $1{,}250$ for the solution $x = 100$, $y = 350$, $v = 150$, $u = 0$, $w = 0$. This completes the example.

Exercises 1.10

1. Prove that a half plane $ax + by \le c$ is a convex region.

2. Prove that the intersection of any finite number of half planes is a convex region.

3. Assuming that the feasible region for a linear programming problem in the xy plane is the boundary plus the interior of a bounded convex polygon, prove that max $(ax + by)$ occurs at at least one vertex of the polygon.

4. Solve the following linear programming problem: maximize $x + y + z$ subject to $x \ge 0$, $y \ge 0$, $z \ge 0$ and $3x + z \le 120$, $z \le 60$, $4y + 3z \le 270$, and $3x + 2y \le 150$.

References

Birkhoff, Garrett, and Saunders MacLane: "A Survey of Modern Algebra," 4th ed., The Macmillan Company, New York, 1977.

Courant, Richard, and David Hilbert: "Methods of Mathematical Physics," Interscience Publishers (Division of John Wiley & Sons, Inc.), New York, 1953, vol. I; 1962, vol. II.

Finkbeiner, Daniel T.: "Introduction to Matrices and Linear Transformations," 3d ed., W. H. Freeman and Company, San Francisco, 1978.

Friedman, Bernard: "Principles and Techniques of Applied Mathematics," John Wiley & Sons, Inc., New York, 1956.

Hadley, G: "Linear Programming," Addison-Wesley Publishing Company, Inc., Reading, Mass., 1962.

Halmos, Paul R.: "Finite-dimensional Vector Spaces," 2d ed., Springer-Verlag, New York, 1958.

Macduffee, Cyrus C.: "Theory of Equations," John Wiley & Sons, Inc., New York, 1954.

Noble, Benjamin: "Applied Linear Algebra," Prentice-Hall, Inc., Englewood Cliffs, N.J., 1969.

Thrall, Robert M., and Leonard Thornheim: "Vector Spaces and Matrices," John Wiley & Sons, Inc., New York, 1957.

Chapter 2. Hilbert Spaces

2.1 Infinite-dimensional Vector Spaces. Function Spaces

Except for the space of continuous functions mentioned in Sec. 1.5 and the space of square integrable functions mentioned in Sec. 1.6, we have considered only finite-dimensional vector spaces so far. The easiest way to construct an infinite-dimensional vector space is to generalize the space of n-tuples of complex numbers to the space of infinite sequences of complex numbers. We shall define a vector in this space as follows:

$$\mathbf{U} = (u_1, u_2, u_3, \ldots)$$

If we try to define a norm in this space as a direct extension of the definition in the space of n-tuples, we have

$$\|\mathbf{U}\|^2 = \sum_{i=1}^{\infty} |u_i|^2$$

Since this involves an infinite sum, not all sequences will define vectors with finite norms. Therefore, we shall include in our space only those sequences for which

$$\sum_{i=1}^{\infty} |u_i|^2 < \infty$$

We still must define addition and multiplication by a scalar and show that the postulates of a vector space are satisfied. If $\mathbf{U} = (u_1, u_2, u_3, \ldots)$ and $\mathbf{V} = (v_1, v_2, v_3, \ldots)$ are vectors such that $\sum_{i=1}^{\infty} |u_i|^2 < \infty$ and $\sum_{i=1}^{\infty} |v_i|^2 < \infty$, then the vector sum is $\mathbf{U} + \mathbf{V} = (u_1 + v_1, u_2 + v_2, u_3 + v_3, \ldots)$. To prove closure, we have to show that $\sum_{i=1}^{\infty} |u_i + v_i|^2 < \infty$. The proof follows.

For every i

$$
\begin{aligned}
|u_i + v_i|^2 &= (u_i^* + v_i^*)(u_i + v_i) \\
&= |u_i|^2 + |v_i|^2 + u_i^* v_i + u_i v_i^* \\
&\leq |u_i|^2 + |v_i|^2 + 2\,|u_i^* v_i| \\
&\leq |u_i|^2 + |v_i|^2 + 2\,|u_i|\,|v_i|
\end{aligned}
$$

We also have

$$(|u_i| - |v_i|)^2 = |u_i|^2 + |v_i|^2 - 2|u_i||v_i| \geq 0$$

so that

$$|u_i|^2 + |v_i|^2 \geq 2|u_i||v_i|$$

Therefore,

$$|u_i + v_i|^2 \leq 2\{|u_i|^2 + |v_i|^2\}$$

and

$$\sum_{i=1}^{\infty} |u_i + v_i|^2 \leq 2\sum_{i=1}^{\infty}|u_i|^2 + 2\sum_{i=1}^{\infty}|v_i|^2 < \infty$$

Multiplication by a complex scalar a is defined as

$$a\mathbf{U} = (au_1, au_2, au_3, \ldots)$$

Closure under this operation is easily verified.

$$\sum_{i=1}^{\infty}|au_i|^2 = \sum_{i=1}^{\infty}|a|^2|u_i|^2$$

$$= |a|^2 \sum_{i=1}^{\infty}|u_i|^2 < \infty$$

The zero vector is defined as

$$\mathbf{0} = (0,0,0,\ldots)$$

and the negative

$$-\mathbf{U} = (-u_1, -u_2, -u_3, \ldots)$$

The other eight postulates are easily verified. Obviously one cannot find a finite number of vectors which span the space, and therefore it is infinite-dimensional.

A scalar product can be defined in this space as follows:

$$(\mathbf{U},\mathbf{V}) = \sum_{i=1}^{\infty} u_i^* v_i$$

We can show that for every pair of vectors in the space this scalar product exists. For every i

$$|u_i^* v_i| = |u_i||v_i| \leq \tfrac{1}{2}\{|u_i|^2 + |v_i|^2\}$$

Therefore, $\sum_{i=1}^{\infty} u_i^* v_i$ converges, since absolute convergence implies convergence.

The set of vectors $\mathbf{X}_1 = (1,0,0,\ldots)$, $\mathbf{X}_2 = (0,1,0,\ldots)$, $\mathbf{X}_3 = (0,0,1,\ldots)$, $\ldots$ is orthonormal, and it spans the space in the following sense. If

$$\mathbf{U}_n = \sum_{i=1}^{n} u_i \mathbf{X}_i$$

then

$$\|\mathbf{U}_n - \mathbf{U}\|^2 = \sum_{i=n+1}^{\infty} |u_i|^2 \to 0$$

as $n \to \infty$. We write

$$U = \sum_{i=1}^{\infty} u_i X_i$$

and say that u_i is the component of U relative to the orthonormal basis X_i; that is, $u_i = (X_i, U)$.

A sequence of vectors $\{X^n\}$ in a metric space is said to be a **Cauchy sequence** if for every $\epsilon > 0$ there exists an $N(\epsilon)$ such that

$$d(X^n, X^m) < \epsilon$$

whenever $n > N$ and $m > N$. A sequence $\{X^n\}$ has a **strong limit**[1] in the space if there exists a vector X in the space such that for every $\epsilon > 0$ there exists an $N(\epsilon)$ such that

$$d(X, X^n) < \epsilon$$

whenever $n > N$. If a sequence has a strong limit, then it is a Cauchy sequence, for given $\epsilon/2$ there exists an $N(\epsilon)$ such that

$$d(X, X^n) < \frac{\epsilon}{2} \qquad d(X, X^m) < \frac{\epsilon}{2}$$

whenever $n > N$ and $m > N$. Then by the triangle inequality we have

$$d(X^n, X^m) \le d(X, X^n) + d(X, X^m) < \epsilon$$

whenever $n > N$ and $m > N$. However, the converse may not be true; that is, a Cauchy sequence may not have a strong limit, unless the space is complete.

A metric space is said to be **complete** if every Cauchy sequence has a strong limit in the space. As is known from analysis, the space of real numbers is complete where the metric is $d(x,y) = |x - y|$. With this fact it is not difficult to show that the space of k-tuples of complex numbers is a complete metric space under the metric

$$d(X, Y) = \sqrt{(x_i^* - y_i^*)(x_i - y_i)} \qquad i = 1, 2, \ldots, k$$

We now show that the space of infinite sequences of complex numbers is a complete metric space.

Let $\{X^n\}$ be a Cauchy sequence. Then for every $\epsilon > 0$ there exists an $N(\epsilon)$ such that

$$d(X^n, X^m) = \left(\sum_{i=1}^{\infty} |x_i^n - x_i^m|^2 \right)^{\frac{1}{2}} < \epsilon$$

whenever $n > N$ and $m > N$. This implies that for each i

$$|x_i^n - x_i^m| < \epsilon$$

---

[1] Later we shall define weak limit.

or that each sequence of components is a Cauchy sequence.   Hence for every i there exists an x_i such that

$$\lim_{n \to \infty} x_i^n = x_i$$

Now define

$$\mathbf{X} = (x_1, x_2, \ldots)$$

It remains to show that $d(\mathbf{X}, \mathbf{X}^n) \to 0$ as $n \to \infty$ and that $\mathbf{X}$ is in the space; that is, $\sum_{i=1}^{\infty} |x_i|^2 < \infty$.

For some fixed M consider $\sum_{i=1}^{M} |x_i - x_i^n|^2$.   Now

$$\sum_{i=1}^{M} |x_i - x_i^n|^2 = \sum_{i=1}^{M} |x_i - x_i^m + x_i^m - x_i^n|^2 \leq 2 \sum_{i=1}^{M} |x_i - x_i^m|^2 + 2 \sum_{i=1}^{M} |x_i^m - x_i^n|^2$$

We can determine an $N(\epsilon)$ such that

$$\sum_{i=1}^{M} |x_i^m - x_i^n|^2 < \frac{\epsilon^2}{4}$$

$$|x_i - x_i^m|^2 < \frac{\epsilon^2}{4M}$$

when $n > N$ and $m > N$.   Then

$$\sum_{i=1}^{M} |x_i - x_i^n|^2 < \epsilon^2$$

This is possible for arbitrary M.   Therefore we can let $M \to \infty$, and we have

$$d(\mathbf{X}, \mathbf{X}^n) = \left(\sum_{i=1}^{\infty} |x_i - x_i^n|^2 \right)^{\frac{1}{2}} < \epsilon$$

when $n > N$.   Finally

$$\sum_{i=1}^{\infty} |x_i|^2 \leq 2 \sum_{i=1}^{\infty} |x_i - x_i^n|^2 + 2 \sum_{i=1}^{\infty} |x_i^n|^2 < \infty$$

so that $\mathbf{X}$ is in the space.

We have seen that the space of infinite sequences of complex numbers described above is an infinite-dimensional complete metric space with a scalar product.   Such a space is known as a **Hilbert space** and is of basic importance in applied mathematics, particularly in the study of quantum mechanics.   We have seen that normed spaces do not necessarily have scalar products (see Exercises 1.6).   However, if a norm is defined, we can define Cauchy sequences and discuss completeness.   A complete normed vector space is called a **Banach space.**

Now let us turn our attention to another infinite-dimensional vector space, the space of square integrable functions defined in Sec. 1.6.   We have already seen that this is a vector space. That it is infinite-dimensional will become apparent as we proceed. We say that a set of functions ϕ_1, ϕ_2, ϕ_3, ... is orthonormal on the interval $a \le x \le b$ if

$$(\boldsymbol{\phi}_i, \boldsymbol{\phi}_j) = \int_a^b \phi_i^* \phi_j \, dx = \delta_{ij}$$

Consider the problem of approximating by the method of least squares a square integrable function $f(x)$ by a linear combination of the functions from the orthonormal set; that is, we wish to choose $c_1, c_2, \ldots, c_n$ to minimize

$$\int_a^b \left| f - \sum_{i=1}^n c_i \phi_i \right|^2 dx$$

Expanding this quantity, we have

$$\int_a^b |f|^2 \, dx + \sum_{i=1}^n |c_i|^2 - \sum_{i=1}^n c_i \int_a^b f^* \phi_i \, dx - \sum_{i=1}^n c_i^* \int_a^b \phi_i^* f \, dx$$

The minimum occurs when

$$\int_a^b |f|^2 \, dx + \sum_{i=1}^n |c_i|^2 - \sum_{i=1}^n c_i \int_a^b f^* \phi_i \, dx - \sum_{i=1}^n c_i^* \int_a^b f \phi_i^* \, dx$$
$$+ \sum_{i=1}^n \int_a^b f^* \phi_i \, dx \int_a^b f \phi_i^* \, dx = \int_a^b |f|^2 \, dx + \sum_{i=1}^n \left| c_i - \int_a^b f \phi_i^* \, dx \right|^2$$

is minimum. Clearly this can be minimized by choosing

$$c_i = \int_a^b \phi_i^* f \, dx = (\boldsymbol{\phi}_i, \mathbf{f})$$

The actual minimum value is then

$$\int_a^b |f|^2 \, dx - \sum_{i=1}^n |c_i|^2 \ge 0$$

The quantity on the left is a positive nonincreasing function of n. This does not imply, however, that, as n increases without bound, this quantity necessarily goes to zero. In any case, we do have **Bessel's inequality**

$$\int_a^b |f|^2 \, dx \ge \sum_{i=1}^\infty |c_i|^2$$

so that the series $\sum_{i=1}^\infty |c_i|^2$ always converges. Therefore, corresponding to every square integrable function $f(x)$ there is a vector in the space of infinite sequences.

Bessel's inequality always becomes an equality if and only if, for every square integrable function,

$$\lim_{n \to \infty} \int_a^b \left| f - \sum_{i=1}^n c_i \phi_i \right|^2 dx = 0$$

In this case we say that ϕ_1, ϕ_2, ϕ_3, ... is a **complete set of functions** and $\sum_{i=1}^{\infty} c_i \phi_i$ **converges in mean** to $f(x)$. Convergence in mean does not imply convergence at each point of the interval. Consider, for example, a square integrable function $g(x)$ which differs from $f(x)$ on a set of points which does not contribute to the integral. Then

$$c_i = \int_a^b \phi_i^* f \, dx = \int_a^b \phi_i^* g \, dx$$

and the series which converge in mean to the two functions are indistinguishable. In general, it will require stronger conditions on $f(x)$ than merely square integrability to prove pointwise convergence. For example, if $f(x)$ and $\phi_i(x)$ are continuous and $\sum_{i=1}^{\infty} c_i \phi_i$ converges uniformly, then

$$\lim_{n \to \infty} \int_a^b \left| f - \sum_{i=1}^n c_i \phi_i \right|^2 dx$$

$$= \int_a^b \left| f - \sum_{i=1}^{\infty} c_i \phi_i \right|^2 dx$$

$$= \int_a^b |f - g|^2 \, dx = 0$$

where $g = \sum_{i=1}^{\infty} c_i \phi_i$ and g is continuous. But $f - g$ is continuous, and therefore

$$f = g = \sum_{i=1}^{\infty} c_i \phi_i$$

everywhere in the interval $a \leq x \leq b$.

Unfortunately, $f(x)$ will not always be continuous, and $\sum_{i=1}^{\infty} c_i \phi_i$ will not necessarily converge uniformly. Therefore, the problem of finding expansions of arbitrary square integrable functions in terms of orthogonal sets of functions needs further discussion. We shall return to this problem in the next section and in Chaps. 3, 4, and 6.

If the orthonormal set of functions ϕ_1, ϕ_2, ϕ_3, ... is complete, then Bessel's inequality becomes an equality:

$$\int_a^b |f|^2 \, dx = \sum_{i=1}^{\infty} |c_i|^2$$

This is called the **completeness relation**. It can be stated more generally; that is, if

$$b_i = \int_a^b \phi_i^* f \, dx = (\boldsymbol{\phi}_i, \mathbf{f})$$

$$c_i = \int_a^b \phi_i^* g \, dx = (\boldsymbol{\phi}_i, \mathbf{g})$$

then
$$\left| \int_a^b f^* g \, dx - \sum_{i=1}^n b_i^* c_i \right| = \left| \int_a^b f^* \left(g - \sum_{i=1}^n c_i \phi_i \right) dx \right|$$

$$\leq \|\mathbf{f}\| \left(\int_a^b \left| g - \sum_{i=1}^n c_i \phi_i \right|^2 dx \right)^{\frac{1}{2}}$$

By the completeness relation

$$\int_a^b \left| g - \sum_{i=1}^n c_i \phi_i \right|^2 dx$$

approaches zero as n approaches infinity. This proves the **Parseval relation**

$$\int_a^b f^* g \, dx = \sum_{i=1}^\infty b_i^* c_i$$

The content of the **Riesz-Fischer theorem**,[1] formulated in terms of the Lebesgue integral, is that the space $L_2(a,b)$ is complete; that is, every Cauchy sequence has a strong limit in the space and $L_2(a,b)$ is therefore a Hilbert space. Let $\phi_1, \phi_2, \phi_3, \ldots$ be a complete orthonormal set of functions in $L_2(a,b)$ and let $f(x)$ be any function in the space. If

$$f_n(x) = \sum_{i=1}^n c_i \phi_i$$

where
$$c_i = (\boldsymbol{\phi}_i, \mathbf{f})$$

then
$$\|\mathbf{f} - \mathbf{f}_n\|^2 = \int_a^b |f - f_n|^2 \, dx \to 0$$

as $n \to \infty$. Therefore, f is a strong limit of the sequence $\{\mathbf{f}_n\}$ and the sequence is a Cauchy sequence. As a Cauchy sequence $\{\mathbf{f}_n\}$ has a strong limit $\mathbf{g}$ in $L_2(a,b)$. But

$$\|\mathbf{f} - \mathbf{g}\| = \|\mathbf{f} - \mathbf{f}_n - \mathbf{g} + \mathbf{f}_n\|$$

$$\leq \|\mathbf{f} - \mathbf{f}_n\| + \|\mathbf{g} - \mathbf{f}_n\| \to 0$$

as $n \to \infty$. Therefore $\mathbf{f} = \mathbf{g}$ almost everywhere. On the other hand, suppose $\phi_1, \phi_2, \phi_3 \ldots$ is an arbitrary set of orthonormal functions in $L_2(a,b)$ and $c_1, c_2, c_3, \ldots$ is any sequence of complex numbers such that

$$\sum_{i=1}^\infty |c_i|^2 < \infty$$

[1] See E. C. Titchmarsh, "The Theory of Functions," Oxford University Press, Fair Lawn, N.J., 1939, pp. 386–388.

Let
$$f_n(x) = \sum_{i=1}^{n} c_i \phi_i$$

Then f_n is in $L_2(a,b)$ for any n.   Furthermore,

$$\|\mathbf{f}_n - \mathbf{f}_m\|^2 = \sum_{i=m+1}^{n} |c_i|^2 \to 0$$

as m and $n \to \infty$.   Hence $\{\mathbf{f}_n\}$ is a Cauchy sequence.   Therefore there exists a function $f(x)$ in $L_2(a,b)$ such that

$$\lim_{n \to \infty} \|\mathbf{f}_n - \mathbf{f}\|^2 = \lim_{n \to \infty} \int_a^b |f - f_n|^2 \, dx = 0$$

Furthermore, by Schwarz's inequality

$$|(\mathbf{f} - \mathbf{f}_n, \boldsymbol{\phi}_k)| \le \|\mathbf{f} - \mathbf{f}_n\| \to 0$$

as $n \to \infty$, so that

$$c_k = \lim_{n \to \infty} (\boldsymbol{\phi}_k, \mathbf{f}_n) = (\boldsymbol{\phi}_k, \mathbf{f})$$

This is not to say that the set of functions $\phi_1, \phi_2, \phi_3, \ldots$ is complete, for although there is a limit function associated with every sequence $c_1, c_2, c_3, \ldots$, some functions in $L_2(a,b)$ may not be producible in this way.   There is, however, an equivalent definition to completeness for a set of orthonormal functions which we shall now state.

A set of orthonormal functions $\phi_1, \phi_2, \phi_3, \ldots$ is said to be **closed** if no normalized function is orthogonal to every function in the set.   *If a set of functions is complete, then it is closed.*   If it is not closed, then there exists a normalized function $f(x)$ such that

$$c_i = \int_a^b \phi_i^* f \, dx = 0$$

for all i.   Furthermore,

$$\lim_{n \to \infty} \int_a^b \left| f - \sum_{i=1}^{n} c_i \phi_i \right|^2 dx = \int_a^b |f|^2 \, dx = 1$$

which contradicts the completeness assumption.   Therefore, the set must be closed.   The converse is also true; that is, *if a set is closed, it is complete.*   For if it is not complete, there exists a function $f(x)$ such that

$$\int_a^b |f|^2 \, dx - \sum_{i=1}^{\infty} |c_i|^2 > 0$$

where
$$c_i = (\boldsymbol{\phi}_i, \mathbf{f})$$

However, the function $g_n = \sum_{i=1}^{n} c_i \phi_i$ converges in mean to a $g(x)$ such that

$$c_i = (\boldsymbol{\phi}_i, \mathbf{g}) = (\boldsymbol{\phi}_i, \mathbf{f})$$

Therefore, the function $h = g - f$ is orthogonal to all the ϕ_i and

$$\|\mathbf{h}\| = \|\mathbf{g} - \mathbf{f}\| \geq \big| \|\mathbf{g} - \mathbf{g}_n\| - \|\mathbf{f} - \mathbf{g}_n\| \big| > 0$$

and h is a normalizable function. Hence, the set cannot be closed.

Suppose we have two complete sets of orthonormal functions, $\phi_1, \phi_2, \phi_3, \cdots$ and $\Psi_1, \Psi_2, \Psi_3, \cdots$. The expansion coefficients of $f(x)$ relative to ϕ_i are $c_i = \int_a^b \phi_i^* f \, dx$, and the expansion coefficients relative to Ψ_i are $d_i = \int_a^b \Psi_i^* f \, dx$. If we apply Parseval's equation to ϕ_i and f, we have

$$c_i = (\boldsymbol{\phi}_i, \mathbf{f}) = \sum_{j=1}^{\infty} (\boldsymbol{\phi}_i, \boldsymbol{\Psi}_j)(\boldsymbol{\Psi}_j, \mathbf{f}) = \sum_{j=1}^{\infty} a_{ij} \, d_j$$

where

$$a_{ij} = (\boldsymbol{\phi}_i, \boldsymbol{\Psi}_j)$$

By applying the Parseval relation to Ψ_i and f, we also have

$$d_i = (\boldsymbol{\Psi}_i, \mathbf{f}) = \sum_{j=1}^{\infty} (\boldsymbol{\phi}_j, \boldsymbol{\Psi}_i)^* (\boldsymbol{\phi}_j, \mathbf{f}) = \sum_{j=1}^{\infty} a_{ji}^* c_j = \sum_{j=1}^{\infty} (a^*)'_{ij} c_j$$

Furthermore,

$$\delta_{ij} = (\boldsymbol{\phi}_i, \boldsymbol{\phi}_j) = \sum_{k=1}^{\infty} (\boldsymbol{\phi}_i, \boldsymbol{\Psi}_k)(\boldsymbol{\Psi}_k, \boldsymbol{\phi}_j) = \sum_{k=1}^{\infty} (\boldsymbol{\phi}_i, \boldsymbol{\Psi}_k)(\boldsymbol{\phi}_j, \boldsymbol{\Psi}_k)^*$$

$$= \sum_{k=1}^{\infty} a_{ik} a_{jk}^* = \sum_{k=1}^{\infty} a_{ik} (a^*)'_{kj}$$

Summarizing in terms of infinite matrices,

$$C = AD$$

$$D = (A^*)'C$$

$$A(A^*)' = I$$

which is to say that the change of representation going from one complete orthonormal set of functions to another corresponds to a unitary transformation of the vector space of infinite sequences of complex numbers. It is this fundamental fact that makes unitary transformations of basic importance in quantum mechanics.

Exercises 2.1

*1. Show that the set of functions $\phi_n = (1/\sqrt{\pi}) \sin nx$, $n = 1, 2, 3, \ldots$, is not a complete set on the interval $0 \leq x \leq 2\pi$.

*2. Let $f(x)$ be continuous on the interval $0 \leq x \leq 2\pi$, and $f(0) = f(2\pi)$, and have a piecewise continuous derivative $f'(x)$. Show that series $\sum_{n=1}^{\infty} n^2(a_n^2 + b_n^2)$, where

$$a_n = \frac{1}{\pi} \int_0^{2\pi} f(x) \cos nx \, dx$$

$$b_n = \frac{1}{\pi} \int_0^{2\pi} f(x) \sin nx \, dx$$

converges. HINT: Apply Bessel's inequality to the function $f'(x)$.

*3. Show that the set of functions $\phi_n = (1/\sqrt{2\pi})e^{inx}$, $n = 0, \pm1, \pm2, \ldots$, is an orthonormal set on the interval $0 \le x \le 2\pi$.

*4. If the functions of exercise 3 are a complete set, show that a series representation of a square integrable function $f(x)$ is

$$\frac{a_0}{2} + \sum_{n=1}^{\infty} (a_n \cos nx + b_n \sin nx)$$

where $a_n = \dfrac{1}{\pi} \displaystyle\int_0^{2\pi} f(x) \cos nx \, dx$ and $b_n = \dfrac{1}{\pi} \displaystyle\int_0^{2\pi} f(x) \sin nx \, dx$.

5. If $f(z)$ is an analytic function of the complex variable z in the region $R_1 < |z| < R_2$, show that $f(re^{i\theta}) = \displaystyle\sum_{n=-\infty}^{\infty} c_n e^{in\theta}$ where

$$c_n = \frac{1}{2\pi} \int_0^{2\pi} f(re^{i\theta})e^{-in\theta} \, d\theta \qquad R_1 < r < R_2$$

HINT: Start with the Laurent expansion for $f(z)$.

6. Consider the space of continuous functions on the interval $a \le x \le b$. Let $\phi_1, \phi_2, \phi_3, \ldots$ be a complete set of orthonormal functions. Let $c_i = \displaystyle\int_a^b \phi_i^* f \, dx$, and assume that $\displaystyle\sum_{i=1}^{\infty} c_i \phi_i$ converges uniformly. Prove that the correspondence $f \sim (c_1, c_2, c_3, \ldots)$ is an isomorphism between the space of continuous functions and the space of infinite sequences, which preserves scalar product.

7. Prove that the space of infinite sequences of complex numbers with

$$\|\mathbf{U}\| = \sum_{i=1}^{\infty} |u_i| < \infty$$

is a Banach space.

2.2   Fourier Series

Perhaps the best known example of a complete set of orthonormal functions is that afforded by the trigonometric functions $1/\sqrt{2\pi}$, $(1/\sqrt{\pi}) \cos x$, $(1/\sqrt{\pi}) \sin x$, $(1/\sqrt{\pi}) \cos 2x$, $(1/\sqrt{\pi}) \sin 2x$, $\ldots$. We shall show in two steps that this set is complete with respect to real-valued functions defined on the interval $0 \le x \le 2\pi$ which are piecewise continuous and have a piecewise continuous derivative. Our first theorem will handle the completeness for the subspace of functions which are continuous and for which $f(0) = f(2\pi)$. The second theorem will extend the result to piecewise continuous functions. In the process of proving completeness we shall obtain the conventional pointwise convergence theorems of Fourier analysis.

We first must show that the set of functions is orthonormal. This is easily established as follows:

$$\frac{1}{2\pi} \int_0^{2\pi} dx = 1$$

$$\frac{1}{\pi} \int_0^{2\pi} \cos^2 kx \, dx = \frac{1}{2\pi} \int_0^{2\pi} (1 + \cos 2kx) \, dx = 1 + \left[\frac{\sin 2kx}{4k\pi}\right]_0^{2\pi} = 1$$

$$\frac{1}{\pi} \int_0^{2\pi} \sin^2 kx \, dx = \frac{1}{2\pi} \int_0^{2\pi} (1 - \cos 2kx) \, dx = 1 - \left[\frac{\sin 2kx}{4k\pi}\right]_0^{2\pi} = 1$$

If $m \neq n$,

$$\frac{1}{\pi} \int_0^{2\pi} \cos mx \cos nx \, dx = \frac{1}{2\pi} \int_0^{2\pi} [\cos (m + n)x + \cos (m - n)x] \, dx$$

$$= \frac{1}{2\pi} \left[\frac{\sin (m + n)x}{m + n} + \frac{\sin (m - n)x}{m - n}\right]_0^{2\pi} = 0$$

$$\frac{1}{\pi} \int_0^{2\pi} \sin mx \sin nx \, dx = \frac{1}{2\pi} \int_0^{2\pi} [\cos (m - n)x - \cos (m + n)x] \, dx$$

$$= \frac{1}{2\pi} \left[\frac{\sin (m - n)x}{m - n} - \frac{\sin (m + n)x}{m + n}\right]_0^{2\pi} = 0$$

$$\frac{1}{\pi} \int_0^{2\pi} \sin mx \cos nx \, dx = \frac{1}{2\pi} \int_0^{2\pi} [\sin (m + n)x + \sin (m - n)x] \, dx$$

$$= -\frac{1}{2\pi} \left[\frac{\cos (m + n)x}{m + n} + \frac{\cos (m - n)x}{m - n}\right]_0^{2\pi} = 0$$

$$\frac{1}{\pi} \int_0^{2\pi} \sin mx \cos mx \, dx = \frac{1}{2\pi} \int_0^{2\pi} \sin 2mx \, dx$$

$$= -\left[\frac{1}{4\pi m} \cos 2mx\right]_0^{2\pi} = 0$$

Theorem 1. The set of functions $1/\sqrt{2\pi}$, $(1/\sqrt{\pi}) \cos kx$, $(1/\sqrt{\pi}) \sin kx$, $k = 1, 2, 3, \ldots$ is a complete set with respect to functions $f(x)$, $0 \leq x \leq 2\pi$, which are continuous and have a piecewise continuous first derivative, and for which $f(0) = f(2\pi)$.

Proof. We first note that $f'(x)$ is integrable and square integrable. Hence, by Bessel's inequality

$$\sum_{k=1}^{\infty} (\alpha_k^2 + \beta_k^2) \leq \frac{1}{\pi} \int_0^{2\pi} |f'(x)|^2 \, dx$$

where
$$\alpha_k = \frac{1}{\pi} \int_0^{2\pi} f'(x) \cos kx \, dx$$

$$= \frac{1}{\pi}\Big[f(x) \cos kx \Big]_0^{2\pi} + \frac{k}{\pi} \int_0^{2\pi} f(x) \sin kx \, dx$$

$$= kb_k$$

$$\beta_k = \frac{1}{\pi} \int_0^{2\pi} f'(x) \sin kx \, dx$$

$$= \frac{1}{\pi}\Big[f(x) \sin kx \Big]_0^{2\pi} - \frac{k}{\pi} \int_0^{2\pi} f(x) \cos kx \, dx$$

$$= -ka_k$$

where a_k and b_k are the **Fourier coefficients** of $f(x)$; that is

$$a_k = \frac{1}{\pi} \int_0^{2\pi} f(x) \cos kx \, dx$$

$$b_k = \frac{1}{\pi} \int_0^{2\pi} f(x) \sin kx \, dx$$

Therefore,
$$\sum_{k=1}^{\infty} (k^2 a_k^2 + k^2 b_k^2) \le \frac{1}{\pi} \int_0^{2\pi} |f'(x)|^2 \, dx$$

We can show that $a_0/2 + \sum_{k=1}^{\infty} (a_k \cos kx + b_k \sin kx)$ converges uniformly. Consider $|S_n - S_m|^2$ where $S_n = a_0/2 + \sum_{k=1}^{n} (a_k \cos kx + b_k \sin kx)$.   Then

$$|S_n - S_m|^2 = \left| \sum_{k=m+1}^{n} (a_k \cos kx + b_k \sin kx) \right|^2$$

$$= \left| \sum_{k=m+1}^{n} \Big[(ka_k) \frac{\cos kx}{k} + (kb_k) \frac{\sin kx}{k} \Big] \right|^2$$

$$\le \sum_{k=m+1}^{n} (k^2 a_k^2 + k^2 b_k^2) \sum_{k=m+1}^{n} \frac{1}{k^2} \le \frac{1}{\pi} \int_0^{2\pi} |f'(x)|^2 \, dx \sum_{k=m+1}^{n} \frac{1}{k^2}$$

The series $\sum_{k=1}^{\infty} \frac{1}{k^2}$ converges.   Hence, $\sum_{k=m+1}^{n} \frac{1}{k^2} \to 0$ as m and $n \to \infty$.   Therefore, $|S_n - S_m|^2 \to 0$ as m and $n \to \infty$ uniformly in x, and we have the Cauchy criterion for uniform convergence of the series.   Next we show that the series converges to $f(x)$.

$$S_n = \frac{a_0}{2} + \sum_{k=1}^{n}(a_k \cos kx + b_k \sin kx)$$

$$= \frac{1}{\pi}\int_0^{2\pi} f(t)[\tfrac{1}{2} + \cos(x-t) + \cos 2(x-t) + \cdots + \cos n(x-t)]\,dt$$

$$= \frac{1}{\pi}\int_0^{2\pi} f(t)\,\operatorname{Re}\,(1 + e^{i(x-t)} + e^{2i(x-t)} + \cdots + e^{ni(x-t)} - \tfrac{1}{2})\,dt$$

$$= \frac{1}{\pi}\int_0^{2\pi} f(t)\,\operatorname{Re}\,\left(\frac{e^{i(n+1)(x-t)}-1}{e^{i(x-t)}-1} - \frac{1}{2}\right)dt$$

$$= \frac{1}{\pi}\int_0^{2\pi} f(t)\,\operatorname{Re}\,\left(\frac{e^{i(n+\frac{1}{2})(x-t)} - e^{-i(x-t)/2}}{e^{i(x-t)/2} - e^{-i(x-t)/2}} - \frac{1}{2}\right)dt$$

$$= \frac{1}{2\pi}\int_0^{2\pi} f(t)\,\frac{\sin(n+\frac{1}{2})(x-t)}{\sin[(x-t)/2]}\,dt$$

$$= \frac{1}{2\pi}\int_{-x}^{2\pi-x} f(u+x)\,\frac{\sin(n+\frac{1}{2})u}{\sin u/2}\,du$$

In the last integral we have made the change of variable $u = t - x$. At this point we must extend the definition of $f(x)$ outside the interval $0 \le x \le 2\pi$. We do this by the periodic extension; that is, for any $-\infty < x < \infty$ we define $f(x) = f(x + 2\pi p)$ where p is the appropriate integer chosen so that $0 \le x + 2p\pi < 2\pi$. We note that by the condition $f(0) = f(2\pi)$ the periodic extension is continuous and also that the derivative is piecewise continuous in any finite interval. We can now write

$$S_n = \frac{1}{2\pi}\int_0^{2\pi} f(u+x)\,\frac{\sin(n+\frac{1}{2})u}{\sin u/2}\,du$$

since the integrand is periodic with period 2π. We also note that

$$\frac{1}{2\pi}\int_0^{2\pi} f(x)\,\frac{\sin(n+\frac{1}{2})u}{\sin u/2}\,du$$

$$= f(x)\left[\frac{1}{\pi}\int_0^{2\pi}(\tfrac{1}{2} + \cos u + \cos 2u + \cdots + \cos nu)\,du\right]$$

$$= f(x)$$

Hence,

$$S_n - f(x) = \frac{1}{2\pi}\int_0^{2\pi}\left[\frac{f(x+u)-f(x)}{\sin u/2}\right]\sin(n+\tfrac{1}{2})u\,du$$

$$= \frac{1}{2\pi}\int_0^{2\pi}\left[\frac{f(x+u)-f(x)}{\sin u/2}\cos\frac{u}{2}\right]\sin nu\,du$$

$$+ \frac{1}{2\pi}\int_0^{2\pi}[f(x+u)-f(x)]\cos nu\,du$$

Now $f(x + u) - f(x)$ is continuous in u and $\dfrac{f(x + u) - f(x)}{\sin u/2} \cos \dfrac{u}{2}$ is piecewise continuous.   The latter follows from

$$\lim_{u \to 0+} 2\frac{f(x + u) - f(x)}{u} \cos \frac{u}{2} \frac{u/2}{\sin u/2} = 2f'(x+)$$

$$\lim_{u \to 0-} 2\frac{f(x + u) - f(x)}{u} \cos \frac{u}{2} \frac{u/2}{\sin u/2} = 2f'(x-)$$

Therefore, $S_n - f(x)$ can be expressed as the sum of Fourier coefficients of a continuous and a piecewise continuous function.   Hence, by Bessel's inequality

$$\lim_{n \to \infty} [S_n - f(x)] = \lim_{n \to \infty} \frac{1}{2\pi} \int_0^{2\pi} \left[\frac{f(x + u) - f(x)}{\sin u/2} \cos \frac{u}{2}\right] \sin nu \, dx$$

$$+ \lim_{n \to \infty} \frac{1}{2\pi} \int_0^{2\pi} [f(x + u) - f(x)] \cos nu \, du$$

$$= 0$$

and we have

$$f(x) = \frac{a_0}{2} + \sum_{k=1}^{\infty} (a_k \cos kx + b_k \sin kx)$$

and the convergence is uniform.   We can multiply the series by $(1/\pi)f(x)$ and integrate term by term, thus obtaining the completeness relation

$$\frac{1}{\pi} \int_0^{2\pi} |f(x)|^2 \, dx = \frac{a_0^2}{2} + \sum_{k=1}^{\infty} (a_k^2 + b_k^2)$$

We note that in proving theorem 1, we have obtained the following corollary.

Corollary 1.1.   If $f(x)$ is a continuous periodic function with period 2π, with a piecewise continuous derivative, it can be expanded in a uniformly convergent series

$$f(x) = \frac{a_0}{2} + \sum_{k=1}^{\infty} (a_k \cos kx + b_k \sin kx)$$

where

$$a_k = \frac{1}{\pi} \int_0^{2\pi} f(x) \cos kx \, dx$$

$$b_k = \frac{1}{\pi} \int_0^{2\pi} f(x) \sin kx \, dx$$

Next we show that the same orthonormal set of trigonometric functions is complete with respect to a larger class of functions.

Theorem 2.   The set of functions $1/\sqrt{2\pi}$, $(1/\sqrt{\pi}) \cos kx$, $(1/\sqrt{\pi}) \sin kx$, $k = 1, 2, 3, \ldots$ is a complete orthonormal set with respect to functions $f(x)$,

$0 \leq x \leq 2\pi$, which are piecewise continuous and have a piecewise continuous derivative. The Fourier series converges pointwise as follows:

$$\tfrac{1}{2}[f(x+) + f(x-)] = \frac{a_0}{2} + \sum_{k=1}^{\infty} (a_k \cos kx + b_k \sin kx)$$

where $f(0-) = f(2\pi-)$ and $f(2\pi+) = f(0+)$. The convergence is uniform in any closed interval not containing a discontinuity of the function.

Proof. We shall prove the pointwise convergence part of the theorem first. To this end we again extend the function by periodicity; that is, $f(x) = f(x + 2\pi p)$ where p is the appropriate integer chosen so that $0 \leq x + 2\pi p < 2\pi$. With this extension we have $f(x)$ defined as a periodic function with period 2π which has but a finite number of discontinuities in each period. Suppose that $f(x)$ has but one discontinuity in each period. Let $f(x)$ have discontinuities at $\xi \pm 2m\pi$, $m = 0, 1, 2, \ldots$ with $0 \leq \xi \leq 2\pi$. Then $\lim\limits_{x \to \xi+} f(x) = f(\xi+)$, $\lim\limits_{x \to \xi-} f(x) = f(\xi-)$, and $f(\xi+) \neq f(\xi-)$.

We can put the discontinuity at the origin by translating the x axis; that is, $t = x - \xi$. This will not affect the Fourier coefficients of $f(x)$ since it is periodic. Let $F(t) = f(t + \xi)$ and

$$g(t) = F(t) - \frac{1}{2}[f(\xi+) - f(\xi-)]h(t)$$

where $h(t) = (1/\pi)(\pi - t)$ for $0 \leq t < 2\pi$ and is extended periodically for other values of t. $h(t)$ is continuous except at $0, \pm 2\pi, \pm 4\pi, \ldots$, and therefore $g(t)$ is continuous except possibly at these points. Actually $g(t)$ is continuous everywhere, since

$$\lim_{t \to 0+} g(t) = F(0+) - \tfrac{1}{2}[f(\xi+) - f(\xi-)]$$
$$= f(\xi+) - \tfrac{1}{2}[f(\xi+) - f(\xi-)]$$
$$= \tfrac{1}{2}[f(\xi+) + f(\xi-)]$$

$$\lim_{t \to 0-} g(t) = F(0-) + \tfrac{1}{2}[f(\xi+) - f(\xi-)]$$
$$= f(\xi-) + \tfrac{1}{2}[f(\xi+) - f(\xi-)]$$
$$= \tfrac{1}{2}[f(\xi+) + f(\xi-)]$$

By periodicity $g(t)$ is continuous at $t = \pm 2\pi, \pm 4\pi, \ldots$. Therefore, the Fourier series representation for $g(t)$ converges uniformly. It remains to show that the Fourier series for $h(t)$ converges and hence that the series for $F(t)$ and $f(x)$ converge. We shall show that the series for $h(t)$ converges to zero at $t = 0$, which will show that the series for $f(x)$ converges to

$$g(0) = \tfrac{1}{2}[f(\xi+) + f(\xi-)]$$

at $x = \xi$.

The function $h(t)$ is odd and, therefore,

$$c_k = \frac{1}{\pi} \int_0^{2\pi} h(t) \cos kt \, dt$$

$$= \frac{1}{\pi} \int_{-\pi}^{\pi} h(t) \cos kt \, dt = 0$$

Its other Fourier coefficients are

$$d_k = \frac{1}{\pi^2} \int_0^{2\pi} (\pi - t) \sin kt \, dt$$

$$= \frac{1}{\pi^2} \left[-\frac{\pi}{k} \cos kt + \frac{t}{k} \cos kt - \frac{\sin kt}{k^2} \right]_0^{2\pi}$$

$$= \frac{2}{k\pi}$$

Now consider the function $H(t) = (1 - \cos t)h(t)$. This function is continuous, odd, and periodic. Therefore, it has a uniformly convergent Fourier series with coefficients

$$\gamma_k = \frac{1}{\pi} \int_0^{2\pi} H(t) \cos kt \, dt$$

$$= \frac{1}{\pi} \int_{-\pi}^{\pi} H(t) \cos kt \, dt = 0$$

$$\delta_k = \frac{1}{\pi} \int_0^{2\pi} (1 - \cos t)h(t) \sin kt \, dt$$

$$= d_k - \frac{1}{\pi} \int_0^{2\pi} h(t) \cos t \sin kt \, dt$$

$$= d_k - \frac{1}{2\pi} \int_0^{2\pi} h(t) \sin (k + 1)t \, dt - \frac{1}{2\pi} \int_0^{2\pi} h(t) \sin (k - 1)t \, dt$$

$$= d_k - \tfrac{1}{2}(d_{k+1} + d_{k-1}) \qquad k = 2, 3, 4, \ldots$$

For $k = 1$, $\delta_1 = d_1 - \tfrac{1}{2}d_2$. Let

$$S_n = \sum_{k=1}^n \delta_k \sin kt = (d_1 - \tfrac{1}{2}d_2) \sin t + [d_2 - \tfrac{1}{2}(d_3 + d_1)] \sin 2t$$

$$+ \cdots + [d_{n-1} - \tfrac{1}{2}(d_n + d_{n-2})] \sin (n - 1)t$$

$$+ [d_n - \tfrac{1}{2}(d_{n+1} + d_{n-1})] \sin nt$$

and $\qquad \sigma_n = \sum_{k=1}^n d_k \sin kt = d_1 \sin t + d_2 \sin 2t + \cdots + d_n \sin nt$

Then

$$(1 - \cos t)\sigma_n = (d_1 - \tfrac{1}{2}d_2) \sin t + [d_2 - \tfrac{1}{2}(d_3 + d_1)] \sin 2t$$

$$+ \cdots + [d_n - \tfrac{1}{2}(d_{n+1} + d_{n-1})] \sin nt$$

$$+ \tfrac{1}{2}d_{n+1} \sin nt - \tfrac{1}{2}d_n \sin (n + 1)t$$

$$= S_n + \frac{1}{(n + 1)\pi} \sin nt - \frac{1}{n\pi} \sin (n + 1)t$$

We know that $\lim\limits_{n \to \infty} S_n = H(t) = (1 - \cos t)h(t)$, uniformly in t. Furthermore,

$$|1 - \cos t|\, |\sigma_n - h(t)| = |S_n - H(t) + \frac{1}{(n + 1)\pi} \sin nt - \frac{1}{n\pi} \sin (n + 1)t|$$

Therefore,

$$|\sigma_n - h(t)| \leq \frac{|S_n - H(t)| + 2/n\pi}{|1 - \cos t|}$$

Suppose t lies in some closed interval not containing $t = 0, \pm 2\pi, \pm 4\pi, \ldots$; then $|1 - \cos t| \geq M > 0$. Therefore, in such an interval $|\sigma_n - h(t)| \to 0$ as $n \to \infty$, uniformly in t. Hence, the Fourier series for $h(t)$ converges uniformly to the function in every closed interval not containing $t = 0, \pm 2\pi, \pm 4\pi, \ldots$. At the exceptional points the series converges to zero, since it contains only terms in $\sin kt$. This completes the proof for one discontinuity. It can clearly be extended to include the possibility of a finite number of discontinuities in a given period.

In the present case we do not have uniform convergence in the interval $0 \leq x \leq 2\pi$. Therefore, we cannot make use of the termwise integration of the series to obtain the completeness relation. However, we can show that the set is closed, and if the set is closed, it is complete.

Let $f(x)$ be a function in the space which is orthogonal to all the functions in the given set; that is,

$$a_k = \frac{1}{\pi} \int_0^{2\pi} f(x) \cos kx \, dx = 0$$

$$b_k = \frac{1}{\pi} \int_0^{2\pi} f(x) \sin kx \, dx = 0$$

Hence, the Fourier series $a_0/2 + \sum\limits_{k=1}^{\infty} (a_k \cos kx + b_k \sin kx)$ converges everywhere to zero. Therefore, $f(x)$ is equal to zero almost everywhere. For the space of piecewise continuous functions, "almost everywhere" means everywhere except at a finite number of points. Hence, $\frac{1}{\pi} \int_0^{2\pi} |f(x)|^2 \, dx = 0$. This shows that the set is closed, since there are no normalizable functions in the space which are orthogonal to every member of the given set of orthonormal functions.

Exercises 2.2

1. Obtain the Fourier series for $f(x) = x/2\pi$, $0 \leq x \leq 2\pi$. What does the series converge to at $x = 0$ and $x = 2\pi$?

2. Show that, if $f(x)$ is an even periodic function with period 2π satisfying the conditions of theorem 2, its Fourier series contains no terms in $\sin kx$. Also show that, if it is odd, its Fourier series contains no $\cos kx$ terms.

3. Obtain a Fourier series for $f(x) = x/2\pi$, $0 \leq x \leq \pi$, which contains no terms in $\sin kx$. Obtain a Fourier series for the same function which contains no terms in $\cos kx$.

4. If $f(x)$ satisfies the conditions of theorem 1, show that $f(x) = \sum\limits_{k=-\infty}^{\infty} c_k e^{-ikx}$ in which $c_k = \dfrac{1}{2\pi} \displaystyle\int_0^{2\pi} f(x) e^{ikx}\, dx$ and the convergence is uniform in x.

5. Show that, if $f(x)$ is periodic with period $2L$, is continuous, and has a piecewise continuous derivative in any period,

$$f(x) = \frac{a_0}{2} + \sum_{k=1}^{\infty} \left(a_k \cos \frac{k\pi x}{L} + b_k \sin \frac{k\pi x}{L} \right)$$

$$a_k = \frac{1}{L} \int_0^{2L} f(x) \cos \frac{k\pi x}{L}\, dx$$

where

$$b_k = \frac{1}{L} \int_0^{2L} f(x) \sin \frac{k\pi x}{L}\, dx$$

Hint: Make the change of variables $x = Lt/\pi$.

6. Prove that, if $f(x)$ is periodic with period 2π and has continuous derivatives up to the $(n-1)$st and a piecewise continuous nth derivative, then its Fourier coefficients have the property

$$\lim_{k \to \infty} k^n a_k = 0$$

$$\lim_{k \to \infty} k^n b_k = 0$$

7. Prove that, if $f(x)$ is periodic and continuous and has piecewise continuous first and second derivatives, then except at points of discontinuity of $f'(x)$ its derivative can be computed by termwise differentiation of its Fourier series.

2.3 Separable Hilbert Spaces

We have defined a Hilbert space as an infinite-dimensional vector space with a scalar product which is complete with respect to the norm defined by the scalar product. For many applications it is sufficient to consider those Hilbert spaces which have a complete orthonormal set, sometimes referred to as **separable Hilbert spaces**. As we have seen in Sec. 2.1, the space of infinite sequences of complex numbers $(u_1, u_2, u_3, \ldots)$ such that $\sum_{i=1}^{\infty} |u_i|^2 < \infty$ is a Hilbert space with a complete orthonormal set

$$\mathbf{X}_1 = (1,0,0, \ldots) \qquad \mathbf{X}_2 = (0,1,0, \ldots) \qquad \mathbf{X}_3 = (0,0,1,0, \ldots)$$

etc. This space plays a very important role since there is a natural isomorphism between it and any other separable Hilbert space over the complex numbers. We show this as follows.

Let $\boldsymbol{\phi}_1, \boldsymbol{\phi}_2, \boldsymbol{\phi}_3, \ldots$ be a complete orthonormal set in the given Hilbert space and let $\mathbf{f}$ be an arbitrary vector. Then $u_i = (\boldsymbol{\phi}_i, \mathbf{f})$ is defined and by the completeness relation

$$\sum_{i=1}^{\infty} |u_i|^2 = \|\mathbf{f}\|^2 < \infty$$

Therefore, $(u_1, u_2, u_3, \ldots)$ is in the space of infinite sequences. Conversely, let $(u_1, u_2, u_3, \ldots)$ be a sequence of complex numbers such that $\sum_{i=1}^{\infty} |u_i|^2 < \infty$. Then

$$\mathbf{f}_n = \sum_{i=1}^{n} u_i \boldsymbol{\phi}_i$$

is a Cauchy sequence since $\|\mathbf{f}_n - \mathbf{f}_m\|^2 = \sum_{i=m+1}^{n} |u_i|^2$ goes to zero as m and n approach infinity. Therefore, since the Hilbert space is complete, $\{\mathbf{f}_n\}$ has a strong limit $\mathbf{f}$ and $u_i = (\boldsymbol{\phi}_i, \mathbf{f})$. We have thus shown that there is a one-to-one correspondence between the vectors in the given Hilbert space and the infinite sequences.

Since, when $u_i = (\boldsymbol{\phi}_i, \mathbf{f})$

$$\lim_{n \to \infty} \|\mathbf{f} - \sum_{i=1}^{n} u_i \boldsymbol{\phi}_i\| = 0$$

we write $\mathbf{f} = \sum_{i=1}^{\infty} u_i \boldsymbol{\phi}_i$. This representation is unique since

$$\sum_{i=1}^{\infty} u_i \boldsymbol{\phi}_i = \sum_{i=1}^{\infty} \bar{u}_i \boldsymbol{\phi}_i$$

implies that

$$\sum_{i=1}^{\infty} (u_i - \bar{u}_i) \boldsymbol{\phi}_i = \mathbf{0}$$

which implies that $\sum_{i=1}^{\infty} |u_i - \bar{u}_i|^2 = \|\mathbf{0}\|^2 = 0$. Hence, $u_i = \bar{u}_i$ for all i.

The correspondence obviously preserves sum and multiplication by a scalar. It also preserves scalar product, since by the Parseval relation

$$(\mathbf{f}, \mathbf{g}) = \sum_{i=1}^{\infty} u_i^* v_i$$

where $u_i = (\boldsymbol{\phi}_i, \mathbf{f})$ and $v_i = (\boldsymbol{\phi}_i, \mathbf{g})$.

If there are two complete orthonormal sets $\boldsymbol{\phi}_1$, $\boldsymbol{\phi}_2$, $\boldsymbol{\phi}_3$, ... and $\boldsymbol{\Psi}_1$, $\boldsymbol{\Psi}_2$, $\boldsymbol{\Psi}_3$, ..., if U denotes the sequence $(u_1, u_2, u_3, \ldots)$ where $u_i = (\boldsymbol{\phi}_i, \mathbf{f})$, and if V denotes the sequence $(v_1, v_2, v_3, \ldots)$ where $v_i = (\boldsymbol{\Psi}_i, \mathbf{f})$, then

$$U = AV$$

where A is the infinite matrix with elements $a_{ij} = (\boldsymbol{\phi}_i, \boldsymbol{\Psi}_j)$ and A is unitary; that is,

$$A(A^*)' = I$$

We conclude this section with a demonstration that the space $L_2(a,b)$ is separable by showing that the trigonometric functions form a complete orthonormal set in the space. We will actually do this for $L_2(0,2\pi)$ but it is clear that there is no essential difference since the change of variable

$$x = \frac{(b-a)t + 2\pi a}{2\pi}$$

will change the interval $a \leq x \leq b$ into $0 \leq t \leq 2\pi$. We shall use the functions $\phi_k = (1/\sqrt{2\pi})e^{ikt}$, $k = 0, \pm 1, \pm 2, \ldots$ and remind the reader that for $c_k = \dfrac{1}{\sqrt{2\pi}} \displaystyle\int_0^{2\pi} f(t)e^{-ikt}\, dt$

$$f(t) = \frac{1}{\sqrt{2\pi}} \sum_{k=-\infty}^{\infty} c_k e^{ikt}$$

if f is continuous and has a piecewise continuous derivative and $f(0) = f(2\pi)$. We must first establish the following lemma.

Lemma 1. Let $f(t)$ be continuous for $0 \leq t \leq 2\pi$ and $f(0) = f(2\pi)$. Then given any $\epsilon > 0$ there exists a trigonometric polynomial $P(t) = \dfrac{1}{\sqrt{2\pi}} \displaystyle\sum_{k=-n}^{n} a_k e^{ikt}$ such that $|f(t) - P(t)| < \epsilon$ for all $0 \leq t \leq 2\pi$.

Proof. Since $f(t)$ is continuous on the closed interval, it is uniformly continuous and $|f(t) - f(t')| < \epsilon/4$ provided $|t - t'| < \delta$ for some δ depending only on ϵ. We subdivide the interval as follows: $0 = t_0 < t_1 < t_2 < \cdots < t_m = 2\pi$ so that $|t_j - t_{j-1}| < \delta$ for $j = 1, 2, 3, \ldots, m$. We construct the graph of a function $g(t)$ by joining $(0, f(0))$ to $(t_1, f(t_1))$ by a straight-line segment, $(t_1, f(t_1))$ to $(t_2, f(t_2))$ by a straight-line segment, and so on until we reach $(2\pi, f(2\pi))$. Now $g(t)$ is continuous and has a piecewise continuous derivative and $g(0) = f(0) = f(2\pi) = g(2\pi)$. Also $|f(t) - g(t)| < \epsilon/2$ for all $0 \leq t \leq 2\pi$. By theorem 1 of Sec. 2.2, $g(t)$ has a uniformly convergent Fourier series $\dfrac{1}{\sqrt{2\pi}} \displaystyle\sum_{k=-\infty}^{\infty} a_k e^{ikt}$, where $a_k = \dfrac{1}{\sqrt{2\pi}} \displaystyle\int_0^{2\pi} g(t)e^{-ikt}\, dt$. There exists n

such that

$$\left| g(t) - \frac{1}{\sqrt{2\pi}} \sum_{k=-n}^{n} a_k e^{ikt} \right| < \frac{\epsilon}{2}$$

for all $0 \le t \le 2\pi$. Hence,

$$\left| f(t) - \frac{1}{\sqrt{2\pi}} \sum_{k=-n}^{n} a_k e^{ikt} \right| < \epsilon$$

and the proof of the lemma is complete.

Now to prove that $L_2(0,2\pi)$ is separable.[1] Let $f(t)$ be in $L_2(0,2\pi)$, and let

$$\int_0^{2\pi} f(t) e^{ikt}\, dt = 0$$

for $k = 0, \pm 1, \pm 2, \ldots$. Let

$$F(t) = \int_0^t f(\tau)\, d\tau$$

then $F(0) = F(2\pi) = 0$. Integrating by parts, we have

$$\int_0^{2\pi} [F(t) - c] e^{ikt}\, dt = \frac{[F(t) - c] e^{ikt}}{ik} \Big|_0^{2\pi} - \frac{1}{ik} \int_0^{2\pi} f(t) e^{ikt}\, dt = 0$$

for $k = \pm 1, \pm 2, \pm 3, \ldots$. This is true for arbitrary c, so we pick

$$c = \frac{1}{2\pi} \int_0^{2\pi} F(t)\, dt$$

which makes $\int_0^{2\pi} [F(t) - c]\, dt = 0$. Now let $G(t) = F(t) - c$. Then $G(0) = G(2\pi)$ and $G(t)$ is continuous for $0 \le t \le 2\pi$. By the lemma, given $\epsilon > 0$ there is a trigonometric polynomial $(1/\sqrt{2\pi}) \sum_{k=-n}^{n} a_k e^{ikt}$ such that for $0 \le t \le 2\pi$

$$\left| G(t) - \frac{1}{\sqrt{2\pi}} \sum_{k=-n}^{n} a_k e^{ikt} \right| < \epsilon$$

By the orthogonality

$$\int_0^{2\pi} |G(t)|^2\, dt = \int_0^{2\pi} G^*(t) \left[G(t) - \frac{1}{\sqrt{2\pi}} \sum_{k=-n}^{n} a_k e^{ikt} \right] dt \le \sqrt{2\pi}\, \epsilon \left[\int_0^{2\pi} |G(t)|^2\, dt \right]^{\frac{1}{2}}$$

If $\int_0^{2\pi} |G(t)|^2\, dt = 0$, then $F(t) = c$, which implies that $f(t) = 0$ almost everywhere. If $\int_0^{2\pi} |G(t)|^2\, dt \ne 0$, then we can square both sides of the

---

[1] We actually use the Lebesgue integral, but the reader should have no difficulty appreciating the proof because all the properties used should be familiar from his experience with the Riemann integral.

inequality and divide through by the integral, proving that

$$\int_0^{2\pi} |G(t)|^2\, dt \leq 2\pi\epsilon^2$$

But ϵ is arbitrary and hence $\displaystyle\int_0^{2\pi} |G(t)|^2\, dt = 0,$ from which we draw the same conclusion, $f(t) = 0$ almost everywhere. We have shown that the set of trigonometric functions is closed, which shows that it is complete. This shows that $L_2(0,2\pi)$ is separable.

There are a lot of other complete sets of functions in $L_2(a,b)$ for various choices of a and b. We shall encounter some of these later in the book.

Exercises 2.3

1. A subset of vectors S in a Hilbert space H is said to be **dense** in H if given $\mathbf{f}$ in H and any $\epsilon > 0$ there is a vector $\mathbf{g}$ in S such that $\|\mathbf{f} - \mathbf{g}\| < \epsilon$. Prove that if $\boldsymbol{\phi}_1, \boldsymbol{\phi}_2, \boldsymbol{\phi}_3, \ldots$ is a complete orthonormal set in H, then the vectors of the form $\displaystyle\sum_{i}^{n} c_i \boldsymbol{\phi}_i$ (n not fixed) are dense in H.

2. Prove that a separable Hilbert space has a countable dense subset. HINT: Consider linear combinations of a complete orthonormal set with rational coefficients.

3. Prove that a Hilbert space with a countable dense subset is separable. HINT: Construct from the countable dense set an orthonormal set and show that it is complete.

4. Prove the **Weierstrass approximation theorem:** if $f(x)$ is continuous on $a \leq x \leq b$ and $\epsilon > 0$ is arbitrary, then there exists a polynomial $P(x)$ such that $|f(x) - P(x)| < \epsilon$ for all $a \leq x \leq b$. HINT: Use the lemma of this section and approximate the trigonometric functions by their Taylor expansions. Be sure to take into account the fact that $f(a) \neq f(b)$.

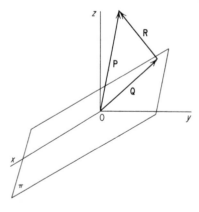

FIGURE 3

2.4   The Projection Theorem

In three-dimensional euclidean geometry there is the familiar idea of the projection of a vector on a plane or on a line. For example, see Fig. 3. The projection of the vector $\mathbf{P}$ on the plane is the vector $\mathbf{Q}$. The vector $\mathbf{R}$ is perpendicular to $\mathbf{Q}$ and we have the relation $\mathbf{P} = \mathbf{Q} + \mathbf{R}$. The same kind of result holds in Hilbert spaces in general. We must first define a subspace of Hilbert space which will play the role of the plane in our simple example.

A **subspace** of Hilbert space is a vector space lying in the Hilbert space equipped with the same scalar product and complete with respect to the norm derivable from that scalar product. If the subspace is finite-dimensional, then it is spanned by a finite number of vectors of the Hilbert space. We have already seen that a finite-dimensional vector space is complete. If the subspace is infinite-dimensional, then it is a Hilbert space because it is complete. In other words, a subspace is either finite-dimensional or a Hilbert space. We now state the **projection theorem**.

Theorem 1. If $\mathbf{h}$ is a given vector of a Hilbert space H and G is a subspace, then there is a unique vector $\mathbf{g}$ in G, called the projection of $\mathbf{h}$ on G, and a unique vector $\mathbf{f}$ orthogonal to every vector in G such that $\mathbf{h} = \mathbf{g} + \mathbf{f}$, with $(\mathbf{u}, \mathbf{f}) = 0$ for all $\mathbf{u}$ in G.

Proof. We consider the collection of norms $\|\mathbf{h} - \mathbf{u}\|$, where $\mathbf{u}$ is in G. Since $0 \leq \|\mathbf{h} - \mathbf{u}\|$, we are dealing with a set of real numbers with a lower bound. Therefore, this set has a greatest lower bound β such that $0 \leq \beta \leq \|\mathbf{h} - \mathbf{u}\|$ for all $\mathbf{u}$ in G. If $\mathbf{h}$ is in G, then $\beta = 0$ because $\|\mathbf{h} - \mathbf{u}\| = 0$ when $\mathbf{u} = \mathbf{h}$, and in this case $\mathbf{h}$ is its own projection on G, $\mathbf{f} = \mathbf{0}$, and the theorem holds. Otherwise let us assume that $\mathbf{h}$ is not in G. There exists a sequence of vectors $\{\mathbf{u}_n\}$ in G such that $\|\mathbf{h} - \mathbf{u}_n\| \to \beta$ as $n \to \infty$. Now

$$\|\mathbf{h} - \tfrac{1}{2}(\mathbf{u}_n + \mathbf{u}_m)\| = \|\tfrac{1}{2}(\mathbf{h} - \mathbf{u}_n) + \tfrac{1}{2}(\mathbf{h} - \mathbf{u}_m)\|$$
$$\leq \tfrac{1}{2}\|\mathbf{h} - \mathbf{u}_n\| + \tfrac{1}{2}\|\mathbf{h} - \mathbf{u}_m\| \to \beta$$

as $n \to \infty$ and $m \to \infty$. On the other hand, $\tfrac{1}{2}(\mathbf{u}_n + \mathbf{u}_m)$ is in G and therefore $\|\mathbf{h} - \tfrac{1}{2}(\mathbf{u}_n + \mathbf{u}_m)\| \geq \beta$. Therefore, $\|\mathbf{h} - \tfrac{1}{2}(\mathbf{u}_n + \mathbf{u}_m)\| \to \beta$ as $n \to \infty$ and $m \to \infty$. It follows from the parallelogram rule

$$\|\mathbf{w} + \mathbf{v}\|^2 + \|\mathbf{w} - \mathbf{v}\|^2 = 2\|\mathbf{w}\|^2 + 2\|\mathbf{v}\|^2$$

with $\mathbf{w} = \mathbf{h} - \mathbf{u}_n$ and $\mathbf{v} = \mathbf{h} - \mathbf{u}_m$ that

$$\|\mathbf{u}_n - \mathbf{u}_m\|^2 = 2\|\mathbf{h} - \mathbf{u}_n\|^2 + 2\|\mathbf{h} - \mathbf{u}_m\|^2 - 4\|\mathbf{h} - \tfrac{1}{2}(\mathbf{u}_n + \mathbf{u}_m)\|^2$$

Hence $\|\mathbf{u}_n - \mathbf{u}_m\| \to 0$ as $n \to \infty$ and $m \to \infty$. This proves that $\{\mathbf{u}_n\}$ is a Cauchy sequence, and since G is complete it has a strong limit $\mathbf{g}$ in G. Now $\|\mathbf{g} - \mathbf{u}_n\| \to 0$ and $\|\mathbf{h} - \mathbf{u}_n\| \to \beta$ as $n \to \infty$, and

$$\|\mathbf{h} - \mathbf{g}\| \leq \|\mathbf{h} - \mathbf{u}_n\| + \|\mathbf{g} - \mathbf{u}_n\| \to \beta$$

as $n \to \infty$. Therefore $\|\mathbf{h} - \mathbf{g}\| \leq \beta$ and $\|\mathbf{h} - \mathbf{g}\| \geq \beta$. This proves that $\|\mathbf{h} - \mathbf{g}\| = \beta$. If $\beta = 0$, then $\mathbf{g} = \mathbf{h}$ and this implies that $\mathbf{h}$ is in G. Therefore, if $\mathbf{h}$ is not in G, $\beta > 0$.

Now let $\mathbf{f} = \mathbf{h} - \mathbf{g}$. We shall prove that $\mathbf{f}$ is orthogonal to every vector in G. Let $\mathbf{u}$ be in G. Then $\mathbf{g} + a\mathbf{u}$ is also in G, and

$$\beta^2 = \|\mathbf{h} - \mathbf{g}\|^2 \leq \|\mathbf{h} - (\mathbf{g} + a\mathbf{u})\|^2$$
$$\beta^2 \leq \|\mathbf{h} - \mathbf{g}\|^2 + |a|^2\|\mathbf{u}\|^2 - a(\mathbf{h} - \mathbf{g}, \mathbf{u}) - a^*(\mathbf{u}, \mathbf{h} - \mathbf{g})$$

If we let $a = (\mathbf{u}, \mathbf{h} - \mathbf{g})/\|\mathbf{u}\|^2$ we have

$$\|\mathbf{h} - \mathbf{g}\|^2 \le \|\mathbf{h} - \mathbf{g}\|^2 - \frac{|(\mathbf{h} - \mathbf{g}, \mathbf{u})|^2}{\|\mathbf{u}\|^2}$$

which is impossible unless $(\mathbf{h} - \mathbf{g}, \mathbf{u}) = 0$ for every $\mathbf{u}$ in G.

It remains to be shown that $\mathbf{g}$ and $\mathbf{f}$ are unique. If $\mathbf{g}$ and $\bar{\mathbf{g}}$ both have the property that $\|\mathbf{h} - \mathbf{g}\| = \|\mathbf{h} - \bar{\mathbf{g}}\| = \beta$, with $\mathbf{g}$ and $\bar{\mathbf{g}}$ in G, then

$$\|\mathbf{h} - \tfrac{1}{2}(\mathbf{g} + \bar{\mathbf{g}})\| \le \tfrac{1}{2}\|\mathbf{h} - \mathbf{g}\| + \tfrac{1}{2}\|\mathbf{h} - \bar{\mathbf{g}}\| = \beta$$

But $\tfrac{1}{2}(\mathbf{g} + \bar{\mathbf{g}})$ is in G and therefore $\|\mathbf{h} - \tfrac{1}{2}(\mathbf{g} + \bar{\mathbf{g}})\| \ge \beta$. This proves that

$$\beta = \|\mathbf{h} - \tfrac{1}{2}(\mathbf{g} + \bar{\mathbf{g}})\| = \tfrac{1}{2}\|\mathbf{h} - \mathbf{g}\| + \tfrac{1}{2}\|\mathbf{h} - \bar{\mathbf{g}}\|$$

which shows that the triangle inequality is an equality. From exercise 4, Sec. 1.6, $\mathbf{h} - \mathbf{g} = \gamma(\mathbf{h} - \bar{\mathbf{g}})$ with $\gamma \ge 0$. If $\gamma = 1$, $\mathbf{g} = \bar{\mathbf{g}}$. If $\gamma \ne 1$, then $\mathbf{h} = (\mathbf{g} - \gamma\mathbf{g})/(1 - \gamma)$ which shows that $\mathbf{h}$ is in G. If $\mathbf{h}$ is in G, then $\mathbf{h}$ is its own projection and no other vector satisfies $\|\mathbf{h} - \mathbf{u}\| = 0$. If $\mathbf{g}$ is unique, then so is $\mathbf{f} = \mathbf{h} - \mathbf{g}$. This completes the proof.

Corollary 1.1. If G is a proper subspace of a Hilbert space H, then there is a subspace F such that H is the direct sum of G and F; that is, $H = G + F$, which means that each vector of H has a unique representation $\mathbf{h} = \mathbf{g} + \mathbf{f}$ where $\mathbf{g}$ is in G and $\mathbf{f}$ is in F.

Proof. Consider each $\mathbf{h}$ in H. By the projection theorem, $\mathbf{h} = \mathbf{g} + \mathbf{f}$ where $\mathbf{g}$ is in G and $\mathbf{f}$ is orthogonal to every vector of G. Let F be the collection of vectors orthogonal to G. It remains to be shown that F is a subspace. It is easy to show that F is a vector space (exercise 1, Sec. 2.4). To show that F is complete, consider a Cauchy sequence $\{\mathbf{f}_n\}$ in F with a strong limit $\mathbf{f}$. We must show that $\mathbf{f}$ is orthogonal to every vector $\mathbf{u}$ in G. Now

$$|(\mathbf{f}, \mathbf{u})| = |(\mathbf{f} - \mathbf{f}_n + \mathbf{f}_n, \mathbf{u})| = |(\mathbf{f} - \mathbf{f}_n, \mathbf{u})|$$
$$\le \|\mathbf{f} - \mathbf{f}_n\| \|\mathbf{u}\|$$

which approaches zero as n approaches infinity. Therefore, $(\mathbf{f}, \mathbf{u}) = 0$. The subspace F is called the orthogonal complement of G in H and is expressed by $F = H - G$.

Exercises 2.4

1. Prove that the collection of vectors orthogonal to a given subspace of a Hilbert space forms a vector space.

2. Given a set of orthonormal vectors $\boldsymbol{\phi}_1, \boldsymbol{\phi}_2, \boldsymbol{\phi}_3, \ldots$ in a Hilbert space H, form a subspace G_n by taking all linear combinations of $\boldsymbol{\phi}_1, \boldsymbol{\phi}_2, \ldots, \boldsymbol{\phi}_n$. What is the projection of a given vector $\mathbf{h}$ on G_n? HINT: What vector in G_n is closest to $\mathbf{h}$?

3. Referring to exercise 2, write $H = G_n + F_n$. In terms of F_n what does it mean to say that the system of orthonormal vectors is complete in H?

2.5   Linear Functionals

In this section we concern ourselves with linear functionals. A **linear functional** assigns to each vector $\mathbf{h}$ in a vector space a complex number $f(\mathbf{h})$ such that $f(a\mathbf{h} + b\mathbf{g}) = af(\mathbf{h}) + bf(\mathbf{g})$. As an example consider $f(\mathbf{h}) = (\mathbf{w},\mathbf{h})$, where $\mathbf{w}$ is a given vector in the space and $\mathbf{h}$ ranges over the whole space. Clearly

$$f(a\mathbf{h} + b\mathbf{g}) = (\mathbf{w}, a\mathbf{h} + b\mathbf{g}) = (\mathbf{w}, a\mathbf{h}) + (\mathbf{w}, b\mathbf{g})$$

$$= a(\mathbf{w},\mathbf{h}) + b(\mathbf{w},\mathbf{g}) = af(\mathbf{h}) + bf(\mathbf{g})$$

so that this is a linear functional.

A linear functional is **bounded** if there exists a real constant c such that $|f(\mathbf{h})| \leq c\,\|\mathbf{h}\|$ for all $\mathbf{h}$ in the vector space. The example given above is bounded since

$$|f(\mathbf{h})| = |(\mathbf{w},\mathbf{h})| \leq \|\mathbf{w}\|\,\|\mathbf{h}\|$$

and we may take $c = \|\mathbf{w}\|$.

A linear functional is **continuous** if $\lim\limits_{n \to \infty} \|\mathbf{u} - \mathbf{u}_n\| = 0$ implies that $\lim\limits_{n \to \infty} f(\mathbf{u}_n) = f(\mathbf{u})$. The example is continuous since

$$|f(\mathbf{u}_n) - f(\mathbf{u})| = |(\mathbf{w},\mathbf{u}_n) - (\mathbf{w},\mathbf{u})| = |(\mathbf{w}, \mathbf{u}_n - \mathbf{u})|$$

$$\leq \|\mathbf{w}\|\,\|\mathbf{u} - \mathbf{u}_n\| \to 0 \qquad \text{as } n \to \infty$$

Theorem 1. A linear functional is continuous if and only if it is bounded.

Proof. Let the functional be bounded. Then there exists a c such that $|f(\mathbf{h})| \leq c\,\|\mathbf{h}\|$. Let $\|\mathbf{u} - \mathbf{u}_n\| \to 0$ as $n \to \infty$. Then

$$|f(\mathbf{u}_n) - f(\mathbf{u})| = |f(\mathbf{u}_n - \mathbf{u})| \leq c\,\|\mathbf{u}_n - \mathbf{u}\| \to 0$$

as $n \to \infty$, which shows that f is continuous.

If $\|\mathbf{h}\| > 0$, then $\mathbf{g} = \mathbf{h}/\|\mathbf{h}\|$ is a unit vector and $f(\mathbf{g}) = f(\mathbf{h})/\|\mathbf{h}\|$. Therefore, to say that f is not bounded is to say that for every positive integer n there is a unit vector $\mathbf{u}_n$ such that $|f(\mathbf{u}_n)| > n$. Now the vectors $\mathbf{v}_n = \mathbf{u}_n/n$ converge strongly to zero. Therefore, if f is continuous $f(\mathbf{v}_n) \to 0$ as $n \to \infty$. However, $f(\mathbf{v}_n) = f(\mathbf{u}_n)/n > 1$. Hence, if f is continuous it is bounded. This completes the proof.

We next show that in a Hilbert space the example we gave at the beginning of this section is really the most general continuous linear functional because of the following theorem.

Theorem 2. If f is a continuous linear functional defined on a Hilbert space H, then there exists a unique vector $\mathbf{w}$ in H such that $f(\mathbf{h}) = (\mathbf{w},\mathbf{h})$.

Proof. Consider the collection of vectors G in H such that for $\mathbf{u}$ in G, $f(\mathbf{u}) = 0$. This collection is not empty since $f(\mathbf{0}) = 0$. Also it is a subspace since it is a complete vector space (exercise 2, Sec. 2.5). If $G = H$, let $\mathbf{w} = \mathbf{0}$, then

$$f(\mathbf{h}) = (\mathbf{0,h}) = 0$$

If G is not the whole space, then there exists a vector $\mathbf{v}$ such that $f(\mathbf{v}) \neq 0$. By the projection theorem, $\mathbf{v}$ has a projection $\mathbf{g}$ in G such that $\mathbf{z} = \mathbf{v} - \mathbf{g}$ is orthogonal to G. Now

$$f(\mathbf{z}) = f(\mathbf{v}) - f(\mathbf{g}) = f(\mathbf{v}) \neq 0$$

We define

$$\mathbf{w} = \frac{f^*(\mathbf{z})\mathbf{z}}{\|\mathbf{z}\|^2}$$

Then

$$f(\mathbf{w}) = \frac{f^*(\mathbf{z})f(\mathbf{z})}{\|\mathbf{z}\|^2} = \frac{|f(\mathbf{z})|^2\|\mathbf{z}\|^2}{\|\mathbf{z}\|^4} = \|\mathbf{w}\|^2$$

Let $\mathbf{h}$ be any vector in H. Then

$$f\left(\mathbf{h} - \frac{f(\mathbf{h})\mathbf{w}}{f(\mathbf{w})}\right) = f(\mathbf{h}) - \frac{f(\mathbf{h})f(\mathbf{w})}{f(\mathbf{w})} = 0$$

which shows that $\mathbf{h} - f(\mathbf{h})\mathbf{w}/f(\mathbf{w})$ is in G. Now $\mathbf{w}$ is orthogonal to G and hence

$$0 = \left(\mathbf{w},\ \mathbf{h} - \frac{f(\mathbf{h})\mathbf{w}}{f(\mathbf{w})}\right) = (\mathbf{w,h}) - \frac{f(\mathbf{h})\|\mathbf{w}\|^2}{f(\mathbf{w})} = (\mathbf{w,h}) - f(\mathbf{h})$$

This shows that $f(\mathbf{h}) = (\mathbf{w,h})$. It remains to show that $\mathbf{w}$ is unique. Suppose there are two vectors $\mathbf{w}$ and $\overline{\mathbf{w}}$ such that $f(\mathbf{h}) = (\mathbf{w,h}) = (\overline{\mathbf{w}},\mathbf{h})$. Then $(\mathbf{w} - \overline{\mathbf{w}},\ \mathbf{h}) = 0$ for all $\mathbf{h}$ in H. But then $\mathbf{h} = \mathbf{w} - \overline{\mathbf{w}}$ implies that $\|\mathbf{w} - \overline{\mathbf{w}}\|^2 = 0$ or that $\mathbf{w} = \overline{\mathbf{w}}$. This completes the proof.

Exercises 2.5

1. Prove that for a linear functional $f(\mathbf{0}) = 0$.

2. Prove that in a Hilbert space with a continuous linear functional f the set of vectors $\mathbf{u}$ such that $f(\mathbf{u}) = 0$ is a subspace.

3. Let f be a bounded linear functional defined on a Hilbert space H. Let the norm of f be the least upper bound of $|f(\mathbf{h})|$ where $\|\mathbf{h}\| = 1$. Show that the norm of f, denoted by $\|f\|$, is defined. Furthermore, show that

$$\|f\| = \max\left|f\left(\frac{\mathbf{u}}{\|\mathbf{u}\|}\right)\right|$$

HINT: Since $f(\mathbf{h}) = (\mathbf{w,h})$, show that $\|f\| \leq \|\mathbf{w}\|$ and that $f(\mathbf{w}/\|\mathbf{w}\|) = \|\mathbf{w}\|$. What about the case $\mathbf{w} = \mathbf{0}$?

2.6 Weak Convergence

We now come to a second kind of convergence in Hilbert space known as **weak convergence.** A sequence of vectors $\{\mathbf{u}_n\}$ is said to have a weak limit

$\mathbf{u}$ in a Hilbert space H if $\lim_{n \to \infty} (\mathbf{h}, \mathbf{u}_n) = (\mathbf{h}, \mathbf{u})$ for every $\mathbf{h}$ in H. In a separable Hilbert space, this might be called componentwise convergence because if $\boldsymbol{\phi}_1, \boldsymbol{\phi}_2, \boldsymbol{\phi}_3, \ldots$ is a complete orthonormal set, then

$$\lim_{n \to \infty} (\boldsymbol{\phi}_i, \mathbf{u}_n) = (\boldsymbol{\phi}_i, \mathbf{u})$$

for every i. This says that the ith component of $\mathbf{u}_n$ converges to the ith component of $\mathbf{u}$ for all i.

Strong convergence implies weak convergence but not vice versa. Let $\{\mathbf{u}_n\}$ be a sequence with a strong limit $\mathbf{u}$. Then for every $\mathbf{h}$ in the Hilbert space

$$|(\mathbf{h}, \mathbf{u}_n) - (\mathbf{h}, \mathbf{u})| = |(\mathbf{h}, \mathbf{u}_n - \mathbf{u})| \leq \|\mathbf{h}\| \, \|\mathbf{u}_n - \mathbf{u}\| \to 0$$

as $n \to \infty$. To show that weak convergence does not imply strong convergence, consider the sequence

$$\mathbf{u}_1 = (1,0,0,0, \ldots) \qquad \mathbf{u}_2 = (0,1,0,0, \ldots) \qquad \mathbf{u}_3 = (0,0,1,0, \ldots)$$

etc., in the Hilbert space of infinite sequences. Let $\mathbf{h}$ be any vector. Then by Bessel's inequality

$$\sum_{n=1}^{\infty} |(\mathbf{h}, \mathbf{u}_n)|^2 \leq \|\mathbf{h}\|^2 < \infty$$

Therefore, $\lim_{n \to \infty} (\mathbf{h}, \mathbf{u}_n) = 0 = (\mathbf{h}, \mathbf{0})$, which shows that $\{\mathbf{u}_n\}$ has the weak limit $\mathbf{0}$. However, $\|\mathbf{u}_n\| = 1$ for all n and therefore the sequence cannot have the strong limit $\mathbf{0}$.

On the other hand, if $\{\mathbf{u}_n\}$ has the weak limit $\mathbf{u}$ and $\|\mathbf{u}_n\| \to \|\mathbf{u}\|$, then the sequence has the strong limit $\mathbf{u}$. This follows from the fact that

$$\|\mathbf{u}_n - \mathbf{u}\|^2 = \|\mathbf{u}_n\|^2 + \|\mathbf{u}\|^2 - (\mathbf{u}_n, \mathbf{u}) - (\mathbf{u}, \mathbf{u}_n) \to 0$$

since $\|\mathbf{u}_n\|^2 \to \|\mathbf{u}\|^2$ and $(\mathbf{u}, \mathbf{u}_n) \to (\mathbf{u}, \mathbf{u}) = \|\mathbf{u}\|^2$.

Theorem 1. The norms of a weakly convergent sequence are uniformly bounded; that is, if $\{\mathbf{u}_n\}$ converges weakly, then there exists a K such that $\|\mathbf{u}_n\| \leq K$ for all n.

Proof. Since $(\mathbf{h}, \mathbf{u}_n) \to (\mathbf{h}, \mathbf{u})$, where $\mathbf{u}$ is the weak limit of the sequence, for a given $\mathbf{h}$, $|(\mathbf{h}, \mathbf{u}_n)|$ is uniformly bounded. We shall use this fact to show that for $\|\mathbf{h} - \mathbf{h}_0\| \leq \delta$ there exists a C such that $|(\mathbf{h}, \mathbf{u}_n)| \leq C$. Here $\mathbf{h}_0$ is a fixed vector. If this were not so, then there would exist a vector $\mathbf{h}_1$ and an n_1 such that $|(\mathbf{h}_1, \mathbf{u}_{n_1})| > 1$. Now the scalar product $(\mathbf{h}, \mathbf{u}_{n_1})$ is continuous in $\mathbf{h}$. Therefore, we can find $\mathbf{h}_1$ satisfying $\|\mathbf{h}_1 - \mathbf{h}_0\| < \delta$ and, furthermore, a neighborhood $N_1 = \{\mathbf{h} \mid \|\mathbf{h} - \mathbf{h}_1\| \leq \delta_1 < \delta\}$ such that for all $\mathbf{h}$ in $N_1 |(\mathbf{h}, \mathbf{u}_{n_1})| > 1$. Similarly we can find an n_2, an $\mathbf{h}_2$, and an $N_2 = \{\mathbf{h} \mid \|\mathbf{h} - \mathbf{h}_2\| \leq \delta_2 < \delta_1 < \delta\}$ such that N_2 is contained in N_1, and for all $\mathbf{h}$ in $N_2 |(\mathbf{h}, \mathbf{u}_{n_2})| > 2$. In this way we construct a collection of nested closed neighborhoods, and by picking the sequence $\{\delta_n\}$ so that it converges to zero we ensure that the sequence of vectors $\{\mathbf{h}_n\}$ converges to a unique vector $\mathbf{v}$ which is in every neighborhood. Therefore $|(\mathbf{v},\mathbf{u}_{n_1})| > 1$, $|(\mathbf{v},\mathbf{u}_{n_2})| > 2$, etc. This contradicts the fact that $|(\mathbf{v},\mathbf{u}_n)|$ is uniformly bounded. Therefore, we have proved that $|(\mathbf{h},\mathbf{u}_n)| \leq C$ for all $\mathbf{h}$ satisfying $\|\mathbf{h} - \mathbf{h}_0\| \leq \delta$. Now for any $\mathbf{h}$

$$(\mathbf{h},\mathbf{u}_n) = \frac{\|\mathbf{h}\|}{\delta}\left(\frac{\delta\mathbf{h}}{\|\mathbf{h}\|} + \mathbf{h}_0, \mathbf{u}_n\right) - \frac{\|\mathbf{h}\|}{\delta}(\mathbf{h}_0,\mathbf{u}_n)$$

The vector $(\delta\mathbf{h}/\|\mathbf{h}\|) + \mathbf{h}_0$ is a distance δ from $\mathbf{h}_0$. Therefore

$$|(\mathbf{h},\mathbf{u}_n)| \leq \frac{2C}{\delta}\|\mathbf{h}\|$$

holds for all $\mathbf{h}$ in the Hilbert space. Let $\mathbf{h} = \mathbf{u}_n$. Then

$$\|\mathbf{u}_n\| \leq \frac{2C}{\delta} = K$$

which completes the proof.

Theorem 2. A Hilbert space is weakly complete; that is, when $(\mathbf{h},\mathbf{u}_n - \mathbf{u}_m) \to 0$ as $n \to \infty$ and $m \to \infty$ for every $\mathbf{h}$ in the Hilbert space, then the sequence $\{\mathbf{u}_n\}$ has a weak limit.

Proof. By hypothesis the complex numbers $(\mathbf{h},\mathbf{u}_n)$ converge for every $\mathbf{h}$ in the Hilbert space. This implies that there exists a K such that $|(\mathbf{h},\mathbf{u}_n)| \leq K\|\mathbf{h}\|$ for every $\mathbf{h}$.† We define a linear functional $f(\mathbf{h}) = \lim_{n\to\infty}(\mathbf{h},\mathbf{u}_n)^*$. This is bounded. Hence there exists a unique vector $\mathbf{u}$ such that

$$(\mathbf{h},\mathbf{u})^* = f(\mathbf{h}) = \lim_{n\to\infty}(\mathbf{h},\mathbf{u}_n)^*$$

and, therefore, $\mathbf{u}$ is the weak limit of $\{\mathbf{u}_n\}$.

Theorem 3. Every bounded sequence in Hilbert space contains a weakly convergent subsequence.

Proof. Let $\{\mathbf{u}_n\}$ be a sequence such that $\|\mathbf{u}_n\| \leq K$ for all n. If $\{\mathbf{u}_n\}$ does not contain an infinite number of distinct vectors, then it must repeat for sufficiently large n. Clearly, in this case, the repeating vector is the weak limit of the sequence. Therefore, we assume that the given sequence contains infinitely many distinct vectors. Consider the sequence of complex numbers $(\mathbf{u}_1,\mathbf{u}_n)$, $n = 1, 2, 3, \ldots$. Since $|(\mathbf{u}_1,\mathbf{u}_n)| \leq \|\mathbf{u}_1\|\|\mathbf{u}_n\| \leq K^2$, this sequence is bounded. There is hence a subsequence $\{\mathbf{u}_{n_1}\}$ such that $(\mathbf{u}_1,\mathbf{u}_{n_1})$ converges. Next consider the sequence of numbers $(\mathbf{u}_2,\mathbf{u}_{n_1})$. This sequence is bounded and hence there is a subsequence of $\{\mathbf{u}_{n_1}\}$, namely, $\{\mathbf{u}_{n_2}\}$ such that

† The previous theorem proved this under the assumption that $\mathbf{u}_n$ has a weak limit. Actually, the proof only depends on the fact that the numbers $(\mathbf{h},\mathbf{u}_n)$ converge for all $\mathbf{h}$ in the Hilbert space.

$(\mathbf{u}_1,\mathbf{u}_{n_2})$ and $(\mathbf{u}_2,\mathbf{u}_{n_2})$ converge. Similarly we establish the existence of a subsequence $\{\mathbf{u}_{n_3}\}$ of both $\{\mathbf{u}_{n_1}\}$ and $\{\mathbf{u}_{n_2}\}$ such that $(\mathbf{u}_1,\mathbf{u}_{n_3})$, $(\mathbf{u}_2,\mathbf{u}_{n_3})$, $(\mathbf{u}_3,\mathbf{u}_{n_3})$ all converge. With a little relabeling we have the following array

$$\{\mathbf{u}_{n_1}\} = \mathbf{v}_{11}, \mathbf{v}_{12}, \mathbf{v}_{13}, \cdots$$
$$\{\mathbf{u}_{n_2}\} = \mathbf{v}_{21}, \mathbf{v}_{22}, \mathbf{v}_{23}, \cdots$$
$$\{\mathbf{u}_{n_3}\} = \mathbf{v}_{31}, \mathbf{v}_{32}, \mathbf{v}_{33}, \cdots$$
$$\cdots = \cdots\cdots\cdots\cdots$$

Now the sequence $\mathbf{v}_{11}, \mathbf{v}_{22}, \mathbf{v}_{33}, \ldots$ is a subsequence of $\{\mathbf{u}_n\}$ and we have that $\lim\limits_{k\to\infty}(\mathbf{u}_n,\mathbf{v}_{kk})$ exists for each n. Next we form a subspace G as follows: take all possible linear combinations of $\mathbf{u}_1$, $\mathbf{u}_2$, $\mathbf{u}_3$, $\ldots$ and all the vectors in the Hilbert space which are strong limits of Cauchy sequences of linear combinations of $\mathbf{u}_1$, $\mathbf{u}_2$, $\mathbf{u}_3$, $\ldots$. Let F be the orthogonal complement of G in H; that is, $H = G + F$. Let $\mathbf{h}$ be any vector in H. Then $\mathbf{h} = \mathbf{g} + \mathbf{f}$, where $\mathbf{g}$ is in G and $\mathbf{f}$ is in F, and

$$(\mathbf{h},\mathbf{v}_{kk}) = (\mathbf{g} + \mathbf{f}, \mathbf{v}_{kk}) = (\mathbf{g},\mathbf{v}_{kk})$$

since $\mathbf{v}_{kk}$ is in G for all k. Now $\mathbf{g}$ is the strong limit of a Cauchy sequence $\{\mathbf{g}_n\}$ in G, where each $\mathbf{g}_n$ is a finite linear combination of $\mathbf{u}_1$, $\mathbf{u}_2$, $\mathbf{u}_3$, $\ldots$. From the above we have

$$\lim_{k\to\infty} (\mathbf{g}_n,\mathbf{v}_{kk}) = l_n$$

$$\lim_{k\to\infty} (\mathbf{g}_m,\mathbf{v}_{kk}) = l_m$$

and $\|\mathbf{g}_n - \mathbf{g}_m\| < \epsilon$ for n and m sufficiently large. Having picked ϵ we fix n and m so that $\|\mathbf{g}_n - \mathbf{g}_m\| < \epsilon$. Then we take k so large that $|l_n - (\mathbf{g}_n,\mathbf{v}_{kk})| < \epsilon$ and $|l_m - (\mathbf{g}_m,\mathbf{v}_{kk})| < \epsilon$. Consequently

$$|l_n - l_m| = |l_n - (\mathbf{g}_n,\mathbf{v}_{kk}) + (\mathbf{g}_n,\mathbf{v}_{kk}) - l_m + (\mathbf{g}_m,\mathbf{v}_{kk}) - (\mathbf{g}_m,\mathbf{v}_{kk})|$$
$$< 2\epsilon + |(\mathbf{g}_n - \mathbf{g}_m, \mathbf{v}_{kk})|$$
$$< 2\epsilon + \|\mathbf{g}_n - \mathbf{g}_m\| \, \|\mathbf{v}_{kk}\|$$
$$< 2\epsilon + K\epsilon$$

Therefore $\{l_n\}$ is a Cauchy sequence of complex numbers, which must have a limit l. Clearly $l = \lim\limits_{k\to\infty}(\mathbf{g},\mathbf{v}_{kk})$. We have thus shown that $\lim\limits_{k\to\infty}(\mathbf{h},\mathbf{v}_{kk})$ exists for all $\mathbf{h}$ in H. By the previous theorem the subsequence $\{\mathbf{v}_{kk}\}$ converges weakly.

Exercises 2.6

1. Prove that a weakly convergent sequence has a unique limit.

2. Prove that the functional introduced in the theorem on weak completeness of Hilbert space is linear and bounded.

3. Show that a bounded sequence does not necessarily contain a strongly convergent subsequence. HINT: Consider any orthonormal sequence.

4. Prove that in a finite-dimensional space strong and weak convergence are equivalent.

*5. Let $\{u_k\}$ and $\{v_k\}$ be sequences of complex numbers such that $\sum_{k=1}^{\infty} |v_k|^2 < \infty$. Prove that the convergence of $\sum_{k=1}^{\infty} u_k v_k$ for all sequences $\{v_k\}$ satisfying the given property implies that $\sum_{k=1}^{\infty} |u_k|^2 < \infty$. HINT: Define vectors $\mathbf{f}_1 = (u_1^*, 0, 0, 0, \ldots)$, $\mathbf{f}_2 = (u_1^*, u_2^*, 0, 0, 0, \ldots)$, $\mathbf{f}_3 = (u_1^*, u_2^*, u_3^*, 0, 0, \ldots)$, etc., and the numerical sequence $\{(\mathbf{f}_k, \mathbf{v})\}$ which converges for all $\mathbf{v}$, and follow the proof about boundedness of weakly convergent sequences.

*6. Let $\{\mathbf{u}_n\}$ converge weakly to $\mathbf{u}$ where $\|\mathbf{u}_n\| \leq C$. Prove that $\|\mathbf{u}\| \leq C$.

2.7   Linear Operators

We now define more precisely what is meant by a linear operator in a Hilbert space. Let A be a **linear operator** in H. Then A assigns to each vector $\mathbf{h}$ in some vector space contained in H a vector $\mathbf{g} = A\mathbf{h}$ in H such that

$$A(a\mathbf{h} + b\mathbf{f}) = aA\mathbf{h} + bA\mathbf{f}$$

The **domain** of the operator is the vector space on which it is defined, and the **range** of the operator is the collection of vectors which are values of the operator. In other words, the operator defines a mapping from its domain in H onto its range, which is also in H. The domain must be a vector space and because of the linearity the range is also a vector space; that is,

$$A(a\mathbf{h} + b\mathbf{f}) = aA\mathbf{h} + bA\mathbf{f}$$

which establishes closure under addition, and the rest of the properties follow.

A linear operator is **bounded** if there exists a real number C such that $\|A\mathbf{h}\| \leq C \|\mathbf{h}\|$ for all $\mathbf{h}$ in its domain. If a linear operator is bounded it has a **norm** which we define by

$$\|A\| = \text{l.u.b.} \frac{\|A\mathbf{h}\|}{\|\mathbf{h}\|}$$

where the least upper bound is determined for all nonzero vectors in the domain of the operator.

A linear operator is **continuous** if $\lim_{n \to \infty} \|\mathbf{u} - \mathbf{u}_n\| = 0$ implies that $\lim_{n \to \infty} \|A\mathbf{u} - A\mathbf{u}_n\| = 0$. It will be left to the reader to show that *a linear operator defined on a Hilbert space is continuous if and only if it is bounded.* If a linear operator A is bounded on a vector space which is dense in a Hilbert space H, then its definition can be extended to the whole space and its extension will be continuous. If $\mathbf{u}$ is any vector in H, then there is a Cauchy sequence $\{\mathbf{u}_n\}$ of vectors in the domain of A such that $\|\mathbf{u} - \mathbf{u}_n\| \to 0$ as $n \to \infty$. Hence, $\|\mathbf{u}_n - \mathbf{u}_m\| \to 0$ as $n \to \infty$ and $m \to \infty$, and

$$\|A\mathbf{u}_n - A\mathbf{u}_m\| = \|A(\mathbf{u}_n - \mathbf{u}_m)\| \leq C \|\mathbf{u}_n - \mathbf{u}_m\|$$

Therefore, $\{A\mathbf{u}_n\}$ is a Cauchy sequence in H with a strong limit $\mathbf{v}$. We define $A\mathbf{u} = \mathbf{v}$. It is then easy to show that A is continuous on H.

Let us form the scalar product $(A\mathbf{h},\mathbf{g})$, where A is defined and bounded on a Hilbert space H. Given $\mathbf{g}$ we have a linear functional $f(\mathbf{h}) = (\mathbf{g},A\mathbf{h})$, which is bounded since

$$|f(\mathbf{h})| \leq \|A\mathbf{h}\| \|\mathbf{g}\| \leq \|\mathbf{g}\| \|A\| \|\mathbf{h}\|$$

Therefore, there exists a unique vector $\mathbf{v}$ in H such that

$$f(\mathbf{h}) = (\mathbf{g},A\mathbf{h}) = (\mathbf{v},\mathbf{h})$$

We let $\mathbf{v} = \tilde{A}\mathbf{g}$. Then $\tilde{A}\mathbf{g}$ is uniquely defined for each $\mathbf{g}$ in H and

$$(A\mathbf{h},\mathbf{g}) = (\mathbf{h},\tilde{A}\mathbf{g})$$

$\tilde{A}$ is a linear operator since

$$(A\mathbf{h}, a_1\mathbf{g}_1 + a_2\mathbf{g}_2) = a_1(A\mathbf{h},\mathbf{g}_1) + a_2(A\mathbf{h},\mathbf{g}_2)$$

$$[\mathbf{h}, \tilde{A}(a_1\mathbf{g}_1 + a_2\mathbf{g}_2)] = a_1(\mathbf{h},\tilde{A}\mathbf{g}_1) + a_2(\mathbf{h},\tilde{A}\mathbf{g}_2)$$

$$[\mathbf{h}, \tilde{A}(a_1\mathbf{g}_1 + a_2\mathbf{g}_2)] = (\mathbf{h}, a_1\tilde{A}\mathbf{g}_1 + a_2\tilde{A}\mathbf{g}_2)$$

for all $\mathbf{h}$ in H. Therefore, $\tilde{A}(a_1\mathbf{g}_1 + a_2\mathbf{g}_2) = a_1\tilde{A}\mathbf{g}_1 + a_2\tilde{A}\mathbf{g}_2$. $\tilde{A}$ is called the **adjoint** of A.

Furthermore, $\tilde{A}$ is bounded since

$$\|\tilde{A}\mathbf{g}\|^2 = |(\tilde{A}\mathbf{g},\tilde{A}\mathbf{g})| = |(A\tilde{A}\mathbf{g},\mathbf{g})| \leq \|A\tilde{A}\mathbf{g}\| \|\mathbf{g}\| \leq \|A\| \|\tilde{A}\mathbf{g}\| \|\mathbf{g}\|$$

that is, $\|\tilde{A}\mathbf{g}\| \leq \|A\| \|\mathbf{g}\|$ for all $\mathbf{g}$ in H. This shows that $\tilde{A}$ has a norm and since $\|A\mathbf{h}\| \leq \|\tilde{A}\| \|\mathbf{h}\|$ for all $\mathbf{h}$ in H it follows that $\|A\| = \|\tilde{A}\|$.

Theorem 1. If there exist two linear operators A and B on a Hilbert space H such that $(A\mathbf{h},\mathbf{g}) = (\mathbf{h},B\mathbf{g})$ for all $\mathbf{h}$ and $\mathbf{g}$ in H, then A and B are bounded and $B = \tilde{A}$.

Proof. It suffices to prove that A is bounded, for then A has a bounded adjoint which must be equal to B. If A is not bounded, then there exists a sequence $\{\mathbf{u}_n\}$ such that $\|\mathbf{u}_n\| = 1$ and $\|A\mathbf{u}_n\| > n$ for all n. Now the sequence contains a weakly convergent subsequence, so without loss of generality we can assume that $\{\mathbf{u}_n\}$ converges weakly to $\mathbf{u}$. Then $(A\mathbf{u}_n,\mathbf{g}) = (\mathbf{u}_n,B\mathbf{g}) \to (\mathbf{u},B\mathbf{g})$, and hence $(A\mathbf{u}_n,\mathbf{g})$ converges for all $\mathbf{g}$ in H. Therefore, $\|A\mathbf{u}_n\|$ is uniformly bounded, which is a contradiction, proving that A is bounded.

If A is a linear operator, then $(A\mathbf{f},\mathbf{g})$ defines a **bilinear form** for any vector $\mathbf{f}$ in the domain of A and any $\mathbf{g}$ in the Hilbert space. Clearly,

$$[A(a_1\mathbf{f}_1 + a_2\mathbf{f}_2), \mathbf{g}] = (a_1 A\mathbf{f}_1 + a_2 A\mathbf{f}_2, \mathbf{g})$$
$$= a_1^*(A\mathbf{f}_1,\mathbf{g}) + a_2^*(A\mathbf{f}_2,\mathbf{g})$$
$$(A\mathbf{f},b_1\mathbf{g}_1 + b_2\mathbf{g}_2) = b_1(A\mathbf{f},\mathbf{g}_1) + b_2(A\mathbf{f},\mathbf{g}_2)$$

A bilinear form is **bounded** if $|(A\mathbf{f},\mathbf{g})| \leq K \|\mathbf{f}\| \|\mathbf{g}\|$ for some K for all $\mathbf{f}$ and $\mathbf{g}$ for which the form is defined. Clearly, if A is a bounded operator the bilinear form is bounded, since

$$|(A\mathbf{f},\mathbf{g})| \leq \|A\mathbf{f}\| \|\mathbf{g}\| \leq \|A\| \|\mathbf{f}\| \|\mathbf{g}\|$$

Conversely if the bilinear form is bounded then the operator is bounded, since

$$\|A\mathbf{f}\|^2 = (A\mathbf{f},A\mathbf{f}) \leq K \|\mathbf{f}\| \|A\mathbf{f}\|$$
$$\|A\mathbf{f}\| \leq K \|\mathbf{f}\|$$

for all $\mathbf{f}$ in the domain of A.

We shall say that an infinite matrix with elements a_{ij} is **bounded** if the bilinear form $\sum\limits_{i=1}^{\infty} \sum\limits_{j=1}^{\infty} a_{ij}^* x_j^* y_i$ is bounded for all sequences $\{x_i\}$ and $\{y_i\}$ such that $\sum\limits_{i=1}^{\infty} |x_i|^2 < \infty$ and $\sum\limits_{i=1}^{\infty} |y_i|^2 < \infty$; that is,

$$\left| \sum_{i=1}^{\infty} \sum_{j=1}^{\infty} a_{ij}^* x_j^* y_i \right| \leq K \sqrt{ \sum_{i=1}^{\infty} |x_i|^2 \sum_{i=1}^{\infty} |y_i|^2 }$$

for some constant K.

Theorem 2. A linear operator A defined on a separable Hilbert space H has a bounded matrix representation if and only if it is bounded.

Proof. Let A be bounded. To say that A has a matrix representation we mean that for any vector $\mathbf{f} = \sum\limits_{i=1}^{\infty} x_i \boldsymbol{\phi}_i$ in terms of an orthonormal basis $\boldsymbol{\phi}_1, \boldsymbol{\phi}_2, \boldsymbol{\phi}_3, \ldots$, we can express the components of

$$A\mathbf{f} = \sum_{i=1}^{\infty} z_i \boldsymbol{\phi}_i$$

in terms of x_i and an infinite matrix with elements a_{ij} as follows:

$$z_i = \sum_{j=1}^{\infty} a_{ij} x_j$$

Clearly $z_i = \left(\boldsymbol{\phi}_i, \sum\limits_{j=1}^{\infty} z_j \boldsymbol{\phi}_j \right) = (\boldsymbol{\phi}_i, A\mathbf{f})$. Let $\mathbf{f}_n = \sum\limits_{j=1}^{n} x_j \boldsymbol{\phi}_j$. Then $A\mathbf{f}_n = \sum\limits_{j=1}^{n} x_j A\boldsymbol{\phi}_j$ and $(\boldsymbol{\phi}_i, A\mathbf{f}_n) = \sum\limits_{j=1}^{n} a_{ij} x_j$, where

$$a_{ij} = (\boldsymbol{\phi}_i, A\boldsymbol{\phi}_j)$$

Now if A is bounded on H it is continuous and hence $A\mathbf{f}_n \to A\mathbf{f}$. Therefore, $(\boldsymbol{\phi}_i, A\mathbf{f}_n) \to (\boldsymbol{\phi}_i, A\mathbf{f})$ and

$$z_i = (\boldsymbol{\phi}_i, A\mathbf{f}) = \sum_{j=1}^{\infty} a_{ij} x_j$$

Since A is bounded the bilinear form $(A\mathbf{f}, \mathbf{g})$ is bounded. Let $\mathbf{g} = \sum\limits_{i=1}^{\infty} y_i \boldsymbol{\phi}_i$. Then

$$|A\mathbf{f}, \mathbf{g}| = \left| \sum_{i=1}^{\infty} \sum_{j=1}^{\infty} a_{ij}^* x_j^* y_i \right| \le \|A\| \sqrt{\sum_{i=1}^{\infty} |x_i|^2 \sum_{i=1}^{\infty} |y_i|^2}$$

which shows that the matrix is bounded.

Now let us assume that a matrix with elements a_{ij} is bounded. Given a vector $\mathbf{g} = \sum\limits_{i=1}^{\infty} y_i \boldsymbol{\phi}_i$, we consider the following special case: $x_j = \delta_{jk}$ and $\mathbf{v} = (0, 0, \ldots, 0, y_{m+1}, y_{m+2}, \ldots, y_n, 0, 0, \ldots)$. Then

$$\left| \sum_{i=m+1}^{n} a_{ik}^* y_i \right| \le K \sqrt{\sum_{i=m+1}^{n} |y_i|^2} \to 0$$

as m and $n \to \infty$. Therefore, $\sum\limits_{i=1}^{\infty} a_{ij}^* y_i$ converges for all $\mathbf{g}$ in H. Using exercise 5, Sec. 2.6, it follows that

$$\sum_{i=1}^{\infty} |a_{ij}|^2 < \infty$$

We define an operator on the basis vectors as follows

$$A\boldsymbol{\phi}_j = \sum_{i=1}^{\infty} a_{ij} \boldsymbol{\phi}_i$$

Then $a_{ij} = (\boldsymbol{\phi}_i, A\boldsymbol{\phi}_j)$. This operator is linear and bounded on the vector space of vectors consisting of finite linear combinations of $\boldsymbol{\phi}_1, \boldsymbol{\phi}_2, \boldsymbol{\phi}_3, \ldots$ since $\mathbf{f} = \sum\limits_{i=1}^{n} x_i \boldsymbol{\phi}_i$, $g = \sum\limits_{i=1}^{\infty} y_i \boldsymbol{\phi}_i$ implies

$$|(A\mathbf{f}, \mathbf{g})| = \left| \sum_{i=1}^{\infty} \sum_{j=1}^{n} a_{ij}^* x_j^* y_i \right| \le K \sqrt{\sum_{i=1}^{n} |x_i|^2 \sum_{i=1}^{\infty} |y_i|^2} = K \|\mathbf{f}\| \|\mathbf{g}\|$$

for all $\mathbf{f}$ with a finite number of components, and since A is bounded on a vector space dense in H we can extend A to all of H as a bounded operator. This completes the proof.

If A is bounded on a separable Hilbert space, it has a matrix representation $a_{ij} = (\boldsymbol{\phi}_i, A\boldsymbol{\phi}_j)$ relative to some orthonormal basis $\boldsymbol{\phi}_1, \boldsymbol{\phi}_2, \boldsymbol{\phi}_3, \ldots$. Also the adjoint $\tilde{A}$ exists and is bounded. Therefore, $\tilde{A}$ has a matrix representation

$$b_{ij} = (\boldsymbol{\phi}_i, \tilde{A}\boldsymbol{\phi}_j) = (A\boldsymbol{\phi}_i, \boldsymbol{\phi}_j) = a_{ji}^*$$

which means that the matrix of the adjoint is the conjugate transpose of the matrix of A.

Exercises 2.7

1. Prove that the range of a linear operator is a vector space.

2. Prove that a linear operator is continuous on a Hilbert space if and only if it is bounded.

3. Prove the following about the norm of a bounded linear operator A defined on a Hilbert space:

a. $\|A\| = \text{l.u.b.} \|A\mathbf{h}\|$ for all unit vectors.

b. $\|A\mathbf{h}\| < \|A\|$ if $\|\mathbf{h}\| < 1$ and $\|A\| > 0$.

c. $\|A\| = \|\tilde{A}\|$.

4. Let A and B be bounded linear operators defined on a Hilbert space. We define AB by $A(B\mathbf{h})$. Prove that AB is bounded and that $\|AB\| \leq \|A\| \|B\|$.

5. Let A be a bounded linear operator on a vector space dense in a Hilbert space H. Prove that the extension by continuity to all of H is a bounded linear operator with the same norm.

6. Prove the following about the adjoints of bounded linear operators defined on a Hilbert space:

a. The adjoint of the adjoint is the original operator.

b. The adjoint of a product is the product of the adjoints in reverse order.

c. The adjoint of a sum is the sum of the adjoints.

7. Prove that every bounded bilinear form $F(\mathbf{f}, \mathbf{g})$ defined for all $\mathbf{f}$ and $\mathbf{g}$ in a Hilbert space H such that

$$F(a_1\mathbf{f}_1 + a_2\mathbf{f}_2, \mathbf{g}) = a_1^* F(\mathbf{f}_1, \mathbf{g}) + a_2^* F(\mathbf{f}_2, \mathbf{g})$$
$$F(\mathbf{f}, b_1\mathbf{g}_1 + b_2\mathbf{g}_2) = b_1 F(\mathbf{f}, \mathbf{g}_1) + b_2 F(\mathbf{f}, \mathbf{g}_2)$$
$$|F(\mathbf{f}, \mathbf{g})| \leq K \|\mathbf{f}\| \|\mathbf{g}\|$$

defines a unique bounded linear operator A on H such that

$$F(\mathbf{f}, \mathbf{g}) = (A\mathbf{f}, \mathbf{g})$$

8. Prove that a necessary condition for a matrix with elements a_{ij}, $i = 1, 2, 3, \ldots$; $j = 1, 2, 3, \ldots$ to be bounded is $\sum_{i=1}^{\infty} |a_{ij}|^2 < \infty$ and $\sum_{j=1}^{\infty} |a_{ij}|^2 < \infty$.

9. Prove that a sufficient condition for a matrix with elements a_{ij}, $i = 1, 2, 3, \ldots$; $j = 1, 2, 3, \ldots$ to be bounded is $\sum_{i=1}^{\infty} |a_{ij}| \leq M$ and $\sum_{j=1}^{\infty} |a_{ij}| \leq M$. HINT: This result is known as Schur's theorem.[1]

---

[1] See Angus E. Taylor, "Introduction to Functional Analysis," pp. 327–329, John Wiley & Sons, Inc., New York, 1958.

10. A bounded linear operator A is said to be **self-adjoint** if $\tilde{A} = A$. In a separable Hilbert space, A has a matrix representation $a_{ij} = (\boldsymbol{\phi}_i, A\boldsymbol{\phi}_j)$. Prove that A is self-adjoint if and only if the matrix is hermitian.

*11. Let A be self-adjoint. Prove that $\|A\| = $ l.u.b. $|(A\mathbf{h},\mathbf{h})|$ for all vectors in the domain of A such that $\|\mathbf{h}\| = 1$. HINT: Let $F(\mathbf{h},\mathbf{g}) = (A\mathbf{h},\mathbf{g})$ and $G(\mathbf{h}) = (A\mathbf{h},\mathbf{h})$; show that $\operatorname{Re} F(\mathbf{h},\mathbf{g}) = G(\frac{1}{2}(\mathbf{h} + \mathbf{g})) - G(\frac{1}{2}(\mathbf{h} - \mathbf{g}))$ and that there exists a scalar a such that $|F(\mathbf{h},\mathbf{g})| = aF(\mathbf{h},\mathbf{g}) = F(\mathbf{h},a\mathbf{g}) = G(\frac{1}{2}(\mathbf{h} + a\mathbf{g})) - G(\frac{1}{2}(\mathbf{h} - a\mathbf{g}))$, where $|a| = 1$.

2.8 Completely Continuous Operators

We now come to a very important class of linear operators known as **completely continuous operators.** A linear operator A defined on a Hilbert space H is completely continuous if it maps weakly convergent sequences into strongly convergent sequences; that is, if $\{\mathbf{u}_n\}$ has the weak limit $\mathbf{u}$, then $\{A\mathbf{u}_n\}$ has the strong limit $A\mathbf{u}$.

Theorem 1. A completely continuous operator is bounded.

Proof. Let A be the operator and assume it is not bounded. Therefore, there exists a sequence of vectors $\{\mathbf{u}_n\}$ such that $\|\mathbf{u}_n\| = 1$ and $\|A\mathbf{u}_n\| > n$. This sequence is bounded and hence it has a weakly convergent subsequence. We may as well assume that this is the subsequence and that $\{\mathbf{u}_n\}$ converges weakly to $\mathbf{u}$. But then $A\mathbf{u}_n$ converges strongly to $A\mathbf{u}$ which implies that $\|A\mathbf{u}_n\| \to \|A\mathbf{u}\|$ as $n \to \infty$. This is a contradiction, which proves the theorem.

If a linear operator A has vector $\mathbf{y}$ and an associated scalar λ such that $\|\mathbf{y}\| \neq 0$ and $A\mathbf{y} = \lambda\mathbf{y}$, then $\mathbf{y}$ is said to be an **eigenvector** of A associated with the **eigenvalue** λ. If A is completely continuous it is bounded and hence has an adjoint $\tilde{A}$. We shall now prove that a completely continuous self-adjoint operator has at least one eigenvalue.

Theorem 2. If A is completely continuous and self-adjoint it has either $\|A\|$ or $-\|A\|$ as an eigenvalue.

Proof. Consider the set of vectors $\mathbf{x}$ such that $A\mathbf{x} = \mathbf{0}$. This set is not empty since it contains the zero vector. If this set is the whole space, then $\|A\mathbf{x}\| = 0$ for all $\mathbf{x}$ and therefore $\|A\| = 0$ and the theorem is proved. Therefore, assume that there is a vector $\mathbf{h}$ such that $A\mathbf{h} \neq \mathbf{0}$ and hence that $\|A\| > 0$. Now there exists a sequence $\{\mathbf{z}_n\}$ such that $\|\mathbf{z}_n\| = 1$ and[1]

$$\|A\| = \lim_{n \to \infty} |(A\mathbf{z}_n, \mathbf{z}_n)|$$

This means that there is a subsequence $\{\mathbf{v}_n\}$ of $\{\mathbf{z}_n\}$ such that $\lim_{n \to \infty} (A\mathbf{v}_n, \mathbf{v}_n)$ is either $\|A\|$ or $-\|A\|$. Let us assume for definiteness that it is $\|A\|$. The

---

[1] See exercise 11, Sec. 2.7, where the reader is asked to prove that $\|A\| = $ l.u.b. $|(A\mathbf{h},\mathbf{h})|$ for all unit vectors in the Hilbert space.

sequence $\{\mathbf{v}_n\}$ is bounded and therefore it has a weakly convergent subsequence $\{\mathbf{u}_n\}$. If $\{\mathbf{u}_n\}$ converges weakly to $\mathbf{u}$, then $A\mathbf{u}_n$ converges strongly to $A\mathbf{u}$, and

$$|(A\mathbf{u}_n,\mathbf{u}_n) - (A\mathbf{u},\mathbf{u})| = |(A\mathbf{u}_n,\mathbf{u}_n) - (A\mathbf{u},\mathbf{u}_n) + (A\mathbf{u},\mathbf{u}_n) - (A\mathbf{u},\mathbf{u})|$$
$$\leq \|A\mathbf{u}_n - A\mathbf{u}\| + |(A\mathbf{u}, \mathbf{u}_n - \mathbf{u})| \to 0$$

as $n \to \infty$. Therefore $\lim\limits_{n\to\infty} (A\mathbf{u}_n,\mathbf{u}_n) = (A\mathbf{u},\mathbf{u}) = \|A\|$. By exercise 6, Sec. 2.6, $\|\mathbf{u}\| \leq 1$. If $\|\mathbf{u}\| < 1$, then there is a real scalar $a > 1$ such that $\|a\mathbf{u}\| = 1$ and $(Aa\mathbf{u},a\mathbf{u}) = a^2(A\mathbf{u},\mathbf{u}) > \|A\|$, contradicting the fact that $\|A\| = $ l.u.b. $|(A\mathbf{h},\mathbf{h})|$ for $\|\mathbf{h}\| = 1$. Therefore, $\|\mathbf{u}\| = 1$. This shows that $\{\mathbf{u}_n\}$ converges strongly to $\mathbf{u}$. We also have, letting $\lambda = \|A\|$,

$$\|A\mathbf{u}_n - \lambda\mathbf{u}_n\|^2 = (A\mathbf{u}_n,A\mathbf{u}_n) - 2\lambda(A\mathbf{u}_n,\mathbf{u}_n) + \lambda^2$$
$$\leq \|A\|^2 - 2\lambda(A\mathbf{u}_n,\mathbf{u}_n) + \lambda^2 \to 0$$

as $n \to \infty$. Hence

$$\|A\mathbf{u} - \lambda\mathbf{u}\| = \|A\mathbf{u} - A\mathbf{u}_n + A\mathbf{u}_n - \lambda\mathbf{u}_n + \lambda\mathbf{u}_n - \lambda\mathbf{u}\|$$
$$\leq \|A\mathbf{u} - A\mathbf{u}_n\| + \|A\mathbf{u}_n - \lambda\mathbf{u}_n\| + \lambda\|\mathbf{u}_n - \mathbf{u}\| \to 0$$

as $n \to \infty$. This shows that $A\mathbf{u} = \lambda\mathbf{u}$ and completes the proof.

Now that we have shown that a completely continuous self-adjoint operator has at least one eigenvalue, we consider the problem of finding others. Assume that $\|A\| > 0$ and that $\lambda_1 = \pm\|A\|$ is an eigenvalue associated with the eigenvector $\mathbf{y}_1$. Consider the subspace orthogonal to $\mathbf{y}_1$. A maps this subspace into itself because if $(\mathbf{x},\mathbf{y}_1) = 0$, then $(A\mathbf{x},\mathbf{y}_1) = (\mathbf{x},A\mathbf{y}_1) = (\mathbf{x},\lambda_1\mathbf{y}_1) = \lambda_1(\mathbf{x},\mathbf{y}_1) = 0$. Therefore, we can consider a new problem of finding an eigenvalue of A in this subspace. Let

$$N_1 = \text{l.u.b. } |(A\mathbf{x},\mathbf{x})|$$

for all $\mathbf{x}$ such that $\|\mathbf{x}\| = 1$ and $(\mathbf{x},\mathbf{y}_1) = 0$. This is the norm of A in the subspace orthogonal to $\mathbf{y}_1$. If $N_1 = 0$, then $\|A\mathbf{x}\| = 0$ for all $\mathbf{x}$ in the subspace, which shows that $A\mathbf{x} = \mathbf{0}$. This shows that the subspace orthogonal to $\mathbf{y}_1$ is the null space of the operator, that is, the subspace which maps into the zero vector. In this case, we can decompose the Hilbert space into the null space plus a one-dimensional subspace consisting of multiples of $\mathbf{y}_1$. If $N_1 > 0$, then the theorem just proved shows the existence of an eigenvalue $\lambda_2 = \pm N_1$ and an associated eigenvector $\mathbf{y}_2$ such that $A\mathbf{y}_2 = \lambda_2\mathbf{y}_2$, $(\mathbf{y}_1,\mathbf{y}_2) = 0$, and $\|\mathbf{y}_2\| = 1$. Also since $\|A\| = $ l.u.b. $|(A\mathbf{h},\mathbf{h})|$ for $\|\mathbf{h}\| = 1$ over the whole space, whereas $N_1 = $ l.u.b. $|(A\mathbf{x},\mathbf{x})|$ for $\|\mathbf{x}\| = 1$ over a subspace, we have $N_1 \leq \|A\|$ or $|\lambda_2| \leq |\lambda_1|$.

We go on from here by an iterative procedure. We arrive at a set of eigenvalues $\lambda_1, \lambda_2, \lambda_3, \ldots$ such that $|\lambda_1| \geq |\lambda_2| \geq |\lambda_3| \geq \cdots$, and orthonormal eigenvectors $\mathbf{y}_1, \mathbf{y}_2, \mathbf{y}_3, \ldots$. If at step $n + 1$ we define $N_n = $ l.u.b. $|(A\mathbf{x},\mathbf{x})|$ for $\|\mathbf{x}\| = 1$ and $(\mathbf{x},\mathbf{y}_i) = 0$, $i = 1, 2, 3, \ldots, n$ and find that

$N_n = 0$, then the subspace orthogonal to the n-dimensional subspace spanned by $y_1, y_2, \ldots, y_n$ is the null space of the operator. Hence, the Hilbert space can be decomposed into the null space plus an n-dimensional subspace spanned by the eigenvectors $y_1, y_2, \ldots, y_n$. If $N_n \neq 0$ for every n, then we have a different situation. In this case, $y_1, y_2, y_3, \ldots$ is an orthonormal set of vectors. Let h be a vector in the Hilbert space and let $h_i = (y_i, h)$. By Bessel's inequality, $\sum_{i=1}^{\infty} |h_i|^2 < \|h\|^2$ and hence $(y_i, h) \to 0$. This shows that $\{y_i\}$ converges weakly to zero and that $Ay_i = \lambda_i y_i$ converges strongly to zero. Therefore $\|\lambda_i y_i\|^2 = |\lambda_i|^2 \to 0$ as $i \to \infty$. This shows that $N_n \to 0$ as $n \to \infty$, which shows that the subspace orthogonal to the null space is spanned by $y_1, y_2, y_3, \ldots$, because if y is not in the null space but is orthogonal to $y_1, y_2, y_3, \ldots$, then

$$|(Ay, y)| \leq N_n \|y\|^2 \to 0$$

as $n \to \infty$. Therefore $\|Ay\| = 0$. This is a contradiction which shows that there are no nonzero vectors orthogonal to the null space which are orthogonal to all the eigenvectors $y_1, y_2, y_3, \ldots$. This proves that these eigenvectors form a complete set in the subspace orthogonal to the null space. We summarize all this in the following theorem:

Theorem 3. Let A be a completely continuous self-adjoint operator defined on a Hilbert space H. Let $H = G + F$, where F is the null space of the operator (x in F implies $Ax = 0$) and G is orthogonal to F. If F is not the whole space, then G contains eigenvectors of A, $y_1, y_2, y_3, \ldots$ such that $Ay_i = \lambda_i y_i$, $(y_i, y_j) = \delta_{ij}$, $|\lambda_1| \geq |\lambda_2| \geq |\lambda_3| \geq \cdots$ and the eigenvectors $y_1, y_2, y_3, \ldots$ are complete in G. If G is not finite-dimensional, then $\lim_{i=\infty} \lambda_i = 0$.

We next turn to the question of solving an equation of the form

$$Ax - \lambda x = \mathfrak{g}$$

in a Hilbert space where A is completely continuous and self-adjoint, λ is a given nonzero scalar, and $\mathfrak{g}$ is a given vector. Let us assume that the null space of the operator consists of the zero vector only. If the equation has a solution x, then in terms of the eigenvectors

$$x = \sum_{i=1}^{\infty} x_i y_i$$

where $x_i = (y_i, x)$. Now $\mathfrak{g} = \sum_{i=1}^{\infty} g_i y_i$, where

$$g_i = (y_i, \mathfrak{g}) = (y_i, Ax - \lambda x) = (Ay_i, x) - \lambda(y_i, x)$$
$$= (\lambda_i y_i, x) - \lambda(y_i, x) = (\lambda_i - \lambda)x_i$$

In other words, a necessary condition for a solution to exist is that $g_i = (\lambda_i - \lambda)x_i$. Now if λ is equal to one of the eigenvalues λ_k, then $g_k = (\mathbf{y}_k, \mathbf{g})$ must be zero for a solution to exist.[1] If $g_k \neq 0$, then no solution exists in this case. If λ is not equal to an eigenvalue, then

$$x_i = \frac{g_i}{\lambda_i - \lambda}$$

Conversely, suppose λ is not equal to an eigenvalue and let

$$\mathbf{x} = \sum_{i=1}^{\infty} \frac{g_i}{\lambda_i - \lambda} \, \mathbf{y}_i$$

This is a vector in H since $|\lambda_i - \lambda| \geq |\,|\lambda| - |\lambda_i|\,| \geq \frac{1}{2}|\lambda|$ for all λ_i such that $|\lambda_i| \leq \frac{1}{2}|\lambda|$. Therefore, $\sum_{i=1}^{\infty} |g_i/(\lambda_i - \lambda)|^2$ converges since $\sum_{i=1}^{\infty} |g_i|^2 < \infty$. We show that $\mathbf{x}$ satisfies the equation. Let

$$\mathbf{x}_n = \sum_{i=1}^{n} \frac{g_i}{\lambda_i - \lambda} \, \mathbf{y}_i$$

The $\lambda\mathbf{x}_n$ and $A\mathbf{x}_n$ converge strongly to $\lambda\mathbf{x}$ and $A\mathbf{x}$, respectively. Furthermore,

$$A\mathbf{x}_n - \lambda\mathbf{x}_n = \sum_{i=1}^{n} g_i \mathbf{y}_i$$

which converges strongly to $\mathbf{g}$. We can also show that this solution is unique. Suppose $\bar{\mathbf{x}}$ is also a solution. Then

$$A(\mathbf{x} - \bar{\mathbf{x}}) - \lambda(\mathbf{x} - \bar{\mathbf{x}}) = \mathbf{0}$$

but λ is not an eigenvalue so this equation has only the trivial solution $\mathbf{x} - \bar{\mathbf{x}} = \mathbf{0}$.

If $\lambda = \lambda_k$ for some eigenvalue and $g_k = (\mathbf{y}_k, \mathbf{g}) = 0$ for all eigenvectors associated with λ_k, then we let

$$\mathbf{x} = \sum_{\substack{i=1 \\ i \neq k}}^{\infty} \frac{g_i}{\lambda_i - \lambda_k} \, \mathbf{y}_i$$

and this is a solution. In this case, however, the solution is not unique because we can add to it any linear combination of the eigenvectors associated with λ_k. We can summarize this discussion in a theorem known as **Fredholm's theorem.**

Theorem 4. Let A be a completely continuous self-adjoint operator defined on a Hilbert space H. Let the null space of the operator consist of

---

[1] This may represent a number of conditions if λ_k is an eigenvalue with degeneracy (see exercise 6, Sec. 2.8).

the zero vector only. Then if $\lambda \neq 0$ and λ is not equal to an eigenvalue of A, the equation $A\mathbf{x} - \lambda\mathbf{x} = \mathbf{g}$ has a unique solution given by

$$\mathbf{x} = \sum_{i=1}^{\infty} \frac{g_i}{\lambda_i - \lambda} \, \mathbf{y}_i$$

where λ_i is the eigenvalue of A corresponding to the eigenvector $\mathbf{y}_i$ and $g_i = (\mathbf{y}_i, \mathbf{g})$. If λ is equal to an eigenvalue λ_k, then there is no solution unless $\mathbf{g}$ is orthogonal to every eigenvector corresponding to that eigenvalue. In the latter case, there is a nonunique solution given by

$$\mathbf{x} = \sum_{\substack{i=1 \\ i \neq k}}^{\infty} \frac{g_i}{\lambda_i - \lambda_k} \, \mathbf{y}_i + \mathbf{h}$$

where $\mathbf{h}$ is an arbitrary linear combination of the eigenvectors corresponding to λ_k.

We conclude this section with some examples of completely continuous operators. Consider a matrix with elements a_{ij} such that $\sum_{i=1}^{\infty} \sum_{j=1}^{\infty} |a_{ij}|^2 < \infty$. Let $\{x_i\}$ be a sequence such that $\sum_{i=1}^{\infty} |x_i|^2 < \infty$. Then clearly

$$y_i = \sum_{j=1}^{\infty} a_{ij} x_j$$

converges for all i and

$$\sum_{i=1}^{\infty} |y_i|^2 \leq \sum_{i=1}^{\infty} \left(\sum_{j=1}^{\infty} |a_{ij}| \, |x_j| \right)^2 \leq \sum_{i=1}^{\infty} \sum_{j=1}^{\infty} |x_j|^2 \sum_{j=1}^{\infty} |a_{ij}|^2$$

$$\leq \sum_{j=1}^{\infty} |x_j|^2 \sum_{i=1}^{\infty} \sum_{j=1}^{\infty} |a_{ij}|^2 < \infty$$

Now consider $\sum_{i=1}^{\infty} y_i^* z_i$, where $\sum_{i=1}^{\infty} |z_i|^2 < \infty$. We have

$$\left| \sum_{i=1}^{\infty} y_i^* z_i \right|^2 \leq \sum_{i=1}^{\infty} |z_i|^2 \sum_{i=1}^{\infty} |y_i|^2 \leq \sum_{i=1}^{\infty} |x_i|^2 \sum_{i=1}^{\infty} |z_i|^2 \sum_{i=1}^{\infty} \sum_{j=1}^{\infty} |a_{ij}|^2$$

which shows that there is an M such that

$$\left| \sum_{i=1}^{\infty} \sum_{j=1}^{\infty} a_{ij}^* x_j^* z_i \right| \leq M \sqrt{ \sum_{i=1}^{\infty} |x_i|^2 \sum_{i=1}^{\infty} |z_i|^2 }$$

Hence the matrix is bounded and represents a bounded operator A. We shall further show that A is completely continuous. Let $\{\mathbf{u}_n\}$ converge weakly to zero with $x_i^{(n)} = (\boldsymbol{\phi}_i, \mathbf{u}_n)$ relative to some orthonormal basis $\boldsymbol{\phi}_1, \boldsymbol{\phi}_2, \boldsymbol{\phi}_3, \ldots$. Let $A\mathbf{u}_n$ be defined by the components

$$y_i^{(n)} = \sum_{j=1}^{\infty} a_{ij} x_j^{(n)}$$

We shall show that $\{A\mathbf{u}_n\}$ converges strongly to zero. Now $\|\mathbf{u}_n\| \leq K$ for some K. Hence $\sum_{i=1}^{\infty} |x_i^{(n)}|^2 \leq K^2$ for all n, and $y_i^{(n)} = (\boldsymbol{\phi}_i, A\mathbf{u}_n) = (\tilde{A}\boldsymbol{\phi}_i, \mathbf{u}_n) \rightarrow (\tilde{A}\boldsymbol{\phi}_i, 0) = 0$. Furthermore,

$$\sum_{i=n+1}^{\infty} |y_i^{(n)}|^2 \leq \sum_{i=n+1}^{\infty} \sum_{j=1}^{\infty} |a_{ij}|^2 \sum_{i=1}^{\infty} |x_i^{(n)}|^2 \leq \epsilon K^2$$

where ϵ can be made arbitrarily small for N sufficiently large. Having picked ϵ we determine N and then take n so large that $\sum_{i=1}^{\infty} |y_i^{(n)}|^2 < \epsilon$ and then

$$\sum_{i=1}^{\infty} |y_i^{(n)}|^2 = \sum_{i=1}^{n} |y_i^{(n)}|^2 + \sum_{i=n+1}^{\infty} |y_i^{(n)}|^2 < \epsilon + \epsilon K^2$$

which proves that $A\mathbf{u}_n$ converges strongly to zero and proves the complete continuity of A. Clearly, in this case A is self-adjoint if and only if $a_{ij} = a_{ji}^*$.

Finally, let H be $L_2(a,b)$ and let $K(s,t)$ satisfy $K(s,t) = K^*(t,s)$ and be square integrable; that is,

$$\int_a^b \int_a^b |K(s,t)|^2 \, ds \, dt < \infty$$

The space is separable so it contains an orthonormal basis $\boldsymbol{\phi}_1, \boldsymbol{\phi}_2, \boldsymbol{\phi}_3, \ldots$. If $x(t) = \sum_{i=1}^{\infty} x_i \phi_i(t)$, where $x_i = \int_a^b \phi_i^*(t) x(t) \, dt$ and

$$y(s) = \int_a^b K(s,t) x(t) \, dt$$

then

$$y(s) = \sum_{i=1}^{\infty} y_i \phi_i(s)$$

We compute matrix elements as follows:

$$a_{ij} = \int_a^b \int_a^b K(s,t) \phi_i^*(s) \phi_j(t) \, ds \, dt$$

Then $y_i = \sum_{j=1}^{\infty} a_{ij} x_j$. The functions

$$\psi_{ij}(s,t) = \phi_i^*(s) \phi_j(t)$$

are orthonormal over the rectangle $R = \{(s,t) \mid a \leq s \leq b, a \leq t \leq b\}$. Therefore, by Bessel's inequality

$$\sum_{i=1}^{\infty} \sum_{j=1}^{\infty} |a_{ij}|^2 \leq \int_a^b \int_a^b |K(s,t)|^2 \, ds \, dt < \infty$$

This shows that the operator is completely continuous. The operator is self-adjoint since

$$a_{ji}^* = \int_a^b \int_a^b K^*(s,t)\phi_j(s)\phi_i^*(t)\,ds\,dt$$

$$= \int_a^b \int_a^b K(t,s)\phi_i^*(t)\phi_j(s)\,dt\,ds$$

$$= a_{ij}$$

Exercises 2.8

1. Prove that A is completely continuous if and only if $\{u_n\}$ bounded implies $\{Au_n\}$ contains a strongly convergent subsequence.

2. Let A and B be completely continuous operators on a Hilbert space H, and let T be a bounded operator. Prove that

a. αA is completely continuous where α is a scalar.

b. $A + B$ is completely continuous.

c. AT and TA are completely continuous.

d. $\tilde{A}A$ is completely continuous.

e. $\tilde{A}$ is completely continuous. HINT: Prove that $\tilde{A}A$ completely continuous implies A is completely continuous.

3. Prove that the identity operator is not completely continuous on a Hilbert space.

4. Prove that every linear operator on a finite-dimensional vector space is completely continuous.

5. Prove that the orthogonal complement of the null space of a self-adjoint operator A is invariant; that is, if $H = G + F$, where F is the null subspace and $\mathfrak{g}$ is in G, then $A\mathfrak{g}$ is in G.

6. An eigenvalue λ is said to be m-fold degenerate if there are exactly m linearly independent eigenvectors associated with it. Prove that the nonzero eigenvalues of a completely continuous operator have finite degeneracy.

7. Prove that the eigenvalues of a self-adjoint operator are real.

8. Prove that eigenvectors corresponding to different eigenvalues of a self-adjoint operator are orthogonal.

9. Prove that if a completely continuous self-adjoint operator exists on a Hilbert space H such that its null space consists of the zero vector only, then H is separable.

10. Generalize the Fredholm theorem in the case where the null space of the operator is a separable Hilbert subspace.

11. Let $\mathbf{x} = (x_1, x_2, x_3, \ldots)$ and $A\mathbf{x} = (x_1, 0, 0, \ldots)$. Show that A is completely continuous and self-adjoint. What is the null space of A? What are the eigenvalues and eigenvectors of A?

References

Akhiezer, N. I., and I. M. Glazman: "Theory of Linear Operators in Hilbert Space," vols. I and II, Pitman Publishing, Ltd., London, 1981.

Berberian, Sterling K.: "Introduction to Hilbert Space," 2d ed., Chelsea Publishers, Inc., New York, 1976.

Churchill, Ruel V., and James Brown: "Fourier Series and Boundary Value Problems," 3d ed., McGraw-Hill Book Company, New York, 1978.

Friedman, Bernard: "Principles and Techniques of Applied Mathematics," John Wiley & Sons, Inc., New York, 1956.

Schmeidler, Werner: "Linear Operators in Hilbert Space," Academic Press Inc., New York, 1965.

Stone, Marshall H.: "Linear Transformations in Hilbert Space and Their Applications to Analysis," reprint of American Mathematical Society Colloquium Publication, vol. 15, New York, 1932, American Mathematical Society, Providence, Rhode Island, 1979.

Taylor, Angus E., and David C. Lay: "Introduction to Functional Analysis," 2d ed., John Wiley & Sons, Inc., New York, 1980.

Chapter 3. Calculus of Variations

3.1 Maxima and Minima of Functions. Lagrange Multipliers

One of the fundamental theorems of analysis tells us that *a continuous function defined over a bounded closed region takes on a maximum and minimum value somewhere in the region.* This maximum and minimum may occur on the boundary of the region or at an interior point. At an interior point we may have a **relative maximum** or a **relative minimum.** The condition for a relative maximum of the function $f(x_1, x_2, \ldots, x_n)$ at a point with coordinates $(x_1, x_2, \ldots, x_n)$ is that there exists some neighborhood of the point for which

$$f(x_1 + \Delta x_1, x_2 + \Delta x_2, \ldots, x_n + \Delta x_n) - f(x_1, x_2, \ldots, x_n) \leq 0$$

for arbitrary $\Delta x_1, \Delta x_2, \ldots, \Delta x_n$. For a relative minimum there must exist some neighborhood of the point throughout which

$$f(x_1 + \Delta x_1, x_2 + \Delta x_2, \ldots, x_n + \Delta x_n) - f(x_1, x_2, \ldots, x_n) \geq 0$$

If the function $f(x_1, x_2, \ldots, x_n)$ has first partial derivatives at a relative maximum or relative minimum, then it is easy to find necessary conditions for such a point. For example, in some neighborhood of a relative maximum

$$\frac{f(x_1, x_2, \ldots, x_k + \Delta x_k, \ldots, x_n) - f(x_1, x_2, \ldots, x_k, \ldots, x_n)}{\Delta x_k} \leq 0$$

for positive Δx_k. Also

$$\frac{f(x_1, x_2, \ldots, x_k + \Delta x_k, \ldots, x_n) - f(x_1, x_2, \ldots, x_k, \ldots, x_n)}{\Delta x_k} \geq 0$$

for Δx_k negative. But the existence of $\partial f / \partial x_k$ at the point $(x_1, x_2, \ldots, x_n)$ implies that

$$\lim_{\Delta x_k \to 0} \frac{f(x_1, x_2, \ldots, x_k + \Delta x_k, \ldots, x_n) - f(x_1, x_2, \ldots, x_k, \ldots, x_n)}{\Delta x_k}$$

exists and is independent of whether Δx_k approaches zero through positive or through negative values. Therefore, $\partial f/\partial x_k = 0$ at a relative maximum point, and this is true for all k. The necessary condition for a relative minimum point is the same, and the proof is quite similar.

The fact that $\partial f/\partial x_k = 0$ for all k is not a sufficient condition for a relative maximum or minimum, as is seen by the example $f(x_1,x_2) = x_1 x_2$.

$$\frac{\partial f}{\partial x_1} = \frac{\partial f}{\partial x_2} = 0 \qquad \text{at} \qquad x_1 = x_2 = 0$$

but $f(x_1,x_2)$ takes on positive values in the first and third quadrants and negative values in the second and fourth quadrants. Hence the origin can be neither a maximum nor a minimum point. In this case, the origin is called a **saddle point.** In any case, we say that $\partial f/\partial x_k = 0$ is a necessary condition for a **stationary value** of the function. The sufficient conditions for relative maxima and minima involve, in addition to the first partial derivatives being zero, inequalities containing higher-order derivatives.[1]

The necessary conditions for a relative maximum or minimum can be phrased in another way. Assuming that $f(x_1,x_2, \ldots ,x_n)$ has continuous first partial derivatives in a neighborhood of the extremum point, we can write

$$\Delta f = f(x_1 + \Delta x_1, x_2 + \Delta x_2, \ldots , x_n + \Delta x_n) - f(x_1,x_2, \ldots ,x_n)$$

$$= \frac{\partial f}{\partial x_1}\Delta x_1 + \frac{\partial f}{\partial x_2}\Delta x_2 + \cdots + \frac{\partial f}{\partial x_n}\Delta x_n + \xi_1\Delta x_1 + \xi_2\Delta x_2 + \cdots + \xi_n\Delta x_n$$

$$= \frac{\partial f}{\partial x_i}\Delta x_i + \xi_i\Delta x_i$$

where $\xi_i \to 0$ as $\Delta x_i \to 0$. The first-order change, or more simply, the **first variation** of f, is $df = (\partial f/\partial x_i)\,dx_i$, where we have replaced Δx_i by the differentials dx_i. *The necessary condition for a relative maximum or minimum of a function with continuous first partial derivatives is that the first variation df vanish for arbitrary changes dx_i in the independent variables x_i.*

If the variables $x_1, x_2, \ldots , x_n$ are not independent but satisfy some condition or conditions of constraint, the extremum points can usually be found by a straightforward procedure of differentiation and solution of simultaneous equations, or by the method of **Lagrange multipliers.** To illustrate the equivalence of these two methods, we consider first the problem of extremizing a function of two variables subject to one constraint. We shall then generalize to the case of a function of n variables subject to k constraints.

Suppose we are asked to locate a relative maximum or minimum of $f(x_1,x_2)$ subject to the constraining condition $g(x_1,x_2) = 0$. In principle we can solve

---

[1] See R. Creighton Buck, "Advanced Calculus," McGraw-Hill Book Company, New York, 1956.

$g(x_1,x_2) = 0$ for x_2 in terms of x_1; that is, $x_2 = G(x_1)$. Substituting in $f(x_1,x_2)$, we have

$$f[x_1,G(x_1)] = F(x_1)$$

a function of the single variable x_1 to be maximized or minimized, a necessary condition for which is $F'(x_1) = 0$. Hence, we can locate the stationary value by solving simultaneously

$$F'(x_1) = \frac{\partial f}{\partial x_1} + \frac{\partial f}{\partial x_2} G'(x_1) = 0$$

$$g(x_1,x_2) = 0$$

From the constraint we have

$$\frac{\partial g}{\partial x_1} dx_1 + \frac{\partial g}{\partial x_2} dx_2 = 0$$

from which we have

$$G'(x_1) = \frac{dx_2}{dx_1} = -\frac{\partial g/\partial x_1}{\partial g/\partial x_2}$$

provided $\partial g/\partial x_2 \neq 0$. Hence the equations we are left to solve are

$$\frac{\partial f}{\partial x_1} \frac{\partial g}{\partial x_2} - \frac{\partial f}{\partial x_2} \frac{\partial g}{\partial x_1} = 0$$

$$g(x_1,x_2) = 0$$

Consider the problem of maximizing or minimizing the function

$$H(x_1,x_2,\lambda) = f(x_1,x_2) + \lambda g(x_1,x_2)$$

as a function of the three variables x_1, x_2, λ subject to no constraint. The necessary conditions for a stationary value are

$$\frac{\partial H}{\partial x_1} = \frac{\partial f}{\partial x_1} + \lambda \frac{\partial g}{\partial x_1} = 0$$

$$\frac{\partial H}{\partial x_2} = \frac{\partial f}{\partial x_2} + \lambda \frac{\partial g}{\partial x_2} = 0$$

$$\frac{\partial H}{\partial \lambda} = g(x_1,x_2) = 0$$

Eliminating λ, we arrive back at the system of equations

$$\frac{\partial f}{\partial x_1} \frac{\partial g}{\partial x_2} - \frac{\partial f}{\partial x_2} \frac{\partial g}{\partial x_1} = 0$$

$$g(x_1,x_2) = 0$$

Thus the two methods are equivalent. The second is called the method of Lagrange multipliers, and the λ is called the **Lagrange multiplier.**

More generally we can consider the problem of maximizing or minimizing a function $f(x_1, x_2, \ldots, x_n)$ subject to constraints

$$g_1(x_1, x_2, \ldots, x_n) = 0$$

$$g_2(x_1, x_2, \ldots, x_n) = 0$$

$$\cdots\cdots\cdots\cdots\cdots$$

$$g_k(x_1, x_2, \ldots, x_n) = 0$$

with $k < n$. In principle we could solve the constraints for k of the variables in terms of the other $n - k$, substitute into $f(x_1, x_2, \ldots, x_n)$, and arrive at a function of $n - k$ independent variables to be extremized. Alternatively, we can proceed as follows. The first variation $df = (\partial f / \partial x_i) \, dx_i$ must vanish, but now, because of the k constraints, only $n - k$ of the differentials dx_i are independent. The differentials must satisfy k conditions of constraint; that is,

$$\frac{\partial g_j}{\partial x_i} \, dx_i = 0 \qquad i = 1, 2, \ldots, n$$

$$j = 1, 2, \ldots, k$$

Multiplying these equations by λ_j and summing, we have

$$\left(\frac{\partial f}{\partial x_i} + \lambda_j \frac{\partial g_j}{\partial x_i} \right) dx_i = 0$$

which is true for arbitrary λ's and arbitrary values of $n - k$ of the differentials. We pick the λ's so that k of the expressions in the parentheses vanish. The remaining expressions in parentheses must also vanish, since the $n - k$ remaining differentials are arbitrary. We arrive at the following necessary conditions for a relative maximum or minimum:

$$\frac{\partial f}{\partial x_i} + \lambda_j \frac{\partial g_j}{\partial x_i} = 0 \qquad i = 1, 2, \ldots, n$$

$$g_j(x_1, x_2, \ldots, x_n) = 0 \qquad j = 1, 2, \ldots, k$$

This gives $n + k$ equations to solve for $x_1, x_2, \ldots, x_n$ and $\lambda_1, \lambda_2, \ldots, \lambda_k$. These equations are the necessary conditions for a relative maximum or minimum of

$$H(x_1, x_2, \ldots, x_n, \lambda_1, \lambda_2, \ldots, \lambda_k) = f(x_1, x_2, \ldots, x_n) + \lambda_j g_j(x_1, x_2, \ldots, x_n)$$

To illustrate the use of Lagrange multipliers, consider the problem of finding the minimum distance from the plane $ax_1 + bx_2 + cx_3 + d = 0$ to the origin. Geometrically it is clear that such a minimum exists. The problem is to minimize $f(x_1, x_2, x_3) = x_1^2 + x_2^2 + x_3^2$ subject to the constraint

$$g(x_1, x_2, x_3) = ax_1 + bx_2 + cx_3 + d = 0$$

In the method of Lagrange multipliers, we minimize

$$H(x_1, x_2, x_3, \lambda) = x_1^2 + x_2^2 + x_3^2 + \lambda(ax_1 + bx_2 + cx_3 + d)$$

The necessary conditions for the minimum are

$$\frac{\partial H}{\partial x_1} = 2x_1 + a\lambda = 0$$

$$\frac{\partial H}{\partial x_2} = 2x_2 + b\lambda = 0$$

$$\frac{\partial H}{\partial x_3} = 2x_3 + c\lambda = 0$$

$$\frac{\partial H}{\partial \lambda} = ax_1 + bx_2 + cx_3 + d = 0$$

Solving for λ, we have

$$\lambda = \frac{2d}{a^2 + b^2 + c^2}$$

The minimum occurs at the point

$$x_1 = -ad/(a^2 + b^2 + c^2) \qquad x_2 = -bd/(a^2 + b^2 + c^2)$$
$$x_3 = -cd/(a^2 + b^2 + c^2)$$

and the actual minimum is equal to $|d|/(a^2 + b^2 + c^2)^{\frac{1}{2}}$.

As another example, consider the real quadratic form

$$Q(x_1, x_2, \ldots, x_n) = X'AX$$

where A is real and symmetric. If we divide Q by $X'X$, which is positive unless $X = 0$, we have a function

$$f(X) = f(x_1, x_2, \ldots, x_n) = \frac{a_{ij}x_i x_j}{x_k x_k} = \frac{X'AX}{X'X}$$

If $\mathbf{Y}$ is an eigenvector of A corresponding to an eigenvalue λ, then

$$f(Y) = \frac{Y'AY}{Y'Y} = \frac{\lambda Y'Y}{Y'Y} = \lambda$$

and, furthermore, f is stationary with respect to small changes of $\mathbf{X}$ about $\mathbf{Y}$; that is,

$$df = \left[\frac{(y_k y_k)(2a_{ij}y_j) - (a_{kj}y_k y_j)(2y_i)}{(y_m y_m)^2} \right] dx_i$$

$$= \left[\frac{(y_k y_k)(2\lambda y_i) - (\lambda y_j y_j)(2y_i)}{(y_m y_m)^2} \right] dx_i$$

$$= 0$$

This suggests that the eigenvalues of A can be found by finding the stationary values of f.

Recall that the eigenvalues of A are all real and that there are a finite number of them. Hence they can be ordered as follows:

$$\lambda_1 \leq \lambda_2 \leq \lambda_3 \leq \cdots \leq \lambda_n$$

assuming A is $n \times n$. Any vector $\mathbf{X}$ can be expressed as a linear combination of the normalized eigenvectors $\mathbf{Y}_i$; that is, $\mathbf{X} = c_i \mathbf{Y}_i$, so that

$$f(X) = \frac{(c_i Y_i') A (c_j Y_j)}{(c_k Y_k')(c_m Y_m)} = \frac{\displaystyle\sum_{i=1}^{n} \sum_{j=1}^{n} (c_i Y_i') c_j (\lambda_j Y_j)}{c_k c_m Y_k' Y_m}$$

$$= \frac{\displaystyle\sum_{i=1}^{n} \sum_{j=1}^{n} \lambda_j c_i c_j \delta_{ij}}{c_k c_m \delta_{km}} = \frac{\displaystyle\sum_{i=1}^{n} \lambda_i c_i^2}{\displaystyle\sum_{k=1}^{n} c_k^2}$$

Now

$$\lambda_1 = \frac{\displaystyle\sum_{i=1}^{n} \lambda_1 c_i^2}{\displaystyle\sum_{k=1}^{n} c_k^2} \quad \text{and} \quad \lambda_n = \frac{\displaystyle\sum_{i=1}^{n} \lambda_n c_i^2}{\displaystyle\sum_{k=1}^{n} c_k^2}$$

so that

$$\lambda_1 - f(X) = \frac{\displaystyle\sum_{i=1}^{n} (\lambda_1 - \lambda_i) c_i^2}{\displaystyle\sum_{k=1}^{n} c_k^2} \leq 0$$

$$\lambda_n - f(X) = \frac{\displaystyle\sum_{i=1}^{n} (\lambda_n - \lambda_i) c_i^2}{\displaystyle\sum_{k=1}^{n} c_k^2} \geq 0$$

or

$$\lambda_1 \leq f(X) \leq \lambda_n$$

for an arbitrary vector $\mathbf{X}$. This inequality allows one very quickly to obtain an upper bound for λ_1 and a lower bound for λ_n, and also proves that one can find λ_1 by minimizing $f(X)$ and λ_n by maximizing $f(X)$. If we wish to find λ_2, we can proceed as follows. Minimize $f(X)$ over the subspace of vectors orthogonal to $\mathbf{Y}_1$. In this case $\mathbf{X}$ can be expressed as a linear combination of $\mathbf{Y}_2, \mathbf{Y}_3, \ldots, \mathbf{Y}_n$, and we have

$$f(X) = \frac{\displaystyle\sum_{i=2}^{n} \lambda_i c_i^2}{\displaystyle\sum_{k=2}^{n} c_k^2}$$

and the inequality

$$\lambda_2 \leq f(X) \leq \lambda_n$$

Therefore, the minimum of the function over the subspace yields λ_2 and the corresponding eigenvector $\mathbf{Y}_2$. The next eigenvalue λ_3 can be found by minimizing $f(X)$ over the subspace orthogonal to $\mathbf{Y}_1$ and $\mathbf{Y}_2$. This iterative procedure yields all the eigenvalues and eigenvectors as solutions to problems in the calculus of variations.

The method of Lagrange multipliers yields all the eigenvalues and eigenvectors in one step. We attempt to find the stationary values of $Q(X)$ subject to the constraint $g(X) = X'X - 1 = 0$. The problem is to extremize

$$H(X) = Q(X) - \lambda g(X)$$

The necessary conditions are

$$2a_{ij}x_j - 2\lambda x_i = 0 \qquad x_i x_i = 1$$

The first of these is the eigenvalue problem

$$AX = \lambda X$$

which leads to the characteristic equation

$$|A - \lambda I| = 0$$

which yields the n eigenvalues $\lambda_1, \lambda_2, \ldots, \lambda_n$. The eigenvectors $\mathbf{Y}_1, \mathbf{Y}_2, \ldots,$ $\mathbf{Y}_n$ are determined by solving

$$AY_\alpha = \lambda_\alpha Y_\alpha$$

subject to $(\mathbf{Y}_\alpha, \mathbf{Y}_\alpha) = 1$. That the eigenvectors are orthogonal follows as in Sec. 1.8.

There is still another way to characterize the eigenvalues of A as a variational problem. Let λ_k be the kth eigenvalue from the sequence of ordered eigenvalues; then

$$\lambda_k - f(X) = \frac{\sum\limits_{i=1}^{n} (\lambda_k - \lambda_i)c_i^2}{\sum\limits_{j=1}^{n} c_j^2}$$

This expression can be made nonnegative by choosing

$$c_{k+1} = c_{k+2} = \cdots = c_n = 0$$

We choose the remaining c's to satisfy the $k - 1$ conditions

$$(\mathbf{X}, \mathbf{V}_1) = (\mathbf{X}, \mathbf{V}_2) = \cdots = (\mathbf{X}, \mathbf{V}_{k-1}) = 0$$

where $\mathbf{V}_1, \mathbf{V}_2, \ldots, \mathbf{V}_{k-1}$ is any set of linearly independent vectors. These are $k - 1$ homogeneous linear algebraic equations in k unknowns. These equations always have a nontrivial solution, which is determined to within a constant of proportionality. This constant does not affect $\lambda_k - f(X)$, since it appears as a common factor in both the numerator and the denominator of the above expression. Alternatively one could require that

$$\|\mathbf{X}\|^2 = x_i x_i = \sum_{i=1}^{k} c_i^2 = 1$$

which would determine $c_1, c_2, \ldots, c_k$ exactly. Since $c_1, c_2, \ldots, c_k$ are determined for a given set of vectors $\mathbf{V}_1, \mathbf{V}_2, \ldots, \mathbf{V}_{k-1}$, and since choosing $c_{k+1}, c_{k+2}, \ldots, c_n$ different from zero cannot increase $\lambda_k - f(X)$, we have

$$\min f(X) \leq \lambda_k$$

where $\mathbf{X}$ is orthogonal to $\mathbf{V}_1, \mathbf{V}_2, \ldots, \mathbf{V}_{k-1}$. Now we know from previous considerations that if $\mathbf{V}_1 = \mathbf{Y}_1, \mathbf{V}_2 = \mathbf{Y}_2, \ldots, \mathbf{V}_{k-1} = \mathbf{Y}_{k-1}$, then $\min f(X) = \lambda_k$. Therefore,

$$\lambda_k = \max \left[\min f(X) \right]$$

as the set $\mathbf{V}_1, \mathbf{V}_2, \ldots, \mathbf{V}_{k-1}$ is allowed to change. This is called the **minimax** definition of the eigenvalue. The vector which defines the kth eigenvalue is the kth eigenvector, for then

$$(\mathbf{X}, \mathbf{Y}_j) = \left(\sum_{i=1}^{k} c_i \mathbf{Y}_i, \mathbf{Y}_j \right) = c_j = \begin{cases} 0 & \text{for } j < k \\ 1 & \text{for } j = k \end{cases}$$

The advantage of this definition is that we can define the kth eigenvalue conceptually without first knowing the $k - 1$ preceding ones.

Exercises 3.1

1. Using the method of Lagrange multipliers, find the points on the surface of the ellipsoid

$$9x_1^2 + 12x_2^2 + 9x_3^2 - 6x_1 x_2 - 6x_2 x_3 - 25 = 0$$

which make the distance from the origin stationary. Show that these points are on mutually orthogonal axes through the origin.

2. Let $Q(X) = X'AX$ be a positive-definite quadratic form with A real and symmetric, and $N(X) = X'X$. We have already shown that the eigenvalues of A are the stationary values of $Q(X)$ subject to $N(X) = 1$. Show that the eigenvalues of A can be found by finding the stationary values of $N(X)$ subject to $Q(X) = 1$. What is the relation between the stationary values of N and the eigenvalues of A? $N(X)$ is just the square of the distance from the origin to the surface $Q(X) = 1$.

*3. Let $Q(X) = X'AX$ be a real quadratic form and $E(X) = X'BX$ be a positive-definite real quadratic form. A and B are both $n \times n$ real symmetric matrices. Show that the stationary values of $Q(X)$ subject to $E(X) = 1$ are the solutions of the characteristic equation $|A - \lambda B| = 0$. How is this problem related to the problem of simultaneous reduction of E and Q to diagonal form discussed in Sec. 1.8?

*4. If $Q(X) = X'AX$ is a real quadratic form, $Q(X,Y) = X'AY$ is called the **associated bilinear form.** Show the following:

a. $Q(X,Y) = Q(Y,X)$.

b. $Q(X,Y) = \frac{1}{4}[Q(X + Y) - Q(X - Y)]$.

    c. If λ is an eigenvalue of A and Y is the corresponding eigenvector,

$$Q(X,Y) - \lambda N(X,Y) = 0$$

for arbitrary X. Here $N(X,Y)$ refers to the bilinear form $N(X,Y) = X'Y$ associated with $N(X) = X'X$.

    d. $N(U,V) = N(U,Y_i)N(V,Y_i)$ where $\mathbf{Y}_i$, $i = 1, 2, 3, \ldots, n$, are the n eigenvectors and $\mathbf{U}$ and $\mathbf{V}$ are any vectors in the n-dimensional vector space.

    5. Let $H(Z) = (Z^*)'AZ$ be a hermitian form, with A a hermitian matrix with eigenvalues $\lambda_1 \leq \lambda_2 \leq \lambda_3 \leq \cdots \leq \lambda_n$. Show that

$$\lambda_1 \leq f(Z) \leq \lambda_n$$

where $f(Z) = H(Z)/(Z^*)'Z$, and that the stationary values of $H(Z)$ subject to $(Z^*)'Z = 1$ are the eigenvalues of A. HINT: In the second part let

$$a_{kj} = \alpha_{kj} + i\beta_{kj} \quad \text{and} \quad z_j = x_j + iy_j$$

3.2 Maxima and Minima of Functionals. Euler's Equation

The types of problems considered in Sec. 3.1 can be handled by the ordinary calculus. The subject matter of the calculus of variations deals, on the other hand, with the problem of maximizing or minimizing **functionals** (usually integrals) which depend on the definition of some function or functions. In Sec. 3.1, we extremized functions in some finite-dimensional vector space. In this section, we shall treat the problem of extremizing functionals in an infinite-dimensional function space. Some examples of this kind of problem are the following:

    1. Find the shortest curve $y = y(x)$ in the xy plane connecting the points $(0,0)$ and (a,b). To solve this we must minimize

$$s = \int_{(0,0)}^{(a,b)} ds = \int_0^a \left[1 + \left(\frac{dy}{dx}\right)^2 \right]^{\frac{1}{2}} dx$$

subject to the conditions $y(0) = 0$ and $y(a) = b$.

    2. Find the shortest curve lying in the surface $z = z(x,y)$ and connecting the points (x_1,y_1,z_1) and (x_2,y_2,z_2) on the surface. Here we must minimize

$$s = \int_{(x_1,y_1,z_1)}^{(x_2,y_2,z_2)} ds = \int_{(x_1,y_1,z_1)}^{(x_2,y_2,z_2)} [(dx)^2 + (dy)^2 + (dz)^2]^{\frac{1}{2}}$$

$$= \int_{(x_1,y_1)}^{(x_2,y_2)} \left[(dx)^2 + (dy)^2 + \left(\frac{\partial z}{\partial x} dx + \frac{\partial z}{\partial y} dy\right)^2 \right]^{\frac{1}{2}}$$

Suppose we seek parametric equations of the curve

$$x = x(t) \qquad y = y(t) \qquad z = z[x(t),y(t)] \qquad 0 \leq t \leq 1$$

such that $x(0) = x_1$ $y(0) = y_1$ $z(x_1,y_1) = z_1$

and $x(1) = x_2$ $y(1) = y_2$ $z(x_2,y_2) = z_2$

Then the problem is to minimize

$$\int_0^1 \left\{ \left[1 + \left(\frac{\partial z}{\partial x} \right)^2 \right] \dot{x}^2 + \left[1 + \left(\frac{\partial z}{\partial y} \right)^2 \right] \dot{y}^2 + 2 \frac{\partial z}{\partial x} \frac{\partial z}{\partial y} \dot{x} \dot{y} \right\}^{\frac{1}{2}} dt$$

3. A mass slides on a frictionless wire connecting points P and Q in a vertical plane. For what shape of the wire is the time of descent a minimum? Energy is conserved, so that the potential energy lost in the descent is equal to the kinetic energy gained. Taking the point P to be the origin of the plane and measuring y positively down, we have $\frac{1}{2}mv^2 = mgy$ or $v = \sqrt{2gy}$. We wish to find the curve $y = y(x)$ which minimizes

$$\int_P^Q dt = \int_P^Q \frac{ds}{v} = \int_0^a \left[\frac{1 + (y')^2}{2gy} \right]^{\frac{1}{2}} dx$$

and which passes through the points $(0,0)$ and (a,b).

4. Find the positive function $y = y(x) > 0$ whose graph passes through the points (a,b) and (c,d) which produces the surface of revolution about the x axis with least area. We wish to minimize

$$\int_a^c 2\pi y \left[1 + \left(\frac{dy}{dx} \right)^2 \right]^{\frac{1}{2}} dx$$

subject to $y(a) = b$ and $y(c) = d$.

5. Fermat's principle states that a light ray passes through a medium along a path which minimized the transit time. Let $\eta(x,y)$ be the index of refraction in a two-dimensional medium; that is, $v(x,y) = c/\eta(x,y)$, where c is a constant (velocity of light in a vacuum). Then the time of transit between points $(0,0)$ and (a,b) is given by

$$\int_0^a \frac{ds}{v} = \frac{1}{c} \int_0^a \eta(x,y) \left[1 + \left(\frac{dy}{dx} \right)^2 \right]^{\frac{1}{2}} dx$$

where $y = y(x)$ is the curve along which light is transmitted. The problem is to find the function $y(x)$ which minimizes this integral.

6. Find the shape of a uniform hanging cable of a given length. The cable will hang in a position which minimizes the potential energy. The potential energy is proportional to the coordinate of the centroid in the direction of the gravitational force. Take the y axis in the vertical direction. The problem is then to minimize

$$\bar{y} = \frac{1}{L} \int_0^a y \, ds = \frac{1}{L} \int_0^a y \left[1 + \left(\frac{dy}{dx} \right)^2 \right]^{\frac{1}{2}} dx$$

subject to the condition

$$L = \int_0^a \left[1 + \left(\frac{dy}{dx} \right)^2 \right]^{\frac{1}{2}} dx$$

and with $y(0) = 0$ and $y(a) = b$.

7. Find the closed curve of given length in the xy plane which encloses the largest area. The parametric equations of the curve are $x = x(t)$, $y = y(t)$, $0 \leq t \leq 1$, with $x(0) = x(1)$ and $y(0) = y(1)$. The area enclosed is

$$\iint\limits_{R} dx\, dy = \tfrac{1}{2} \oint_{C} x\, dy - y\, dx$$

by use of Green's lemma. Thus we wish to maximize

$$\tfrac{1}{2} \int_{0}^{1} (x\dot{y} - y\dot{x})\, dt$$

subject to
$$L = \int_{0}^{1} (\dot{x}^2 + \dot{y}^2)^{\frac{1}{2}}\, dt$$

It is understood that certain continuity conditions on the functions and/or their derivatives must be prescribed in each case in order to make the extremum meaningful. This serves to define the function space in which the extremum is sought. It may be that there exists no solution of the variational problem within the **class of admissible functions,** that is, the function space in which the extremum is sought. For example, if we seek the positive continuous function $y = y(x)$ passing through (a,c) and (b,d) which minimizes the area between the curve and the x axis, there is no solution. This is because the function corresponding to the lower bound

$$\int_{a}^{b} y\, dx = 0$$

is $y(x) = 0$, $a < x < b$, $y(a) = c$, $y(b) = d$, and this is not a continuous function as prescribed by the variational problem. Continuous functions can be found which will make the area arbitrarily close to zero, but it cannot be made exactly zero for any continuous function.

The problem of finding sufficient conditions for the solution of a variational problem is one of the more difficult aspects of the calculus of variations. We shall be content to deal only with necessary conditions, that is, assuming that a maximizing or a minimizing function exists, to find the conditions it must satisfy, and hence to find a function which satisfies these conditions. One can usually show that it is the required extremizing function within the context of the individual problem.

Problems 1, 3, 4, and 5 can all be put in the form of the following problem: to minimize

$$\int_{a}^{b} F\left(x, y, \frac{dy}{dx}\right) dx$$

by the proper choice of $y = y(x)$, which goes through the end points $y(a) = c$ and $y(b) = d$. Under suitable restrictions on F and on the class of functions admitted to competition, we shall derive necessary conditions for the minimizing function, assuming that it exists in the class of admissible functions. Let F have continuous second partial derivatives with respect to its arguments x, y, and y'. Let y have a continuous second derivative. Assuming that there is a function $\phi(x)$, which has a continuous second derivative, passes through the given points, and makes the functional take on its minimum value, we have

$$I(\epsilon) = \int_a^b F(x, \phi + \epsilon\eta, \phi' + \epsilon\eta')\, dx \geq I(0) = \int_a^b F(x, \phi, \phi')\, dx$$

where we have set $y(x) = \phi(x) + \epsilon\eta(x)$. ϵ is a small scalar, and $\eta(x)$ is an arbitrary function in the class of admissible functions, that is, has a continuous second derivative. Furthermore,

$$y' = \frac{dy}{dx} = \frac{d\phi}{dx} + \epsilon\frac{d\eta}{dx} = \phi' + \epsilon\eta'$$

Also, since $y(a) = \phi(a)$ and $y(b) = \phi(b)$, then $\eta(a) = \eta(b) = 0$.

$I(\epsilon)$ takes on a minimum value at $\epsilon = 0$. Therefore, $I'(0) = 0$, and we use this to derive necessary conditions for $\phi(x)$ to be the minimizing function.

$$I'(0) = \lim_{\epsilon \to 0} \frac{I(\epsilon) - I(0)}{\epsilon}$$

$$= \lim_{\epsilon \to 0} \int_a^b \left[\frac{F(x, \phi + \epsilon\eta, \phi' + \epsilon\eta') - F(x, \phi, \phi')}{\epsilon} \right] dx$$

Since F has continuous first partial derivatives, we can write

$$I'(0) = \lim_{\epsilon \to 0} \int_a^b [F_y(x, \phi + \theta_1\epsilon\eta, \phi')\eta + F_{y'}(x, \phi, \phi' + \theta_2\epsilon\eta')\eta']\, dx$$

where F_y refers to the partial derivative of F with respect to its second argument and $F_{y'}$ refers to the partial derivative of F with respect to its third argument. Here $0 < \theta_1 < 1$, and $0 < \theta_2 < 1$. The fact that the integral is a continuous function of ϵ allows us to take the limit under the integral sign, and we have

$$I'(0) = \int_a^b [F_y(x, \phi, \phi')\eta + F_{y'}(x, \phi, \phi')\eta']\, dx = 0$$

We can integrate the second term by parts, giving

$$0 = \left[F_{y'}(x, \phi, \phi')\eta \right]_a^b - \int_a^b \left[\frac{d}{dx} F_{y'}(x, \phi, \phi') - F_y(x, \phi, \phi') \right] \eta\, dx$$

Because $\eta(a) = \eta(b) = 0$, we have

$$\int_a^b \left[\frac{d}{dx} F_{y'}(x, \phi, \phi') - F_y(x, \phi, \phi') \right] \eta\, dx = 0$$

The function $(d/dx)F_{y'}(x,\phi,\phi') - F_y(x,\phi,\phi')$ is a continuous function of x. When it is multiplied by an arbitrary continuous function $\eta(x)$, which vanishes at the end points, and is integrated, the result is zero. This means that

$$\frac{d}{dx} F_{y'}(x,\phi,\phi') - F_y(x,\phi,\phi') = 0$$

in the whole interval. If this were not so, the function would be different from zero at some point. The continuity implies that it is of the same sign in some neighborhood of that point. Then η can be chosen with the same sign in all or part of this neighborhood and with the value zero elsewhere. This would imply that the integral of a positive continuous function is zero, which is clearly impossible.

The necessary condition for $y = \phi(x)$ to be the minimizing function is thus

$$\frac{d}{dx} F_{y'}(x,\phi,\phi') - F_y(x,\phi,\phi') = 0$$

which is **Euler's equation.** It is a second-order differential equation for ϕ which, along with the boundary conditions $\phi(a) = c$ and $\phi(b) = d$, generally will yield the desired solution.

As an example, let us solve problem 1. Here

$$F(x,y,y') = \left[1 + \left(\frac{dy}{dx}\right)^2\right]^{\frac{1}{2}}$$

The Euler equation yields

$$\frac{d}{dx} \frac{\phi'}{[1 + (\phi')^2]^{\frac{1}{2}}} = 0$$

Integrating this equation, we have

$$\frac{\phi'}{[1 + (\phi')^2]^{\frac{1}{2}}} = k$$

which can be solved for ϕ' as follows:

$$\phi' = \frac{k}{(1 - k^2)^{\frac{1}{2}}} = K$$

Integrating and using the boundary conditions $\phi(0) = 0$ and $\phi(a) = b$, we have

$$\phi = \frac{b}{a} x$$

which is to say that the straight line is the shortest curve connecting two points in the plane.

Problems 3 and 4 can likewise be solved, but first let us note a useful property of the Euler equation when F does not depend explicitly on the independent variable x. In this case,

$$\frac{d}{dx}(y'F_{y'} - F) = y''F_{y'} + y'\frac{d}{dx}F_{y'} - F_yy' - F_{y'}y'' = y'\left(\frac{d}{dx}F_{y'} - F_y\right) = 0$$

when $y = \phi$, the minimizing function. This implies that

$$\phi'F_{y'}(x,\phi,\phi') - F(x,\phi,\phi') = k$$

where k is a constant.

In problem 3,

$$F = y^{-\frac{1}{2}}[1 + (y')^2]^{\frac{1}{2}}$$

which does not depend explicitly on x. Therefore,

$$(\phi)'^2\{\phi[1 + (\phi')^2]\}^{-\frac{1}{2}} - \phi^{-\frac{1}{2}}[1 + (\phi')^2]^{\frac{1}{2}} = k$$

$$\frac{-1}{\{\phi[1 + (\phi')^2]\}^{\frac{1}{2}}} = k = \frac{1}{K}$$

Solving for dx, we have

$$dx = \frac{\phi^{\frac{1}{2}}\,d\phi}{(K^2 - \phi)^{\frac{1}{2}}}$$

Making the substitution $\phi = K^2 \sin^2 t$, we have

$$x = 2K^2\int_0^t \sin^2 t\,dt = K^2\left(t - \frac{\sin 2t}{2}\right)$$

$$x = \tfrac{1}{2}K^2(2t - \sin 2t)$$

$$\phi = \tfrac{1}{2}K^2(1 - \cos 2t)$$

Therefore, the solution is a cycloid, with the constant K determined by the condition that the curve pass through the point (a,b).

Problem 2 is of the following type: to minimize

$$\int_a^b F(t,x,y,\dot{x},\dot{y})\,dt$$

by choosing properly functions $x(t)$ and $y(t)$ passing through given end values $x(a) = x_1$, $y(a) = y_1$, and $x(b) = x_2$, $y(b) = y_2$. This is a special case of the following problem in n dependent variables: to minimize

$$\int_a^b F(t,x_1,x_2, \ldots ,x_n,\dot{x}_1,\dot{x}_2, \ldots ,\dot{x}_n)\,dt$$

by choosing properly functions $x_1(t)$, $x_2(t)$, $\ldots$, $x_n(t)$ passing through given end points. It is possible to find necessary conditions for the minimizing functions in this more general case. We shall do this in the case of two dependent variables, and the extension to n dependent variables will be immediate.

Let F have continuous second partial derivatives with respect to its arguments t, x, y, $\dot{x}$, and $\dot{y}$, and let $x(t)$ and $y(t)$ have continuous second derivatives. Assuming that $\phi(t)$ and $\psi(t)$ are the actual minimizing functions,

$$I(\epsilon_1, \epsilon_2) = \int_a^b F(t, \phi + \epsilon_1 \eta_1, \psi + \epsilon_2 \eta_2, \dot{\phi} + \epsilon_1 \dot{\eta}_1, \dot{\psi} + \epsilon_2 \dot{\eta}_2)\, dt$$

has a minimum at $\epsilon_1 = \epsilon_2 = 0$. Here ϵ_1 and ϵ_2 are scalars, and $\eta_1(t)$ and $\eta_2(t)$ are arbitrary functions with continuous second derivatives. Under the hypotheses, we can compute partial derivatives of $I(\epsilon_1, \epsilon_2)$ as in the simpler case of only one dependent variable. We have

$$I_{\epsilon_1}(0,0) = \int_a^b [F_x(t,\phi,\psi,\dot{\phi},\dot{\psi})\eta_1 + F_{\dot{x}}(t,\phi,\psi,\dot{\phi},\dot{\psi})\dot{\eta}_1]\, dt = 0$$

$$I_{\epsilon_2}(0,0) = \int_a^b [F_y(t,\phi,\psi,\dot{\phi},\dot{\psi})\eta_2 + F_{\dot{y}}(t,\phi,\psi,\dot{\phi},\dot{\psi})\dot{\eta}_2]\, dt = 0$$

Integrating by parts gives

$$\left[F_{\dot{x}}\eta_1 \right]_a^b - \int_a^b \left(\frac{d}{dt} F_{\dot{x}} - F_x \right)\eta_1\, dt = 0$$

$$\left[F_{\dot{y}}\eta_2 \right]_a^b - \int_a^b \left(\frac{d}{dt} F_{\dot{y}} - F_y \right)\eta_2\, dt = 0$$

Using the fact that η_1 and η_2 vanish at the end points but are otherwise arbitrary, we arrive at the pair of Euler equations

$$\frac{d}{dt} F_{\dot{x}} - F_x = 0 \qquad \frac{d}{dt} F_{\dot{y}} - F_y = 0$$

as necessary conditions for the minimizing functions $\phi(t)$ and $\psi(t)$. They represent a system of second-order differential equations for the unknown functions ϕ and ψ. In the general case with n dependent variables the necessary conditions are

$$\frac{d}{dt} F_{\dot{x}_i} - F_{x_i} = 0 \qquad i = 1, 2, \ldots, n$$

For purposes of illustration, let us solve problem 2 in the case of the cylindrical surface $z = (1 - x^2)^{\frac{1}{2}}$. The problem is to minimize

$$\int_0^1 \left[\dot{x}^2 + \dot{y}^2 + \left(\frac{\partial z}{\partial x} \dot{x} + \frac{\partial z}{\partial y} \dot{y} \right)^2 \right]^{\frac{1}{2}} dt$$

Here
$$F = [(1 - x^2)^{-1}\dot{x}^2 + \dot{y}^2]^{\frac{1}{2}}$$

$$\frac{\partial F}{\partial x} = \frac{1}{F} \frac{x\dot{x}^2}{(1 - x^2)^2} \qquad \frac{\partial F}{\partial y} = 0$$

$$\frac{\partial F}{\partial \dot{x}} = \frac{1}{F} \frac{\dot{x}}{(1 - x^2)} \qquad \frac{\partial F}{\partial \dot{y}} = \frac{1}{F} \dot{y}$$

It is well known that there are many ways of describing a curve parametrically. So far, what we have said holds for a general parameter t. We can simplify the problem by picking a particular parameter s, the length of arc along the shortest curve. With this choice of parameter,

$$F(\phi, \psi, \dot\phi, \dot\psi) = \left[(1 - \phi^2)^{-1} \left(\frac{d\phi}{ds} \right)^2 + \left(\frac{d\psi}{ds} \right)^2 \right]^{\frac{1}{2}} = 1$$

The Euler equations, written in terms of the parameter s, become

$$\frac{d}{ds} \left(\frac{\dot\phi}{1 - \phi^2} \right) - \frac{\phi \dot\phi^2}{(1 - \phi^2)^2} = \frac{\ddot\phi}{1 - \phi^2} + \frac{\phi \dot\phi^2}{(1 - \phi^2)^2} = 0$$

$$\frac{d^2}{ds^2} \psi = 0$$

where $\dot\phi = d\phi/ds$. The second equation can be integrated directly, giving $\psi = as + b$. Also, from the condition

$$\frac{\dot\phi^2}{1 - \phi^2} = 1 - \left(\frac{d\psi}{ds} \right)^2 = 1 - a^2$$

we have

$$\ddot\phi = -\phi \frac{\dot\phi^2}{1 - \phi^2} = -(1 - a^2)\phi$$

Therefore,

$$\phi = c \cos \left(\sqrt{1 - a^2}\, s + d \right)$$

If $a = 1$, $\psi = s + b$ and $\phi = k$, and we have a straight line parallel to the y axis. If $a = 0$, $\psi = b$ and $\phi = c \cos (s + d)$. However,

$$\frac{c^2 \sin^2 (s + d)}{1 - c^2 \cos^2 (s + d)} = 1$$

implies that $c = 1$. In this case we have a circle in a plane perpendicular to the axis of the cylinder. If $0 < a < 1$, we again must have $c = 1$ in order that the condition

$$\frac{(1 - a^2) c^2 \sin^2 \left(\sqrt{1 - a^2}\, s + d \right)}{1 - c^2 \cos^2 \left(\sqrt{1 - a^2}\, s + d \right)} = 1 - a^2$$

be satisfied. The parametric equations of the curve are therefore

$$\phi = \cos \left(\sqrt{1 - a^2}\, s + d \right)$$

$$\psi = as + b$$

$$\zeta = \sin \left(\sqrt{1 - a^2}\, s + d \right)$$

and we have a spiral.

An alternative approach to the solution of the last problem is a technique related to the Lagrange multiplier method of extremizing functions subject to constraints. In the present case we wish to minimize

$$I = \int_a^b F(t,x,y,z,\dot{x},\dot{y},\dot{z})\, dt$$

subject to a condition (constraint)

$$G(x,y,z) = 0$$

Assuming that there are minimizing functions $\phi(t)$, $\psi(t)$, $\zeta(t)$, then the first variation

$$dI = I_{\epsilon_1}(0,0,0)\epsilon_1 + I_{\epsilon_2}(0,0,0)\epsilon_2 + I_{\epsilon_3}(0,0,0)\epsilon_3$$

$$= \int_a^b \left[F_x(t,\phi,\psi,\zeta,\dot{\phi},\dot{\psi},\dot{\zeta}) - \frac{d}{dt} F_{\dot{x}}(t,\phi,\psi,\zeta,\dot{\phi},\dot{\psi},\dot{\zeta}) \right]\epsilon_1\eta_1(t)\, dt$$

$$+ \int_a^b \left[F_y(t,\phi,\psi,\zeta,\dot{\phi},\dot{\psi},\dot{\zeta}) - \frac{d}{dt} F_{\dot{y}}(t,\phi,\psi,\zeta,\dot{\phi},\dot{\psi},\dot{\zeta}) \right]\epsilon_2\eta_2(t)\, dt$$

$$+ \int_a^b \left[F_z(t,\phi,\psi,\zeta,\dot{\phi},\dot{\psi},\dot{\zeta}) - \frac{d}{dt} F_{\dot{z}}(t,\phi,\psi,\zeta,\dot{\phi},\dot{\psi},\dot{\zeta}) \right]\epsilon_3\eta_3(t)\, dt$$

must vanish. In this case, however, the variations $\epsilon_1\eta_1$, $\epsilon_2\eta_2$, $\epsilon_3\eta_3$ are not independent, but must satisfy the constraint

$$G(\phi + \epsilon_1\eta_1,\ \psi + \epsilon_2\eta_2,\ \zeta + \epsilon_3\eta_3) = 0$$

Therefore, $dG = G_x(\phi,\psi,\zeta)\epsilon_1\eta_1 + G_y(\phi,\psi,\zeta)\epsilon_2\eta_2 + G_z(\phi,\psi,\zeta)\epsilon_3\eta_3 = 0$

Multiplying by an arbitrary function $\lambda(t)$ and integrating, we have

$$\int_a^b (\lambda G_x\epsilon_1\eta_1 + \lambda G_y\epsilon_2\eta_2 + \lambda G_z\epsilon_3\eta_3)\, dt = 0$$

Adding this to dI,

$$\int_a^b \left(F_x - \frac{d}{dt} F_{\dot{x}} + \lambda G_x \right)\epsilon_1\eta_1\, dt + \int_a^b \left(F_y - \frac{d}{dt} F_{\dot{y}} + \lambda G_y \right)\epsilon_2\eta_2\, dt$$

$$+ \int_a^b \left(F_z - \frac{d}{dt} F_{\dot{z}} + \lambda G_z \right)\epsilon_3\eta_3\, dt = 0$$

λ is arbitrary, so we choose it to make

$$F_x - \frac{d}{dt} F_{\dot{x}} + \lambda G_x = 0$$

Two of the variations $\epsilon_2\eta_2$ and $\epsilon_3\eta_3$ are arbitrary, so that

$$F_y - \frac{d}{dt} F_{\dot{y}} + \lambda G_y = 0$$

$$F_z - \frac{d}{dt} F_{\dot{z}} + \lambda G_z = 0$$

λ can be eliminated between two of these equations, leaving two equations to be solved, say for ϕ and ψ. ζ is then determined from the condition $G(\phi,\psi,\zeta) = 0$.

To determine the shortest curves on the cylindrical surface of the above example, we must minimize

$$\int_a^b (\dot{x}^2 + \dot{y}^2 + \dot{z}^2)^{\frac{1}{2}} \, dt$$

subject to

$$G(x,y,z) = x^2 + z^2 - 1 = 0$$

Then $F = (\dot{x}^2 + \dot{y}^2 + \dot{z}^2)^{\frac{1}{2}}$, and the equations become

$$-\frac{d}{dt}\frac{\dot{\phi}}{F} + 2\lambda\phi = 0$$

$$-\frac{d}{dt}\frac{\dot{\psi}}{F} = 0$$

$$-\frac{d}{dt}\frac{\dot{\zeta}}{F} + 2\lambda\zeta = 0$$

Changing to the parameter s, the arc length on the shortest curve, we have the equations

$$\ddot{\phi} = 2\lambda\phi \qquad \ddot{\psi} = 0 \qquad \ddot{\zeta} = 2\lambda\zeta$$

where the dot now refers to d/ds. The second equation can be integrated to give

$$\psi = as + b$$

We can show in this case that λ is a constant, for

$$\phi^2 + \zeta^2 = 1$$

$$\phi\dot{\phi} + \zeta\dot{\zeta} = 0$$

$$\phi\ddot{\phi} + \zeta\ddot{\zeta} + \dot{\phi}^2 + \dot{\zeta}^2 = 0$$

$$2\lambda(\phi^2 + \zeta^2) = \dot{\psi}^2 - 1 = a^2 - 1$$

$$\lambda = \frac{a^2 - 1}{2}$$

Therefore, $\ddot{\phi} = (a^2 - 1)\phi$, $\ddot{\zeta} = (a^2 - 1)\zeta$, and we have

$$\phi = \cos\left(\sqrt{1 - a^2}\, s + d\right) \qquad \zeta = \sin\left(\sqrt{1 - a^2}\, s + d\right)$$

the same solution as before.

Still another alternative is to work the problem in cylindrical coordinates, since the calculus-of-variation technique is independent of the particular set of coordinates used to describe the surface. Let us minimize

$$\int_a^b (\dot{r}^2 + r^2\dot{\theta}^2 + \dot{y}^2)^{\frac{1}{2}} \, dt$$

subject to $r = 1$, where $x = r \cos \theta$, $y = y$, $z = r \sin \theta$.†

Because of the constraint, $\dot{r} = 0$, and hence we are to minimize

$$\int_{\theta_1}^{\theta_2} \left[1 + \left(\frac{dy}{d\theta}\right)^2 \right]^{\frac{1}{2}} \, d\theta$$

This is the same as problem 1 above, and therefore

$$\psi = A \cos^{-1} \phi + B$$

$$\phi = \cos \left(\frac{\psi - B}{A}\right)$$

which is the spiral solution obtained by the other methods.

In problems 6 and 7, we are to minimize a certain integral subject to a subsidiary condition. The problem is to minimize

$$\int_a^b F(x,y,y') \, dx$$

subject to

$$\int_a^b G(x,y,y') \, dx = k$$

k being a constant. Let F and G both have continuous second partial derivatives with respect to their arguments, and let y have a continuous second derivative.

If $\phi(x)$ is the solution of the problem, then

$$I(\epsilon_1,\epsilon_2) = \int_a^b F(x, \phi + \epsilon_1\eta_1 + \epsilon_2\eta_2, \phi' + \epsilon_1\eta_1' + \epsilon_2\eta_2') \, dx$$

has a minimum at $\epsilon_1 = \epsilon_2 = 0$, while

$$J(\epsilon_1,\epsilon_2) = \int_a^b G(x, \phi + \epsilon_1\eta_1 + \epsilon_2\eta_2, \phi' + \epsilon_1\eta_1' + \epsilon_2\eta_2') \, dx = k$$

This is equivalent to a problem in Lagrange multipliers, that is, to minimize $I(\epsilon_1,\epsilon_2)$ subject to $J(\epsilon_1,\epsilon_2) = k$. If this has a solution for $\epsilon_1 = \epsilon_2 = 0$, then

$$\left(\frac{\partial I}{\partial \epsilon_1} + \lambda \frac{\partial J}{\partial \epsilon_1}\right)_{\epsilon_1=\epsilon_2=0} = 0$$

$$\left(\frac{\partial I}{\partial \epsilon_2} + \lambda \frac{\partial J}{\partial \epsilon_2}\right)_{\epsilon_1=\epsilon_2=0} = 0$$

$$(J - k)_{\epsilon_1=\epsilon_2=0} = 0$$

† These are not the usual cylindrical coordinates because of the choice of the y axis as the axis of the cylinder.

The first of these equations gives

$$\int_a^b \left[\left(\frac{d}{dx} F_{y'} - F_y \right) + \lambda \left(\frac{d}{dx} G_{y'} - G_y \right) \right] \eta_1 \, dx = 0$$

Unless $(d/dx)G_{y'} - G_y \equiv 0$, η_1 can be chosen so that $\int_a^b \left(\frac{d}{dx} G_{y'} - G_y \right) \eta_1 \, dx \neq$ 0. This equation can then be used to define λ. Using this λ in the second equation, we have

$$\int_a^b \left[\left(\frac{d}{dx} F_{y'} - F_y \right) + \lambda \left(\frac{d}{dx} G_{y'} - G_y \right) \right] \eta_2 \, dx = 0$$

for arbitrary η_2 vanishing at the end points. Hence, our necessary condition is

$$\frac{d}{dx} \frac{\partial}{\partial y'} (F + \lambda G) - \frac{\partial}{\partial y} (F + \lambda G) = 0$$

which is the necessary condition for a minimum of the functional

$$\int_a^b (F + \lambda G) \, dx$$

without regard to the subsidiary condition. In actual practice, the parameter λ is determined from the condition

$$\int_a^b G(x, \phi, \phi') \, dx = k$$

after we have first solved the Euler equation for ϕ in terms of λ.

To illustrate the problem of the calculus of variations with a subsidiary condition, let us solve problem 7. Let

$$H = F + \lambda G = \tfrac{1}{2}(x\dot{y} - y\dot{x}) + \lambda(\dot{x}^2 + \dot{y}^2)^{\frac{1}{2}}$$

Then

$$\frac{\partial H}{\partial x} = \tfrac{1}{2}\dot{y}$$

$$\frac{\partial H}{\partial y} = -\tfrac{1}{2}\dot{x}$$

$$\frac{\partial H}{\partial \dot{x}} = -\tfrac{1}{2}y + \frac{\lambda \dot{x}}{(\dot{x}^2 + \dot{y}^2)^{\frac{1}{2}}}$$

$$\frac{\partial H}{\partial \dot{y}} = \tfrac{1}{2}x + \frac{\lambda \dot{y}}{(\dot{x}^2 + \dot{y}^2)^{\frac{1}{2}}}$$

If we now pick a special parameter s, the arc length along the solution curve, then $(\dot{\phi} + \dot{\psi})^{\frac{1}{2}} = 1$, and the Euler equations become

$$\lambda\ddot{\phi} - \dot{\psi} = 0$$

$$\lambda\ddot{\psi} + \dot{\phi} = 0$$

Integrating these, we have

$$\lambda \dot{\phi} - \psi = c_1$$

$$\lambda \dot{\psi} + \phi = c_2$$

Eliminating ψ, we have

$$\lambda^2 \ddot{\phi} + \phi - c_2 = 0$$

which has the solution

$$\phi = A \sin \frac{1}{\lambda} s + B \cos \frac{1}{\lambda} s + c_2$$

Solving for ψ, we have

$$\psi = A \cos \frac{1}{\lambda} s - B \sin \frac{1}{\lambda} s - c_1$$

Let $\phi(0) = c_2$; then $B = 0$, and $\psi(0) = A - c_1$. Finally, $\phi(0) = \phi(L)$, $\psi(0) = \psi(L)$ implies that $\sin (1/\lambda)L = 0$ and $\cos (1/\lambda)L = 1$, so that $\lambda = L/2\pi$ and the desired maximizing curve is a circle.

Before this section is concluded, a few words about notation are in order. Going back to the first class of problem treated in this section, we asked for necessary conditions for stationary values of the functional

$$\int_a^b F\left(x, y, \frac{dy}{dx}\right) dx$$

Assuming the existence of an extremizing function $\phi(x)$, we let

$$y(x) = \phi(x) + \epsilon \eta(x) = \phi(x) + \delta y(x)$$

Then

$$I(\epsilon) = \int_a^b F(x, \phi + \delta y, \phi' + \delta y') \, dx$$

has a stationary value at $\epsilon = 0$, a necessary condition for which is that the first variation

$$\delta I = I'(0)\epsilon = \int_a^b \left[F_y(x, \phi, \phi') - \frac{d}{dx} F_{y'}(x, \phi, \phi') \right] \epsilon \eta \, dx = 0$$

from which the Euler equation follows, since $\epsilon \eta$ is arbitrary. This argument is often abbreviated as follows:

$$\delta I = \delta \int_a^b F \, dx = \int_a^b \delta F \, dx = \int_a^b \left(F_y - \frac{d}{dx} F_v \right) \delta y \, dx = 0$$

This notation has become standard, so that the reader should be able to use it without losing sight of the basic arguments behind it.

Exercises 3.2

1. Show that if $F(x, y, y')$ does not depend explicitly on y, then the Euler equation has a first integral $F_{y'} = c$.

2. Solve problem 4 of the text.

3. Find the light ray passing through the two-dimensional medium between the points $(0,0)$ and (a,b) if the index of refraction is $\eta(x,y) = k(x + 1)$.

4. Find the **geodesics** (shortest curves) on the surface of a sphere.

5. Find the geodesics on the surface of a cone.

6. Solve problem 6 of the text.

7. What are the necessary conditions for an extreme value of

$$\int_a^b F\left(x, y, \frac{dx}{dx}, \frac{d^2y}{dx^2}\right) dx$$

for functions $y(x)$ with continuous fourth derivatives passing through the points (a,c) and (b,d) and having specified first derivatives at $x = a$ and $x = b$?

8. Solve problem 7 of the text by the following device. Among all simple closed curves expressed in polar coordinates $r = f(\theta)$, $f(0) = f(2\pi)$, such that $f'(\theta)$ is piecewise continuous, minimize $A = \frac{1}{2}\int_0^{2\pi} r^2\, d\theta$ subject to $L = \int_0^{2\pi} r\, d\theta = k$ (constant). HINT: Write the Fourier series for $f(\theta)$ and express A and L in terms of the Fourier coefficients.

3.3 Hamilton's Principle. Lagrange's Equations

One of the reasons the calculus of variations plays such an important role in mathematical physics is that a large number of problems in mechanics can be stated in terms of a variational principle known as **Hamilton's principle.** Consider a system of n particles with masses $m_1, m_2, \ldots, m_n$ and displacement vectors $\mathbf{X}_1, \mathbf{X}_2, \ldots, \mathbf{X}_n$ referred to some inertial (unaccelerating) coordinate system. Under the action of forces $\mathbf{F}_1, \mathbf{F}_2, \ldots, \mathbf{F}_n$, referred to the same coordinate system, Newtonian mechanics tells us that the equations of motion of the system are

$$m_\alpha \ddot{\mathbf{X}}_\alpha = \mathbf{F}_\alpha \qquad \alpha = 1, 2, 3, \ldots, n$$

Each displacement vector consists of a triple of numbers, each of which is, in general, a function of time. Hence, the configuration (the positions of all the particles) of the system can be described at any time by $3n$ coordinates. However, the system may be subject to certain constraints, which would imply that not all the $3n$ coordinates are independent. Suppose that there are k relations of the type

$$\phi_i(\mathbf{X}_1, \mathbf{X}_2, \ldots, \mathbf{X}_n, t) = 0 \qquad i = 1, 2, \ldots, k$$

which must be satisfied by the coordinates due to the constraints.[1] Then there are only $3n - k = p$ independent coordinates, and we can find a set of p

---

[1] When this is the case, the constraints are called **holonomic.** For a discussion of nonholonomic constraints, see Herbert Goldstein, "Classical Mechanics," Addison-Wesley Publishing Company, Inc., Reading, Mass., 1950.

independent **generalized coordinates** q_i, in terms of which the displacement vectors can be written

$$\mathbf{X}_i = \mathbf{X}_i(q_1, q_2, \ldots, q_p, t)$$

The forces acting on the particles consist of applied forces $\mathbf{F}_\alpha^{(a)}$ and forces of constraint $\mathbf{F}_\alpha^{(c)}$, so we may write

$$m_\alpha \ddot{\mathbf{X}}_\alpha = \mathbf{F}_\alpha^{(a)} + \mathbf{F}_\alpha^{(c)}$$

Let the actual path of the system between times t_1 and t_2 be given by the functions $\mathbf{X}_\alpha(t)$. Now consider some virtual (fictitious) path

$$\mathbf{X}_\alpha(t) + \epsilon \mathbf{Y}_\alpha(t) = \mathbf{X}_\alpha(t) + \delta \mathbf{X}_\alpha(t)$$

between t_1 and t_2 slightly different from the actual motion, but having the same configuration at the ends of the time interval; that is,

$$\delta \mathbf{X}_\alpha(t_1) = \delta \mathbf{X}_\alpha(t_2) = \mathbf{0}$$

We shall also require that the virtual displacements satisfy the constraints. We can now write

$$\sum_{\alpha=1}^{n} [m_\alpha(\ddot{\mathbf{X}}_\alpha, \delta \mathbf{X}_\alpha) - (\mathbf{F}_\alpha^{(a)}, \delta \mathbf{X}_\alpha) - (\mathbf{F}_\alpha^{(c)}, \delta \mathbf{X}_\alpha)] = 0$$

If we restrict ourselves to systems in which the forces of constraint do no work,[1] the last term in this expression vanishes, and we have

$$\sum_{\alpha=1}^{n} [m_\alpha(\ddot{\mathbf{X}}_\alpha, \delta \mathbf{X}_\alpha) - (\mathbf{F}_\alpha^{(a)}, \delta \mathbf{X}_\alpha)] = 0$$

for any time t. This is known as **D'Alembert's principle.** Integrating between t_1 and t_2, we have

$$\int_{t_1}^{t_2} \sum_{\alpha=1}^{n} [m_\alpha(\ddot{\mathbf{X}}_\alpha, \delta \mathbf{X}_\alpha) - (\mathbf{F}_\alpha^{(a)}, \delta \mathbf{X}_\alpha)] \, dt = 0$$

Integrating the first term by parts,

$$\left[\sum_{\alpha=1}^{n} m_\alpha(\dot{\mathbf{X}}_\alpha, \delta \mathbf{X}_\alpha) \right]_{t_1}^{t_2} - \int_{t_1}^{t_2} \sum_{\alpha=1}^{n} [m_\alpha(\dot{\mathbf{X}}_\alpha, \delta \dot{\mathbf{X}}_\alpha) + (\mathbf{F}_\alpha^{(a)}, \delta \mathbf{X}_\alpha)] \, dt = 0$$

Making use of the fact that the virtual displacements vanish at t_1 and t_2, we have

$$\int_{t_1}^{t_2} \sum_{\alpha=1}^{n} [m_\alpha(\dot{\mathbf{X}}_\alpha, \delta \dot{\mathbf{X}}_\alpha) + (\mathbf{F}_\alpha^{(a)}, \delta \mathbf{X}_\alpha)] \, dt = 0$$

[1] For example, the normal reaction of a plane on a particle constrained to move on a plane surface, the tension in an inextensible pendulum cord, the internal forces in a rigid body, etc.

We define the kinetic energy T as follows:

$$T = \tfrac{1}{2} \sum_{\alpha=1}^{n} m_\alpha (\dot{\mathbf{X}}_\alpha, \dot{\mathbf{X}}_\alpha)$$

Then the first-order change in the kinetic energy induced by the virtual displacements is

$$\delta T = \sum_{\alpha=1}^{n} m_\alpha (\dot{\mathbf{X}}_\alpha, \delta \dot{\mathbf{X}}_\alpha)$$

Hence we have
$$\int_{t_1}^{t_2} \left[\delta T + \sum_{\alpha=1}^{n} (\mathbf{F}_\alpha^{(a)}, \delta \mathbf{X}_\alpha) \right] dt = 0$$

If the applied forces are derivable from a **potential function** V, that is, if the force field is **conservative**, then[1]

$$\mathbf{F}_\alpha^{(a)} = -\left(\frac{\partial V}{\partial x_{\alpha 1}}, \frac{\partial V}{\partial x_{\alpha 2}}, \frac{\partial V}{\partial x_{\alpha 3}} \right)$$

and
$$\sum_{\alpha=1}^{n} (\mathbf{F}_\alpha^{(a)}, \delta \mathbf{X}_\alpha) = -\delta V$$

Then
$$\int_{t_1}^{t_2} \delta(T - V)\, dt = \delta \int_{t_1}^{t_2} (T - V)\, dt = 0$$

This is **Hamilton's principle** for a conservative system. It may be stated as follows: *The motion of a conservative system between time t_1 and t_2 proceeds in such a way that* $\int_{t_1}^{t_2} (T - V)\, dt$ *is stationary with respect to arbitrary small changes in the motion consistent with the constraints and vanishing at t_1 and t_2.*

From Hamilton's principle we can derive **Lagrange's equations.** We first define the **Lagrangian function**

$$L = T - V$$

which can be defined in terms of the generalized coordinates $q_1, q_2, \ldots, q_p$.

We have
$$\dot{\mathbf{X}}_\alpha = \frac{\partial \mathbf{X}_\alpha}{\partial q_j} \dot{q}_j + \frac{\partial \mathbf{X}_\alpha}{\partial t} \qquad \alpha = 1, 2, \ldots, n$$

$$j = 1, 2, \ldots, p$$

Hence
$$T = \tfrac{1}{2} \sum_{\alpha=1}^{n} m_\alpha (\dot{\mathbf{X}}_\alpha, \dot{\mathbf{X}}_\alpha) = T(q_1, q_2, \ldots, q_p, \dot{q}_1, \dot{q}_2, \ldots, \dot{q}_p, t)$$

Likewise,
$$V = V(\mathbf{X}_1, \mathbf{X}_2, \ldots, \mathbf{X}_n) = V(q_1, q_2, \ldots, q_p, t)$$

---

[1] Here $x_{\alpha 1}$, $x_{\alpha 2}$, $x_{\alpha 3}$ are the three rectangular cartesian coordinates of the αth particle.

so that L is a function of the generalized coordinates q_i, the **generalized velocities** $\dot{q}_i$, and time.   Using the notation of the calculus of variations, we have

$$\delta \int_{t_1}^{t_2} L\, dt = \int_{t_1}^{t_2} \left(\frac{\partial L}{\partial q_i} - \frac{d}{dt} \frac{\partial L}{\partial \dot{q}_i} \right) \delta q_i\, dt = 0$$

Since the δq_i are independent, we have

$$\frac{d}{dt} \frac{\partial L}{\partial \dot{q}_i} - \frac{\partial L}{\partial q_i} = 0 \qquad i = 1, 2, \ldots, p$$

These are the **Lagrange equations** for a conservative system.

As an example of the use of Lagrange's equations, consider the small vibrations of the following system of coupled pendulums, assuming the pendulum rods weightless:

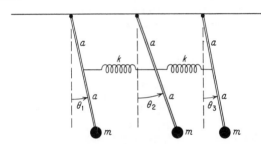

FIGURE 4

The generalized coordinates can be taken as θ_1, θ_2, θ_3.   The kinetic and potential energies are then

$$T = 2ma^2(\dot{\theta}_1^2 + \dot{\theta}_2^2 + \dot{\theta}_3^2)$$
$$V = \tfrac{1}{2}ka^2(\sin\theta_1 - \sin\theta_2)^2 + \tfrac{1}{2}ka^2(\sin\theta_2 - \sin\theta_3)^2$$
$$\quad + 2mga[(1 - \cos\theta_1) + (1 - \cos\theta_2) + (1 - \cos\theta_3)]$$

For small vibrations we can make the approximations $\sin\theta \approx \theta$ and

$$1 - \cos\theta = 2\sin^2\frac{\theta}{2} \approx \frac{\theta^2}{2}$$

Then

$$V = \tfrac{1}{2}ka^2[(\theta_1 - \theta_2)^2 + (\theta_2 - \theta_3)^2] + mga(\theta_1^2 + \theta_2^2 + \theta_3^2)$$

and the equations of motion are

$$4ma^2\ddot{\theta}_1 = -ka^2(\theta_1 - \theta_2) - 2mga\theta_1$$
$$4ma^2\ddot{\theta}_2 = -ka^2(2\theta_2 - \theta_1 - \theta_3) - 2mga\theta_2$$
$$4ma^2\ddot{\theta}_3 = -ka^2(\theta_3 - \theta_2) - 2mga\theta_3$$

The solution can then be found by the methods developed in Sec. 1.9. We shall return to the general theory of small vibrations in the next section.

It is possible to derive Lagrange equations even if the field is not conservative. In the absence of a potential function V, we can write

$$\sum_{\alpha=1}^{n} (\mathbf{F}_{\alpha}^{(a)}, \delta\mathbf{X}_{\alpha}) = Q_k \delta q_k \qquad k = 1, 2, \ldots, p$$

with

$$\delta\mathbf{X}_{\alpha} = \frac{\partial\mathbf{X}_{\alpha}}{\partial q_k} \delta q_k$$

so that

$$Q_k = \sum_{\alpha=1}^{n} \left(\mathbf{F}_{\alpha}^{(a)}, \frac{\partial\mathbf{X}_{\alpha}}{\partial q_k} \right)$$

We call Q_k the **generalized forces.** They do not necessarily have the dimensions of a force, since the δq_k do not necessarily have dimensions of length. However, $Q_k \delta q_k$ must have the dimensions of work.

Introducing the generalized forces, we have

$$\int_{t_1}^{t_2} (\delta T + Q_k \delta q_k)\, dt = 0$$

or

$$\int_{t_1}^{t_2} \left(\frac{\partial T}{\partial q_k} - \frac{d}{dt}\frac{\partial T}{\partial \dot{q}_k} + Q_k \right) \delta q_k\, dt = 0$$

Thus

$$\frac{d}{dt}\frac{\partial T}{\partial \dot{q}_k} - \frac{\partial T}{\partial q_k} = Q_k$$

are the equations of motion when the field is not conservative.

Consider next a conservative system in which the constraints are not functions of time, and hence

$$\mathbf{X}_i = \mathbf{X}_i(q_1, q_2, \ldots, q_p)$$

are not explicitly functions of time. In this case, V is not explicitly a function of time, nor is it a function of the generalized velocities $\dot{q}_i$. Then

$$\frac{dL}{dt} = \frac{\partial L}{\partial q_i}\dot{q}_i + \frac{\partial L}{\partial \dot{q}_i}\ddot{q}_i$$

From Lagrange's equations we have

$$\frac{\partial L}{\partial q_i} = \frac{d}{dt}\frac{\partial L}{\partial \dot{q}_i}$$

so that

$$\frac{dL}{dt} = \left(\frac{d}{dt}\frac{\partial L}{\partial \dot{q}_i} \right)\dot{q}_i + \frac{\partial L}{\partial \dot{q}_i}\ddot{q}_i = \frac{d}{dt}\left(\dot{q}_i \frac{\partial L}{\partial \dot{q}_i} \right)$$

Hence we have

$$\frac{d}{dt}\left(\dot{q}_i \frac{\partial L}{\partial \dot{q}_i} - L \right) = 0$$

which implies that
$$\dot{q}_i \frac{\partial L}{\partial \dot{q}_i} - L = H$$

where H is a constant. This statement can be put in a simpler form, for under the given hypotheses,

$$\frac{\partial L}{\partial \dot{q}_i} = \frac{\partial T}{\partial \dot{q}_i}$$

and we have
$$H = \dot{q}_i \frac{\partial T}{\partial \dot{q}_i} - T + V$$

But
$$T = \tfrac{1}{2} \sum_{\alpha=1}^{n} m_\alpha (\dot{\mathbf{X}}_\alpha, \dot{\mathbf{X}}_\alpha)$$

$$= \tfrac{1}{2} \sum_{\alpha=1}^{n} m_\alpha \left(\frac{\partial \mathbf{X}_\alpha}{\partial q_j}, \frac{\partial \mathbf{X}_\alpha}{\partial q_k} \right) \dot{q}_j \dot{q}_k$$

$$= \tfrac{1}{2} a_{jk}(q_1, q_2, \ldots, q_p) \dot{q}_j \dot{q}_k$$

and
$$\dot{q}_i \frac{\partial T}{\partial \dot{q}_i} = a_{ik} \dot{q}_i \dot{q}_k = 2T$$

Hence, we have the **conservation-of-energy principle**

$$T + V = H$$

a constant.

There are many reasons why the formulation of mechanics as a variational principle is extremely important. First, it gives us a formulation which is independent of the particular coordinate system in terms of which the motion is described. In other words, if we change the generalized coordinates to a new set by a coordinate transformation

$$\bar{q}_i = \bar{q}_i(q_1, q_2, \ldots, q_p) \qquad i = 1, 2, \ldots, p$$

then Hamilton's principle

$$\delta \int (T - V) \, dt = 0$$

will yield the equations of motion

$$\frac{d}{dt} \frac{\partial L}{\partial \dot{\bar{q}}_i} - \frac{\partial L}{\partial \bar{q}_i} = 0$$

and the form of the equations is the same, as it should be if the theory is to be independent of the particular mathematical description.

Second, the variational calculus approach is one which is more easily generalized to situations other than the equations of motion as such. Indeed new theories have been developed in areas unrelated to mechanics by means of variational principles.   For example, various field theories can be based on variational principles.

Third, it gives us numerical techniques for approximating certain physical constants by means of establishing upper and lower bounds, which are easily computed from the functionals involved in the variational problem.   Some of these methods will be illustrated in Sec. 3.9.

Exercises 3.3

1. Rework the example of this section, taking into account the mass of the rods and assuming the linear density of the rods uniform.   The kinetic energy of the rods can be expressed as the rotational kinetic energy $\frac{1}{2}I\dot{\theta}^2$, where I is the moment of inertia about the point of rotation.

2. Assume a change of generalized coordinates $\bar{q}_i = \bar{q}_i(q_1, q_2, \dots, q_p)$ and show the following:

a. $\bar{q}_i = \dfrac{\partial \bar{q}_i}{\partial q_j}\dot{q}_j$ and $\dot{q}_i = \dfrac{\partial q_i}{\partial \bar{q}_j}\dot{\bar{q}}_j$

b. $\dfrac{\partial \dot{q}_i}{\partial \bar{q}_j} = \dfrac{\partial^2 q_i}{\partial \bar{q}_j \partial \bar{q}_k}\dot{\bar{q}}_k$

Hence, show directly that the form of Lagrange's equations is independent of the coordinate system.

3. The notion of generalized coordinates can be used to derive equations of motion in various coordinate systems.   In spherical coordinates

$$T = \tfrac{1}{2}\sum_{\alpha=1}^{n} m_\alpha(\dot{r}_\alpha^2 + r_\alpha^2 \sin^2 \phi_\alpha \dot{\theta}_\alpha^2 + r_\alpha^2 \dot{\phi}_\alpha^2)$$

Use this to derive equations of motion for a system of particles referred to spherical coordinates.

4. Suppose that additional conditions of constraint

$$\phi_i(q_1, q_2, \dots, q_p) = 0 \qquad i = 1, 2, \dots, k < p$$

are imposed on a system.   Derive the equations of motion

$$\frac{d}{dt}\frac{\partial T}{\partial \dot{q}_j} - \frac{\partial T}{\partial q_j} - Q_j = \lambda_i \frac{\partial \phi_i}{\partial q_j} \qquad \begin{matrix} i = 1, 2, \dots, k \\ j = 1, 2, \dots, p \end{matrix}$$

5. Define the **generalized momentum** $p_i = \partial T/\partial \dot{q}_i$ and the **Hamiltonian** for a conservative system $H = p_i\dot{q}_i - L$.   Show that

a. $\dfrac{\partial H}{\partial \dot{q}_i} = 0$

b. $\dfrac{\partial H}{\partial p_i} = \dot{q}_i$

c. $\dfrac{\partial H}{\partial q_i} = -\dot{p}_i$

d. $\dfrac{\partial H}{\partial t} = -\dfrac{\partial L}{\partial t}$

On the basis of these equations give an alternative proof of conservation of energy when the constraints do not depend explicitly on time.

3.4 Theory of Small Vibrations

We next turn our attention to the general theory of small vibrations of conservative systems near a position of stable equilibrium.[1] Consider a conservative system in which the constraints are not functions of time and hence the potential energy is not explicitly a function of time. Suppose the system is in a configuration of stable equilibrium. Equilibrium occurs when the forces on all the particles are zero. Hence, in a small virtual displacement at equilibrium

$$\delta W = \sum_{\alpha=1}^{n} (\mathbf{F}_\alpha, \delta \mathbf{X}_\alpha) = 0$$

$$= \sum_{\alpha=1}^{n} (\mathbf{F}_\alpha^{(a)} + \mathbf{F}_\alpha^{(c)}, \delta \mathbf{X}_\alpha) = 0$$

$$= \sum_{\alpha=1}^{n} (\mathbf{F}_\alpha^{(a)}, \delta \mathbf{X}_\alpha) = 0$$

since the forces of constraint are assumed to do no work. In terms of generalized forces Q_k and generalized coordinates q_k

$$\delta W = Q_k \delta q_k = 0$$

for arbitrary δq_k at equilibrium. Hence the condition for equilibrium can be stated as

$$Q_k = -\frac{\partial V}{\partial q_k} = 0$$

Therefore, V is stationary at equilibrium. That it is a relative minimum follows from the fact that we are assuming stable equilibrium. If not, a small displacement would decrease V and hence increase T, since energy $H = T + V$ is conserved. This would imply that a small displacement would cause the system to run away, contrary to assumption.

If we choose the generalized coordinates so that they vanish in the equilibrium position and choose V so that it vanishes at equilibrium, then

$$V(q_1, q_2, \ldots, q_p) = \frac{1}{2} \frac{\partial^2 V}{\partial q_i \, \partial q_j} q_i q_j$$

$$= \tfrac{1}{2} a_{ij} q_i q_j$$

neglecting higher-order terms in the small displacements from equilibrium.

---

[1] This means that, if the system is slightly displaced, it will tend to return to the equilibrium state rather than run away.

The a_{ij} are constants, since the second partial derivatives are evaluated at the equilibrium position $q_1 = q_2 = \cdots = q_p = 0$. The square matrix A with elements a_{ij} is real and symmetric. The potential energy is, therefore, a quadratic form in the displacements.

Since the constraints do not depend on time, the kinetic energy is

$$T = \tfrac{1}{2}b_{ij}\dot{q}_i\dot{q}_j$$

where

$$b_{ij} = \sum_{\alpha=1}^{n} m_\alpha \left(\frac{\partial \mathbf{X}_\alpha}{\partial q_i}, \frac{\partial \mathbf{X}_\alpha}{\partial q_j} \right)$$

In the case of small vibrations about equilibrium, we can evaluate b_{ij} at the equilibrium position. Hence, the b_{ij} are constants and the matrix B with elements b_{ij} is real and symmetric. From the definition of kinetic energy we see that it is positive unless all the velocities are zero. Therefore, B has only positive eigenvalues.

The Lagrangian is equal to

$$L = T - V = \tfrac{1}{2}b_{ij}\dot{q}_i\dot{q}_j - \tfrac{1}{2}a_{ij}q_iq_j$$

and the Lagrange equations yield

$$b_{ij}\ddot{q}_j + a_{ij}q_j = 0$$

Suppose we attempt to separate out the time-dependent part of the solution; that is, we look for a solution of the form

$$q_j = y_j f(t)$$

Then the equations of motion become

$$b_{ij}y_j\ddot{f} + a_{ij}y_j f = 0$$

For each i we can then write

$$\frac{\ddot{f}}{f} = -\frac{a_{ij}y_j}{b_{ik}y_k} = -\lambda$$

from which it follows that

$$\ddot{f}(t) + \lambda f(t) = 0$$
$$a_{ij}y_j = \lambda b_{ij}y_j$$

We are thus led to the eigenvector problem

$$AY = \lambda BY$$

with characteristic equation

$$|A - \lambda B| = 0$$

which has p real solutions. There are thus p separation constants λ_α corresponding to p **normal modes** of vibration of the system

$$y_{\alpha j}f_\alpha(t) = y_{\alpha j}\sin\left(\sqrt{\lambda_\alpha}\,t + \phi_\alpha\right) \qquad \alpha = 1, 2, \ldots, p$$

The general solution is a linear combination of these normal modes:

$$q_j = \sum_{\alpha=1}^{p} c_\alpha y_{\alpha j} \sin (\sqrt{\lambda_\alpha}\, t + \phi_\alpha)$$

for

$$\ddot{q}_j = -\sum_{\alpha=1}^{p} \lambda_\alpha c_\alpha y_{\alpha j} \sin (\sqrt{\lambda_\alpha}\, t + \phi_\alpha)$$

and

$$b_{kj}\ddot{q}_j = -\sum_{\alpha=1}^{p} \lambda_\alpha c_\alpha b_{kj} y_{\alpha j} \sin (\sqrt{\lambda_\alpha}\, t + \phi_\alpha)$$

$$= -\sum_{\alpha=1}^{p} c_\alpha a_{kj} y_{\alpha j} \sin (\sqrt{\lambda_\alpha}\, t + \phi_\alpha)$$

$$= -a_{kj}q_j$$

Alternatively, we could have performed the coordinate transformation

$$q_i = s_{ij}\bar{q}_j$$
$$\dot{q}_i = s_{ij}\dot{\bar{q}}_j$$

which will simultaneously reduce the potential and kinetic energy quadratic forms to the diagonal form

$$T = \tfrac{1}{2}\delta_{ij}\dot{\bar{q}}_i\dot{\bar{q}}_j$$
$$V = \tfrac{1}{2}d_{ij}\bar{q}_i\bar{q}_j$$

Then the equations of motion give

$$\ddot{\bar{q}}_\alpha + \lambda_\alpha \bar{q}_\alpha = 0$$

The $\bar{q}$'s are called the **normal coordinates** for the system. We solve for them as follows:

$$\bar{q}_\alpha = c_\alpha \sin (\sqrt{\lambda_\alpha}\, t + \phi_\alpha)$$

That the λ's are positive follows from the fact that V is positive-definite.

A third approach is to obtain the normal modes as a solution of a problem in the calculus of variations. The normal modes are characterized by the fact that they are solutions with periodic time dependence. Let

$$q_i = y_i \sin (\sqrt{\lambda}\, t + \phi)$$

Then

$$2T = \lambda b_{ij} y_i y_j \cos^2 (\sqrt{\lambda}\, t + \phi)$$
$$2V = a_{ij} y_i y_j \sin^2 (\sqrt{\lambda}\, t + \phi)$$

If y_i are the components of one of the eigenvectors of the problem

$$A Y = \lambda B Y$$

then

$$2V = \lambda b_{ij} y_i y_j \sin^2 (\sqrt{\lambda}\, t + \phi)$$

and $\bar{V} = \bar{T}$, where $\bar{V}$ and $\bar{T}$ are the mean potential and kinetic energies over one period τ in periodic motion.

$$2\bar{V} = \frac{1}{\tau} \int_0^\tau a_{ij} y_i y_j \sin^2 (\sqrt{\lambda}\, t + \phi)\, dt$$

$$2\bar{T} = \frac{1}{\tau} \int_0^\tau \lambda b_{ij} y_i y_j \cos^2 (\sqrt{\lambda}\, t + \phi)\, dt$$

Then

$$\frac{a_{ij} y_i y_j}{b_{mn} y_m y_n} = \frac{\bar{V}}{\bar{U}} = \lambda$$

where $\bar{T} = \lambda \bar{U}$.   For a normal mode with a natural frequency $\sqrt{\lambda}/2\pi$

$$\bar{V} = \lambda \bar{U}$$

and furthermore, λ is a stationary value of $\bar{V}/\bar{U}$ or, alternatively, a stationary value of $\bar{V}$ subject to $\bar{U} = 1$.   These properties have already been developed in Sec. 3.1.   Thus, the eigenvalues and eigenvectors can be defined by any one of three problems in the calculus of variations:

1. Minimize $\bar{V}/\bar{U} = a_{ij} x_i x_j / b_{mn} x_m x_n$ subject to $X'BY_j = 0, j = 1, 2, \ldots, k - 1$.

2. Minimize $\bar{V} = a_{ij} x_i x_j$ subject to $\bar{U} = b_{ij} x_i x_j = 1$ and $X'BY_j = 0$, $j = 1, 2, \ldots, k - 1$.

3. Find the minimum of $\bar{V}/\bar{U}$ subject to $X'BV_j = 0, j = 1, 2, \ldots, k - 1$, where V_j are any set of linearly independent vectors.   Then find the maximum of the minima as the V's are allowed to change.

Exercises 3.4

1. Find the natural frequencies and normal modes of small vibrations about equilibrium of the double pendulum with identical masses and rods, assuming plane motion.

2. Find the natural frequencies and normal modes of small transverse vibrations of a system consisting of two identical rods hinged together and supported by three identical springs, one at the hinge and the others at the ends of the rods. Assume plane motion.   HINT: The kinetic energy of a rod is the sum of the translational energy of the center of mass and the rotational energy about the center of mass.

3. A simple pendulum is suspended from a point and is free to move so that the mass remains on the surface of a sphere with a radius equal to the length of the pendulum rod.   Analyze small vibrations about equilibrium.

4. A rod of length l and mass m is suspended at one end by a string of length l. Analyze small vibrations about equilibrium in a plane.

3.5   The Vibrating String

Most of the above ideas for systems of particles can be carried over to the case of continuous distributions of mass.   Consider, for example, a string of uniform density ρ stretched between two fixed points $x = 0$ and $x = L$, with a tension σ. The density and tension will be assumed constant, and we shall ignore the weight of the string. The kinetic energy per unit length is $\frac{1}{2}\rho\dot{y}^2$, where y is the displacement from equilibrium.[1] Hence, the total kinetic energy is

$$T = \tfrac{1}{2}\rho \int_0^L \dot{y}^2 \, dx$$

The potential energy can be computed as the negative of the work done in the displacement. Consider an element of string subject to tension σ on the ends.

FIGURE 5

The force in the direction of the displacement is

$$F = \sigma(\sin \theta_2 + \sin \theta_1) \approx \sigma(\tan \theta_2 + \tan \theta_1)$$

Therefore,
$$F \approx \sigma[-y'(x + \Delta x) + y'(x)]$$

$$\approx -\sigma y''(x) \, \Delta x = -\sigma p \frac{\partial p}{\partial y} \Delta x$$

where $p = y'$. The work done per unit length is then

$$-\sigma \int_0^y p \frac{\partial p}{\partial y} \, dy = -\sigma \frac{p^2}{2} = \frac{-\sigma(y')^2}{2}$$

The total potential energy is thus

$$V = \tfrac{1}{2} \int_0^L \sigma(y')^2 \, dx$$

The equilibrium position is a stationary value of V. Therefore,

$$\delta V = \tfrac{1}{2}\delta \int_0^L \sigma(y')^2 \, dx = 0$$

will tell us what the equilibrium configuration is. The Euler equation for this problem is $y'' = 0$, which implies that $y = 0$ for equilibrium, assuming that the ends are on the x axis.

---

[1] y is a function of x and t. We are using the notation $\dot{y} = \partial y / \partial t$ and $y' = \partial y / \partial x$.

Hamilton's principle for this problem states the following:

$$\delta \int_{t_1}^{t_2} \int_0^L (\tfrac{1}{2}\rho\dot{y}^2 - \tfrac{1}{2}\sigma y'^2)\, dx\, dt = 0$$

Let $y(x,t) = \phi(x,t) + \epsilon\eta(x,t)$, where ϕ is assumed to be the solution of the variational problem. The boundary conditions are

$$y(0,t) = \phi(0,t) = y(L,t) = \phi(L,t) = 0$$

Hence, $\eta(0,t) = \eta(L,t) = 0$. Also we must have $\eta(x,t_1) = \eta(x,t_2) = 0$. We know that

$$I(\epsilon) = \int_{t_1}^{t_2} \int_0^L [\tfrac{1}{2}\rho(\dot{\phi} + \epsilon\dot{\eta})^2 - \tfrac{1}{2}\sigma(\phi' + \epsilon\eta')^2]\, dx\, dt$$

has a stationary value at $\epsilon = 0$. Therefore,

$$I'(0) = \int_{t_1}^{t_2} \int_0^L [\rho\dot{\phi}\dot{\eta} - \sigma\phi'\eta']\, dx\, dt = 0$$

Integrating by parts and using the boundary conditions, we have

$$\int_{t_1}^{t_2} \int_0^L (\rho\ddot{\phi} - \sigma\phi'')\eta\, dx\, dt = 0$$

Since η is arbitrary, the Euler equation becomes the following partial differential equation:

$$\frac{\partial^2 \phi}{\partial x^2} = \frac{\rho}{\sigma}\frac{\partial^2 \phi}{\partial t^2}$$

This is the **wave equation** which must be satisfied by the displacement function on the string. To determine the solution, we must also have boundary conditions and initial conditions. In this case the boundary conditions are $\phi(0,t) = \phi(L,t) = 0$. The initial conditions are $\phi(x,0) = g(x)$ and $\dot{\phi}(x,0) = h(x)$, where $g(x)$ and $h(x)$ are some prescribed functions which give the initial displacements and initial velocities along the string.

We can solve the string problem by techniques analogous to those used in the case of systems of particles. We attempt to find fundamental solutions in the form of simple periodic functions by performing a separation of variables. Let $\phi(x,t) = \psi(x)f(t)$. Then

$$f\psi'' = \frac{\rho}{\sigma}\psi\ddot{f}$$

so that

$$\frac{\psi''}{\psi} = \frac{\rho}{\sigma}\frac{\ddot{f}}{f} = -\lambda$$

The time-dependent part $f(t)$ must then satisfy the equation

$$\ddot{f} + \frac{\sigma\lambda}{\rho}f = 0$$

and the space-dependent part $\psi(x)$ must satisfy the equation

$$\psi'' + \lambda\psi = 0$$

If $\lambda = 0$, then $\psi = c_1 x + c_2$, and the boundary conditions would imply that c_1 and c_2 are both zero. If λ were negative, then

$$\psi = c_1 \sinh \sqrt{-\lambda}\,x + c_2 \cosh \sqrt{-\lambda}\,x$$

and again the boundary conditions would imply that $c_1 = c_2 = 0$. For λ positive

$$\psi(x) = c \sin(\sqrt{\lambda}\,x + \theta) = A \sin\sqrt{\lambda}\,x + B \cos\sqrt{\lambda}\,x$$

$\psi(0) = 0$ implies $B = 0$, and $\psi(L) = 0$ implies

$$\lambda = \frac{k^2\pi^2}{L^2}$$

There are, therefore, an infinite number of discrete separation constants possible with $k = 1, 2, 3, \ldots$.

The time-dependent part of the fundamental solution is

$$f_k = \sin(\omega_k t + \theta_k)$$

with $\omega_k = (k\pi/L)\sqrt{\sigma/\rho}$. These are the **natural frequencies** corresponding to the **normal modes**

$$\phi_k = c_k \sin\frac{k\pi}{L}x \sin(\omega_k t + \theta_k)$$

Any linear combination of normal modes will satisfy the partial differential equation and the boundary conditions. We must find out which of all possible such functions satisfies the initial conditions. To this end, we use the theory of Fourier series developed in Sec. 2.2. Consider the series

$$\phi(x,t) = \sum_{k=1}^{\infty} \sin\frac{k\pi x}{L}(a_k \cos\omega_k t + b_k \sin\omega_k t)$$

If this represents the solution, then

$$\phi(x,0) = g(x) = \sum_{k=1}^{\infty} a_k \sin\frac{k\pi x}{L}$$

If this is the Fourier sine series for $g(x)$, then

$$a_k = \frac{2}{L} \int_0^L g(x) \sin \frac{k\pi x}{L} \, dx$$

Also, we must have

$$\dot{\phi}(x,0) = h(x) = \sum_{k=1}^{\infty} \omega_k b_k \sin \frac{k\pi x}{L}$$

and

$$b_k = \frac{2}{\omega_k L} \int_0^L h(x) \sin \frac{k\pi x}{L} \, dx$$

These series will converge to the given functions $g(x)$ and $h(x)$ if both are continuous and have a piecewise continuous derivative.

We have thus obtained an infinite series which converges to the given initial conditions. To show that this is the solution to our problem, we proceed as follows:

$$\phi(x,t) = \sum_{k=1}^{\infty} \frac{a_k}{2} \left[\sin \frac{k\pi}{L} (x - ct) + \sin \frac{k\pi}{L} (x + ct) \right]$$

$$+ \sum_{k=1}^{\infty} \frac{b_k}{2} \left[\cos \frac{k\pi}{L} (x - ct) - \cos \frac{k\pi}{L} (x + ct) \right]$$

where $c = \sqrt{\sigma/\rho}$. Since a_k is the Fourier coefficient of $g(x)$, the first sum is obviously $\frac{1}{2}[g(x - ct) + g(x + ct)]$, where we must extend the definition of $g(x)$ to the interval $-L \leq x \leq 0$ as an odd function and from there to any other interval as a periodic function with period $2L$. The other series can also be summed in closed form. To see this, extend the definition of $h(x)$ first to the interval $-L \leq x \leq 0$ as an odd function and then to any other interval as a periodic function with period $2L$. Define

$$H(x) = \int_0^x h(\xi) \, d\xi$$

Then $H(x)$ is continuous, even, and periodic with period $2L$ and has a continuous derivative and a piecewise continuous second derivative. Now

$$H(x + ct) - H(x - ct) = \int_{x-ct}^{x+ct} h(\xi) \, d\xi$$

$$= \int_{x-ct}^{x+ct} \sum_{k=1}^{\infty} \omega_k b_k \sin \frac{k\pi \xi}{L} \, d\xi$$

$$= c \sum_{k=1}^{\infty} b_k \left[\cos \frac{k\pi}{L} (x - ct) - \cos \frac{k\pi}{L} (x + ct) \right]$$

Hence, $\phi(x,t) = \frac{1}{2}[g(x - ct) + g(x + ct)] + \dfrac{1}{2c} \displaystyle\int_{x-ct}^{x+ct} h(\xi)\, d\xi$

It can be shown by direct substitution that this is a solution of the problem if $g(x)$ is twice differentiable and $h(x)$ is once differentiable.

We can analyze the natural frequencies and normal modes of the string problem, using the calculus of variations.

For a fundamental solution ϕ_k the average potential and kinetic energies over one period τ_k are equal; that is,

$$\bar{T} = \frac{1}{\tau_k} \int_0^{\tau_k} \int_0^L \frac{\omega_k^2}{2} \rho \sin^2 \frac{k\pi x}{L} \cos^2 (\omega_k t + \theta_k)\, dx\, dt$$

$$= \frac{\omega_k^2 \rho L}{2}$$

$$\bar{V} = \frac{1}{\tau_k} \int_0^{\tau_k} \int_0^L \left(\frac{k\pi}{L}\right)^2 \frac{\sigma}{2} \cos^2 \frac{k\pi x}{L} \sin^2 (\omega_k t + \theta_k)\, dx\, dt$$

$$= \frac{k^2 \pi^2}{L^2} \frac{\sigma L}{8} = \frac{\omega_k^2 \rho L}{8}$$

Integrating out the time-dependent part, we have

$$\bar{T} = \frac{\omega_k^2 \rho}{4} \int_0^L \psi_k^2\, dx$$

$$\bar{V} = \frac{\sigma}{4} \int_0^L \psi_k'^2\, dx$$

so that

$$\frac{\sigma}{\rho \omega_k^2} \frac{\displaystyle\int_0^L \psi_k'^2\, dx}{\displaystyle\int_0^L \psi_k^2\, dx} = 1$$

or

$$\frac{Q(\psi_k)}{N(\psi_k)} = \omega_k^2 \frac{\rho}{\sigma} = \frac{k^2 \pi^2}{L^2} = \lambda_k$$

where $Q(\psi_k) = \displaystyle\int_0^L \psi_k'^2\, dx$ and $N(\psi_k) = \displaystyle\int_0^L \psi_k^2\, dx$

We are led to speculate that the eigenvalues λ_k are stationary values of $Q(\psi)/N(\psi)$ or, alternatively, stationary values of $Q(\psi)$ subject to $N(\psi) = 1$. This is indeed the case, as we shall now show.

Let F be the function space of real-valued functions defined on the interval $0 \leq x \leq L$, which are continuous and have piecewise continuous first and second derivatives[1] and vanish at $x = 0$ and $x = L$. Then $\lambda_k = k^2\pi^2/L^2$, $k = 1, 2, 3, \ldots$ are stationary values of $Q(f)$ over F subject to $N(f) = 1$ and $N(f,\psi_j) = 0, j = 1, 2, 3, \ldots, k - 1$, where

$$\psi_k = \left(\frac{2}{L}\right)^{\frac{1}{2}} \sin \frac{k\pi x}{L}$$

are the eigenfunctions of the variational problem. We prove this as follows.

We first extend the definition of every function in F into the interval $-L \leq x \leq 0$ as an odd function. Then, since the trigonometric functions $\sin(k\pi x/L)$ are complete with respect to this set of odd functions,

$$\frac{1}{L} \int_{-L}^{L} [f(x)]^2 \, dx = \sum_{k=1}^{\infty} b_k^2$$

where

$$b_k = \frac{1}{L} \int_{-L}^{L} f(x) \sin \frac{k\pi x}{L} \, dx$$

Furthermore, $f'(x)$ is even, and the functions $\cos(k\pi x/L)$, $k = 0, 1, 2, \ldots$ are complete with respect to this set of functions. Hence,

$$\frac{1}{L} \int_{-L}^{L} [f'(x)]^2 \, dx = \frac{a_0^2}{2} + \sum_{k=1}^{\infty} a_k^2$$

where

$$a_k = \frac{1}{L} \int_{-L}^{L} f'(x) \cos \frac{k\pi x}{L} \, dx$$

$$= \frac{1}{L} \left[f(x) \cos \frac{k\pi x}{L} \right]_{-L}^{L} + \frac{k\pi}{L^2} \int_{-L}^{L} f(x) \sin \frac{k\pi x}{L} \, dx$$

$$= \frac{k\pi}{L} b_k$$

Therefore,

$$Q(f) = \int_0^L [f'(x)]^2 \, dx = \tfrac{1}{2} \int_{-L}^{L} [f'(x)]^2 \, dx$$

$$= \frac{L}{2} \sum_{k=1}^{\infty} \frac{k^2\pi^2}{L^2} b_k^2$$

$$N(f) = \int_0^L [f(x)]^2 \, dx = \tfrac{1}{2} \int_{-L}^{L} [f(x)]^2 \, dx$$

$$= \frac{L}{2} \sum_{k=1}^{\infty} b_k^2$$

---

[1] Using more sophisticated methods, we would not have to assume the piecewise continuity of the second derivative.

Now consider
$$\psi_1(x) = \left(\frac{2}{L}\right)^{\frac{1}{2}} \sin \frac{\pi x}{L}$$

$$N(\psi_1) = \frac{2}{L} \int_0^L \sin^2 \frac{\pi x}{L}\, dx$$

$$= \frac{1}{L} \int_0^L \left(1 - \cos \frac{2\pi x}{L}\right) dx = 1$$

$$Q(\psi_1) = \frac{2\pi^2}{L^3} \int_0^L \cos^2 \frac{\pi x}{L}\, dx$$

$$= \frac{\pi^2}{L^3} \int_0^L \left(1 + \cos \frac{2\pi x}{L}\right) dx$$

$$= \frac{\pi^2}{L^2}$$

For any normalized $f(x)$ in F
$$N(f) = \frac{L}{2} \sum_{k=1}^\infty b_k^2 = 1$$

and
$$Q(f) - Q(\psi_1) = \frac{L}{2} \sum_{k=1}^\infty \frac{k^2\pi^2}{L^2} b_k^2 - \frac{L}{2} \sum_{k=1}^\infty \frac{\pi^2}{L^2} b_k^2$$

$$= \frac{L}{2} \sum_{k=2}^\infty \frac{(k^2-1)\pi^2}{L^2} b_k^2 \geq 0$$

This proves that $Q(\psi_1) = \pi^2/L^2$ is the minimum over the whole space subject to $N(f) = 1$.

Next consider the subspace of F such that $N(f) = 1$ and $N(f,\psi_1) = 0$. The latter condition implies that $b_1 = 0$. Let $\psi_2(x) = (2/L)^{\frac{1}{2}} \sin (2\pi x/L)$. Then

$$N(\psi_2) = 1 \qquad Q(\psi_2) = \frac{4\pi^2}{L^2}$$

$$Q(f) - Q(\psi_2) = \frac{L}{2} \sum_{k=2}^\infty \frac{k^2\pi^2}{L^2} b_k^2 - \frac{L}{2} \sum_{k=2}^\infty \frac{4\pi^2}{L^2} b_k^2$$

$$= \frac{L}{2} \sum_{k=3}^\infty \frac{(k^2-4)\pi^2}{L^2} b_k^2 \geq 0$$

This proves the $Q(\psi_2) = 4\pi^2/L^2$ is the minimum over the subspace. More generally, consider the subspace such that $N(f) = 1$ and $N(f,\psi_j) = 0$, $j = 1, 2, 3, \ldots, k-1$, where $\psi_j = (2/L)^{\frac{1}{2}} \sin (j\pi x/L)$. The latter condition implies that $b_1 = b_2 = \cdots = b_{k-1} = 0$. Let $\psi_k(x) = (2/L)^{\frac{1}{2}} \sin (k\pi x/L)$.

Then

$$N(\psi_k) = 1 \qquad Q(\psi_k) = \frac{k^2\pi^2}{L^2}$$

$$Q(f) - Q(\psi_k) = \frac{L}{2} \sum_{m=k}^{\infty} \frac{m^2\pi^2}{L^2} b_m^2 - \frac{L}{2} \sum_{m=k}^{\infty} \frac{k^2\pi^2}{L^2} b_m^2$$

$$= \frac{L}{2} \sum_{m=k+1}^{\infty} \frac{(m^2 - k^2)\pi^2}{L^2} b_m^2 \geq 0$$

This proves that $Q(\psi_k)$ is a solution of the kth minimum problem.

Notice that by the fortuitous choice of the trigonometric functions as the complete set of functions in the space F we obtained the quadratic forms $Q(f)$ and $N(f)$ in diagonal form. For another complete set of functions these forms would not have been diagonal. Hence, the variational problem is essentially equivalent to the simultaneous reduction of $Q(f)$ and $N(f)$ to diagonal form.

It should also be mentioned that the eigenvalues and eigenfunctions of the variational problem can be formulated in terms of a minimax definition.

Exercises 3.5

1. A string of length L and fixed ends is started into oscillation with no initial velocity but with an initial displacement $g(x) = \epsilon x(L - x)$. Find the displacement for any time $t > 0$.

*2. Using direct methods (not Euler's equation), show that $(1/L)^{\frac{1}{2}}$, $(2/L)^{\frac{1}{2}} \cos(k\pi x/L)$, $k = 1, 2, 3, \ldots$ are solutions of the following variational problem: Over the space of real-valued functions defined on the interval $0 \leq x \leq L$, which are continuous and have piecewise continuous first and second derivatives, minimize

$$Q(f) = \int_0^L [f'(x)]^2 \, dx$$

subject to $N(f) = \int_0^L [f(x)]^2 \, dx = 1$ and $N(f,\psi_j) = 0$, $j = 0, 1, 2, \ldots, k - 1$,

where

$$\psi_0 = \left(\frac{1}{L}\right)^{\frac{1}{2}} \qquad \psi_j = \left(\frac{2}{L}\right)^{\frac{1}{2}} \cos \frac{j\pi x}{L}$$

*3. Formulate and establish the minimax definition of the eigenvalues and eigenfunctions for the string problem with fixed ends.

*4. If a force of linear density $f(x)$ is applied to a string stretched between two fixed points with tension σ, show that the total potential energy is

$$\int_0^L \left[\frac{\sigma}{2}\left(\frac{dy}{dx}\right)^2 - f(x)y \right] dx$$

Hence show that the equation for the equilibrium deflection of the string is $\sigma(d\phi^2/dx^2) = -f(x)$.

*5. Obtain the differential equation for the longitudinal displacement y of the cross section of a rod with fixed ends, if the potential energy per unit length is $\frac{1}{2}EA(\partial y/\partial x)^2$, where E is Young's modulus and A is the area of the cross section.

*6. Obtain the differential equation for the transverse displacement y of a beam built in at the ends, if the potential energy per unit length is $\frac{1}{2}EI(\partial^2 y/\partial x^2)^2$ where E is Young's modulus and I is the moment of inertia of the cross section. The boundary conditions are $y(0,t) = y(L,t) = y'(0,t) = y'(L,t) = 0$.

3.6 Boundary-value Problems of Mathematical Physics

In the last section, we arrived at the partial differential equation for the displacement of a stretched string in small vibrations by an application of Hamilton's principle to the continuous system of masses representing the string.   By a similar analysis we can obtain the equation for the displacement of a stretched elastic membrane.   Let us say that the membrane covers the region R in the xy plane, bounded by a simple closed curve[1] C.   The displacement from equilibrium is measured in the z direction.   The membrane is assumed to have uniform and constant density ρ, and, therefore, the total kinetic energy is given by

$$T = \tfrac{1}{2}\rho \iint\limits_{R} \left(\frac{\partial z}{\partial t}\right)^2 dx\,dy$$

The tension σ, in small vibrations, is assumed constant, and by an extension of the argument given for the string the total potential energy is

$$V = \tfrac{1}{2}\sigma \iint\limits_{R} \left[\left(\frac{\partial z}{\partial x}\right)^2 + \left(\frac{\partial z}{\partial y}\right)^2 \right] dx\,dy$$

Hamilton's principle then states

$$\delta \int_{t_1}^{t} \iint\limits_{R} \left\{ \rho\left(\frac{\partial z}{\partial t}\right)^2 - \sigma\left[\left(\frac{\partial z}{\partial x}\right)^2 + \left(\frac{\partial z}{\partial x}\right)^2 \right] \right\} dx\,dy\,dt = 0$$

If we assume that there is a function $\phi(x,y,t)$, with continuous second partial derivatives, which makes this integral stationary, then

$$I(\epsilon) = \int_{t_1}^{t_2} \iint\limits_{R} \left\{ \rho(\phi_t + \epsilon\eta_t)^2 - \sigma[(\phi_x + \epsilon\eta_x)^2 + (\phi_y + \epsilon\eta_y)^2] \right\} dx\,dy\,dt$$

has a stationary value at $\epsilon = 0$.   Let us assume that the membrane is held fixed to the boundary curve C.   Hence, the solution ϕ and the virtual displacement $\phi + \epsilon\eta$ must vanish on C.   Also in the formulation of Hamilton's principle $\eta(x,y,t_1) = \eta(x,y,t_2) = 0$.   Otherwise η is arbitrary, except that it

---

[1] By a simple closed curve we mean a curve consisting of a finite number of arcs with continuously turning tangents, joined at the end points, but not crossing over itself.

must have continuous first partial derivatives. The necessary condition for I to have a stationary value at $\epsilon = 0$ is

$$I'(0) = \int_{t_1}^{t_2} \iint_R \left(\rho \frac{\partial \phi}{\partial t} \frac{\partial \eta}{\partial t} - \sigma \frac{\partial \phi}{\partial x} \frac{\partial \eta}{\partial x} - \sigma \frac{\partial \phi}{\partial y} \frac{\partial \eta}{\partial y} \right) dx\, dy\, dt = 0$$

Integrating the first term by parts with respect to t and using the conditions $\eta(x,y,t_1) = \eta(x,y,t_2) = 0$, we have

$$\int_{t_1}^{t_2} \left[\iint_R \left(\rho \frac{\partial^2 \phi}{\partial t^2} \eta + \sigma \nabla \phi \cdot \nabla \eta \right) dx\, dy \right] dt = 0$$

Next we use the following Green's identity:

$$\iint_R \nabla u \cdot \nabla v\, dx\, dy = - \iint_R v \nabla^2 u\, dx\, dy + \int_C v \frac{du}{dn}\, ds$$

plus the fact that η vanishes on C to allow us to write

$$\int_{t_1}^{t_2} \iint_R \left(\rho \frac{\partial^2 \phi}{\partial t^2} - \sigma \frac{\partial^2 \phi}{\partial x^2} - \sigma \frac{\partial^2 \phi}{\partial y^2} \right) \eta\, dx\, dy\, dt = 0$$

Since η is arbitrary, we have the necessary condition

$$\frac{\rho}{\sigma} \frac{\partial^2 \phi}{\partial t^2} = \frac{\partial^2 \phi}{\partial x^2} + \frac{\partial^2 \phi}{\partial y^2} = \nabla^2 \phi$$

which is the wave equation in two dimensions.

To solve for the displacement of the membrane, we can again try a separation of the time-dependent part of the function and write

$$\phi(x,y,t) = \psi(x,y)f(t)$$

Substituting, we have

$$\frac{\nabla^2 \psi}{\psi} = \frac{\rho}{\sigma} \frac{\ddot{f}}{f} = -\lambda$$

where λ is a constant. If this is to hold

$$\ddot{f} + \frac{\sigma}{\rho} \lambda f = 0$$

$$\nabla^2 \psi + \lambda \psi = 0$$

The first equation is a simple second-order ordinary differential equation which will yield a sinusoidal solution if λ is positive. That λ is positive follows, since

$$
\begin{aligned}
\lambda \iint_R \psi^2 \, dx \, dy &= -\iint_R \psi \nabla^2 \psi \, dx \, dy \\
&= \iint_R \nabla \psi \cdot \nabla \psi \, dx \, dy - \int_C \psi \frac{d\psi}{dn} \, ds \\
&= \iint_R \|\nabla \psi\|^2 \, dx \, dy \geq 0
\end{aligned}
$$

The line integral over the boundary curve C vanishes because of the boundary condition $\psi = 0$ on C. If

$$
\iint_R \|\nabla \psi\|^2 \, dx \, dy = \iint_R \left[\left(\frac{\partial \psi}{\partial x} \right)^2 + \left(\frac{\partial \psi}{\partial y} \right)^2 \right] dx \, dy = 0
$$

then $\partial \psi / \partial x = \partial \psi / \partial y = 0$ in R, which implies that $\psi = k$, a constant, in R. To satisfy the boundary condition, k would then have to be zero. This leads to the trivial solution $\psi \equiv 0$ in R.

The space-dependent functions ψ and the separation constants λ are determined by solving the following **boundary-value problem**:[1]

$$
\nabla^2 \psi + \lambda \psi = 0 \qquad \text{in } R
$$

$$
\psi = 0 \qquad \text{on } C
$$

The solution will depend on the geometry of the region R. We shall find, in general, an infinite set of nontrivial fundamental solutions (**eigenfunctions**) $\psi_1, \psi_2, \psi_3, \ldots$ corresponding to a discrete set of separation constants (**eigenvalues**) $0 < \lambda_1 \leq \lambda_2 \leq \lambda_3 \leq \cdots$. The general solution is then a superposition of these eigenfunctions multiplied by the corresponding time-dependent functions

$$
f_\alpha = A_\alpha \sin \omega_\alpha t + B_\alpha \cos \omega_\alpha t
$$

where

$$
\omega_\alpha = \sqrt{\frac{\sigma}{\rho} \lambda_\alpha}
$$

or

$$
\phi(x,y,t) = \sum_{\alpha=1}^\infty (A_\alpha \sin \omega_\alpha t + B_\alpha \cos \omega_\alpha t) \psi_\alpha(x,y)
$$

A_α and B_α are determined by the initial conditions. Under suitable restrictions on the initial conditions, the infinite series will converge and represent the displacement of the membrane. Before turning to the solution of this

---

[1] The first equation is a partial differential equation known as **Helmholtz's equation**.

problem, we shall cite several other physical problems which lead to this type or related types of boundary-value problems, hence indicating that the analysis to follow is really much more general than we have already indicated.

In a large number of physical situations we seek a solution of the **wave equation** with various boundary conditions prescribed and subject to certain prescribed initial values of the function and its first derivative with respect to time. The wave equation in rectangular coordinates is

$$\nabla^2 \phi = \frac{\partial^2 \phi}{\partial x^2} + \frac{\partial^2 \phi}{\partial y^2} + \frac{\partial^2 \phi}{\partial z^2} = \frac{1}{a^2} \frac{\partial^2 \phi}{\partial t^2}$$

Some of these situations are the following:

1. *The Vibrating String.* We have already seen in Sec. 3.5 that the displacement of the string satisfies the one-dimensional wave equation. The boundary condition for a fixed end is $\phi = 0$. If the end is free,[1] the boundary condition is $\partial \phi / \partial x = 0$, since there will then be no vertical force exerted at this point, and the force is proportional to the first derivative of the displacement. If the end is only partially free, that is, if the end is held by an elastic support, then by Hooke's law for the support, the restoring force is proportional to the displacement, and this is equal to the force on the end of the string, which in turn is proportional to the derivative of displacement. Hence, $\partial \phi / \partial x = \alpha \phi$ is the boundary condition for an elastically supported end.

2. *The Vibrating Membrane.* We have seen in this section that the displacement of the membrane satisfies the two-dimensional wave equation. The boundary condition for a fixed boundary is $\phi = 0$ on C, the boundary curve. If a part of the boundary is free, then there is no vertical force at the boundary, and hence $d\phi / dn = 0$. For an elastically supported boundary, $d\phi / dn = -\alpha \phi$.

3. *Acoustic Problems.* In acoustics the velocity potential for the flow of fluid through which sound is being transmitted is a scalar function whose gradient gives the velocity of the flow. The velocity potential satisfies the three-dimensional wave equation where a is the velocity of sound in the medium. For a rigid boundary there is no flow at right angles to the boundary, and hence the boundary condition $d\phi / dn = 0$ must be satisfied. The opposite of a rigid boundary is a boundary at which all the flow is normal. This would imply that $\phi = k$, a constant, on such a boundary. Then $\Phi = \phi - k$ would satisfy the wave equation and satisfy $\Phi = 0$ on the boundary. For a "soft" boundary, the appropriate boundary condition is $d\phi / dn = -\alpha \phi$.

4. *Longitudinal Vibrations of an Elastic Rod.* The longitudinal displacement of a cross section of an elastic rod in small vibrations satisfies the one-dimensional wave equation

$$\frac{\partial^2 \phi}{\partial x^2} = \frac{\rho}{EA} \frac{\partial^2 \phi}{\partial t^2}$$

---

[1] This condition can be approximated by tying the string to a ring which rides on a vertical rod with little friction.

where ρ is the linear density, E is Young's modulus, and A is the cross-sectional area. At a fixed end $\phi = 0$. At a free end $\partial\phi/\partial x = 0$, and at an elastically supported end $\partial\phi/\partial x = -\alpha\phi$.

5. *Electromagnetic Cavity.* The electric field $\mathbf{E}$ and the magnetic field $\mathbf{H}$ satisfy the three-dimensional wave equation in a vacuum in the absence of charge and dielectric; that is,

$$\nabla^2\mathbf{E} = \frac{1}{c^2}\frac{\partial^2\mathbf{E}}{\partial t^2}$$

$$\nabla^2\mathbf{H} = \frac{1}{c^2}\frac{\partial^2\mathbf{H}}{\partial t^2}$$

where c is the velocity of light in a vacuum. In addition, they must satisfy

$$\nabla \cdot \mathbf{E} = 0$$
$$\nabla \cdot \mathbf{H} = 0$$

On a boundary consisting of a perfect conductor

$$\mathbf{n} \times \mathbf{E} = \mathbf{0}$$
$$\mathbf{n} \cdot \mathbf{H} = 0$$

where $\mathbf{n}$ is a unit vector normal to the boundary.

In the case of the **heat** or **diffusion equation,** the first derivative with respect to time enters; that is,

$$\nabla^2\phi = \frac{1}{a^2}\frac{\partial\phi}{\partial t}$$

However, a separation of the time dependence still yields Helmholtz's equation, with the time-dependent part satisfying the first-order equation

$$\dot{f} + a^2\lambda f = 0$$

yielding a solution $f = Ce^{-a^2\lambda t}$. In the case of heat flow, ϕ represents the temperature. At a boundary of constant temperature, the zero of temperature may be so chosen that $\phi = 0$. Flow of heat takes place in the direction of negative temperature gradient. Therefore, at a perfectly insulated boundary, $d\phi/dn = 0$, since there is no flow of heat across such a boundary. For a partially insulated boundary, the appropriate boundary condition is

$$\frac{d\phi}{dn} = -\alpha\phi$$

The partial differential equation governing the flow of electricity along a cable is the **telegrapher's equation**

$$\frac{\partial^2\phi}{\partial x^2} = a\phi + b\frac{\partial\phi}{\partial t} + c\frac{\partial^2\phi}{\partial t^2}$$

Both the current I and the voltage V satisfy an equation of this type.   At a point where the cable is shorted, the boundary conditions are $V = 0$ and $\partial I/\partial x = 0$.   At a point where the cable is open, the boundary conditions are $I = 0$ and $\partial V/\partial x = 0$.

In each of the above examples, we are attempting to solve a partial differential equation of the general type

$$\nabla^2 \phi = a\phi + b \frac{\partial \phi}{\partial t} + c \frac{\partial^2 \phi}{\partial t^2}$$

subject to boundary conditions of one of the types

$$\phi = 0 \qquad \frac{d\phi}{dn} = 0 \qquad \text{or} \qquad \frac{d\phi}{dn} + \alpha\phi = 0$$

Such a problem is an example of a **homogeneous problem.**[1]   The differential equation and the boundary conditions are said to be homogeneous. Such a problem has the following important property.   If ϕ_1 and ϕ_2 are both solutions of the differential equation satisfying the boundary conditions, then $\gamma_1\phi_1 + \gamma_2\phi_2$, where γ_1 and γ_2 are constants, also satisfies the differential equation and boundary conditions, for

$$\left(\nabla^2 - a - b\frac{\partial}{\partial t} - c\frac{\partial^2}{\partial t^2}\right)(\gamma_1\phi_1 + \gamma_2\phi_2)$$

$$= \gamma_1\left(\nabla^2\phi_1 - a\phi_1 - b\frac{\partial\phi_1}{\partial t} - c\frac{\partial^2\phi_1}{\partial t^2}\right)$$

$$+ \gamma_2\left(\nabla^2\phi_2 - a\phi_2 - b\frac{\partial\phi_2}{\partial t} - c\frac{\partial^2\phi_2}{\partial t^2}\right) = 0$$

and

$$\gamma_1\phi_1 + \gamma_2\phi_2 = 0$$

or

$$\frac{d}{dn}(\gamma_1\phi_1 + \gamma_2\phi_2) = \gamma_1\frac{d\phi_1}{dn} + \gamma_2\frac{d\phi_2}{dn} = 0$$

or

$$\left(\frac{d}{dn} + \alpha\right)(\gamma_1\phi_1 + \gamma_2\phi_2) = \gamma_1\left(\frac{d\phi_1}{dn} + \alpha\phi_1\right) + \gamma_2\left(\frac{d\phi_2}{dn} + \alpha\phi_2\right) = 0$$

Because of this property, we expect to find the general solution of the boundary-value problem as a superposition of fundamental solutions.   To determine the particular solution which fits a given physical situation, one must specify initial conditions on the solution and its first derivative with respect to time.   The fundamental solutions of the boundary-value problem are found by separating out the time dependence.   Let

$$\phi(x,y,z,t) = \psi(x,y,z)f(t)$$

---

[1] Nonhomogeneous boundary-value problems will be treated in Chap. 5.

Then
$$\frac{\nabla^2\psi}{\psi} = \frac{af + bf + cf}{f} = -\lambda$$

or
$$cf + bf + (a + \lambda)f = 0$$

and
$$\nabla^2\psi + \lambda\psi = 0$$

The boundary conditions on ψ will be one of the following types: $\psi = 0$, $d\psi/dn = 0$, $d\psi/dn + \alpha\psi = 0$, or a mixture of these. We shall consider the solution of the boundary-value problem for the space-dependent part in the next section, from the point of view of the calculus of variations.

There are many other problems which do not exactly fit the above general theory but nevertheless are quite closely related. Some of these are listed below according to the basic partial differential equation involved.

1. *Schrödinger's Wave Equation.* In nonrelativistic quantum mechanics the fundamental equation is Schrödinger's wave equation

$$i\hbar \frac{\partial\phi}{\partial t} = -\frac{\hbar^2}{2m}\nabla^2\phi + V(x,y,z)\phi$$

where $\hbar = h/2\pi$, h is Planck's constant, V is the potential energy, and m is the mass of the particle whose wave function is ϕ. ϕ has the interpretation that $|\phi|^2\,dx\,dy\,dz$ is the probability that the particle may be found in the volume element $dx\,dy\,dz$ at any particular time. It must satisfy the normalizing condition

$$\iiint\limits_{\text{all space}} |\phi|^2\,dx\,dy\,dz = 1$$

Although only the first derivative with respect to time enters the equation, by assuming a complex time dependence of the form $e^{-iEt/\hbar}$, where E is constant, we get a separation of variables which leads to the equation

$$\frac{\hbar^2}{2m}\nabla^2\psi + (E - V)\psi = 0$$

for the space-dependent part of the wave function. If V is zero, we have Helmholtz's equation. In any case, the requirement that ψ satisfy certain boundary conditions may lead to discrete values of E, that is, discrete quantum energy states. The eigenvalue problem is hence fundamental in quantum mechanics.

2. *Laplace's Equation.* Laplace's equation, $\nabla^2\phi = 0$, is satisfied by the equilibrium displacement of a membrane when the displacement on the boundary curve is specified, by the temperature in steady-state heat flow in the absence of sources or sinks of heat, by the velocity potential for an incompressible, irrotational, homogeneous fluid in the absence of sources or sinks, by the electrostatic potential in the absence of charge, by the gravitational potential in the absence of mass, and in many other situations. This is a

special case of Helmholtz's equation with λ equal to zero. Hence, it appears as though the eigenvalue problem does not occur here. However, separation of space variables may lead to eigenvalue problems quite similar to those for the Helmholtz equation.

3. *Poisson's Equation.*   Poisson's equation, $\nabla^2\phi = f(x,y,z)$, where $f(x,y,z)$ is a known function, is satisfied by the equilibrium displacement of a membrane under distributed forces, by the electrostatic potential in the presence of distributed charge, by the velocity potential for an incompressible, irrotational, homogeneous fluid in the presence of distributed sources or sinks, by the gravitational potential in the presence of distributed matter, by the steady-state temperature in the presence of distributed sources or sinks of heat, and in many other situations.   This is a nonhomogeneous problem, and its solution will be discussed when we take up the general study of nonhomogeneous problems and their solutions using Green's functions.   It is mentioned here merely for completeness.

4. *Biharmonic Wave Equation.*   In elasticity the biharmonic wave equation is extremely important.   For example, the displacement of a thin flat elastic plate in small vibrations satisfies the equation

$$\nabla^4\phi = \nabla^2\nabla^2\phi = \nabla^2\left(\frac{\partial^2\phi}{\partial x^2} + \frac{\partial^2\phi}{\partial y^2}\right) = -\frac{1}{a^2}\frac{\partial^2\phi}{\partial t^2}$$

The boundary conditions for a clamped plate are $\phi = 0$ and $d\phi/dn = 0$ on the boundary.   The corresponding problem in one space variable is the transverse vibrations of an elastic beam built in at the ends.   Here the differential equation is

$$\frac{\partial^4\phi}{\partial x^4} = -\frac{\rho}{EI}\frac{\partial^2\phi}{\partial t^2}$$

where ρ is the linear density, E Young's modulus, and I the moment of inertia of the cross section.   The end conditions are $\phi = \partial\phi/\partial x = 0$.   If an end were free, the boundary conditions would be $\partial^2\phi/\partial x^2 = \partial^3\phi/\partial x^3 = 0$.   If an end were hinged, the boundary conditions would be $\phi = \partial^2\phi/\partial x^2 = 0$.

Exercises 3.6

*1. Assuming a separation of space variables for a rectangular membrane of length a and width b, write $\psi(x,y) = X(x)Y(y)$ and find the normal modes of vibration and the natural frequencies if the membrane is (a) clamped on the edges and (b) free on the edges.

*2. Consider acoustic vibrations of a rectangular room with dimensions a, b, and c.   Assume the walls to be rigid boundaries.   Find the normal modes and natural frequencies of the room.

*3. Suppose a certain boundary-value problem for a partial differential equation leads to the Helmholtz' equation for the space-dependent part of the function

$$\nabla^2\psi + \lambda\psi = 0$$

with the boundary condition $d\psi/dn + \alpha\psi = 0$ with $\alpha > 0$. Show that the separation constant λ is positive.

*4. If an external force of density $f(x,y,t)$ is acting on an elastic membrane, a term

$$-\iint_R f(x,y,t)z \, dx \, dy$$

must be added to the potential energy. Use Hamilton's principle to derive the differential equation

$$\rho \frac{\partial^2 \phi}{\partial t^2} - \sigma \nabla^2 \phi = f(x,y,t)$$

for forced vibrations of the membrane with clamped edges.

*5. If an external force of density $f(x,y)$ is acting on an elastic membrane, derive the equation for the equilibrium displacement of the membrane

$$\nabla^2 \phi = -\frac{1}{\sigma} f(x,y)$$

from the minimum principle for potential energy.

*6. Suppose the boundary of a membrane is elastically supported. If a part of the boundary of length ds is displaced by an amount z, the restoring force is $kz \, ds$. Add the appropriate term to the potential energy and show that the boundary condition $d\phi/dn = -(k/\sigma)\phi$ can be derived directly from Hamilton's principle. When this is the case, we say that it is a **natural boundary condition**.

7. Consider Hamilton's principle for the vibrating string. By removing the condition that the displacement vanish at one end of the string, show that $\partial\phi/\partial x = 0$ is a natural boundary condition.

8. Consider Hamilton's principle for the vibrating string with an elastically supported end at $x = L$. Add the potential energy of the spring $\frac{1}{2}ky^2(L,t)$, where k is the spring constant, to the Lagrangian and show that $\partial\phi/\partial x + \alpha\phi = 0$ is a natural boundary condition. What is the value of α? HINT: Do not add anything for the kinetic energy of the spring.

3.7 Eigenvalues and Eigenfunctions

In this section, we shall investigate the solution of the boundary-value problem

$$\nabla^2 \psi + \lambda\psi = 0$$

in a three-dimensional region R, which has a boundary[1] sufficiently regular so that the divergence theorem applies for sufficiently well-behaved functions as integrands, subject to the boundary condition $d\psi/dn + \alpha\psi = 0$ on S, the boundary of R. We shall characterize the solutions of this boundary-value problem as solutions of any one of the following problems in the calculus of variations:

1. Among all functions f which are continuous and have piecewise continuous first derivatives in R (that is, R can be subdivided into a finite number

---

[1] For quite general conditions on the boundary see Oliver D. Kellogg, "Foundations of Potential Theory," Dover Publications, Inc., New York, 1953.

of subregions in each of which the first partial derivatives of f are continuous and have limits as the boundary is approached from the interior), minimize the ratio[1]

$$\frac{Q(f)}{N(f)} = \frac{\iiint\limits_{R} (\nabla f \cdot \nabla f)\, dV + \iint\limits_{S} \alpha f^2\, dS}{\iiint\limits_{R} f^2\, dV}$$

The minimum will be equal to the lowest eigenvalue λ_1, and the minimizing function will be the first eigenfunction ψ_1. Next minimize $Q(f)/N(f)$ subject to $\iiint\limits_{R} f\psi_1\, dV = 0$. This minimum will be equal to the next eigenvalue λ_2, and the minimizing function will be the second eigenfunction ψ_2. Since we have added a constraint to the first problem, λ_2 is at least as large as λ_1, and thus $\lambda_1 \leq \lambda_2$. For the kth eigenvalue and eigenfunction we minimize $Q(f)/N(f)$ subject to $\iiint\limits_{R} f\psi_j\, dV = 0, j = 1, 2, \ldots, k - 1$. We therefore arrive at a sequence of eigenvalues $\lambda_1 \leq \lambda_2 \leq \lambda_3 \leq \cdots$ and corresponding eigenfunctions $\psi_1, \psi_2, \psi_3, \ldots$. The eigenfunctions will constitute an orthogonal set but will not necessarily be normalized. However, one can construct from them an orthonormal set, since the problem is homogeneous.

2. Among all functions f which are continuous and have piecewise continuous first derivatives in R, minimize $Q(f)$ subject to $N(f) = 1$. The minimum will be equal to the first eigenvalue λ_1, and the minimizing function will be the first eigenfunction ψ_1, which will be normalized; that is, $N(\psi_1) = 1$. The kth eigenvalue and eigenfunction are found by minimizing $Q(f)$ subject to $N(f) = 1$ and $\iiint\limits_{R} f\psi_j\, dV = 0, j = 1, 2, \ldots, k - 1$. The solution of this problem generates a sequence of eigenvalues $\lambda_1 \leq \lambda_2 \leq \lambda_3 \leq \cdots$ and an orthonormal set of eigenfunctions $\psi_1, \psi_2, \psi_3, \ldots$.

3. Among all functions f which are continuous and have piecewise continuous first derivatives in R, minimize $Q(f)$ subject to $N(f) = 1$ and

$$\iiint\limits_{R} f v_i\, dV = 0$$

where v_i, $i = 1, 2, \ldots, k - 1$, is *any* set of $k - 1$ linearly independent functions which are piecewise continuous in R. For all possible choices of the

---

[1] α can be a function of position on the surface of R but must be nonnegative everywhere on S.

set v_i choose the maximum of the minima of $Q(f)$. This maximum will be the kth eigenvalue, and the function which produces the maximum will be the kth eigenfunction. This is the **minimax definition** of the kth eigenvalue and eigenfunction. It has the advantage that one can characterize the kth eigenvalue directly without going through $k - 1$ previous variational problems, as in the other two cases, and gives us the means to compare the eigenvalues of different problems.

We shall be content here to derive only necessary conditions for the solution of the variational problems. The sufficient conditions for the existence of a solution and the proof that the solution has continuous second partial derivatives in R are beyond the scope of this book.[1]

The solutions of problems 1 and 2 can be handled together. Assume that there exists a solution ψ_1 to the first minimum problem, in each case, with continuous second derivatives in R. Then

$$Q(\psi_1 + \epsilon\eta) \geq \lambda_1 N(\psi_1 + \epsilon\eta)$$

where ϵ is an arbitrary constant and η is an arbitrary function from the class of admissible functions, that is, is continuous and has piecewise continuous first derivatives in R. Since ψ_1 is the solution to the minimum problem, $Q(\psi_1) = \lambda_1 N(\psi_1)$. Expanding the above inequality and using this fact, we have[2]

$$2\epsilon[Q(\psi_1,\eta) - \lambda_1 N(\psi_1,\eta)] + \epsilon^2[Q(\eta) - \lambda_1 N(\eta)] \geq 0$$

This must be true for arbitrary ϵ. Therefore

$$Q(\psi_1,\eta) - \lambda_1 N(\psi_1,\eta) = 0$$

for arbitrary η. Otherwise, given an η, we could choose an ϵ sufficiently small, and of the proper sign, so that

$$2\epsilon[Q(\psi_1,\eta) - \lambda_1 N(\psi_1,\eta)] + \epsilon^2[Q(\eta) - \lambda_1 N(\eta)] < 0$$

contradicting the above inequality. Consequently we have

$$\iiint_R \nabla\psi_1 \cdot \nabla\eta \, dV + \iint_S \alpha\psi_1\eta \, dS - \lambda_1 \iiint_R \psi_1\eta \, dV = 0$$

---

[1] See Richard Courant and David Hilbert, "Methoden der mathematischen Physik," Springer-Verlag, Berlin, 1937, vol. II, chap. 7.

[2] $Q(f,g)$ and $N(f,g)$ are the **associated bilinear forms**

$$Q(f,g) = \iiint_R \nabla f \cdot \nabla g \, dV + \iint_S \alpha fg \, dS \text{ and } N(f,g) = \iiint_R fg \, dV$$

ψ_1 has continuous second derivatives in R, and η has piecewise continuous first derivatives in R; therefore we may use Green's identity, giving us

$$\iint\limits_S \left(\frac{d\psi_1}{dn} + \alpha\psi_1\right)\eta \, dS - \iiint\limits_R (\nabla^2\psi_1 + \lambda_1\psi_1)\eta \, dV = 0$$

for arbitrary η.   Since η is arbitrary, we can first pick it to be zero on the boundary but otherwise arbitrary in the interior of R.   The surface integral then vanishes, and we have

$$\iiint\limits_R (\nabla^2\psi_1 + \lambda_1\psi_1)\eta \, dV = 0$$

implying that $\nabla^2\psi_1 + \lambda_1\psi_1 = 0$ in the interior of R.   But this implies that the volume integral vanishes, and hence for all η

$$\iint\limits_S \left(\frac{d\psi_1}{dn} + \alpha\psi_1\right)\eta \, dS = 0$$

This implies that $d\psi_1/dn + \alpha\psi_1 = 0$ on S.   Therefore, the solution to the first variational problem is also a solution of the boundary-value problem.

Notice that we arrived at the boundary condition $d\psi_1/dn + \alpha\psi_1 = 0$ as a necessary condition for the solution of the variational problem without requiring that this condition be satisfied by all the functions in the class of admissible functions.   Such a boundary condition is called a **natural boundary condition.**   If $\alpha = 0$ everywhere on S, then $d\psi_1/dn = 0$ is the natural boundary condition.   On the other hand, if we want our solution to satisfy the boundary condition $\psi_1 = 0$ on S, we have to require that $f = 0$ on S for every function in the class of admissible functions.   In this case

$$Q(f) = \iiint\limits_R \nabla f \cdot \nabla f \, dV$$

and $\psi_1 = 0$ on S is a consequence of ψ_1 being in this class of functions.   This is then a **prescribed boundary condition,** as opposed to a natural boundary condition.

Next we solve the same minimum problem, but with the further condition that

$$N(\psi_1,f) = \iiint\limits_R \psi_1 f \, dV = 0$$

Let us assume that this problem has a solution ψ_2 with continuous second partial derivatives.   Then $Q(\psi_2) = \lambda_2 N(\psi_2)$, $N(\psi_1,\psi_2) = 0$, and

$$Q(\psi_2 + \epsilon\eta) \geq \lambda_2 N(\psi_2 + \epsilon\eta)$$

This time, however, η is not completely arbitrary, since $N(\psi_1, \eta) = 0$. Therefore, starting with an arbitrary function ζ in the class of admissible functions, we can construct an η by subtracting out the part not orthogonal to ψ_1; that is,

$$\eta = \zeta - c\psi_1$$

where

$$c = \frac{N(\psi_1, \zeta)}{N(\psi_1)}$$

Then we have

$$Q(\psi_2 + \epsilon\zeta - \epsilon c\psi_1) \geq \lambda_2 N(\psi_2 + \epsilon\zeta - \epsilon c\psi_1)$$

or

$$Q(\psi_2, \zeta) - \lambda_2 N(\psi_2, \zeta) - c[Q(\psi_2, \psi_1) - \lambda_2 N(\psi_2, \psi_1)] = 0$$

by the same kind of argument as before. Now we know that $N(\psi_2, \psi_1) = 0$ and that

$$Q(\psi_1, \zeta) - \lambda_1 N(\psi_1, \zeta) = 0$$

for arbitrary ζ, from the previous variational problem. Letting $\zeta = \psi_2$, we have $Q(\psi_1, \psi_2) = 0$. Therefore, $Q(\psi_2, \zeta) - \lambda_2 N(\psi_2, \zeta) = 0$ where ζ is arbitrary. Hence

$$\nabla^2\psi_2 + \lambda_2\psi_2 = 0 \qquad \text{in } R$$

$$\frac{d\psi_2}{dn} + \alpha\psi_2 = 0 \qquad \text{on } S$$

We continue in this way generating successive eigenvalues and eigenfunctions.

To establish the minimax definition of the kth eigenvalue and eigenfunction, we first note that the minimum of $Q(f) = \lambda_k$ when $v_i = \psi_i$, $i = 1, 2, \ldots,$ $k - 1$. Also, if $v_i \neq \psi_i$, the minimum of $Q(f) \leq \lambda_k$ for $f = \sum_{j=1}^{k} c_j\psi_j$ produces a value

$$Q(f) = \sum_{j=1}^{k} c_j^2 \lambda_j \leq \lambda_k$$

because $N(f) = \sum_{j=1}^{k} c_j^2 = 1$. The c's are uniquely determined by the $k - 1$ conditions $N(f, v_i) = 0$ plus $\sum_{j=1}^{k} c_j^2 = 1$. We therefore know that for every choice of v_i, the minimum of Q is less than or equal to λ_k. Yet the minimum of Q is equal to λ_k for the particular choice $v_i = \psi_i$, from which it must follow that λ_k is equal to the maximum of the minima of Q, and this maximum is taken on when $f = \psi_k$.

The characterization of eigenvalues in terms of variational problems gives a powerful tool for comparing eigenvalues for different problems. The basis for these comparisons is in the following two theorems.

Theorem 1. Let λ_k be the kth eigenvalue of the variational problem in which $Q(f)$ is minimized over a certain class of admissible functions F. Let $\bar{\lambda}_k$

be the kth eigenvalue of the variational problem in which $Q(\bar{f})$ is minimized over a class of admissible functions $\bar{F}$ resulting from the addition of certain constraints to F. Then $\lambda_k \leq \bar{\lambda}_k$.

The proof of this theorem follows from the minimax definition of the eigenvalue. Since the v_i need only be piecewise continuous in R, the same sets can be chosen for either problem. Therefore, the minimum of $Q(f)$ is less than or equal to the minimum of $Q(\bar{f})$ for every choice of v_i. From this it follows that the maximum of the minima of $Q(f)$ is less than or equal to the maximum of the minima of $Q(\bar{f})$, or $\lambda_k \leq \bar{\lambda}_k$. Notice that this result is not obtainable from variational problems 1 or 2 because the eigenfunctions of one problem are not the same as for the other. Therefore, except for the lowest eigenvalue, the constraints in the two problems are not directly comparable.

Theorem 2. Let λ_k be the kth eigenvalue of the variational problem in which $Q(f)$ is minimized over a class of admissible functions F. Let $\bar{\lambda}_k$ be the kth eigenvalue of the variational problem in which $\bar{Q}(f)$ is minimized over the same class of functions, but $Q(f) \leq \bar{Q}(f)$ for every function in F. Then $\lambda_k \leq \bar{\lambda}_k$. The proof of this theorem will be left for the exercises.

To illustrate the use of these theorems, consider the following examples. Let $\nu_k = \sqrt{(\sigma/\rho)\lambda_k}$ be a natural frequency of an elastic membrane under tension σ and with density ρ. Suppose the membrane covers the region R in the xy plane and is clamped to the boundary curve C. Let us now take a smaller membrane over a region $\bar{R}$ contained in R and clamped on its boundary $\bar{C}$. The natural frequency $\bar{\nu}_k = \sqrt{(\sigma/\rho)\bar{\lambda}_k}$ of the smaller membrane would tend to be larger, or at least not smaller, than ν_k. This follows from theorem 1. We get λ_k by minimizing

$$Q(f) = \iint_R \nabla f \cdot \nabla f \, dx \, dy$$

over a class of functions which vanish on C. By comparison, we get $\bar{\lambda}_k$ by minimizing Q over a class of functions which vanishes on C, $\bar{C}$, and the region between C and $\bar{C}$. But this class of functions has additional restrictions beyond those imposed on the first class of function. Hence, $\lambda_k \leq \bar{\lambda}_k$.

As another example, consider a membrane over the region R and attached to the boundary curve C by elastic supports. The boundary condition is $d\psi/dn + \alpha\psi = 0$ on C, where α is the modulus of elasticity. Suppose we now increase the modulus of elasticity on some part of the boundary. The natural frequencies will tend to increase, or at least not decrease. This follows from theorem 2. We get λ_k for the first problem by minimizing

$$Q(f) = \iint_R \nabla f \cdot \nabla f \, dx \, dy + \int_C \alpha f^2 \, ds$$

We get $\bar{\bar{\lambda}}_k$ in the second problem by minimizing

$$\bar{\bar{Q}}(f) = \iint\limits_R \nabla f \cdot \nabla f \, dx \, dy + \int_C \bar{\bar{\alpha}} f^2 \, ds$$

over the same class of functions, where $\bar{\bar{\alpha}} \geq \alpha$ on C. Therefore, $\bar{Q} \leq \bar{\bar{Q}}$ for every f, and hence $\bar{\lambda}_k \leq \bar{\bar{\lambda}}_k$.

Theorems 1 and 2 are also useful in telling us what the general behavior of the sequence of eigenvalues is. Consider the boundary-value problems for the same simply connected region R but under three different boundary conditions. Let λ_k, $\bar{\lambda}_k$, and $\bar{\bar{\lambda}}_k$ be the kth eigenvalues for the three problems corresponding to the boundary conditions $\psi = 0$, $d\psi/dn = 0$, and $d\psi/dn + \alpha\psi = 0$, respectively. By theorem 2 we know that $\bar{\lambda}_k \leq \bar{\bar{\lambda}}_k$, since we are minimizing functionals

$$\bar{Q}(f) = \iiint\limits_R \nabla f \cdot \nabla f \, dV \leq \bar{\bar{Q}}(f) = \iiint\limits_R \nabla f \cdot \nabla f \, dV + \iint\limits_S \alpha f^2 \, dS$$

over the same class of admissible functions. By theorem 1 we know that $\bar{\lambda}_k \leq \lambda_k$, since we are minimizing the same functional $\bar{Q}$, but for λ_k we prescribe the condition $f = 0$ on S on the class of admissible functions. Therefore, we have

$$\bar{\lambda}_k \leq \bar{\bar{\lambda}}_k \leq \lambda_k$$

We shall show that $\lim\limits_{k \to \infty} \lambda_k = \infty$ using theorem 1. Let R' be a rectangular parallelepiped containing R with dimensions a, b, and c and surface S'. Consider the eigenvalue problem

$$\frac{\partial^2 \psi}{\partial x^2} + \frac{\partial^2 \psi}{\partial y^2} + \frac{\partial^2 \psi}{\partial z^2} + \mu\psi = 0 \qquad \text{in } R'$$

with $\psi = 0$ on S'. By separating the variables, we obtain solutions

$$\psi_k = \sin\frac{m\pi x}{a} \sin\frac{n\pi y}{b} \sin\frac{p\pi z}{c}$$

with eigenvalues $\mu_k = \pi^2\left(\dfrac{m^2}{a^2} + \dfrac{n^2}{b^2} + \dfrac{p^2}{c^2}\right)$, $m = 1, 2, 3, \ldots$, $n = 1, 2, 3, \ldots$, $p = 1, 2, 3, \ldots$. It can be shown that the completeness of the functions $\sin(m\pi x/a)$, $\sin(n\pi y/b)$, $\sin(p\pi z/c)$ on each of the intervals $0 \leq x \leq a$, $0 \leq y \leq b$, and $0 \leq z \leq c$, implies the completeness of ψ_k on the rectangle R' (see Sec. 4.1 for the extension from one to two dimensions). This means that the sequence $\{\mu_k\}$ contains all the eigenvalues for the problem on R'.

Obviously $\lim_{k \to \infty} \mu_k = \infty$, and since the partial differential equation is the Euler equation for the variational problem

$$\min \iiint_{R'} \nabla f \cdot \nabla f \, dV \qquad f \equiv 0 \qquad \text{on } S'$$

there is a subsequence of the sequence $\{\mu_k\}$ whose members are respectively lower bounds for λ_k. Since every subsequence of $\{\mu_k\}$ is unbounded, $\lim_{k \to \infty} \lambda_k = \infty$.

Actually $\lim_{k \to \infty} \bar{\lambda}_k = \infty$ and $\lim_{k \to \infty} \tilde{\lambda}_k = \infty$, but we shall have to wait until Chap. 6, which deals with integral equations, to show this. The fact that the sequence of eigenvalues approaches infinity is important in the proof of the completeness of the set of eigenfunctions of the variational problem. It also implies that an eigenvalue of the variational problem can have only a finite degeneracy; that is, if a single eigenvalue corresponds to more than one linearly independent eigenfunction, it can correspond to at most a finite number of them.

The analysis of this section is quite a bit more general than we have already indicated. The important features of the variational problem are the following:

1. $Q(f)$ and $N(f)$ must be real **quadratic functionals.** This means that

$$Q(f + g) + Q(f - g) = 2Q(f) + 2Q(g)$$
$$N(f + g) + N(f - g) = 2N(f) + 2N(g)$$

When this is the case, we can define **associated bilinear forms**

$$Q(f,g) = \tfrac{1}{2}[Q(f + g) - Q(f) - Q(g)]$$
$$= \tfrac{1}{4}[Q(f + g) - Q(f - g)]$$
$$N(f,g) = \tfrac{1}{2}[N(f + g) - N(f) - N(g)]$$
$$= \tfrac{1}{4}[N(f + g) - N(f - g)]$$

with the properties[1]

$$Q(f,g) = Q(g,f)$$
$$Q(c_1 f_1 + c_2 f_2, g) = c_1 Q(f_1, g) + c_2 Q(f_2, g)$$

and the corresponding properties for N.

2. $N(f)$ is positive-definite; that is, $N(f) > 0$ unless $f \equiv 0$.

3. $Q(f)$ is bounded from below (or above).

Let us assume that $Q(f)$ is bounded from below. In this case we minimize $Q(f)$ subject to $N(f) = 1$. [If $Q(f)$ is bounded from above, we maximize

---

[1] See exercise 5, Sec. 1.6.

$Q(f)$.] Assuming that there is a minimizing function ψ_1, for which Q takes on the value λ_1, we have

$$Q(\psi_1) = \lambda_1 N(\psi_1)$$

$$Q(\psi_1 + \epsilon\eta) \geq \lambda_1 N(\psi_1 + \epsilon\eta)$$

for arbitrary ϵ and η. Then

$$2\epsilon[Q(\psi_1,\eta) - \lambda_1 N(\psi_1,\eta)] + \epsilon^2[Q(\eta) - \lambda_1 N(\eta)] \geq 0$$

from which it follows that[1]

$$Q(\psi_1,\eta) = \lambda_1 N(\psi_1,\eta)$$

for arbitrary η in the class of admissible functions. Continuing in this way, we define a sequence of problems in the calculus of variations leading to a sequence of eigenvalues $\lambda_1 \leq \lambda_2 \leq \lambda_3 \leq \cdots$ and corresponding eigenfunctions $\psi_1, \psi_2, \psi_3, \ldots$ satisfying

$$Q(\psi_\alpha,\eta) = \lambda_\alpha N(\psi_\alpha,\eta)$$

$$Q(\psi_i) = \lambda_i$$

$$N(\psi_i,\psi_j) = \delta_{ij}$$

In the next section, we shall discuss the completeness of the set of eigenfunctions and the expansion of arbitrary admissible functions in series of eigenfunctions.

Exercises 3.7

1. Prove theorem 2 of this section.

*2. Prove directly from the differential equation $\nabla^2\psi + \lambda\psi = 0$ and boundary condition $d\psi/dn + \alpha\psi = 0$, $\alpha > 0$, that if $\lambda_i \neq \lambda_j$, then $N(\psi_i,\psi_j) = 0$.

3. Show that the eigenvalues and eigenfunctions of the boundary-value problem $\nabla^4\psi - \lambda\psi = 0$ in R, $\psi = d\psi/dn = 0$ on C, arising in the study of vibrations of a clamped elastic plate can be obtained from the solution of the following variational problem: to minimize

$$\frac{Q(f)}{N(f)} = \frac{\displaystyle\iint_R (\nabla^2 f)^2 \, dx \, dy}{\displaystyle\iint_R f^2 \, dx \, dy}$$

over the class of admissible functions f with piecewise continuous second derivatives which satisfy $f = df/dn = 0$ on C, subject to $N(f,\psi_i) = 0$, where ψ_i, $i = 1$, $2, \ldots, k - 1$, are the eigenfunctions already found. Assume that the solution of the variational problem exists and has continuous fourth derivatives in R.

4. An elastic membrane over the region R is clamped on its edge, the boundary curve C. Indicate with reasons whether the following will tend to increase or decrease the natural frequencies of the membrane:

a. Part of the boundary is unclamped.

b. An interior point is held down.

---

[1] Compare with exercise 4, Sec. 3.1.

c. The membrane is cut along an interior curve.

d. A particle of mass m is attached to an interior point. Hint: Consider min Q/N.

5. Consider the transverse vibrations of an elastic beam. The partial differential equation for small displacements is $\dfrac{\partial^4 \phi}{\partial x^4} + \dfrac{1}{a^2} \dfrac{\partial^2 \phi}{\partial t^2} = 0$. The boundary conditions for a built-in end are $\phi = \partial\phi/\partial x = 0$, for a hinged end are $\phi = \partial^2\phi/\partial x^2 = 0$, and for a completely free end are $\partial^2\phi/\partial x^2 = \partial^3\phi/\partial x^3 = 0$. Show that the natural frequencies can be derived from a variational principle based on the functional

$$Q(f) = \int_0^L \left(\frac{d^2 f}{dx^2}\right)^2 dx$$

Which boundary conditions are natural and which must be prescribed? If one end of the beam is built in, compare the natural frequencies for three conditions on the other end: free, hinged, and built in.

3.8   Eigenfunction Expansions

In the last section we showed that eigenvalues and eigenfunctions for certain boundary-value problems can be obtained from variational problems. We have yet to show that all the eigenvalues and eigenfunctions of a given problem can be so obtained. We can do this if we can show that the eigenfunctions of the appropriate variational problem form a complete set. If there exists an eigenfunction of the boundary-value problem corresponding to an eigenvalue not found by the variational problem, then it is orthogonal to every one of a complete set[1] and is therefore identically zero. If there exists an eigenfunction ψ of the boundary-value problem corresponding to an eigenvalue λ found by the variational problem, then a set of constants c_i, $i = 1, 2, \ldots, n$, where n is the multiplicity of this eigenvalue of the variational problem, can be found such that

$$g = \psi - c_i \psi_i \qquad i = 1, 2, \ldots, n$$

is orthogonal to every ψ_i, that is, those eigenfunctions corresponding to the eigenvalue λ. This can be done with

$$c_i = N(\psi, \psi_i) \qquad i = 1, 2, \ldots, n$$

Then g is orthogonal not only to the eigenfunctions corresponding to λ, but to all the eigenfunctions of the variational problem corresponding to different eigenvalues. It is therefore orthogonal to every member of a complete set. It must therefore be identically zero, implying that ψ is a linear combination of the ψ_i. Consequently, the variational problem furnishes all the eigenfunctions. We have therefore to show that the eigenfunctions of the variational problem form a complete set.

[1] See exercise 2, Sec. 3.7.

Let f be an arbitrary function from the class of admissible functions in the variational problem.   If we attempt to get an approximation for f in terms of a linear combination of eigenfunctions $c_i\psi_i$, $i = 1, 2, \ldots, m$, we know from Chap. 2 that we get the best approximation in the least-mean-square sense if we choose the c_i as

$$c_i = N(f,\psi_i)$$

Let
$$f_m = f - c_i\psi_i \qquad i = 1, 2, \ldots, m$$

where the c_i are defined in this way.   Then f_m is an admissible function for the variational problem defining the $(m + 1)$st eigenvalue λ_{m+1}, for

$$N(f_m,\psi_j) = N(f,\psi_j) - c_i\delta_{ij} \qquad j = 1, 2, \ldots, m$$
$$= c_j - c_j = 0$$

Therefore,
$$Q(f_m) \geq \lambda_{m+1}N(f_m)$$

$Q(f_m)$ is bounded, since

$$Q(f) = Q(f_m) + Q(c_i\psi_i) + 2Q(c_i\psi_i,f_m)$$
$$= Q(f_m) + Q(c_i\psi_i)$$

This follows from the relation $Q(\psi_i,\eta) - \lambda_i N(\psi_i,\eta) = 0$, which is true for ψ_i and an arbitrary admissible function η.   If we let $\eta = f_m$, then

$$N(\psi_i,f_m) = Q(\psi_i,f_m) = 0$$

Therefore, we have

$$0 \leq Q(f_m) = Q(f) - Q(c_i\psi_i)$$
$$= Q(f) - \lambda_i c_i^2$$

$\lambda_i c_i^2$ is positive and nondecreasing with m. $Q(f)$ exists and is independent of m.   Thus, $Q(f_m)$ is bounded as m increases.   This implies, since $\lambda_{m+1} \to \infty$ as $m \to \infty$, that

$$N(f_m) \leq \frac{Q(f_m)}{\lambda_{m+1}} \to 0 \text{ as } m \to \infty$$

$$N(f_m) = N(f) - \sum_{i=1}^{m} c_i^2 \to 0 \text{ as } m \to \infty$$

or
$$N(f) = \sum_{i=1}^{\infty} c_i^2$$

which is the **completeness relation**.   The ψ_i therefore are a complete set in the class of admissible functions.

By arguments similar to those given in Sec. 2.1, we get **Parseval's Equation**

$$N(f,g) = \sum_{i=1}^{\infty} c_i b_i$$

where $c_i = N(f, \psi_i)$ and $b_i = N(g, \psi_i)$ and f and g are any pair of functions from the class of admissible functions.

We must remember that completeness of a set of orthonormal functions, which is equivalent to convergence in mean of a series to an arbitrary function, is not the same as pointwise convergence. However, based on completeness, we can often establish pointwise or even uniform convergence. For example, consider the problem of the vibrating string stretched between $x = 0$ and $x = L$. The eigenvalues of the problem are $\lambda_k = k^2 \pi^2 / L^2$ and the eigenfunctions are $\psi_k = (2/L)^{\frac{1}{2}} \sin(k\pi/L)x$. From the variational problem we know that the eigenfunctions are a complete set with respect to continuous functions which vanish at $x = 0$ and $x = L$ and have piecewise continuous first derivatives. Consider the series $\sum_{k=1}^{\infty} c_k \psi_k$, where

$$c_k = \int_0^L f\psi_k \, dx$$

By Schwarz's inequality,

$$\left(\sum_{k=n}^{p} c_k \psi_k \right)^2 \le \sum_{k=n}^{p} c_k^2 \lambda_k \sum_{k=n}^{p} \frac{\psi_k^2}{\lambda_k} \le \sum_{k=1}^{\infty} c_k^2 \lambda_k \sum_{k=n}^{\infty} \frac{\psi_k^2}{\lambda_k}$$

We have already shown that

$$0 \le Q(f_m) = Q(f) - \sum_{k=1}^{m} c_k^2 \lambda_k$$

Therefore, the series $\sum_{k=1}^{\infty} c_k^2 \lambda_k$ converges. Also

$$\sum_{k=1}^{\infty} \frac{\psi_k^2}{\lambda_k} = \frac{2L}{\pi^2} \sum_{k=1}^{\infty} \frac{\sin^2(k\pi/L)x}{k^2}$$

converges uniformly, since $\sin^2 \dfrac{k\pi x}{L} \le 1$ and the series $\sum_{k=1}^{\infty} \dfrac{1}{k^2}$ converges. Therefore, $\sum_{k=n}^{\infty} \dfrac{\psi_k^2}{\lambda_k}$ approaches zero as n approaches infinity uniformly in x. This implies that $\left(\sum_{k=n}^{p} c_k \psi_k \right)^2$ approaches zero as n approaches infinity uniformly in x for every $p > n$, and by the Cauchy criterion for convergence the series $\sum_{k=1}^{\infty} c_k \psi_k$ converges uniformly. The series must therefore converge uniformly to f by the completeness of the eigenfunctions.[1]

The same argument would hold if we replaced $\left(\sum_{k=1}^{p} c_k \psi_k \right)^2$ by $\left(\sum_{k=1}^{p} |c_k| \, |\psi_k| \right)^2$. Therefore, we also have absolute convergence of the series $\sum_{k=1}^{\infty} c_k \psi_k$.

[1] See Sec. 2.1.

Thus we have been able to show that every continuous function, with a piecewise continuous first derivative, which vanishes at $x = 0$ and $x = L$ can be expanded in a uniformly convergent series of sine functions, that is, the eigenfunctions of the string with fixed ends. This, however, depends on the completeness of the set of eigenfunctions, which in turn depends on the existence of the solution of the variational problem.[1] However, if the completeness property has already been shown, then the expansion theorem can be proved directly. We shall return to the problem of expanding arbitrary functions in terms of orthonormal sets of functions in Chap. 4, when we discuss the Sturm-Liouville problem.

Exercises 3.8

1. Obtain a uniformly convergent series of sine functions which converges to

$$f(x) = x \qquad 0 \leq x \leq \tfrac{1}{2}$$
$$f(x) = 1 - x \qquad \tfrac{1}{2} \leq x \leq 1$$

What does the series converge to for $-1 \leq x \leq 0$?

2. Consider the vibrating-string problem with free ends. What are the eigenvalues and eigenfunctions? With respect to what class of functions are the eigenfunctions a complete set? Obtain a uniformly convergent series of cosine functions which converges to the function of problem 1. What does the series converge to for $-1 \leq x \leq 0$?

3.9 Upper and Lower Bounds for Eigenvalues

The definition of eigenvalues of a boundary-value problem by variational principles allows one to develop procedures for approximating eigenvalues when the corresponding eigenfunction is not known. For example, we know that λ_1 is the minimum of the ratio of two functionals in a certain function space. Therefore, if we take *any* function in the space and evaluate this ratio, we shall obtain an upper bound for λ_1. As an example, consider the string problem with fixed ends. The first eigenvalue λ_1 is the minimum of

$$\frac{Q(f)}{N(f)} = \frac{\displaystyle\int_0^1 (f')^2 \, dx}{\displaystyle\int_0^1 f^2 \, dx}$$

over the class of functions which are continuous,[2] vanish at $x = 0$ and $x = 1$, and have a piecewise continuous derivative. Such a function is

$$g_1(x) = x \qquad 0 \leq x \leq \tfrac{1}{2}$$
$$g_1(x) = 1 - x \qquad \tfrac{1}{2} \leq x \leq 1$$

[1] See Sec. 2.2.
[2] For convenience we have taken a string of unit length.

An elementary calculation leads to

$$\frac{Q(g_1)}{N(g_1)} = 12 > \lambda_1 = \pi^2$$

As another example, we might take

$$g_2 = x(1 - x)$$

This function leads to an upper bound

$$\frac{Q(g_2)}{N(g_2)} = 10$$

which is a better approximation for π^2.

What is needed is a procedure for systematically improving the approximation afforded by the upper bound. Because the function space is a linear vector space, any linear combination of functions in the space is also in the space. Therefore, we may take a linear combination $c_1 g_1 + c_2 g_2$ to determine an upper bound for λ_1. However, now c_1 and c_2 are arbitrary and so can be chosen to give the smallest possible upper bound over the subspace spanned by g_1 and g_2. We have

$$\lambda_1 \leq \frac{Q(c_1 g_1 + c_2 g_2)}{N(c_1 g_1 + c_2 g_2)}$$

for any c_1 and c_2. Expanding, we have

$$\frac{Q(c_1 g_1 + c_2 g_2)}{N(c_1 g_1 + c_2 g_2)} = \frac{c_1^2 Q(g_1) + 2c_1 c_2 Q(g_1, g_2) + c_2^2 Q(g_2)}{c_1^2 N(g_1) + 2c_1 c_2 N(g_1, g_2) + c_2^2 N(g_2)}$$

The problem of minimizing the ratio of two quadratic forms where the denominator is positive-definite we have met before in exercise 3, Sec. 3.1. We are led to the characteristic equation

$$\begin{vmatrix} Q(g_1) - \lambda N(g_1) & Q(g_1, g_2) - \lambda N(g_1, g_2) \\ Q(g_1, g_2) - \lambda N(g_1, g_2) & Q(g_2) - \lambda N(g_2) \end{vmatrix} = 0$$

to solve for the stationary values of the ratio. In the present example, the characteristic equation becomes

$$3\lambda^2 - 416\lambda + 3{,}840 = 0$$

The smaller of the roots of this equation is approximately 9.944, as compared with π^2, which is approximately 9.870. The procedure outlined here is known as the **Rayleigh-Ritz method.** We shall now discuss it in a more general framework.

We wish to find an upper bound for the smallest eigenvalue of the boundary-value problem

$$\nabla^2 \psi + \lambda \psi = 0 \qquad \text{in } R$$

$$\frac{d\psi}{dn} + \alpha \psi = 0 \qquad \text{on } S$$

We know from the variational principle that

$$\lambda_1 \leq \frac{\iiint\limits_R \nabla f \cdot \nabla f \, dV + \iint\limits_S \alpha f^2 \, dS}{\iiint\limits_R f^2 \, dV}$$

where f is any continuous function with piecewise continuous first derivatives in R. Let $f = c_i \varphi_i$, $i = 1, 2, \ldots, n$, where φ_i is any set of linearly independent functions in the space of admissible functions. Evaluating the functionals involved, we have

$$\lambda_1 \leq \frac{Q(c_i \varphi_i)}{N(c_k \varphi_k)} = \frac{a_{ij} c_i c_j}{b_{km} c_k c_m}$$

where $a_{ij} = Q(\varphi_i, \varphi_j)$ and $b_{ij} = N(\varphi_i, \varphi_j)$. We note that the matrices A with elements a_{ij} and B with elements b_{ij} are both real and symmetric. $N(f)$ is never negative, and it is zero only if $f = c_i \varphi_i = 0$. But the φ_i are independent, so that $b_{ij} c_i c_j = 0$ only if $c_i = 0$ for all i. Therefore, $b_{ij} c_i c_j$ is a positive-definite quadratic form.

We now pick the c's to give us the "best" upper bound. This is the problem of minimizing the ratio of two quadratic forms, the denominator of which is positive-definite. The stationary values of this ratio are given by the solutions of the characteristic equation

$$|A - \mu B| = 0$$

There are n stationary values, all of which are real. The smallest of these will give us the best upper bound for λ_1 over the subspace spanned by the functions φ_i.

We next ask ourselves what is the relation between the other stationary values and eigenvalues of the boundary-value problem other than λ_1. Let C_α and C_β be eigenvectors of the problem

$$AC = \mu BC$$

corresponding to different eigenvalues μ_α and μ_β. Then

$$AC_\alpha = \mu_\alpha BC_\alpha$$
$$AC_\beta = \mu_\beta BC_\beta$$
$$C'_\beta AC_\alpha = \mu_\alpha C'_\beta BC_\alpha$$
$$C'_\alpha AC_\beta = \mu_\beta C'_\alpha BC_\beta$$
$$0 = (\mu_\alpha - \mu_\beta)C'_\beta BC_\alpha$$

Since $\mu_\alpha \neq \mu_\beta$, $C'_\beta BC_\alpha = 0$. Let $\theta_i = c_{ij}\varphi_j$, where c_{ij} is the jth component of the ith eigenvector. The θ_i are an orthogonal set for

$$N(\theta_i, \theta_j) = N(c_{ik}\varphi_k, c_{jm}\varphi_m) = c_{ik}c_{jm}N(\varphi_k, \varphi_m) = c_{ik}c_{jm}b_{km} = C'_i BC_j = 0$$

Even if there is a repeated eigenvalue, it will correspond to a finite-dimensional subspace in which we can construct an orthogonal basis, and we will have a set of n orthogonal eigenvectors.

We can go another step in the approximation procedure if we add another function to the set already used. Since the θ_i are an orthogonal set in the function space, let us use these and add a new function θ_{n+1}, which we assume is constructed orthogonal to the others. We shall also assume that the θ's are normalized. We then have

$$N(\theta_i, \theta_j) = \delta_{ij} \qquad i = 1, 2, \ldots, n+1$$
$$j = 1, 2, \ldots, n+1$$

Thus at each stage in the approximation the functions which produce stationary values are linear combinations of an orthonormal set. If we consider the kth stationary values in each of the successive approximations, we find that they form a nonincreasing sequence. This can be seen from a minimax definition of the stationary value.

$$\mu_k^{(n)} = \max \min \frac{Q(f)}{N(f)}$$

where f is in the subspace S spanned by $\theta_1, \theta_2, \ldots, \theta_n$, and $N(f, v_i) = 0$, $i = 1, 2, \ldots, k-1$, and v_i are any set of $k-1$ piecewise continuous functions.

$$\mu_k^{(n+1)} = \max \min \frac{Q(\tilde{f})}{N(\tilde{f})}$$

where $\tilde{f}$ is in the subspace $\tilde{S}$ spanned by $\theta_1, \theta_2, \ldots, \theta_{n+1}$ and $N(\tilde{f}, v_i) = 0$, $i = 1, 2, \ldots, k-1$. Now we know that S is contained in $\tilde{S}$, since any function in S can be written as a linear combination of $\theta_1, \theta_2, \ldots, \theta_{n+1}$ with $c_{n+1} = 0$. Therefore, $\mu_k^{(n+1)} \leq \mu_k^{(n)}$, $k = 1, 2, \ldots, n$. We know also that every stationary value is greater than or equal to λ_k, which is the maximum of the minima of Q/N in the whole space. Therefore, we have

$$\lambda_k \leq \cdots \leq \mu_k^{(n+2)} \leq \mu_k^{(n+1)} \leq \mu_k^{(n)}$$

and the set of kth stationary values forms a nonincreasing sequence bounded from below by the kth eigenvalue of the boundary-value problem. Such a sequence always has a limit. We have also shown that the function which produces the kth stationary value can be expressed as a linear combination of a set of orthonormal functions. If this is a complete set, then $\theta_k^{(n)}$ converges in mean to ψ_k, the kth eigenfunction of the boundary-value problem; that is,

$$\lim_{n \to \infty} N(\psi_k - \theta_k^{(n)}) = 0$$

It can be shown that this implies that

$$\lim_{n \to \infty} \mu_k^{(n)} = \lambda_k$$

but this involves some methods of functional analysis which we have not yet taken up. We shall return to this question again in Chap. 6, after we have shown how to convert the eigenvalue problem for the Helmholtz equation to an eigenvalue problem for an integral equation. The operator in the integral equation will be self-adjoint and completely continuous and this will simplify matters somewhat.

The Rayleigh-Ritz procedure has more than just a computational interest, for it can be used as a starting point for a proof of the existence of the solution of the variational problem. This is what Courant and Hilbert refer to as the "direct method" of the calculus of variations.[1] We shall not attempt to discuss these methods further here.

So far we have discussed only upper bounds for eigenvalues. One can also get lower bounds. Assume, for example, that

$$\nabla^2 f = g$$

where f has continuous second partial derivatives in R and satisfies the appropriate boundary condition of the eigenvalue problem. By Parseval's equation we have

$$N(f) = \sum_{k=1}^{\infty} c_k^2$$

$$N(f,g) = \sum_{k=1}^{\infty} c_k b_k$$

where $c_k = N(f,\psi_k)$ and $b_k = N(g,\psi_k)$. From Green's theorem,

$$b_\alpha = \iiint_R \psi_\alpha \nabla^2 f \, dV = \iiint_R \nabla^2 \psi_\alpha f \, dV + \iint_S \left(\psi_\alpha \frac{df}{dn} - f \frac{d\psi_\alpha}{dn} \right) dS$$

$$= -\lambda_\alpha \iiint_R \psi_\alpha f \, dV = -\lambda_\alpha c_\alpha$$

---

[1] See Richard Courant and David Hilbert, "Methods of Mathematical Physics," Interscience Publishers (Division of John Wiley & Sons, Inc.), New York, 1953, vol. I, pp. 174–176. See also Richard Courant and David Hilbert, "Methoden der mathematischen Physik," vol. II, chap. 7.

so that
$$N(f,g) = N(f,\nabla^2 f) = -Q(f) = -\sum_{k=1}^{\infty} \lambda_k c_k^2$$

Furthermore,
$$D(f) = N(\nabla^2 f, \nabla^2 f) = N(g,g) = \sum_{k=1}^{\infty} b_k^2 = \sum_{k=1}^{\infty} \lambda_k^2 c_k^2$$

Let
$$\lambda = \frac{Q(f)}{N(f)} = \frac{\sum_{k=1}^{\infty} \lambda_k c_k^2}{\sum_{k=1}^{\infty} c_k^2}$$

Then
$$\frac{D(f)}{N(f)} - \lambda^2 = \frac{D}{N} - \lambda \frac{Q}{N} = \frac{D - \lambda Q}{N}$$
$$= \frac{D - 2\lambda Q + \lambda Q}{N} = \frac{D - 2\lambda Q + \lambda^2 N}{N}$$
$$= \frac{\sum_{k=1}^{\infty} c_k^2 (\lambda_k^2 - 2\lambda \lambda_k + \lambda^2)}{\sum_{k=1}^{\infty} c_k^2}$$
$$= \frac{\sum_{k=1}^{\infty} c_k^2 (\lambda_k - \lambda)^2}{\sum_{k=1}^{\infty} c_k^2} \geq (\lambda_i - \lambda)^2$$

where λ_i is the eigenvalue closest to λ. Then
$$\lambda - \sqrt{\frac{D}{N} - \frac{Q^2}{N^2}} \leq \lambda_i \leq \lambda + \sqrt{\frac{D}{N} - \frac{Q^2}{N^2}}$$

As an example, consider the function g_2, which gave us an approximation of 10 for $\lambda_1 = \pi^2$. We have $D(g_2) = 4$, $N(g_2) = \frac{1}{30}$, so that
$$10 - \sqrt{20} \leq \lambda_1 = \pi^2 \leq 10 + \sqrt{20}$$

These are not very good bounds in this case. For a procedure which leads to better bounds see Bernard Friedman, "Principles and Techniques of Applied Mathematics," page 212. There is quite an extensive literature on the problem of obtaining bounds for eigenvalues. Those interested in pursuing the subject should refer to the mathematical reviews for the works of Aronszajn, Diaz, Duffin, Payne, Polya, Protter, Szegö, Weinberger, Weinstein, etc.

It may happen that the value of some functional is directly related to some physical quantity whose value is to be determined. Then finding upper and lower bounds for this functional is a good means of approximating that quantity. For example, in electrostatics, for a typical capacitor the electrostatic potential φ satisfies

$$\nabla^2\varphi = 0 \quad \text{in } R$$
$$\varphi = 1 \quad \text{on } S_1$$
$$\varphi = 0 \quad \text{on } S_0$$

where R is the region between two surfaces S_1 and S_0 which are perfect conductors and form the boundaries of R.† This is a Dirichlet problem, and its solution is the solution of the following minimum problem: to minimize

$$Q(f) = \iiint_R \nabla f \cdot \nabla f \, dV$$

over the class of functions continuous in R and on S, with piecewise continuous first derivatives in R, and taking on the given boundary values on S_0 and S_1. $Q(f)$ is positive-definite, for if $Q(f) = 0$, then $\partial f/\partial x = \partial f/\partial y = \partial f/\partial z = 0$ in R. This implies that f is constant in R and on S, but this is not possible if $f = 0$ on S_0 and $f = 1$ on S_1 and is continuous. Therefore,

$$\min Q(f) = \lambda > 0$$

Assume that the problem has a solution φ, and then $Q(\varphi) = \lambda$ and

$$Q(\varphi + \epsilon\eta) \geq \lambda$$

where ϵ and η are arbitrary, except that $\eta = 0$ on S. Hence

$$2\epsilon Q(\varphi,\eta) + \epsilon^2 Q(\eta) \geq 0$$

This implies, since ϵ is arbitrary, that $Q(\varphi,\eta) = 0$. By Green's theorem,

$$Q(\varphi,\eta) = -\iiint_R \eta\nabla^2\varphi \, dV + \iint_S \eta\frac{d\varphi}{dn} \, dS$$

$$= -\iiint_R \eta\nabla^2\varphi \, dV = 0$$

Since η is arbitrary in R, $\nabla^2\varphi = 0$ in R; and since φ is in the function space, it must satisfy the boundary conditions $\varphi = 1$ on S_1 and $\varphi = 0$ on S_0.

Now λ is not an eigenvalue of the differential equation, but in this case it is proportional to the capacity of the capacitor. Therefore, we are interested in its value, which we can approximate by getting upper and lower bounds. The Rayleigh-Ritz procedure for obtaining upper bounds is as follows. Let f_0 be any admissible function which satisfies the boundary conditions, and

---

† Actually, to fit the present formulation of the problem, we have to think of these boundary values as an idealization of a continuous function which makes the transition from 0 to 1 in a small region of the boundary where S_0 and S_1 are joined.

let $f_1, f_2, \ldots, f_n$ be any set of continuous functions with piecewise continuous first derivatives which satisfy the boundary condition $f_i = 0$ on S, $i = 1, 2, \ldots, n$. Then

$$f = f_0 + c_i f_i$$

is admissible, and $Q(f) \geq \lambda$. The c's are arbitrary and hence can be adjusted to give the smallest upper bound.

A method due to Trefftz for getting lower bounds is the following. Let g be any function satisfying the differential equation $\nabla^2 g = 0$ in R. By Schwarz's inequality,

$$[Q(\varphi,g)]^2 \leq Q(\varphi)Q(g)$$

By Green's theorem,

$$Q(\varphi,g) = -\iiint_R \varphi \nabla^2 g \, dV + \iint_S \varphi \frac{dg}{dn} \, dS$$

$$= \iint_{S_1} \frac{dg}{dn} \, dS$$

Therefore,

$$\frac{\left[\iint_{S_1} \frac{dg}{dn} \, dS \right]^2}{\iiint_R \nabla g \cdot \nabla g \, dV} \leq Q(\varphi) = \lambda$$

For a method of getting successively better lower bounds, see the paper by J. B. Diaz, "Upper and Lower Bounds for Quadratic Functionals," *Proceedings of the Symposium on Spectral Theory and Differential Problems*, Oklahoma A. & M., Stillwater, Okla., 1951.

Exercises 3.9

1. Find the best upper bound for the smallest eigenvalue of the string problem for one end ($x = 0$) fixed and one end ($x = 1$) free over the subspace spanned by the functions $g_1 = x$ and $g_2 = x^2$.

2. Prove that $N(\theta_i, \theta_j) = 0$ if $\lim_{n \to \infty} N(\theta_i^{(n)} - \theta_i) = 0$, $\lim_{n \to \infty} N(\theta_j^{(n)} - \theta_j) = 0$, and $N(\theta_i^{(n)}, \theta_j^{(n)}) = 0$ for all n.

3. If φ is the solution of the boundary-value problem $\nabla^2 \varphi = -\rho$, ρ a known function in R, and $\varphi = 0$ on S, the boundary of R, then show that

$$\frac{\left[\iiint_R \rho u \, dV \right]^2}{Q(u)} \leq Q(\varphi) \leq Q(v)$$

where $Q(f) = \iiint_R \nabla f \cdot \nabla f \, dV$, $f \not\equiv 0$ in R, and $\nabla^2 v = -\rho$ in R, and $u = 0$ on S.

HINT: Use Schwarz's inequality.

4. Under the conditions for the lower bound derived in this section, show that

$$\lambda_1 \geq \frac{\beta Q - D}{\beta N - Q}$$

where β is any number less than λ_2 and $\beta N - Q > 0$. Let $\beta = 39 < 4\pi^2 = \lambda_2$ and find a lower bound for $\lambda_1 = \pi^2$ in the string problem using $g_2 = x(1 - x)$ as the comparison function. HINT: $\sum_{k=1}^{\infty} (\lambda_k - \lambda_1)(\lambda_k - \beta)c_k^2 \geq 0$.

References

Akhiezer, N. I.: "The Calculus of Variations," Blaisdell Publishing Company (Division of Ginn and Company), Waltham, Mass., 1962.

Courant, Richard, and David Hilbert: "Methods of Mathematical Physics," Interscience Publishers (Division of John Wiley & Sons, Inc.), New York, 1953, vol. I; 1962, vol. II.

Friedman, Bernard: "Principles and Techniques of Applied Mathematics," John Wiley & Sons, Inc., New York, 1956.

Goldstein, Herbert: "Classical Mechanics," 2d ed., Addison-Wesley Publishing Company, Inc., Reading, Mass., 1980.

Hildebrand, F. B.: "Methods of Applied Mathematics," 2d ed., Prentice-Hall, Inc., Englewood Cliffs, N.J., 1965.

Sagan, Hans: "Introduction to the Calculus of Variations," McGraw-Hill Book Company, New York, 1969.

Temple, George F., and W. G. Brickley: "Rayleigh's Principle," Dover Publications, Inc., New York, 1956.

Weinstock, Robert: "Calculus of Variations," Dover Publications, Inc., New York, 1974.

Chapter 4. Boundary-value Problems.
Separation of Variables

4.1 Orthogonal Coordinate Systems. Separation of Variables

In the last chapter, we characterized the eigenfunctions of the Helmholtz equation as solutions of variational problems.   Although this was very instructive and it led to several interesting theorems, as well as to approximation procedures, it is not very constructive as a method of finding explicit eigenfunctions.   The method of **separation of variables** is one of the most important for finding explicit solutions of the Helmholtz and related partial differential equations.   The Helmholtz equation separates into ordinary differential equations in eleven different **orthogonal coordinate systems,**[1] and these are sufficient to solve many problems of practical significance.   Therefore, we shall restrict our discussion to orthogonal coordinates and avoid the difficulties of completely general coordinate transformations.

We shall assume that ours is a three-dimensional euclidean space, so that there exists a line element

$$(ds)^2 = (dx)^2 + (dy)^2 + (dz)^2$$

in terms of the usual orthogonal euclidean coordinates (x,y,z).   We consider a coordinate transformation to new coordinates (u,v,w) by

$$u = u(x,y,z)$$
$$v = v(x,y,z)$$
$$w = w(x,y,z)$$

where u, v, w are single-valued, continuous, differentiable functions.   At

[1] See Chester H. Page, "Physical Mathematics," D. Van Nostrand Company, Inc., Princeton, N.J., 1955.

points where the Jacobian

$$J = \begin{vmatrix} \dfrac{\partial u}{\partial x} & \dfrac{\partial u}{\partial y} & \dfrac{\partial u}{\partial z} \\[2mm] \dfrac{\partial v}{\partial x} & \dfrac{\partial v}{\partial y} & \dfrac{\partial v}{\partial z} \\[2mm] \dfrac{\partial w}{\partial x} & \dfrac{\partial w}{\partial y} & \dfrac{\partial w}{\partial z} \end{vmatrix} \neq 0$$

the transformation has a unique inverse:

$$x = x(u,v,w)$$
$$y = y(u,v,w)$$
$$z = z(u,v,w)$$

A **coordinate surface** is a surface on which one of the coordinates is constant, so that $u = u_0$, $v = v_0$, and $w = w_0$ define three coordinate surfaces which intersect in the point (u_0,v_0,w_0). A **coordinate curve** is a curve along which only one of the coordinates varies. For example, if we set $v = v_0$ and $w = w_0$ we have the intersection of two coordinate surfaces, which is a coordinate curve along which only u varies. We call this a u-coordinate curve. At the point (u_0,v_0,w_0) we have the intersection of three coordinate curves (see Fig. 6).

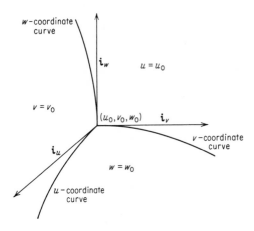

FIGURE 6

Let us draw unit tangent vectors to each of the coordinate curves at the point (u_0,v_0,w_0). An **orthogonal coordinate system** is defined as one in which these three vectors are always mutually perpendicular, or equivalently, one in which the coordinate curves always meet at right angles.

The orthogonality of the coordinate system can be expressed in terms of the unit vectors $\mathbf{i}_u$, $\mathbf{i}_v$, $\mathbf{i}_w$ as follows:

$$\mathbf{i}_u \cdot \mathbf{i}_v = \mathbf{i}_u \cdot \mathbf{i}_w = \mathbf{i}_v \cdot \mathbf{i}_w = 0$$

This can be expressed analytically in at least two different ways. $\mathbf{i}_u$, $\mathbf{i}_v$, $\mathbf{i}_w$ are vectors normal to surfaces $u = u_0$, $v = v_0$, and $w = w_0$, respectively. Therefore, we have

$$\mathbf{i}_u = \frac{\nabla u}{\|\nabla u\|}$$

$$\mathbf{i}_v = \frac{\nabla v}{\|\nabla v\|}$$

$$\mathbf{i}_w = \frac{\nabla w}{\|\nabla w\|}$$

where these gradients are evaluated at (u_0, v_0, w_0). In terms of partial derivatives,

$$\nabla u \cdot \nabla v = \frac{\partial u}{\partial x}\frac{\partial v}{\partial x} + \frac{\partial u}{\partial y}\frac{\partial v}{\partial y} + \frac{\partial u}{\partial z}\frac{\partial v}{\partial z} = 0$$

$$\nabla u \cdot \nabla w = \frac{\partial u}{\partial x}\frac{\partial w}{\partial x} + \frac{\partial u}{\partial y}\frac{\partial w}{\partial y} + \frac{\partial u}{\partial z}\frac{\partial w}{\partial z} = 0$$

$$\nabla v \cdot \nabla w = \frac{\partial v}{\partial x}\frac{\partial w}{\partial x} + \frac{\partial v}{\partial y}\frac{\partial w}{\partial y} + \frac{\partial v}{\partial z}\frac{\partial w}{\partial z} = 0$$

Also, $\mathbf{i}_u$, $\mathbf{i}_v$, $\mathbf{i}_w$ are vectors tangent to the u-coordinate curve, v-coordinate curve, and w-coordinate curve, respectively. Along the u-coordinate curve, for example,

$$x = x(u, v_0, w_0)$$

$$y = y(u, v_0, w_0)$$

$$z = z(u, v_0, w_0)$$

are the parametric equations of the curve using u as the parameter. Hence, the unit tangent is

$$\mathbf{i}_u = \left(\frac{\partial x}{\partial u}, \frac{\partial y}{\partial u}, \frac{\partial z}{\partial u}\right)\frac{du}{ds_u}$$

where

$$\frac{du}{ds_u} = \left[\left(\frac{\partial x}{\partial u}\right)^2 + \left(\frac{\partial y}{\partial u}\right)^2 + \left(\frac{\partial z}{\partial u}\right)^2\right]^{-\frac{1}{2}}$$

and the partial derivatives are evaluated at (u_0, v_0, w_0). Similarly,

$$\mathbf{i}_v = \left(\frac{\partial x}{\partial v}, \frac{\partial y}{\partial v}, \frac{\partial z}{\partial v}\right)\frac{dv}{ds_v}$$

$$\mathbf{i}_w = \left(\frac{\partial x}{\partial w}, \frac{\partial y}{\partial w}, \frac{\partial z}{\partial w}\right)\frac{dw}{ds_w}$$

and we have

$$\frac{\partial x}{\partial u}\frac{\partial x}{\partial v} + \frac{\partial y}{\partial u}\frac{\partial y}{\partial v} + \frac{\partial z}{\partial u}\frac{\partial z}{\partial v} = 0$$

$$\frac{\partial x}{\partial u}\frac{\partial x}{\partial w} + \frac{\partial y}{\partial u}\frac{\partial y}{\partial w} + \frac{\partial z}{\partial u}\frac{\partial z}{\partial w} = 0$$

$$\frac{\partial x}{\partial v}\frac{\partial x}{\partial w} + \frac{\partial y}{\partial v}\frac{\partial y}{\partial w} + \frac{\partial z}{\partial v}\frac{\partial z}{\partial w} = 0$$

We are now in a position to express the line element in terms of the co-ordinates u, v, w. By the chain rule we have

$$dx = \frac{\partial x}{\partial u}\,du + \frac{\partial x}{\partial v}\,dv + \frac{\partial x}{\partial w}\,dw$$

$$dy = \frac{\partial y}{\partial u}\,du + \frac{\partial y}{\partial v}\,dv + \frac{\partial y}{\partial w}\,dw$$

$$dz = \frac{\partial z}{\partial u}\,du + \frac{\partial z}{\partial v}\,dv + \frac{\partial z}{\partial w}\,dw$$

Then

$$(ds)^2 = (dx)^2 + (dy)^2 + (dz)^2$$

$$= \left[\left(\frac{\partial x}{\partial u}\right)^2 + \left(\frac{\partial y}{\partial u}\right)^2 + \left(\frac{\partial z}{\partial u}\right)^2\right](du)^2 + \left[\left(\frac{\partial x}{\partial v}\right)^2 + \left(\frac{\partial y}{\partial v}\right)^2 + \left(\frac{\partial z}{\partial v}\right)^2\right](dv)^2$$

$$+ \left[\left(\frac{\partial x}{\partial w}\right)^2 + \left(\frac{\partial y}{\partial w}\right)^2 + \left(\frac{\partial z}{\partial w}\right)^2\right](dw)^2$$

$$= \left(\frac{ds_u}{du}\,du\right)^2 + \left(\frac{ds_v}{dv}\,dv\right)^2 + \left(\frac{ds_w}{dw}\,dw\right)^2$$

$$= (ds_u)^2 + (ds_v)^2 + (ds_w)^2$$

The cross-product terms in $du\,dv$, $du\,dw$, and $dv\,dw$ have dropped out because of the orthogonality of the coordinate system.   It is common to write the line element as

$$(ds)^2 = h_1^2(du)^2 + h_2^2(dv)^2 + h_3^2(dw)^2$$

where

$$h_1^2 = \left(\frac{\partial x}{\partial u}\right)^2 + \left(\frac{\partial y}{\partial u}\right)^2 + \left(\frac{\partial z}{\partial u}\right)^2$$

$$h_2^2 = \left(\frac{\partial x}{\partial v}\right)^2 + \left(\frac{\partial y}{\partial v}\right)^2 + \left(\frac{\partial z}{\partial v}\right)^2$$

$$h_3^2 = \left(\frac{\partial x}{\partial w}\right)^2 + \left(\frac{\partial y}{\partial w}\right)^2 + \left(\frac{\partial z}{\partial w}\right)^2$$

The magnitude of a gradient is equal to the directional derivative in a direction normal to a surface.   Therefore,

$$\mathbf{i}_u = \frac{\nabla u}{\|\nabla u\|} = \frac{\nabla u}{\dfrac{du}{ds_u}} = h_1 \nabla u$$

$$\mathbf{i}_v = \frac{\nabla v}{\|\nabla v\|} = \frac{\nabla v}{\dfrac{dv}{ds_v}} = h_2 \nabla v$$

$$\mathbf{i}_w = \frac{\nabla w}{\|\nabla w\|} = \frac{\nabla w}{\dfrac{dw}{ds_w}} = h_3 \nabla w$$

We shall now express the gradient, divergence, and curl in terms of the coordinates u, v, w and unit vectors $\mathbf{i}_u$, $\mathbf{i}_v$, $\mathbf{i}_w$.   Then the Laplacian can be expressed as the divergence of a gradient.   By the chain rule we have

$$\frac{\partial \psi}{\partial x} = \frac{\partial \psi}{\partial u}\frac{\partial u}{\partial x} + \frac{\partial \psi}{\partial v}\frac{\partial v}{\partial x} + \frac{\partial \psi}{\partial w}\frac{\partial w}{\partial x}$$

$$\frac{\partial \psi}{\partial y} = \frac{\partial \psi}{\partial u}\frac{\partial u}{\partial y} + \frac{\partial \psi}{\partial v}\frac{\partial v}{\partial y} + \frac{\partial \psi}{\partial w}\frac{\partial w}{\partial y}$$

$$\frac{\partial \psi}{\partial z} = \frac{\partial \psi}{\partial u}\frac{\partial u}{\partial z} + \frac{\partial \psi}{\partial v}\frac{\partial v}{\partial z} + \frac{\partial \psi}{\partial w}\frac{\partial w}{\partial z}$$

Now $\nabla \psi = (\partial \psi / \partial x)\mathbf{i} + (\partial \psi / \partial y)\mathbf{j} + (\partial \psi / \partial z)\mathbf{k}$ in   rectangular   coordinates. Therefore,

$$\nabla \psi = \frac{\partial \psi}{\partial u}\left(\frac{\partial u}{\partial x}\mathbf{i} + \frac{\partial u}{\partial y}\mathbf{j} + \frac{\partial u}{\partial z}\mathbf{k}\right) + \frac{\partial \psi}{\partial v}\left(\frac{\partial v}{\partial x}\mathbf{i} + \frac{\partial v}{\partial y}\mathbf{j} + \frac{\partial v}{\partial z}\mathbf{k}\right)$$

$$+ \frac{\partial \psi}{\partial w}\left(\frac{\partial w}{\partial x}\mathbf{i} + \frac{\partial w}{\partial y}\mathbf{j} + \frac{\partial w}{\partial z}\mathbf{k}\right)$$

$$= \frac{\partial \psi}{\partial u}\nabla u + \frac{\partial \psi}{\partial v}\nabla v + \frac{\partial \psi}{\partial w}\nabla w$$

$$= \frac{1}{h_1}\frac{\partial \psi}{\partial u}\mathbf{i}_u + \frac{1}{h_2}\frac{\partial \psi}{\partial v}\mathbf{i}_v + \frac{1}{h_3}\frac{\partial \psi}{\partial w}\mathbf{i}_w$$

which expresses the gradient entirely in the new coordinate system.

Let $\mathbf{V}$ be a vector expressed in terms of the u, v, w coordinate system; that is,

$$\mathbf{V} = V_u \mathbf{i}_u + V_v \mathbf{i}_v + V_w \mathbf{i}_w$$

If $\mathbf{i}_u$, $\mathbf{i}_v$, $\mathbf{i}_w$ form a right-handed triad, then

$$\mathbf{i}_u = \mathbf{i}_v \times \mathbf{i}_w = h_2 h_3 (\nabla v \times \nabla w)$$
$$\mathbf{i}_v = \mathbf{i}_w \times \mathbf{i}_u = h_1 h_3 (\nabla w \times \nabla u)$$
$$\mathbf{i}_w = \mathbf{i}_u \times \mathbf{i}_v = h_1 h_2 (\nabla u \times \nabla v)$$

and $\quad \mathbf{V} = h_2 h_3 V_u (\nabla v \times \nabla w) + h_1 h_3 V_v (\nabla w \times \nabla u) + h_1 h_2 V_w (\nabla u \times \nabla v)$

The divergence of $\mathbf{V}$ is

$$\begin{aligned}
\nabla \cdot \mathbf{V} &= \nabla \cdot [h_2 h_3 V_u (\nabla v \times \nabla w)] + \nabla \cdot [h_1 h_3 V_v (\nabla w \times \nabla u)] \\
&\quad + \nabla \cdot [h_1 h_2 V_w (\nabla u \times \nabla v)] \\
&= \nabla (h_2 h_3 V_u) \cdot (\nabla v \times \nabla w) + \nabla (h_1 h_3 V_v) \cdot (\nabla w \times \nabla u) \\
&\quad + \nabla (h_1 h_2 V_w) \cdot (\nabla u \times \nabla v) + h_2 h_3 V_u \nabla \cdot (\nabla v + \nabla w) \\
&\quad + h_1 h_3 V_v \nabla \cdot (\nabla w \times \nabla u) + h_1 h_2 V_w \nabla \cdot (\nabla u \times \nabla v)
\end{aligned}$$

The last three terms in this expression are zero because of the identity

$$\nabla \cdot (\nabla f \times \nabla g) = \nabla g \cdot (\nabla \times \nabla f) - \nabla f \cdot (\nabla \times \nabla g)$$

plus the fact that the curl of a gradient is zero. Using the expression for the gradient, which we have already derived, and making use of the identity $\mathbf{A} \cdot (\mathbf{A} \times \mathbf{B}) = 0$ for any pair of vectors, we have

$$\begin{aligned}
\nabla \cdot \mathbf{V} &= \frac{\partial}{\partial u} (h_2 h_3 V_u) \nabla u \cdot (\nabla v \times \nabla w) + \frac{\partial}{\partial v} (h_1 h_3 V_v) \nabla v \cdot (\nabla w \times \nabla u) \\
&\quad + \frac{\partial}{\partial w} (h_1 h_2 V_w) \nabla w \cdot (\nabla u \times \nabla v)
\end{aligned}$$

Finally, we have

$$\begin{aligned}
1 &= \mathbf{i}_u \cdot (\mathbf{i}_v \times \mathbf{i}_w) = \mathbf{i}_v \cdot (\mathbf{i}_w \times \mathbf{i}_u) = \mathbf{i}_w \cdot (\mathbf{i}_u \times \mathbf{i}_v) \\
&= h_1 h_2 h_3 \nabla u \cdot (\nabla v \times \nabla w) = h_1 h_2 h_3 \nabla v \cdot (\nabla w \times \nabla u) \\
&= h_1 h_2 h_3 \nabla w \cdot (\nabla u \times \nabla v)
\end{aligned}$$

so that the divergence in the u, v, w coordinate system is

$$\nabla \cdot \mathbf{V} = \frac{1}{h_1 h_2 h_3} \left[\frac{\partial}{\partial u} (h_2 h_3 V_u) + \frac{\partial}{\partial v} (h_1 h_3 V_v) + \frac{\partial}{\partial w} (h_1 h_2 V_w) \right]$$

For completeness we shall derive the expression for the curl of a vector. Let

$$\mathbf{V} = V_u \mathbf{i}_u + V_v \mathbf{i}_v + V_w \mathbf{i}_w = h_1 V_u \nabla u + h_2 V_v \nabla v + h_3 V_w \nabla w$$

Then $\quad \nabla \times \mathbf{V} = \nabla(h_1 V_u) \times \nabla u + \nabla(h_2 V_v) \times \nabla v + \nabla(h_3 V_w) \times \nabla w$

where we have again made use of the fact that the curl of a gradient is zero. Therefore,

$$\nabla \times \mathbf{V} = \frac{\partial}{\partial v}(h_1 V_u)(\nabla v \times \nabla u) + \frac{\partial}{\partial w}(h_1 V_u)(\nabla w \times \nabla u)$$

$$+ \frac{\partial}{\partial u}(h_2 V_v)(\nabla u \times \nabla v) + \frac{\partial}{\partial w}(h_2 V_v)(\nabla w \times \nabla v)$$

$$+ \frac{\partial}{\partial u}(h_3 V_w)(\nabla u \times \nabla w) + \frac{\partial}{\partial v}(h_3 V_w)(\nabla v \times \nabla w)$$

$$= \frac{1}{h_2 h_3}\left[\frac{\partial}{\partial v}(h_3 V_w) - \frac{\partial}{\partial w}(h_2 V_v)\right]\mathbf{i}_u$$

$$+ \frac{1}{h_1 h_3}\left[\frac{\partial}{\partial w}(h_1 V_u) - \frac{\partial}{\partial u}(h_3 V_w)\right]\mathbf{i}_v$$

$$+ \frac{1}{h_1 h_2}\left[\frac{\partial}{\partial u}(h_2 V_v) - \frac{\partial}{\partial v}(h_1 V_u)\right]\mathbf{i}_w$$

$$= \frac{1}{h_1 h_2 h_3}\begin{vmatrix} h_1\mathbf{i}_u & h_2\mathbf{i}_v & h_3\mathbf{i}_w \\ \dfrac{\partial}{\partial u} & \dfrac{\partial}{\partial v} & \dfrac{\partial}{\partial w} \\ h_1 V_u & h_2 V_v & h_3 V_w \end{vmatrix}$$

Our main interest here is in the expression for the Laplacian, $\nabla^2 \psi$. We shall write this as $\nabla \cdot \nabla \psi$, and hence in the expression for the divergence we put $V_u = (1/h_1)(\partial \psi/\partial u)$, $V_v = (1/h_2)(\partial \psi/\partial v)$, and $V_w = (1/h_3)(\partial \psi/\partial w)$. Then

$$\nabla^2 \psi = \frac{1}{h_1 h_2 h_3}\left[\frac{\partial}{\partial u}\left(\frac{h_2 h_3}{h_1}\frac{\partial \psi}{\partial u}\right) + \frac{\partial}{\partial v}\left(\frac{h_1 h_3}{h_2}\frac{\partial \psi}{\partial v}\right) + \frac{\partial}{\partial w}\left(\frac{h_1 h_2}{h_3}\frac{\partial \psi}{\partial w}\right)\right]$$

As an example, let us take spherical coordinates, one of the coordinate systems in which the Helmholtz equation is separable. The transformation to spherical coordinates is

$$x = r \sin \theta \cos \phi$$
$$y = r \sin \theta \sin \phi$$
$$z = r \cos \theta$$

with $0 \leq r$, $0 \leq \theta \leq \pi$, and $0 \leq \phi < 2\pi$, where r, θ, ϕ are the coordinates shown in Fig. 7.

The line element is

$$(ds)^2 = (dr)^2 + r^2(d\theta)^2 + r^2 \sin^2 \theta(d\phi)^2$$

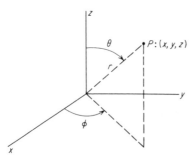

FIGURE 7

Hence, $h_1 = 1$, $h_2 = r$, and $h_3 = r \sin \theta$. The Laplacian is

$$\nabla^2 \psi = \frac{1}{r^2 \sin \theta} \left[\frac{\partial}{\partial r} \left(r^2 \sin \theta \frac{\partial \psi}{\partial r} \right) + \frac{\partial}{\partial \theta} \left(\sin \theta \frac{\partial \psi}{\partial \theta} \right) + \frac{\partial}{\partial \phi} \left(\frac{1}{\sin \theta} \frac{\partial \psi}{\partial \phi} \right) \right]$$

$$= \frac{\partial^2 \psi}{\partial r^2} + \frac{2}{r} \frac{\partial \psi}{\partial r} + \frac{1}{r^2} \frac{\partial^2 \psi}{\partial \theta^2} + \frac{\cot \theta}{r^2} \frac{\partial \psi}{\partial \theta} + \frac{1}{r^2 \sin^2 \theta} \frac{\partial^2 \psi}{\partial \phi^2}$$

In Helmholtz's equation let us assume that $\psi(r,\theta,\phi) = U(r)V(\theta)W(\phi)$; then

$$\nabla^2 \psi + \lambda \psi = \frac{VW}{r^2} \frac{d}{dr} \left(r^2 \frac{dU}{dr} \right) + \frac{UW}{r^2 \sin \theta} \frac{d}{d\theta} \left(\sin \theta \frac{dV}{d\theta} \right)$$
$$+ \frac{UV}{r^2 \sin^2 \theta} \frac{d^2 W}{d\phi^2} + \lambda UVW = 0$$

Dividing through by UVW/r^2 and transposing, we have

$$\frac{1}{U} \frac{d}{dr} \left(r^2 \frac{dU}{dr} \right) + \lambda r^2 = -\frac{1}{V \sin \theta} \frac{d}{d\theta} \left(\sin \theta \frac{dV}{d\theta} \right) - \frac{1}{W \sin^2 \theta} \frac{d^2 W}{d\phi^2}$$

The left-hand side is a function of r only, whereas the right-hand side is a function of θ and ϕ only. The only way this can be an identity in r, θ, and ϕ is for both sides to be equal to a constant which we shall call α. Therefore, we have

$$\frac{d}{dr} \left(r^2 \frac{dU}{dr} \right) + \lambda r^2 U - \alpha U = 0$$

$$\frac{\sin \theta}{V} \frac{d}{d\theta} \left(\sin \theta \frac{dV}{d\theta} \right) + \alpha \sin^2 \theta = -\frac{1}{W} \frac{d^2 W}{d\phi^2}$$

In the last line the left-hand side is a function of θ only, and the right-hand side is a function of ϕ only. This can be an identity in θ and ϕ only if both sides are equal to a constant β. Then we have

$$\frac{d}{d\theta} \left(\sin \theta \frac{dV}{d\theta} \right) + \alpha V \sin \theta - \frac{\beta}{\sin \theta} V = 0$$

$$\frac{d^2 W}{d\phi^2} + \beta W = 0$$

Thus, the partial differential equation has been reduced to three ordinary differential equations to be solved. If there exist values of the separation constants α, β, and λ for which these ordinary differential equations have solutions, then

$$\psi(r,\theta,\phi) = U(r,\lambda,\alpha) V(\theta,\alpha,\beta) W(\phi,\beta)$$

is an eigenfunction of Helmholtz's equation in spherical coordinates. Of course the problem is not completely stated until boundary conditions and continuity conditions on the solution are specified.   It is precisely these conditions which allow us to determine the admissible values of α, β, and λ.   For example, if we were attempting to solve Helmholtz's equation on the interior of a sphere of radius a subject to one of the boundary conditions $\psi = 0$, $\partial\psi/\partial r = 0$, or $\partial\psi/\partial r + \sigma\psi = 0$ at $r = a$, then these become conditions on $U(r)$ at $r = a$; that is, $U(a) = 0$, $U'(a) = 0$, or $U'(a) + \sigma U(a) = 0$.   The requirement that ψ be continuous and have continuous first derivatives at $r = 0$ means that $U(0)$ and $U'(0)$ must be finite.   If ψ is to be single-valued in the sphere and have a continuous derivative, then $W(0) = W(2\pi)$, $W'(0) = W'(2\pi)$, which means that $\beta = n^2$, n an integer. $V(0)$, $V'(0)$, $V(\pi)$, $V'(\pi)$ must all be finite if ψ is to be continuous and have continuous first derivatives on the polar axis of the sphere.

We notice that each of the separated equations can be written in the form

$$(py')' - qy + \lambda\rho y = 0$$

The solution of this equation subject to boundary conditions of the following types is known as the **Sturm-Liouville problem**:

1. $y(a) = 0$ or $y(b) = 0$
2. $y'(a) = 0$ or $y'(b) = 0$
3. $y'(a) - \sigma_1 y(a) = 0$ or $y'(b) + \sigma_2 y(b) = 0$, $\sigma_1 > 0$, $\sigma_2 > 0$
4. $y(a) = y(b)$ and $p(a)y'(a) = p(b)y'(b)$
5. $y(a)$ and $y'(a)$ finite, with $p(a) = 0$ or
    $y(b)$ and $y'(b)$ finite, with $p(b) = 0$

The first three are conditions imposed at physical boundaries; the fourth is a requirement of periodicity; the fifth is a requirement imposed to guarantee that ψ is sufficiently well-behaved in the interior of the region.   We shall discuss this problem in detail in the next section, but first let us show that the solutions of a Sturm-Liouville problem form an orthonormal set of functions.

We first note that the Sturm-Liouville problem is homogeneous.   Therefore, we can assume that any nontrivial solution has been normalized as follows:[1]

$$\int_a^b \rho y^2 \, dx = 1$$

Next assume that y_i is a solution corresponding to λ_i and that y_j is a solution corresponding to λ_j with $\lambda_i \neq \lambda_j$.   Then

$$(py_i')' - qy_i + \lambda_i \rho y_i = 0$$
$$(py_j')' - qy_j + \lambda_j \rho y_j = 0$$

Multiplying the first equation by y_j and the second by y_i, subtracting, and

---

[1] We assume that ρ is a nonnegative function.   In actual practice this is the case.

integrating, we have

$$(\lambda_i - \lambda_j) \int_a^b \rho y_i y_j \, dx = \int_a^b [(py_j')' y_i - (py_i')' y_j] \, dx$$
$$= [py_j' y_i - py_i' y_j]_a^b$$
$$= 0$$

for any combination of the above boundary conditions. Summarizing, we have

$$\int_a^b \rho y_i y_j \, dx = \delta_{ij}$$

which is to say that the solutions form an orthonormal set with respect to the **weighting function** $\rho(x)$.

We have stated that the technique of separation of variables leads to eigenfunctions of the Helmholtz equation under appropriate boundary conditions. It remains to show that it leads to all possible eigenfunctions of the boundary-value problem. This can be established if we can show that the solutions obtained by separation of variables form a complete set. Hence, any other solution of the boundary-value problem would be orthogonal to a complete set of functions and would, therefore, be identically zero. We shall give the argument for a two-dimensional problem in which it is assumed that Helmholtz's equation separates into a pair of ordinary differential equations leading to two Sturm-Liouville problems:

$$(py')' - qy + \lambda r y = 0$$
$$(PY')' - QY + \mu R Y = 0$$

We shall assume that there exists a denumerable set of solutions for each problem and that they are complete sets.[1] We therefore have completeness relations

$$\int_a^b rfg \, dx = \sum_{i=1}^{\infty} a_i b_i$$
$$\int_c^d RFG \, dx = \sum_{i=1}^{\infty} A_i B_i$$

where

$$a_i = \int_a^b rfy_i \, dx$$
$$b_i = \int_a^b rgy_i \, dx$$
$$A_i = \int_c^d RFY_i \, dx$$
$$B_i = \int_c^d RGY_i \, dx$$

---

[1] These facts will be proved in the next section.

We shall prove that the functions $y_i(x)\,Y_j(t)$ are a complete set; that is,

$$\int_a^b \int_c^d r(x)R(t)f^2(x,t)\,dt\,dx = \sum_{i=1}^{\infty} \sum_{j=1}^{\infty} c_{ij}^2$$

where

$$c_{ij} = \int_a^b \int_c^d r(x)R(t)f(x,t)y_i(x)\,Y_j(t)\,dt\,dx$$

First we note that the functions $y_i(x)\,Y_j(t)$ are an orthonormal set; that is,

$$\int_a^b \int_c^d r(x)R(t)y_i(x)y_j(x)\,Y_k(t)\,Y_m(t)\,dt\,dx = \delta_{ij}\,\delta_{km}$$

For each t, such that $c \le t \le d$,

$$\int_a^b r(x)f^2(x,t)\,dx = \sum_{i=1}^{\infty} g_i^2(t)$$

where

$$g_i(t) = \int_a^b r(x)f(x,t)y_i(x)\,dx$$

The series $\displaystyle\sum_{i=1}^{\infty} g_i^2(t)$ converges uniformly.   If this were not so, then there would exist a positive number ϵ, a sequence $\{n_k\}$, and a sequence $\{t_k\}$ such that

$$S_{n_k}(t_k) < S(t_k) - \epsilon$$

where

$$S(t) = \sum_{i=1}^{\infty} g_i^2(t) \qquad \text{and} \qquad S_n(t) = \sum_{i=1}^{n} g_i^2(t)$$

However, for a fixed N,

$$S_N(t_k) \le S_{n_k}(t_k)$$

for $n_k > N$.   Therefore,

$$S_N(t_k) < S(t_k) - \epsilon$$

Also we know that the sequence $\{t_k\}$ must have a limit τ in the interval. Letting t_k pass to the limit and using the continuity of S_N and S, we have

$$S_N(\tau) < S(\tau) - \epsilon$$

But this is not possible, since we have convergence at $t = \tau$ and $S_N(\tau)$ can be made arbitrarily close to $S(\tau)$.   Therefore, the series must converge uniformly.

We multiply the uniformly convergent series by a continuous function $R(t)$ and integrate term by term:

$$\int_a^b \int_c^d r(x)R(t)f^2(x,t)\,dt\,dx = \sum_{i=1}^{\infty} \int_c^d R(t)g_i^2(t)\,dt$$

Using the completeness relation for $R(t)g_i^2(t)$, we have

$$\int_a^b \int_c^d r(x)R(t)f^2(x,t)\,dt\,dx = \sum_{i=1}^\infty \sum_{j=1}^\infty c_{ij}^2$$

where $\quad c_{ij} = \int_c^d R(t)g_i(t)\,Y_j(t)\,dt = \int_a^b \int_c^d r(x)R(t)f(x,t)y_i(x)\,Y_j(t)\,dt\,dx$

We know that any linear combination of solutions of the Helmholtz equation satisfying the physical boundary conditions will also satisfy the original partial differential equation and the boundary conditions. We are therefore led to seek the general solution of the boundary-value–initial-value problem as a finite or infinite linear combination of fundamental solutions. The exact solution will be determined when specific initial values are imposed. This will be possible if we can expand the functions representing the initial values in series of eigenfunctions. Therefore, we shall investigate the problem of expanding functions in series of solutions obtained by separation of variables, but, as we have seen, this ultimately depends on the completeness of the set of solutions of a Sturm-Liouville problem. Hence, we turn our attention in the next section to the Sturm-Liouville problem.

Exercises 4.1

*1. Discuss cylindrical coordinates. Describe the coordinate surfaces and curves. Separate the variables in the Helmholtz equation. Discuss the boundary conditions on the solutions of the separated equations, assuming the region is the interior of a cylinder of radius a and height h.

2. Discuss parabolic coordinates

$$x = \sqrt{uv}\cos w$$

$$y = \sqrt{uv}\sin w$$

$$z = \frac{u-v}{2}$$

$0 \le u,\ 0 \le v,\ 0 \le w \le 2\pi$. Describe the coordinate surfaces and curves. Separate the variables in the Helmholtz equation.

*3. Assuming that $p(x)$, $q(x)$, and $\rho(x)$ are all greater than zero for $a \le x \le b$, prove that the eigenvalue λ in the Sturm-Liouville equation is positive under any of the boundary conditions 1 to 5.

*4. A circular membrane of radius 1 is in equilibrium with a displacement of $\phi(1,\theta) = a\theta(2\pi - \theta)$ on the boundary. Find the displacement at interior points. HINT: Separate Laplace's equation in cylindrical coordinates.

4.2 Sturm-Liouville Problems

As pointed out in the previous section, the Sturm-Liouville equation is

$$(py')' - qy + \lambda\rho y = 0$$

We shall assume that $p(x) \geq 0$ and $\rho(x) \geq 0$ over the interval $a \leq x \leq b$ in which the differential equation is to be satisfied.   If either vanishes, this will occur at an end point or at an isolated point inside the interval.   We sometimes write the equation

$$L(y) + \lambda \rho y = 0$$

where L is the linear operator $\dfrac{d}{dx}\left(p\dfrac{d}{dx}\right) - q$.   The operator has the following property.   If u and v are both functions which satisfy the same boundary conditions, then

$$\int_a^b [uL(v) - vL(u)]\,dx = \int_a^b [u(pv')' - v(pu')']\,dx$$
$$= [puv' - pvu']_b^a = 0$$

If we call the scalar product $(f,g) = \displaystyle\int_b^a fg\,dx$, then we can write

$$(u,Lv) = (Lu,v)$$

Therefore, the operator L is self-adjoint.

The fact that L is self-adjoint has some interesting consequences.   For example, let λ_i and λ_j be two different eigenvalues of the Sturm-Liouville equation, corresponding to solutions y_i and y_j; then

$$L(y_i) + \lambda_i \rho y_i = 0$$
$$L(y_i) + \lambda_j \rho y_j = 0$$

and
$$(\lambda_i - \lambda_j)\int_a^b \rho y_i y_j\,dx = (y_j,Ly_i) - (Ly_j,y_i) = 0$$

Since $\lambda_i \neq \lambda_j$, $\displaystyle\int_a^b \rho y_i y_j\,dx = 0$.   We say that solutions corresponding to different eigenvalues are orthogonal with respect to the weighting function ρ, or

$$(\sqrt{\rho}y_i, \sqrt{\rho}y_j) = 0$$

Except for the weighting factor, this is similar to all the other self-adjoint operators we have encountered.

If there exist two different solutions u and v corresponding to the same eigenvalue λ, then

$$uL(v) - vL(u) = u(pv')' - v(pu')' = 0$$

$$= \frac{d}{dx}[p(uv' - vu')]$$

This implies that $p(uv' - vu')$ is constant.   For every boundary condition except condition 4, the condition for periodicity, the expression vanishes at the end points of the interval. However, p is not identically zero over any subinterval; therefore, $uv' - vu' = 0$. Hence,

$$\frac{u'}{u} = \frac{v'}{v}$$

which implies that $u = cv$. This shows that λ cannot correspond to two linearly independent solutions except in the case of the boundary condition for periodicity. In the latter case we have, for example,

$$y'' + \lambda y = 0$$

and with $\lambda = n^2$ we have independent solutions $\cos n\theta$ and $\sin n\theta$ for the interval $0 \le \theta \le 2\pi$.

The solutions of the Sturm-Liouville problem can be characterized as solutions of problems in the calculus of variations. We shall begin by considering boundary conditions 1 to 3, corresponding to physical boundaries at the ends of the interval $a \le x \le b$. As we shall see in Sec. 4.3, if $p(x)$ vanishes at any points in the interval $a \le x \le b$ or if $q(x)$ or $\rho(x)$ become unbounded in the interval, the solution of the equation may become unbounded. We avoid this situation for now by making the following assumptions: $p(x)$, $q(x)$, and $\rho(x)$ are continuous functions in the interval $a \le x \le b$, and $p > 0$ and $\rho > 0$ in the interval. We shall call this the **regular case**. For boundary condition 3, the variational principle defining the kth eigenvalue and eigenfunction is

$$\lambda_k = \min \frac{Q(f)}{N(f)}$$

subject to $N(f, y_i) = 0$, $i = 1, 2, \ldots, k - 1$, where

$$Q(f) = \int_a^b [p(f')^2 + qf^2]\, dx + \sigma_1 p(a)[f(a)]^2 + \sigma_2 p(b)[f(b)]^2$$

$$N(f) = \int_a^b \rho f^2\, dx$$

and f is in the space of functions continuous and having a piecewise continuous derivative in the interval $a \le x \le b$.

We shall derive necessary conditions for the first eigenvalue and eigenfunction. The extension to the kth eigenvalue and eigenfunction is straightforward. We notice that Q and N are both quadratic functionals and that N is positive-definite since $\rho > 0$. Let y_1 be the first eigenfunction and λ_1 be the first eigenvalue; then

$$Q(y_1 + \epsilon\eta) \ge \lambda_1 N(y_1 + \epsilon\eta)$$

where ϵ is an arbitrary constant and η is an arbitrary function. Expanding, we have

$$Q(y_1) + 2\epsilon Q(y_1,\eta) + \epsilon^2 Q(\eta) \geq \lambda_1[N(y_1) + 2\epsilon N(y_1,\eta) + \epsilon^2 N(\eta)]$$

Since $Q(y_1) = \lambda_1 N(y_1)$,

$$2\epsilon[Q(y_1,\eta) - \lambda_1 N(y_1,\eta)] + \epsilon^2[Q(\eta) - \lambda_1 N(\eta)] \geq 0$$

This is to be true for arbitrary ϵ; therefore

$$Q(y_1,\eta) - \lambda_1 N(y_1,\eta) = 0$$

for arbitrary η. Integrating by parts, we have

$$\int_a^b [-(py_1')' + qy_1 - \lambda_1\rho y_1]\eta \, dx + p(b)[y_1'(b) + \sigma_2 y_1(b)]\eta(b)$$
$$- p(a)[y_1'(a) - \sigma_1 y_1(a)]\eta(a) = 0$$

Since η is arbitrary, we have the following necessary conditions for the minimizing function:

$$(py_1')' - qy_1 + \lambda_1\rho y_1 = 0 \qquad a \leq x \leq b$$

$$y_1'(b) + \sigma_2 y_1(b) = 0$$

$$y_1'(a) - \sigma_1 y_1(a) = 0$$

We see that the boundary conditions are natural boundary conditions. If we wanted the boundary conditions $y'(a) = y'(b) = 0$, we would merely set $\sigma_1 = \sigma_2 = 0$. The conditions $y(a) = y(b) = 0$ are not natural boundary conditions. In this case, we would prescribe $f(a) = f(b) = 0$. We could have one boundary condition satisfied at one end point and another at the other end point by properly specifying σ_1 and σ_2 and the conditions on the function space of admissible functions.

We recall from Sec. 3.8 that to show that all the solutions of the differential equation are obtained from the variational problem, we have to show that the eigenfunctions of the variational problem form a complete set. This is easily done if we can show that the eigenvalues increase without bound. Let λ_k, $\tilde{\lambda}_k$, and $\bar{\lambda}_k$ be the eigenvalues corresponding to the boundary conditions $y(a) = y(b) = 0$, $y'(a) - \sigma_1 y(a) = y'(b) + \sigma_2 y(b) = 0$, and $y'(a) = y'(b) = 0$, respectively. Then

$$\bar{\lambda}_k \leq \tilde{\lambda}_k \leq \lambda_k$$

This follows from theorems 1 and 2 of Sec. 3.7. The inequality on the left follows from theorem 2, for the minimum principles use the same class of functions, but the functionals differ in a predetermined sense for every function in the class. The inequality on the right follows from theorem 1, since, in this case, the functionals are the same, but the class of admissible functions is restricted for the prescribed boundary condition. For mixed boundary conditions, the eigenvalues can be interspersed in this inequality. However, $\bar{\lambda}_k$ will be the smallest and $\bar{\lambda}_k$ the largest for any choice of boundary conditions in the regular case. Therefore, in this equality $\bar{\lambda}_k$ could be interpreted as the eigenvalue for any mixed set of boundary conditions as well.

We are assuming that p, q, and ρ are all continuous functions. Hence,

$$0 < p_m \le p(x) \le p_M$$
$$0 < \rho_m \le \rho(x) \le \rho_M$$
$$q_m \le q(x) \le q_M$$

where p_m, ρ_m, q_m are the minimum values of p, ρ, and q, and p_M, ρ_M, and q_M are the maximum values of p, ρ, and q in the interval $a \le x \le b$. Consider the following variational principle:

$$\mu_k = \min \frac{\displaystyle\int_a^b [p_m(f')^2 + q_m f^2] \, dx}{\displaystyle\int_a^b \rho_M f^2 \, dx}$$

subject to $\displaystyle\int_b^a \rho_M f u_i \, dx = 0$, $i = 1, 2, \ldots, k - 1$, where $u_1, u_2, \ldots, u_{k-1}$ are the solutions of $k - 1$ preceding problems. The necessary conditions for the minimizing function are

$$p_m u_k'' - q_m u_k + \mu_k \rho_M u_k = 0$$
$$u_k'(a) = u_k'(b) = 0$$

The differential equation can be written

$$u_k'' + \alpha_k^2 u_k = 0$$

with $\alpha_k^2 = (\rho_M \mu_k - q_m)/p_m$. The solutions are

$$u_k = \cos \frac{(k-1)\pi(x-a)}{(b-a)} \qquad k = 1, 2, \ldots$$

Thus, we have $\qquad \mu_k = \dfrac{p_m[(k-1)^2 \pi^2/(b-a)^2] + q_m}{\rho_M}$

Also, from theorem 2, Sec. 3.7, we have $\mu_k \le \bar{\lambda}_k$.

Next consider the variational principle

$$\gamma_k = \min \frac{\displaystyle\int_a^b [p_M(f')^2 + q_M f^2] \, dx}{\displaystyle\int_a^b \rho_m f^2 \, dx}$$

subject to $f(a) = f(b) = 0$ and $\displaystyle\int_a^b \rho_m f v_i \, dx = 0$, $i = 1, 2, \ldots, k - 1$, where

$v_1, v_2, \ldots, v_{k-1}$ are the solutions of $k-1$ preceding minimum problems.   The necessary conditions for the solution are

$$p_M v_k'' - q_M v_k + \gamma_k \rho_m v_k = 0$$

$$v_k(a) = v_k(b) = 0$$

The differential equation can be written

$$v_k'' + \beta_k^2 v_k = 0$$

with $\beta_k^2 = (\rho_m \gamma_k - q_M)/p_M$.   The solutions are

$$v_k = \sin \frac{k\pi(x-a)}{b-a} \qquad k = 1, 2, \ldots$$

Thus, we have
$$\gamma_k = \frac{p_M[k^2\pi^2/(b-a)^2] + q_M}{\rho_m}$$

From theorem 2, Sec. 3.7, we have $\lambda_k \le \gamma_k$.   The following inequalities result:

$$\frac{p_m[(k-1)^2\pi^2/(b-a)^2] + q_m}{\rho_M} \le \bar{\lambda}_k \le \tilde{\lambda}_k \le \lambda_k \le \frac{p_M[k^2\pi^2/(b-a)^2] + q_M}{\rho_m}$$

This implies that the sequence of eigenvalues increases without bound regardless of what combination of boundary conditions 1 to 3 we take.   It also implies that, although q may be negative and there may be negative eigenvalues, there are at most a finite number of them.   Following the method of Sec. 3.8, we can show that the eigenfunctions of the successive variational problems are a complete set.   This implies that eigenfunctions of the variational problem represent all the eigenfunctions of the Sturm-Liouville problem.

We can use the completeness of the set of eigenfunctions to prove an expansion theorem for arbitrary functions in the class of admissible functions for the variational principle.   First, however, we must prove that the eigenfunctions are bounded independent of x and k.   To do this we must transform the differential equations as follows.   Let $v = fy$ and $t = \displaystyle\int_a^b \left(\frac{\rho}{p}\right)^{\frac{1}{2}} dx$, where $f = (\rho p)^{\frac{1}{4}}$.   Then

$$y' = \frac{d}{dt}\left(\frac{v}{f}\right)\frac{dt}{dx} = \left(\frac{\dot{v}}{f} - \frac{v}{f^2}\dot{f}\right)\left(\frac{\rho}{p}\right)^{\frac{1}{2}}$$

$$py' = f\dot{v} - v\dot{f}$$

$$(py')' = (f\ddot{v} - v\ddot{f})\left(\frac{\rho}{p}\right)^{\frac{1}{2}} = \frac{\rho}{f}\ddot{v} - v\ddot{f}\frac{\rho}{f^2}$$

$$(py')' - qy + \lambda\rho y = \frac{\rho}{f}\left[\ddot{v} - \left(\frac{\ddot{f}}{f} + \frac{q}{\rho}\right)v + \lambda v\right] = 0$$

The equation satisfied by v is then

$$\ddot{v} - rv + \lambda v = 0$$

where $r = \ddot{f}/f + q/\rho$, which is a continuous function.   The equation is to hold in the interval $0 \le t \le T$ where $T = \int_b^a \left(\frac{\rho}{p}\right)^{\frac{1}{2}} dx$.

Multiplying by $\dot{v}$ and integrating from 0 to t, we have

$$[\dot{v}(t)]^2 + \lambda[v(t)]^2 - 2\int_0^t rv\dot{v}\, d\tau = [\dot{v}(0)]^2 + \lambda[v(0)]^2$$

We next integrate the equation from 0 to T to show that

$$T\{[\dot{v}(0)]^2 + \lambda[v(0)]^2\} = \int_0^T \dot{v}^2\, dt - 2\int_0^T \int_0^t rv\dot{v}\, d\tau\, dt + \lambda$$

We are assuming that the function has been normalized as follows:

$$\int_a^b \rho y^2\, dx = \int_0^T v^2\, dt = 1$$

As a result, we have

$$\lambda[v(t)]^2 \le \lambda[v(t)]^2 + [\dot{v}(t)]^2 = 2\int_0^t rv\dot{v}\, d\tau + [\dot{v}(0)]^2 + \lambda[v(0)]^2$$

or $$\lambda[v(t)]^2 \le \left| 2\int_0^t rv\dot{v}\, d\tau \right| + \frac{\lambda}{T} + \frac{1}{T}\int_0^T \dot{v}^2\, dt + \frac{2}{T}\left| \int_0^T \int_0^t rv\dot{v}\, d\tau\, dt \right|$$

By Schwarz's inequality,

$$\left| 2\int_0^t rv\dot{v}\, d\tau \right| \le 2\left[\int_0^t r^2 v^2\, d\tau\right]^{\frac{1}{2}}\left[\int_0^t \dot{v}^2\, d\tau\right]^{\frac{1}{2}}$$

$$\le C_1\left[\int_0^T v^2\, d\tau\right]^{\frac{1}{2}}\left[\int_0^T \dot{v}^2\, d\tau\right]^{\frac{1}{2}}$$

$$\le C_1\left[\int_0^T \dot{v}^2\, d\tau\right]^{\frac{1}{2}}$$

where C_1 is a constant independent of t and λ.   Also

$$\left| \int_0^T \int_0^t rv\dot{v}\, d\tau\, dt \right| \le \left[\int_0^T \int_0^t r^2 v^2\, d\tau\, dt\right]^{\frac{1}{2}}\left[\int_0^T \int_0^t \dot{v}^2\, d\tau\, dt\right]^{\frac{1}{2}}$$

$$\le \left[\int_0^T \int_0^T r^2 v^2\, d\tau\, dt\right]^{\frac{1}{2}}\left[\int_0^T \int_0^T \dot{v}^2\, d\tau\, dt\right]^{\frac{1}{2}}$$

$$\le T\left[\int_0^T r^2 v^2\, d\tau\right]^{\frac{1}{2}}\left[\int_0^T \dot{v}^2\, d\tau\right]^{\frac{1}{2}}$$

$$\le C_2\left[\int_0^T \dot{v}^2\, d\tau\right]^{\frac{1}{2}}$$

Since $v = (\rho p)^{\frac{1}{2}} y = fy$ and $\dot{v} = (f'y + fy')(p/\rho)^{\frac{1}{2}}$,

$$\int_0^T \dot{v}^2 \, dt = \int_a^b \frac{(f')^2}{\rho} \left(\frac{p}{\rho}\right)^{\frac{1}{2}} \rho y^2 \, dx + 2\int_a^b \frac{ff'}{\rho} \, \rho^{\frac{1}{2}} y p^{\frac{1}{2}} y' \, dx + \int_a^b py'^2 \, dx$$

Therefore,

$$\int_0^T \dot{v}^2 \, dt \leq C_3 \int_a^b py^2 \, dx + 2\left[\int_a^b \left(\frac{ff'}{\rho}\right)^2 \rho y^2 \, dx\right]^{\frac{1}{2}} \left[\int_a^b py'^2 \, dx\right]^{\frac{1}{2}} + \int_a^b py'^2 \, dx$$

$$\leq C_3 + C_4 \left[\int_a^b py'^2 \, dx\right]^{\frac{1}{2}} + \int_a^b py'^2 \, dx$$

Starting from the Sturm-Liouville equation, we have

$$\int_a^b [(py')'y - qy^2 + \lambda \rho y^2] \, dx = 0$$

which implies that

$$\int_a^b py'^2 \, dx + \sigma_2 p(b)[y(b)]^2 + \sigma_1 p(a)[y(a)]^2 = \lambda - \int_a^b qy^2 \, dx$$

Thus, $$\int_a^b py'^2 \, dx \leq \int_a^b py'^2 \, dx + \sigma_2 p(b)[y(b)]^2 + \sigma_1 p(a)[y(a)]^2$$

$$\leq \lambda + C_5$$

Combining this with the above,

$$\int_0^T \dot{v}^2 \, dt \leq C_6 + C_7 \sqrt{\lambda} + \lambda$$

and $$\lambda[v(t)]^2 \leq C_8 + C_9 \sqrt{\lambda} + C_{10}\lambda$$

$$[v(t)]^2 \leq C_{10} + \frac{C_9}{\sqrt{\lambda}} + \frac{C_8}{\lambda}$$

where the C's are all independent of t and λ.   We have therefore shown that $v(t)$ is bounded independent of t and λ, which in turn implies that $y(x)$ is bounded independent of x and λ.   We have assumed the third boundary condition, but it is clear that the same result holds for boundary conditions 1 and 2.

We are now ready to prove that any function f in the class of admissible functions appropriate to the variational principle defining the Sturm-Liouville functions for each of the three homogeneous boundary conditions can be expanded in a uniformly convergent series of the appropriate eigenfunctions.   Consider the series $\sum_{k=1}^{\infty} c_k y_k$ where

$$c_k = \int_a^b \rho f y_k \, dx$$

and
$$\int_a^b \rho y_k^2 \, dx = 1$$

By Schwarz's inequality, using the fact that $\lambda_k > 0$ for large k,

$$\left(\sum_{k=n}^p c_k y_k\right)^2 \le \sum_{k=n}^p c_k^2 \lambda_k \sum_{k=n}^p \frac{y_k^2}{\lambda_k} \le \sum_{k=n}^\infty c_k^2 \lambda_k \sum_{k=n}^\infty \frac{y_k^2}{\lambda_k}$$

We know from the proof of completeness of the eigenfunctions that for sufficiently large m

$$0 \le Q(f_m) = Q(f) - \sum_{k=1}^m c_k^2 \lambda_k$$

where $f_m = f - \sum_{k=1}^m c_k y_k$. Therefore, the series $\sum_{k=1}^\infty c_k^2 \lambda_k$ converges. We have shown above that $y_k^2(x)$ is bounded independent of x and k. Therefore, the series $\sum_{k=1}^\infty \frac{y_k^2}{\lambda_k}$ converges uniformly. This follows because $\lambda_k \ge M(k-1)^2$ for k sufficiently large and $\sum_{k=2}^\infty \frac{1}{(k-1)^2}$ converges. Therefore, $\sum_{k=n}^\infty \frac{y_k^2}{\lambda_k}$ approaches zero as n approaches infinity uniformly in x. This implies that $\left(\sum_{k=n}^p c_k y_k\right)^2$ approaches zero uniformly in x for every $p > n$, and by the Cauchy criterion for convergence the series $\sum_{k=1}^\infty c_k y_k$ converges uniformly. The series must converge uniformly to $f(x)$ by the completeness of the eigenfunctions. The same result would hold if we use $\left(\sum_{k=n}^p |c_k y_k|\right)^2$ in the above argument. Therefore, we also have absolute convergence.

If we do not have the regular case considered above, certain modifications in the proof have to be made. For example, consider Bessel's equation, which occurs in the separation of the Helmholtz equation in cylindrical coordinates.[1] The equation has the form

$$(xy')' - \frac{m^2}{x} y + \lambda xy = 0$$

where m is commonly a nonnegative integer. We see that $p = x$ is zero at $x = 0$, $\rho = x$ is zero at $x = 0$, and $q = m^2/x$ is not continuous at $x = 0$. Therefore, if our independent variable lies in the interval $0 \le x \le b$, then we do not have the regular case. Let us take m, an integer greater than or equal to 1, and the boundary conditions $y(0)$ finite and $y(b) = 0$. We shall see later that Bessel's equation has two linearly independent solutions $J_m(\sqrt{\lambda}x)$, the Bessel function of the first kind of order m, and $Y_m(\sqrt{\lambda}x)$, the Bessel function of the second kind of order m. The second of these is not finite at $x = 0$;

---

[1] See exercise 1, Sec. 4.1.

therefore, we discard it. To find the eigenvalues λ_k we would then have to solve the equation $J_m(\sqrt{\lambda}b) = 0$.

We can get some idea of the behavior of the sequence of eigenvalues if we consider the problem from the point of view of the calculus of variations. The solution of the above Sturm-Liouville problem can be characterized by the following variational principle.

$$\lambda_k = \min \frac{\displaystyle\int_0^b [x(f')^2 + (m^2/x)f^2]\, dx}{\displaystyle\int_0^b xf^2\, dx}$$

subject to $\displaystyle\int_0^b xfJ_m(\sqrt{\lambda_j}x)\, dx = 0$ and $f(b) = 0$, where λ_j are the first $k-1$ solutions of $J_m(\sqrt{\lambda}b) = 0$. Before proceeding, we first transform the differential equation by introducing the new dependent variable $v = \sqrt{x}y$. The equation then becomes

$$v'' - \frac{4m^2 - 1}{4x^2}\, v + \lambda v = 0$$

The fact that $y(0)$ is finite implies that $v(0) = 0$. The variational principle corresponding to this differential equation is

$$\lambda_k = \min \frac{\displaystyle\int_0^b \{(f')^2 + [(4m^2 - 1)/4x^2]f^2\}\, dx}{\displaystyle\int_0^b f^2\, dx}$$

where $f(0) = f(b) = 0$, and $\displaystyle\int_a^b fv_i\, dx = 0$, $i = 1, 2, \ldots, k-1$. By theorem 2, Sec. 3.7,

$$\lambda_k \geq \mu_k = \min \frac{\displaystyle\int_0^b (f')^2\, dx}{\displaystyle\int_0^b f^2\, dx}$$

subject to $\displaystyle\int_0^b f \sin \sqrt{\mu_j}x\, dx = 0$ and $f(0) = f(b) = 0$, where

$$\mu_j = \frac{j^2\pi^2}{b^2} \qquad j = 1, 2, \ldots, k-1$$

$\mu_k = k^2\pi^2/b^2$, and therefore we have the inequality $k^2\pi^2/b^2 \leq \lambda_k$. There is, therefore, an infinite sequence of zeros of the Bessel function which increases without bound. This immediately implies that the set of functions $J_m(\sqrt{\lambda_k}x)$ is a complete set. Also, if we exclude an arbitrarily small neighborhood of $x = 0$, we can prove that the eigenfunctions are bounded independent of x and

λ_k. It is then possible to prove that $\sum\limits_{k=1}^{\infty} c_k J_m(\sqrt{\lambda_k}x)$ converges uniformly to $f(x)$ in the interval $0 < \epsilon \leq x \leq b$ if f is a continuous function with a piecewise continuous first derivative and c_k is defined as follows:

$$c_k = \frac{\displaystyle\int_0^b xf J_m(\sqrt{\lambda_k}x)\, dx}{\displaystyle\int_0^b x J_m^2(\sqrt{\lambda_k}x)\, dx}$$

The denominator is easily evaluated. Multiplying the left-hand side of the differential equation by xJ'_m and integrating, we have

$$\int_0^x [(xJ'_m)'xJ'_m - m^2 J_m J'_m + \lambda_k x^2 J_m J'_m]\, dx = 0$$

$$[xJ'_m(\sqrt{\lambda_k}x)]^2 - m^2 J_m^2(\sqrt{\lambda_k}x) + \lambda_k x^2[J_m(\sqrt{\lambda_k}x)]^2$$

$$-\int_0^x xJ'_m \left[(xJ'_m)' - \frac{m^2}{x}J_m + \lambda_k xJ_m\right] dx - 2\lambda_k \int_0^x xJ_m^2\, dx = 0$$

or, letting $x = b$ and making use of $J_m(\sqrt{\lambda_k}b) = 0$, we have

$$\int_0^b xJ_m^2(\sqrt{\lambda_k}x)\, dx = \frac{b^2[J'_m(\sqrt{\lambda_k}b)]^2}{2\lambda_k}$$

If $m = 0$, we cannot use the same comparison procedure for the eigenvalues as above, so we must use a different means of showing that the sequence of zeros of $J_0(\sqrt{\lambda}b)$ increases without bound. Making the change of variable $z = \sqrt{\lambda}x$, we have

$$zJ_0''(z) + J_0'(z) + zJ_0(z) = 0$$

Differentiating this equation, we have

$$zJ_0''' + 2J_0'' + zJ_0' + J_0 = 0$$

$$zJ_0''' + J_0'' - \frac{J_0'}{z} + zJ_0' = 0$$

The latter equation satisfied by $J_0'(z)$ is the equation satisfied by $J_1(z)$. Therefore, $J_0'(z) = cJ_1(z)$. This means that the zeros of $J_1(z)$ are the zeros of $J_0'(z)$. We have already shown that the zeros of $J_1(z)$ form an unbounded sequence; therefore, the same is true of the zeros of $J_0'(z)$. We shall infer from this that the zeros of $J_0(z)$ increase without bound.

Let ξ be a zero of $J_0'(z)$; then $\xi J_0''(\xi) + \xi J_0(\xi) = 0$. This implies that $J_0''(\xi)$ and $J_0(\xi)$ are of opposite sign. Otherwise $J_0(\xi) = J_0'(\xi) = J_0''(\xi) = 0$, and the differential equation would imply that all higher derivatives of J_0 would vanish at ξ.   Thus the power series expansion about ξ would vanish identically.   Now let ξ_1 and ξ_2 be adjacent zeros of $J_0'(z)$.   Hence, $J_0'(z)$ is all of one sign between ξ_1 and ξ_2.   By Rolle's theorem $J_0''(z)$ must have an odd number of zeros between ξ_1 and ξ_2, where a zero of order n is counted n times.   This implies that $J_0''(\xi_1)$ and $J_0''(\xi_2)$ are of opposite sign and, therefore, that $J_0(\xi_1)$ and $J_0(\xi_2)$ are of opposite sign.   This implies that $J_0(z)$ has at least one zero between ξ_1 and ξ_2.   It cannot have more than one zero, for if it did, Rolle's theorem would imply that $J_0'(z)$ has a zero between ξ_1 and ξ_2.   We have thus shown that the zeros of $J_0'(z)$ bracket the zeros of $J_0(z)$ and, therefore, the zeros of $J_0(z)$ increase without bound.   The set of functions $J_0(\sqrt{\lambda_k}x)$, where the λ_k are the solutions of $J_0(\sqrt{\lambda}b) = 0$, is a complete set, and we have an expansion theorem in terms of Bessel functions of order zero.

The trigonometric functions $\sin nx$ and $\cos nx$ are special cases of Sturm-Liouville functions which come from the differential equation

$$y'' + \lambda y = 0$$

to be satisfied for $-\pi \le x \le \pi$, under periodic boundary conditions

$$y(\pi) = y(-\pi) \qquad \text{and} \qquad y'(\pi) = y'(-\pi)$$

The eigenvalues are $\lambda = n^2$, $n = 0, 1, 2, \ldots$.   The expansion theorem leads to Fourier series expansions of functions which satisfy suitable continuity requirements.   We can arrive at the expansion theorem by the following procedure.

Consider the solution of the following variational problem:

$$\lambda_k = \min \frac{\displaystyle\int_0^\pi (f')^2\, dx}{\displaystyle\int_0^\pi f^2\, dx}$$

over all continuous functions f which have a piecewise continuous derivative, vanish at the end points, and satisfy $\int_0^\pi f \sin jx\, dx = 0, j = 1, 2, 3, \ldots, k-1$. This is a regular Sturm-Liouville problem with normalized solutions

$$\sqrt{2/\pi}\, \sin kx$$

We therefore know that any admissible function can be expanded in a uniformly convergent series

$$f(x) = \sum_{k=1}^{\infty} b_k \sin kx$$

where $b_k = \dfrac{2}{\pi} \displaystyle\int_0^\pi f \sin kx \, dx$.   Next consider the variational problem

$$\lambda_k = \min \frac{\displaystyle\int_0^\pi (f')^2 \, dx}{\displaystyle\int_0^\pi f^2 \, dx}$$

over all continuous functions f which have a piecewise continuous first derivative and satisfy $\displaystyle\int_0^\pi f \cos jx \, dx = 0$, $j = 0, 1, 2, \ldots, k-1$. This is also a regular Sturm-Liouville problem with normalized solutions $\sqrt{1/\pi}$, $\sqrt{2/\pi} \cos kx$, $k = 1, 2, 3, \ldots$. We have an expansion theorem which states that any admissible function can be expanded in a uniformly convergent series

$$f(x) = \frac{a_0}{2} + \sum_{k=1}^\infty a_k \cos kx$$

where
$$a_k = \frac{2}{\pi} \int_0^\pi f \cos kx \, dx$$

Now consider any continuous function with a piecewise continuous derivative which is periodic with period 2π. Such a function can be written as follows:

$$f(x) = \tfrac{1}{2}[f(x) + f(-x)] + \tfrac{1}{2}[f(x) - f(-x)]$$

It is obvious that $\tfrac{1}{2}[f(x) + f(-x)]$ is even and $\tfrac{1}{2}[f(x) - f(-x)]$ is odd. Hence, we can write any such function as the sum of an odd and an even function; that is,

$$f(x) = f_e(x) + f_o(x)$$

$f_e(x)$ and $f_o(x)$ are continuous and have piecewise continuous first derivatives. Furthermore,

$$f_o(0) = \tfrac{1}{2}[(f0) - f(0)] = 0$$
$$f_o(\pi) = \tfrac{1}{2}[f(\pi) - f(-\pi)] = 0$$

since $f(x)$ is periodic. Therefore, $f_o(x)$ is admissible in the first variational principle and $f_e(x)$ is admissible in the second variational principle. Hence,

$$f_e(x) = \frac{a_0}{2} + \sum_{k=1}^\infty a_k \cos kx$$

$$f_o(x) = \sum_{k=1}^\infty b_k \sin kx$$

in the interval $0 \le x \le \pi$. These series also represent the functions in the interval $-\pi \leq x \leq 0$, for

$$f_e(-x) = f_e(x) = \frac{a_0}{2} + \sum_{k=1}^{\infty} a_k \cos kx$$

$$f_o(-x) = -f_o(x) = -\sum_{k=1}^{\infty} b_k \sin kx$$

Therefore, $f(x) = f_e(x) + f_o(x) = \frac{a_0}{2} + \sum_{k=1}^{\infty} (a_k \cos kx + b_k \sin kx)$

valid for $-\pi \leq x \leq \pi$, and the convergence is uniform.   The expansion coefficients can be expressed in terms of $f(x)$ as follows:

$$a_k = \frac{2}{\pi} \int_0^{\pi} f_e \cos kx \, dx$$

$$= \frac{2}{\pi} \int_0^{\pi} \tfrac{1}{2}[f(x) + f(-x)] \cos kx \, dx$$

$$= \frac{1}{\pi} \left[\int_0^{\pi} f(x) \cos kx \, dx + \int_{-\pi}^0 f(x) \cos kx \, dx \right]$$

$$= \frac{1}{\pi} \int_{-\pi}^{\pi} f(x) \cos kx \, dx$$

$$b_k = \frac{2}{\pi} \int_0^{\pi} f_o \sin kx \, dx$$

$$= \frac{2}{\pi} \int_0^{\pi} \tfrac{1}{2}[f(x) - f(-x)] \sin kx \, dx$$

$$= \frac{1}{\pi} \left[\int_0^{\pi} f(x) \sin kx \, dx + \int_{-\pi}^0 f(x) \sin kx \, dx \right]$$

$$= \frac{1}{\pi} \int_{-\pi}^{\pi} f(x) \sin kx \, dx$$

We therefore have the following theorem: *Any continuous function which is periodic with period 2π and has a piecewise continuous first derivative can be expanded in a uniformly convergent Fourier series.*

$$f(x) = \frac{a_0}{2} + \sum_{k=1}^{\infty} (a_k \cos kx + b_k \sin kx)$$

where

$$a_k = \frac{1}{\pi} \int_{-\pi}^{\pi} f(x) \cos kx \, dx$$

$$b_k = \frac{1}{\pi} \int_{-\pi}^{\pi} f(x) \sin kx \, dx$$

This is the same as theorem 1 of Sec. 2.2.   Here the approach was from the point of view of the calculus of variations, as contrasted with the more algebraic approach of Sec. 2.2.   It also should be remembered that in Sec. 3.5 we showed the existence of the solutions of a problem in the calculus of variations using the completeness of the set of trigonometric functions. Hence, all these methods are intimately connected. The present expansion theorem can be extended to the class of piecewise continuous functions, as was done in theorem 2 of Sec. 2.2.

Exercises 4.2

*1. Prove that the zeros of $J'_m(z)$ increase without bound and hence that

$$\phi_k(x) = J_m(\sqrt{\lambda_k} x)$$

where λ_k is a solution of $J'_m(\sqrt{\lambda} b) = 0$, forms a complete set of functions. HINT: Consider the eigenvalue problem for the Bessel equation with the boundary condition $y'(b) = 0$.

*2. Show that the Legendre polynomials $P_n(x)$, which are solutions of

$$[(1 - x^2)y']' + \lambda y = 0$$

$-1 \leq x \leq 1$, $y(-1)$, $y(1)$ finite, can be derived from the variational principle

$$\lambda_k = \min \frac{\displaystyle\int_{-1}^{1} (1 - x^2)(f')^2 \, dx}{\displaystyle\int_{-1}^{1} f^2 \, dx}$$

subject to $\displaystyle\int_{-1}^{1} fP_n(x) \, dx = 0$, $n = 1, 2, \ldots, k - 1$.

3. A tapered homogeneous elastic rod with density ρ, length L, and Young's modulus E has a circular cross section with diameter a at one end and b at the other. It is built in at both ends. Set up Hamilton's principle for small longitudinal vibrations of the cross sections, derive the Euler partial differential equation, and show that the method of separation of variables leads to a Sturm-Liouville problem.

4. Prove that for any of the regular Sturm-Liouville problems

$$\frac{p_m}{(b - a)^2 \rho_M} \leq \lim_{k \to \infty} \frac{\lambda_k}{k^2 \pi^2} \leq \frac{p_M}{(b - a)^2 \rho_m}$$

5. State the variational principle for the eigenvalue problem $\ddot{v} - rv + \lambda v = 0$, $v(0) = v(T) = 0$. Relate this problem to $(py')' - qy + \lambda \rho y = 0$, $y(a) = y(b) = 0$. Prove that

$$\lim_{k \to \infty} \frac{k^2 \pi^2}{\lambda_k} = \left(\int_a^b \sqrt{\frac{\rho}{p}} \, dx \right)^2$$

4.3   Series Solutions of Ordinary Differential Equations

The general second-order linear homogeneous ordinary differential equation can be written

$$y'' + f(x)y' + g(x)y = 0$$

Clearly the Sturm-Liouville equation can be put in this form. For the purpose of finding explicit solutions of the Sturm-Liouville equation, we want to investigate the problem of finding series solutions of this equation.

We say that x_0 is an **ordinary point** of the differential equation if there exists an interval $|x - x_0| < R$ in which $f(x)$ and $g(x)$ have convergent power series representations

$$f(x) = \sum_{k=0}^{\infty} a_k (x - x_0)^k$$

$$g(x) = \sum_{k=0}^{\infty} b_k (x - x_0)^k$$

We shall prove that in this case the differential equation has a power series solution

$$y(x) = \sum_{k=0}^{\infty} c_k (x - x_0)^k$$

which converges in the interval $|x - x_0| < R$.

We can assume, without loss of generality, that $x_0 = 0$. If it is not, we can introduce a new independent variable, $\xi = x - x_0$, and then the equation will have an ordinary point at the origin. Let us assume a solution

$$y(x) = \sum_{k=0}^{\infty} c_k x^k$$

then

$$y'(x) = \sum_{k=0}^{\infty} k c_k x^{k-1}$$

$$y''(x) = \sum_{k=0}^{\infty} k(k-1) c_k x^{k-2}$$

Substituting in the differential equation and equating coefficients of various powers of x to zero, we have

$$2c_2 + a_0 c_1 + b_0 c_0 = 0$$

$$6c_3 + 2a_0 c_2 + a_1 c_1 + b_0 c_1 + b_1 c_0 = 0$$

$$12c_4 + 3a_0 c_3 + 2a_1 c_2 + a_2 c_1 + b_0 c_2 + b_1 c_1 + b_2 c_0 = 0$$

etc. The coefficients c_0 and c_1 can be assigned arbitrarily, giving us the two required arbitrary constants in the general solution of the second-order differential equation. The first equation then determines c_2 in terms of c_0 and c_1. The second equation determines c_3 in terms of c_0, c_1, and c_2. The next equation determines c_4, etc. We therefore have the general solution of the differential equation, provided we can justify the operations performed on the series, that is, differentiation, multiplication, addition, etc. This can be done provided we can find some interval of convergence for the power series we have formally determined. We can show that it will converge for

$|x| < R$, which is the common interval of convergence for the power series representations of $f(x)$ and $g(x)$.

We know that, by absolute convergence,

$$\sum_{k=0}^{\infty} |a_k|\, |x|^k \qquad \text{and} \qquad \sum_{k=0}^{\infty} |b_k|\, |x|^k$$

converge for $|x| \le r < R$. Therefore,

$$\lim_{k \to \infty} |a_k|\, r^k = \lim_{k \to \infty} |b_k|\, r^k = 0$$

Hence, there exist constants M and N such that

$$|a_k| \le Mr^{-k} \qquad \text{and} \qquad |b_k| \le Nr^{-k}$$

for all k. Let K be the larger of M and Nr; then

$$|a_k| \le Kr^{-k}$$
$$|b_k| \le Kr^{-k-1}$$

If $|c_0| = \beta_0$ and $|c_1| = \beta_1$, then

$$2\,|c_2| \le \beta_1 |a_0| + \beta_0 |b_0|$$
$$\le 2K\beta_1 + K\beta_0 r^{-1}$$

so that $|c_2| \le \beta_2$, where $2\beta_2 = K(2\beta_1 + \beta_0 r^{-1})$. Also

$$2 \cdot 3\,|c_3| \le 2\beta_2 |a_0| + \beta_1 |a_1| + \beta_1 |b_0| + \beta_0 |b_1|$$
$$\le 3\beta_2 K + 2\beta_1 Kr^{-1} + \beta_0 Kr^{-2}$$

and $\quad 3 \cdot 4\,|c_4| \le 3\beta_3 |a_0| + 2\beta_2 |a_1| + \beta_1 |a_2| + \beta_2 |b_0| + \beta_1 |b_1| + \beta_0 |b_2|$
$$\le 4\beta_3 K + 3\beta_2 Kr^{-1} + 2\beta_1 Kr^{-2} + \beta_0 Kr^{-3}$$

Therefore, $|c_3| \le \beta_3$ and $|c_4| \le \beta_4$, where

$$2 \cdot 3\beta_3 = K(3\beta_2 + 2\beta_1 r^{-1} + \beta_0 r^{-2})$$
$$3 \cdot 4\beta_4 = K(4\beta_3 + 3\beta_2 r^{-1} + 2\beta_1 r^{-2} + \beta_0 r^{-3})$$

Continuing this way, we have $|c_k| \le \beta_k$, where

$$(k-1)k\beta_k = K[k\beta_{k-1} + (k-1)\beta_{k-2}r^{-1} + \cdots + 2\beta_1 r^{-k+2} + \beta_0 r^{-k+1}]$$

Next we write

$$(k-2)(k-1)\beta_{k-1}r^{-1} = K[(k-1)\beta_{k-2}r^{-1} + \cdots + 2\beta_1 r^{-k+2} + \beta_0 r^{-k+1}]$$

and subtract. This gives us

$$(k-1)k\beta_k - (k-2)(k-1)\beta_{k-1}r^{-1} = Kk\beta_{k-1}$$

from which it follows that

$$\frac{\beta_k}{\beta_{k-1}} = \frac{k-2}{kr} + \frac{K}{k-1}$$

Now consider the series $\sum\limits_{k=0}^{\infty} \beta_k x^k$. Applying the ratio test, we see that this converges for $|x| < r$, for

$$\lim_{k \to \infty} \frac{\beta_k |x|}{\beta_{k-1}} = \frac{|x|}{r} < 1$$

implies convergence for $|x| < r$. By a comparison test, therefore, we have absolute convergence of $\sum\limits_{k=0}^{\infty} c_k x^k$ in the same interval, because $|c_k x^k| \leq \beta_k |x|^k$. But r is any positive number less than R. Therefore, we have convergence of

$$y(x) = \sum_{k=0}^{\infty} c_k x^k$$

for $|x| < R$. This completes the proof.

Returning to the differential equation, if $f(x)$ or $g(x)$ are not continuous at x_0, then x_0 is called a singularity of the differential equation. If $(x - x_0)f(x)$ and $(x - x_0)^2 g(x)$ have power series representations in some interval $|x - x_0| < R$, then we say that x_0 is a **regular singular point**. In the neighborhood of a regular singular point we can always get a series solution of the differential equation as follows.[1]

We shall again assume that $x_0 = 0$. We first multiply the differential equations through by x^2, and then we have to solve

$$x^2 y'' + x F(x) y' + G(x) y = 0$$

where

$$F(x) = x f(x) = \sum_{k=0}^{\infty} a_k x^k$$

$$G(x) = x^2 g(x) = \sum_{k=0}^{\infty} b_k x^k$$

and the series converge for $|x| < R$. We assume a solution of the form

$$y(x) = \sum_{k=0}^{\infty} c_k x^{k+\alpha}$$

Then

$$x^2 y'' = \sum_{k=0}^{\infty} (k+\alpha)(k+\alpha-1) c_k x^{k+\alpha}$$

$$xy' = \sum_{k=0}^{\infty} (k+\alpha) c_k x^{k+\alpha}$$

Substituting in the differential equation and equating coefficients of the

[1] This is usually referred to as **Frobenius's method**.

various powers of x to zero, we have

$$c_0[\alpha(\alpha - 1) + \alpha a_0 + b_0] = 0$$

$$c_k[(\alpha + k)(\alpha + k - 1) + a_0(\alpha + k) + b_0]$$

$$+ \sum_{j=0}^{k-1} c_j[(\alpha + j)a_{k-j} + b_{k-j}] = 0 \qquad k = 1, 2, \ldots$$

The first equation indicates that c_0 is arbitrary and that α must be a solution of the **indicial equation**

$$\alpha(\alpha - 1) + \alpha a_0 + b_0 = 0$$

This equation is quadratic in α and therefore has two solutions, not necessarily distinct. Let α_1 and α_2 be the roots of the indicial equation. If $\alpha_1 \neq \alpha_2$ and $\alpha_2 - \alpha_1 \neq$ an integer, then the second equation yields the coefficients in each of two independent solutions, each of which contains an arbitrary constant. The sum of the two then gives us the general solution of the differential equation, provided the series can be shown to converge.

If $\alpha_1 = \alpha_2$ then the method yields only one solution. However, if we know one solution, we can always get another independent solution by a substitution which reduces the problem to the solution of a linear first-order equation. We shall return to this problem presently. If $\alpha_2 = \alpha_1 + n$, where n is a positive integer, then for $k = n$

$$(\alpha_1 + n)(\alpha_1 + n - 1) + a_0(\alpha_1 + n) + b_0 = 0$$

In this case, the solution corresponding to the root α_1 fails unless

$$\sum_{j=0}^{n-1} c_j[(\alpha_1 + j)a_{n-j} + b_{n-j}] = 0$$

In the latter case c_n is arbitrary. Otherwise the method fails to yield two solutions. We will, however, always get one series using the root α_2. This series can be shown to converge for $|x| < R$.

We can write

$$c_k I(\alpha + k) = -\sum_{j=0}^{k-1} c_j[(\alpha + j)a_{k-j} + b_{k-j}]$$

where

$$I(\alpha) = \alpha(\alpha - 1) + \alpha a_0 + b_0 = (\alpha - \alpha_1)(\alpha - \alpha_2)$$

Then $\quad I(\alpha_1 + k) = k(k + \alpha_1 - \alpha_2) \quad$ and $\quad I(\alpha_2 + k) = k(k + \alpha_2 - \alpha_1)$

We can write the inequality

$$k(k - |\alpha_1 - \alpha_2|) |c_k| \leq |I(\alpha_2 + k)| |c_k| \leq \sum_{j=0}^{k-1} |c_j| [(|\alpha_2| + j) |a_{k-j}| + |b_{k-j}|]$$

for all $k \geq |\alpha_1 - \alpha_2|$. Let $|c_j| = \beta_j$ for $j < m$, where m is some integer greater than $|\alpha_1 - \alpha_2|$. Then

$$m(m - |\alpha_1 - \alpha_2|) |c_m| \leq \sum_{j=0}^{m-1} \beta_j[(|\alpha_2| + j) |a_{m-j}| + |b_{m-j}|]$$

As in the analysis of the ordinary point there exists a constant K such that

$$|a_k| \leq Kr^{-k}$$
$$|b_k| \leq Kr^{-k}$$

Then $\qquad m(m - |\alpha_1 - \alpha_2|)\,|c_m| \leq K \sum_{j=0}^{m-1} \beta_j(|\alpha_2| + j + 1)r^{-m+j}$

and $|c_m| \leq \beta_m$, where

$$m(m - |\alpha_1 - \alpha_2|)\beta_m = K \sum_{j=0}^{m-1} \beta_j(|\alpha_2| + j + 1)r^{-m+j}$$

Furthermore, for $n \geq m$,

$$n(n - |\alpha_1 - \alpha_2|)\,|c_n| \leq K \sum_{j=0}^{n-1} \beta_j(|\alpha_2| + j + 1)r^{-n+j}$$

and $|c_n| \leq \beta_n$, where

$$n(n - |\alpha_1 - \alpha_2|)\beta_n = K \sum_{i=0}^{n-1} \beta_j(|\alpha_2| + j + 1)r^{-n+j}$$

Replacing n by $n - 1$ and dividing by r, we have

$$(n - 1)(n - 1 - |\alpha_1 - \alpha_2|)\beta_{n-1}r^{-1} = K \sum_{j=0}^{n-2} \beta_j(|\alpha_2| + j + 1)r^{-n+j}$$

Subtracting,

$$n(n - |\alpha_1 - \alpha_2|)\beta_n - (n - 1)(n - 1 - |\alpha_1 - \alpha_2|)\beta_{n-1}r^{-1}$$
$$= K\beta_{n-1}(n + |\alpha_2|)r^{-1}$$

or $\qquad \dfrac{\beta_n}{\beta_{n-1}} = \dfrac{(n - 1)(n - 1 - |\alpha_1 - \alpha_2|)}{n(n - |\alpha_1 - \alpha_2|)r} + \dfrac{K(n + |\alpha_2|)}{n(n - |\alpha_1 - \alpha_2|)r}$

Consider the series $\sum_{k=0}^{\infty} \beta_k x^k$. This series converges absolutely for $|x| < r$ by the ratio test, since

$$\lim_{n \to \infty} \frac{\beta_n |x|}{\beta_{n-1}} = \frac{|x|}{r} < 1$$

when $|x| < r$. By comparison $\sum_{k=0}^{\infty} c_k x^k$ converges absolutely in the same interval, but this implies that it converges absolutely for $|x| < R$ since r is any positive number less than R. This justifies all the operations involved in the substitution of the series in the differential equation and shows that

$$y(x) = x^{\alpha_2} \sum_{k=0}^{\infty} c_k x^k$$

where the coefficients c_k are determined by the above recurrence formulas, is a solution of the differential equation valid for $|x| < R$.

If $\alpha_2 - \alpha_1$ is not zero or a positive integer, then we get a second independent solution in the Frobenius method using the root α_1, and the proof of convergence of the series is essentially the same as that for α_2. If the method fails to give two independent solutions, then we proceed as follows. Let $y(x) = u(x)v(x)$, where $u(x)$ is a known solution of the differential equation. Upon substituting we find that the differential equation which v must satisfy is

$$v'' + \left(\frac{2u'}{u} + f\right)v' = 0$$

This is a first-order linear equation in v' with the solution

$$v' = Au^{-2} \exp\left[-\int f(x)\,dx\right]$$

Then

$$v = A\int\left\{u^{-2}\exp\left[-\int f(x)\,dx\right]\right\}dx + B$$

$$y = uv = Au\int\left\{u^{-2}\exp\left[-\int f(x)\,dx\right]\right\}dx + Bu$$

This is the general solution.

To illustrate some of these ideas, let us consider the series solution of the Legendre equation

$$[(1 - x^2)y']' + \lambda y = 0$$

The equation can be written as

$$y'' - \frac{2x}{1 - x^2}y' + \frac{\lambda}{1 - x^2}y = 0$$

and we see that $x = 0$ is an ordinary point, while $x = \pm 1$ are regular singular points. The indicial equation at either of the singular points is $\alpha^2 = 0$, so that $\alpha_1 = \alpha_2 = 0$. Hence, we can get a power series solution valid in the neighborhood of either of the singularities. However, these solutions are not finite at the other singularity, unless λ has certain integer values. We can best determine these eigenvalues by seeking solutions valid near the origin as series in powers of x. Hence, we assume

$$y = \sum_{k=0}^{\infty} c_k x^k$$

Substituting in the differential equation, we have

$$(1 - x^2)\sum_{k=2}^{\infty} k(k - 1)c_k x^{k-2} - 2x\sum_{k=1}^{\infty} kc_k x^{k-1} + \lambda\sum_{k=0}^{\infty} c_k x^k = 0$$

or

$$\sum_{m=0}^{\infty}[\lambda c_m + (m + 2)(m + 1)c_{m+2}]x^m - \sum_{m=0}^{\infty} 2mc_m x^m - \sum_{m=0}^{\infty} m(m - 1)c_m x^m = 0$$

Setting the coefficients of the various powers of x equal to zero, we have

$$c_{m+2} = \frac{m(m+1) - \lambda}{(m+2)(m+1)} c_m \qquad m = 0, 1, 2, \ldots$$

If we let $c_1 = 0$ while $c_0 \neq 0$, we obtain a series in even powers of x.   If we let $c_0 = 0$ while $c_1 \neq 0$, we obtain a series in odd powers of x.   These solutions are obviously linearly independent, and therefore a linear combination of them will be the general solution of the differential equation.   If $\lambda \neq n(n+1)$ for every nonnegative integer n, these solutions will be infinite series converging for $|x| < 1$.   However, it can be shown that they diverge for $x = \pm 1$.†   If we are seeking solutions finite at $x = \pm 1$, λ will have to equal $n(n+1)$, where n is a nonnegative integer.   In this case, either the series in even powers or the one in odd powers terminates, depending on whether n is even or odd, giving us polynomial solutions.   If c_0 or c_1 is adjusted so that the solution takes on the value 1 at $x = 1$, we have the Legendre polynomials $P_n(x)$.   The first few of these are $P_0(x) = 1$, $P_1(x) = x$, $P_2(x) = \frac{1}{2}(3x^2 - 1)$, $P_3(x) = \frac{1}{2}(5x^3 - 3x)$, etc.   More generally,

$$P_n(x) = \sum_{k=0}^{K} \frac{(-1)^k (2n - 2k)!}{2^n k! \, (n-k)! \, (n-2k)!} x^{n-2k}$$

where $K = \frac{1}{2}n$ or $K = \frac{1}{2}(n-1)$, depending on whether n is even or odd. We rephrase the definition of $P_n(x)$ as follows:

$$P_n(x) = \sum_{k=0}^{K} \frac{(-1)^k}{2^n k! \, (n-k)!} \frac{d^n}{dx^n} x^{2n-2k}$$

$$= \frac{1}{2^n n!} \frac{d^n}{dx^n} \sum_{k=0}^{K} \frac{(-1)^k n!}{k! \, (n-k)!} x^{2n-2k}$$

$$= \frac{1}{2^n n!} \frac{d^n}{dx^n} \sum_{k=0}^{n} \frac{(-1)^k n!}{k! \, (n-k)!} x^{2n-2k}$$

$$= \frac{1}{2^n n!} \frac{d^n}{dx^n} (x^2 - 1)^n$$

The last line is known as **Rodrigues' formula.**

The Legendre polynomials are solutions of the Legendre equation

$$(1 - x^2)y'' - 2xy' + n(n+1)y = 0$$

for $n = 0, 1, 2, \ldots$.   Since they are polynomials, they are finite in any finite interval.   They are, therefore, solutions of the Sturm-Liouville problem

$$[(1 - x^2)y']' + \lambda y = 0$$

---

† See Richard Courant and David Hilbert, "Methods of Mathematical Physics," Interscience Publishers (Division of John Wiley & Sons, Inc.), New York, 1953, vol. I, pp. 325 and 326.

with $y(1)$ and $y(-1)$ finite for $\lambda_n = n(n + 1)$. It follows from the differential equation that the Legendre polynomials are mutually orthogonal in the following sense:

$$\int_{-1}^{1} P_n(x)P_m(x)\, dx = 0$$

for $n \neq m$. That they are a complete set of functions follows from the fact that they are solutions of a variational problem with an unbounded infinite sequence of eigenvalues.[1] This tells us that the Legendre polynomials are the only solutions of the differential equation finite at $x = \pm 1$.

The Legendre polynomials are not normalized, but it can be shown, by use of Rodrigues' formula, that

$$\int_{-1}^{1} P_n^2(x)\, dx = \frac{2}{2n + 1}$$

The **associated Legendre equation** is

$$(1 - x^2)y'' - 2xy' + \left(\lambda - \frac{m^2}{1 - x^2}\right)y = 0$$

where in most applications m is an integer. For example, in the separation of the Helmholtz equation in spherical coordinates,[2] m is a separation constant which must be an integer in order to make the ϕ-dependent part of the solution single-valued. We shall consider only the case where m is an integer. Let us perform the following transformation on the differential equation:

$$y = (1 - x^2)^{\frac{1}{2}m}v$$

Then
$$y' = (1 - x^2)^{\frac{1}{2}m}v' - xm(1 - x^2)^{\frac{1}{2}m-1}v$$
$$y'' = (1 - x^2)^{\frac{1}{2}m}v'' - 2xm(1 - x^2)^{\frac{1}{2}m-1}v'$$
$$-m(1 - x^2)^{\frac{1}{2}m-1}v + x^2m(m - 2)(1 - x^2)^{\frac{1}{2}m-2}v$$

Substituting in the differential equation, we have

$$(1 - x^2)v'' - 2(m + 1)xv' + (\lambda - m - m^2)v = 0$$

We next let $v = (d^m/dx^m)u$. Then the differential equation becomes

$$(1 - x^2)\frac{d^{m+2}u}{dx^{m+2}} - 2(m + 1)x\frac{d^{m+1}u}{dx^{m+1}} + (\lambda - m - m^2)\frac{d^mu}{dx^m} = 0$$

This can be written as follows:

$$\frac{d^m}{dx^m}\left[(1 - x^2)\frac{d^2u}{dx^2} - 2x\frac{du}{dx} + \lambda u\right] = 0$$

[1] See exercise 2, Sec. 4.2.
[2] See Sec. 4.1.

for $\dfrac{d^m}{dx^m}\left[(1 - x^2)\dfrac{d^2u}{dx^2}\right] = \displaystyle\sum_{k=0}^{m} \dfrac{m!}{(m - k)!\, k!} \dfrac{d^k}{dx^k}(1 - x^2)\dfrac{d^{m-k+2}u}{dx^{m-k+2}}$

$$= (1 - x^2)\dfrac{d^{m+2}u}{dx^{m+2}} - 2xm\dfrac{d^{m+1}u}{dx^{m+1}} - m(m - 1)\dfrac{d^m u}{dx^m}$$

and $\dfrac{d^m}{dx^m}\left(-2x\dfrac{du}{dx}\right) = \displaystyle\sum_{k=0}^{m} \dfrac{m!}{(m - k)!\, k!} \dfrac{d^k(-2x)}{dx^k} \dfrac{d^{m+k-1}u}{dx^{m-k+1}}$

$$= -2\dfrac{d^{m+1}u}{dx^{m+1}} - 2m\dfrac{d^m u}{dx^m}$$

We know that if $\lambda = n(n + 1)$, with n a nonnegative integer, the equation

$$(1 - x^2)u'' - 2xu' + n(n + 1)u = 0$$

has a polynomial solution $P_n(x)$ which is finite at $x = \pm 1$. Therefore, the associated Legendre equation has solutions

$$P_n^m(x) = (1 - x^2)^{\frac{1}{2}m}\dfrac{d^m}{dx^m}P_n(x)$$

$$= \dfrac{(1 - x^2)^{\frac{1}{2}m}}{2^n n!}\dfrac{d^{n+m}}{dx^{n+m}}(x^2 - 1)^n$$

which are finite at $x = \pm 1$. These functions are called the **associated Legendre functions.** One can prove that for a given integer m these are the only solutions finite at $x = \pm 1$ by showing that these solutions are obtainable from a variational principle with eigenvalues $\lambda = n(n + 1)$, $n = m$, $m + 1$, $m + 2, \ldots \to \infty$ and, therefore, are a complete set of functions satisfying the differential equation. It can be shown by direct integration that

$$\int_{-1}^{1}[P_n^m(x)]^2\, dx = \dfrac{2}{2n + 1}\dfrac{(n + m)!}{(n - m)!}$$

In the separation of the Helmholtz equation in spherical coordinates we have the equation

$$\sin\theta\dfrac{\partial}{\partial\theta}\left(\sin\theta\dfrac{\partial S}{\partial\theta}\right) + \dfrac{\partial^2 S}{\partial\phi^2} + \alpha S \sin^2\theta = 0$$

where $S(\theta,\phi)$ is the angular dependent part of the eigenfunction. This equation also occurs in the solution of the steady-state Schrödinger equation in quantum mechanics when the potential $V(r)$ is spherically symmetric. We require that the solution be single-valued, and finite for $\theta = 0$ and π. A further separation of variables $S = V(\theta)W(\phi)$ yields

$$\dfrac{d^2 W}{d\phi^2} + \beta W = 0$$

with solutions $\sin \sqrt{\beta}\phi$, $\cos \sqrt{\beta}\phi$. $\beta = m^2$, with m an integer, is required to make W periodic with period 2π, and hence continuous. The other equation is then

$$\sin^2 \theta \frac{d^2V}{d\theta^2} + \sin \theta \cos \theta \frac{dV}{d\theta} + (\alpha \sin^2 \theta - m^2)V = 0$$

We seek a solution of this equation which is finite for $\theta = 0$ and π. Let $\cos \theta = x$, then

$$\frac{dV}{d\theta} = -\sin \theta \frac{dV}{dx}$$

$$\frac{d^2V}{d\theta^2} = \sin^2 \theta \frac{d^2V}{dx^2} - \cos \theta \frac{dV}{dx}$$

and the equation becomes

$$(1 - x^2) \frac{d^2V}{dx^2} - 2x \frac{dV}{dx} + \left(\alpha - \frac{m^2}{1 - x^2}\right)V = 0$$

This is the associated Legendre equation which has solutions $P_n^m(x)$ which are finite at $x = \pm 1$, or $\cos \theta = \pm 1$. The solutions

$$S(\theta, \phi) = AP_n^m(\cos \theta) \sin m\phi + BP_n^m(\cos \theta) \cos m\phi$$

are called **spherical harmonics**. The discussion at the end of Sec. 4.1 shows that they are a complete set of functions over the region $0 \leq \theta \leq \pi$, $0 \leq \phi \leq 2\pi$.

In the separation of the Helmholtz equation in spherical coordinates, the equation satisfied by the radial dependent part of the solution is

$$r^2 \frac{d^2U}{dr^2} + 2r \frac{dU}{dr} + \lambda r^2 U - n(n + 1)U = 0$$

Here the separation constant α is set equal to $n(n + 1)$ as required to make the angular dependent part of the solution finite at $\theta = 0$ and π. We make the change of variables:

$$x = \lambda^{\frac{1}{2}}r$$

$$U = r^{-\frac{1}{2}}y$$

Then
$$\frac{dU}{dr} = -\tfrac{1}{2}r^{-\frac{3}{2}}y + \lambda^{\frac{1}{2}}r^{-\frac{1}{2}} \frac{dy}{dx}$$

$$\frac{d^2U}{dr^2} = \tfrac{3}{4}r^{-\frac{5}{2}}y - r^{-\frac{3}{2}}\lambda^{\frac{1}{2}} \frac{dy}{dx} + \lambda r^{-\frac{1}{2}} \frac{d^2y}{dx^2}$$

The differential equation becomes

$$x^2y'' + xy' + [x^2 - (n + \tfrac{1}{2})^2]y = 0$$

This is **Bessel's differential equation.** It has a regular singular point at $x = 0$. We shall consider the solution of the equation

$$x^2 y'' + xy' + (x^2 - \nu^2)y = 0$$

If we seek a series solution valid in the neighborhood of $x = 0$, the indicial equation is $\alpha^2 - \nu^2 = 0$. The roots are $\alpha_1 = -\nu$ and $\alpha_2 = \nu$. If ν is not an integer, we get the two independent solutions

$$y_1 = x^\nu \sum_{k=0}^{\infty} a_k x^k$$

$$y_2 = x^{-\nu} \sum_{k=0}^{\infty} b_k x^k$$

Substituting the first into the differential equation, we have

$$\sum_{k=0}^{\infty} k(k + 2\nu)a_k x^{k+\nu} + \sum_{k=0}^{\infty} a_k x^{k+\nu+2} = 0$$

The coefficient of $x^{\nu+1}$ is $(1 + 2\nu)a_1$. Since this must be zero, $a_1 = 0$ if $\nu \neq -\frac{1}{2}$. The other coefficients are determined by the recurrence relation

$$a_{k+2} = - \frac{a_k}{(k + 2)(k + 2 + 2\nu)}$$

Since $a_1 = 0$, all the coefficients with odd subscripts are also zero. a_0 is arbitrary, and the other coefficients are determined in terms of a_0 as follows:

$$a_{2n} = \frac{(-1)^n a_0}{2^{2n} n! \, (\nu + 1)(\nu + 2) \cdots (\nu + n)}$$

Let $\Gamma(z)$ be the gamma function.[1] Then

$$\Gamma(\nu + n + 1) = (\nu + n)(\nu + n - 1) \cdots (\nu + 1)\Gamma(\nu + 1)$$

and if we let

$$a_0 = [2^\nu \Gamma(\nu + 1)]^{-1}$$

we can write

$$J_\nu(x) = \sum_{n=0}^{\infty} \frac{(-1)^n (x/2)^{2n+\nu}}{n! \, \Gamma(\nu + n + 1)}$$

as a solution of the Bessel differential equation. This is the **Bessel function of the first kind of order ν.** If ν is not an integer and $\nu \neq \frac{1}{2}$ another solution corresponding to the root $\alpha_1 = -\nu$ is

$$J_{-\nu}(x) = \sum_{n=0}^{\infty} \frac{(-1)^n (x/2)^{2n-\nu}}{n! \, \Gamma(-\nu + n + 1)}$$

---

[1] See David V. Widder, "Advanced Calculus," 2d ed., Prentice-Hall, Inc., Englewood Cliffs, N.J., 1961, pp. 367–373.

This is the Bessel function of the first kind of order $-\nu$. These two solutions are linearly independent and so the general solution is

$$y = AJ_\nu(x) + BJ_{-\nu}(x)$$

If $\nu = 0$, then the two above solutions are obviously the same. If $\nu = m$, $m = 1, 2, 3, \ldots$, then $\alpha_2 - \alpha_1 = 2m$ is an integer and we have the situation described above where the Frobenius type series with $\alpha = \alpha_1$ may not yield a solution. The recurrence formula is

$$b_k + (k + 2)(k + 2 - 2m)b_{k+2} = 0$$

When $k = 2m - 2$ we cannot use this to determine b_{2m}. However, suppose we set $b_0 = 0$. Then $b_0 = b_2 = b_4 = \cdots = b_{2m-2} = 0$ and the recurrence relation for $k = 2m - 2$ is satisfied for arbitrary b_{2m}. Hence,

$$b_{k+2} = -\frac{b_k}{(k + 2)(k + 2 - 2m)}$$

$k = 2m, 2m + 2, 2m + 4, \ldots$. From this we obtain

$$b_{2m+2n} = \frac{(-1)^n b_{2m}}{2^{2n}n!\,(m + 1)(m + 2)\cdots(m + n)}$$

and we have a solution

$$y = x^{-m}b_{2m}\sum_{n=0}^{\infty}\frac{(-1)^n x^{2m+2n}}{2^{2n}n!\,(m + 1)(m + 2)\cdots(m + n)}$$

$$= 2^m m!\,b_{2m}\sum_{n=0}^{\infty}\frac{(-1)^n (x/2)^{2n+m}}{2^{2n}n!\,\Gamma(m + n + 1)}$$

$$= 2^m m!\,b_{2m}J_m(x)$$

But this solution is proportional to $J_m(x)$. Therefore, if $\nu = 1, 2, 3, \ldots$ we do not get independent solutions out of the two series. In this case, a second independent solution is given by

$$Y_m(x) = \frac{2}{\pi}J_m(x)\int\left\{[J_m(x)]^{-2}\exp\left[-\int(1/x)\,dx\right]\right\}dx$$

$$= \frac{2}{\pi}J_m(x)\int\frac{1}{x[J_m(x)]^2}\,dx$$

There is of course ambiguity in the definition of the indefinite integral. Without being very specific, we shall merely note that when the proper multiple of $J_m(x)$ is added this second solution is called the **Bessel function of the second kind of order** m. It is important to observe that $Y_m(x)$

becomes unbounded as x approaches zero.   Since

$$J_m(x) = x^m \left(\frac{1}{2^m m!} + \cdots \right)$$

$$\frac{1}{x[J_m(x)]^2} = \frac{1}{x^{2m+1}} [2^{2m}(m!)^2 + \cdots]$$

Integrating and multiplying by $(2/\pi)J_m(x)$ we have the leading term in the expansion of $Y_m(x)$, which turns out to be

$$-\frac{1}{\pi}(m-1)! \left(\frac{x}{2} \right)^{-m}$$

This shows that $Y_m(x)$ becomes unbounded as x approaches zero.   If $m = 0$, we have

$$Y_0(x) = \frac{2}{\pi} J_0(x) \int \frac{1}{x[J_0(x)]^2} \, dx$$

and it is easy to show that $Y_0(x)$ contains a term of the form $(2/\pi) \ln x$.   This shows that $Y_0(x)$ is unbounded as x approaches zero.

If ν is not an integer we define the **Bessel function of the second kind of order** ν as

$$Y_\nu(x) = \frac{J_\nu(x) \cos \nu\pi - J_{-\nu}(x)}{\sin \nu\pi}$$

It can be shown that for m a nonnegative integer

$$Y_m(x) = \lim_{\nu \to m} Y_\nu(x)$$

It is also useful to have another set of independent solutions of the Bessel differential equation known as the **Hankel functions** of the first and second kind, respectively,

$$H_\nu^{(1)}(x) = J_\nu(x) + i Y_\nu(x)$$
$$H_\nu^{(2)}(x) = J_\nu(x) - i Y_\nu(x)$$

These play an important role in the solution of the wave equation in two-dimensional problems, as we shall see in Chap. 5.

In the Sturm-Liouville problem

$$(xy')' - \frac{\nu^2}{x} y + \lambda x y = 0$$

with y finite at $z = 0$, we have the solution $J_\nu(\sqrt{\lambda}x)$, since the second solution is not finite at $x = 0$.   This is true whether ν is an integer or not.   The eigenvalue λ is determined by the requirement on the solution to satisfy the other boundary condition.   We have seen in the previous section that the Bessel functions form a complete set of functions.   In the above separation-of-variables problem in spherical coordinates, the solution of

$$r^2 \frac{d^2 U}{dr^2} + 2r \frac{dU}{dr} + \lambda r^2 U - n(n+1)U = 0$$

which is finite at $r = 0$, is $r^{-\frac{1}{2}}J_{n+\frac{1}{2}}(\sqrt{\lambda}r)$, and the eigenvalue λ is determined by the physical boundary condition at $r = a$.

Exercises 4.3

1. Find the bounded solution $J_0(x)$ of the zeroth-order Bessel equation $xy'' + y' + xy = 0$ valid at the origin. Show that the other solution $Y_0(x)$ becomes unbounded as $(2/\pi)\ln x$ as x approaches zero.

2. Show that the method of Frobenius yields two independent solutions of Bessel's equation of order $1/2$ valid near the origin, even though the roots of the identical equation differ by an integer. Show that the general solution can be written as $y = (A \sin x + B \cos x)/\sqrt{x}$. Find $J_{1/2}(x)$ and $J_{-1/2}(x)$ and express the general solution in terms of Bessel functions.

3. Show that the method of Frobenius fails to give a nontrivial solution of $x^3y'' + x^2y' + y = 0$ valid near the origin. Note that the singularity at $x = 0$ is not a regular singularity.

4. Find the general solution of the Legendre differential equation $(1 - x^2)y'' - 2xy' + n(n + 1)y = 0$, with n a nonnegative integer, valid near $x = 1$. Show that it becomes unbounded as a constant times $\ln |x - 1|$ as x approaches 1.

5. Show that $\int_{-1}^{1} P_n^2(x)\,dx = 2/(2n + 1)$, where $P_n(x)$ is a Legendre polynomial.

6. Show that $\int_{-1}^{1} [P_n^m(x)]^2\,dx = [2/(2n + 1)][(n + m)!/(n - m)!]$, where $P_n^m(x)$ is an associated Legendre function.

7. Sometimes we are interested in solutions of $y'' + f(x)y' + g(x)y = 0$ for large values of $|x|$. Make the transformation $x = 1/\xi$ and state criteria in terms of $f(x)$ and $g(x)$ for an ordinary point and a regular singular point at $x = \infty$. Develop a method for obtaining solutions of the form $y(x) = x^{-\alpha} \sum_{k=0}^{\infty} c_k x^{-k}$ valid for $|x| > R$ when there is a regular singular point at $x = \infty$.

8. Consider the equation $y'' - xy' + ny = 0$. For what values of n does this equation have polynomial solutions? Find several of these polynomials. If the highest power has the coefficient 1 they are called **Hermite polynomials.**

9. Consider the equation $xy'' + (1 - x)y' + ny = 0$. For what values of n does this equation have polynomial solutions? Find several of these polynomials. If the constant term in each of them is 1, they are called **Laguerre polynomials.**

10. The Bessel functions satisfy a large number of identities just as the trigonometric functions do. Prove, for example, the following recurrence relations:

$$J_{\nu-1}(x) + J_{\nu+1}(x) = \frac{2\nu}{x}J_\nu(x)$$

$$J_{\nu-1}(x) - J_{\nu+1}(x) = 2J_\nu'(x)$$

$$\frac{\nu}{x}J_\nu(x) + J_\nu'(x) = J_{\nu-1}(x)$$

$$\frac{\nu}{x}J_\nu(x) - J_\nu'(x) = J_{\nu+1}(x)$$

11. Prove the following recurrence relations for the Legendre polynomials:

$$nP_n(x) - (2n - 1)xP_{n-1}(x) + (n - 1)P_{n-2}(x) = 0$$

$$xP'_n(x) - P'_{n-1}(x) = nP_n(x)$$

$$P'_n(x) - xP'_{n-1}(x) = nP_{n-1}(x)$$

$$P'_{n-1}(x) - xP'_{n-1}(x) = (2n + 1)P_n(x)$$

4.4   Series Solutions of Boundary-value Problems

We conclude this chapter by solving some boundary-value problems which lend themselves nicely to the separation-of-variables technique and solution using eigenfunction expansions.

We consider first the vibrations of a circular membrane.   Here we must solve the two-dimensional wave equation $\nabla^2\phi = (1/a^2)(\partial^2\phi/\partial t^2)$ in the interior of the membrane; that is, $0 \leq r < b$, $0 \leq \theta \leq 2\pi$, subject to the boundary condition $\phi = 0$ on the edge, $r = b$.   We shall specify an initial displacement

$$\phi(r,\theta,0) = f(r,\theta)$$

but no initial velocity; that is, $\dot{\phi}(r,\theta,0) = 0$.   With the introduction of polar coordinates, the wave equation becomes

$$\frac{\partial^2\phi}{\partial r^2} + \frac{1}{r}\frac{\partial\phi}{\partial r} + \frac{1}{r^2}\frac{\partial^2\phi}{\partial\theta^2} = \frac{1}{a^2}\frac{\partial^2\phi}{\partial t^2}$$

Assuming a solution of this equation of the form

$$\phi(r,\theta,0) = \psi(r,\theta)F(t)$$

we obtain the separated equations

$$\ddot{F} + a^2\lambda F = 0$$

$$\frac{\partial^2\psi}{\partial r^2} + \frac{1}{r}\frac{\partial\psi}{\partial r} + \frac{1}{r^2}\frac{\partial^2\psi}{\partial\theta^2} + \lambda\psi = 0$$

The first of these has the solution $F = C \cos a\sqrt{\lambda}t$, since we are assuming no initial velocity.   We next assume a separation of the space variables, that is,

$$\psi(r,\theta) = U(r)V(\theta)$$

giving us
$$r^2\frac{d^2U}{dr^2} + r\frac{dU}{dr} + (r^2\lambda - n^2)U = 0$$

$$\frac{d^2V}{d\theta^2} + n^2V = 0$$

The separation constant must be n^2 with $n = 0, 1, 2, 3, \ldots$ in order to guarantee that

$$V_n(\theta) = A_n \cos n\theta + B_n \sin n\theta$$

and its derivatives are continuous at $\theta = 0$. In the equation satisfied by U we make the substitution $x = \sqrt{\lambda}\,r$, and the equation becomes

$$x^2 \frac{d^2U}{dx^2} + x \frac{dU}{dx} + (x^2 - n^2)U = 0$$

This is Bessel's equation of order n. The solution which is finite at $x = 0$ is $J_n(x)$. Therefore,

$$U_n(r) = J_n(\sqrt{\lambda}\,r)$$

To satisfy the boundary condition identically in θ, we set $J_n(\sqrt{\lambda}\,b) = 0$. We know that the Bessel function of order n has an infinite sequence of zeros increasing to infinity. Let these zeros be $\sqrt{\lambda_{nm}}\,b$, thus determining the eigenvalues λ_{nm}. The eigenfunctions are $J_n(\sqrt{\lambda_{nm}}\,r) \cos n\theta$ and $J_n(\sqrt{\lambda_{nm}}\,r) \sin n\theta$. A formal series solution of the problem is then

$$\phi(r,\theta,t) = \sum_{n=0}^{\infty} \sum_{m=1}^{\infty} (a_{nm} \cos n\theta + b_{nm} \sin n\theta)J_n(\sqrt{\lambda_{nm}}\,r) \cos a\sqrt{\lambda_{nm}}\,t$$

If this is to be a solution, the initial condition must be satisfied; that is,

$$\phi(r,\theta,0) = f(r,\theta) = \sum_{n=0}^{\infty} \sum_{m=1}^{\infty} (a_{nm} \cos n\theta + b_{nm} \sin n\theta)J_n(\sqrt{\lambda_{nm}}\,r)$$

The coefficients must therefore be

$$a_{0m} = \frac{(1/2\pi)\int_0^{2\pi}\int_0^b rJ_0(\sqrt{\lambda_{0m}}\,r)f(r,\theta)\,dr\,d\theta}{\int_0^b rJ_0^2(\sqrt{\lambda_{0m}}\,r)\,dr}$$

$$a_{nm} = \frac{(1/\pi)\int_0^{2\pi}\int_0^b rJ_n(\sqrt{\lambda_{nm}}\,r)f(r,\theta)\cos n\theta\,dr\,d\theta}{\int_0^b rJ_n^2(\sqrt{\lambda_{nm}}\,r)\,dr}$$

$$b_{nm} = \frac{(1/\pi)\int_0^{2\pi}\int_0^b rJ_n(\sqrt{\lambda_{nm}}\,r)f(r,\theta)\sin n\theta\,dr\,d\theta}{\int_0^b rJ_n^2(\sqrt{\lambda_{nm}}\,r)\,dr}$$

The integral in the denominators can be evaluated simply as follows; using a result of Sec. 4.2 and exercise 10, Sec. 4.3,

$$\int_0^b r J_n^2(\sqrt{\lambda_{nm}}r)\,dr = \frac{b^2[J_n'(\sqrt{\lambda_{nm}}b)]^2}{2\lambda_{nm}}$$

$$= \frac{b^2}{2}\left[\frac{n}{\sqrt{\lambda_{nm}}b}J_n(\sqrt{\lambda_{nm}}b) - J_{n+1}(\sqrt{\lambda_{nm}}b)\right]^2$$

$$= \frac{b^2}{2}J_{n+1}^2(\sqrt{\lambda_{nm}}b)$$

We know that the eigenfunctions form a complete set of solutions of Helmholtz's equation. Thus, if $f(r,\theta)$ is continuous in the closed region and has piecewise continuous first partial derivatives, then the series converges in mean to the function. We shall later prove, in connection with our study of integral equations in Chap. 6, that if $f(r,\theta)$ is continuous along with its first and second partial derivatives, the series will converge absolutely and uniformly to the function.

We next turn our attention to a problem in heat conduction. Suppose a homogeneous hollow cylinder which occupies the region $a \le r \le b$, $0 \le \theta \le 2\pi$, $0 \le z \le h$ has its ends $z = 0$ and $z = h$ maintained at temperatures 0 and 100°C, respectively. The faces at $r = a$ and $r = b$ are insulated against the flow of heat. Assuming an initial temperature distribution $f(r,\theta,z)$, we have to solve the following boundary-value problem to determine the temperature $\phi(r,\theta,z,t)$ in the solid:

$$\frac{\partial^2\phi}{\partial r^2} + \frac{1}{r}\frac{\partial\phi}{\partial r} + \frac{1}{r^2}\frac{\partial^2\phi}{\partial\theta^2} + \frac{\partial^2\phi}{\partial z^2} = \frac{1}{c^2}\frac{\partial\phi}{\partial t}$$

for $a < r < b$, $0 \le \theta \le 2\pi$, $0 < z < h$, subject to

$$\phi(r,\theta,0,t) = 0$$

$$\phi(r,\theta,h,t) = 100$$

$$\phi_r(a,\theta,z,t) = \phi_r(b,\theta,z,t) = 0$$

$$\phi(r,\theta,z,0) = f(r,\theta,z)$$

Here we have a nonhomogeneous boundary condition at $z = h$. However, we can solve the problem by taking the following point of view. Let $\phi = \phi^{(S)} + \phi^{(T)}$, where $\phi^{(S)}$ is the **steady-state** solution, which does not depend on t, satisfying $\nabla^2\phi^{(S)} = 0$, subject to $\phi^{(S)}(r,\theta,0) = 0$, $\phi^{(S)}(r,\theta,h) = 100$, $\phi_r^{(S)}(a,\theta,z) = \phi_r^{(S)}(b,\theta,z) = 0$; and $\phi^{(T)}$ is the **transient solution** satisfying $\nabla^2\phi^{(T)} = (1/c^2)(\partial\phi^{(T)}/\partial t)$, subject to

$$\phi^{(T)}(r,\theta,0,t) = 0, \qquad \phi^{(T)}(r,\theta,h,t) = 0, \qquad \phi_r^{(T)}(a,\theta,z,t) = \phi_r^{(T)}(b,\theta,z,t) = 0$$

Obviously, the sum $\phi^{(S)} + \phi^{(T)}$ satisfies the differential equation and the boundary conditions.   The initial condition for $\phi^{(T)}$ becomes

$$\phi^{(T)}(r,\theta,z,0) = f(r,\theta,z) - \phi^{(S)}(r,\theta,z)$$

Because of the symmetry of the problem and the boundary conditions on $\phi^{(S)}$, we see that the steady-state solution is just a uniform flow of heat from the high-temperature end to the low-temperature end, and hence

$$\phi^{(S)} = 100 \frac{z}{h}$$

For the transient solution we attempt a separation of variables. Let $\phi^{(T)} = U(r)V(\theta)W(z)F(t)$.   Then we have

$$\dot{F} + \lambda c^2 F = 0$$

$$\frac{d^2W}{dz^2} + (\lambda - \alpha)W = 0$$

$$\frac{d^2V}{d\theta^2} + n^2 V = 0$$

$$r^2 \frac{d^2U}{dr^2} + r \frac{dU}{dr} + (\alpha r^2 - n^2)U = 0$$

The solutions are

$$F = c_1 e^{-\lambda c^2 t}$$

$$W = c_2 \cos \sqrt{\lambda - \alpha} z + c_3 \sin \sqrt{\lambda - \alpha} z$$

$$V = c_4 \cos n\theta + c_5 \sin n\theta$$

$$U = c_6 J_n(\sqrt{\alpha} r) + c_7 Y_n(\sqrt{\alpha} r)$$

$W(0) = 0$ implies that $c_2 = 0$ and $W(h) = 0$ implies that $\lambda - \alpha = k^2\pi^2/h^2$, $k = 1, 2, 3, \ldots$.   The boundary conditions at $r = a$ and $r = b$ are to hold independently of θ and z.   Therefore,

$$c_6 J_n'(\sqrt{\alpha} a) + c_7 Y_n'(\sqrt{\alpha} a) = 0$$

$$c_6 J_n'(\sqrt{\alpha} b) + c_7 Y_n'(\sqrt{\alpha} b) = 0$$

This requires that α be a solution of the equation

$$J_n'(\sqrt{\alpha} b) Y_n'(\sqrt{\alpha} a) - J_n'(\sqrt{\alpha} a) Y_n'(\sqrt{\alpha} b) = 0$$

We let $\quad U_{mn}(r) = \dfrac{J_n(\sqrt{\alpha_{nm}} r) Y_n'(\sqrt{\alpha_{nm}} a) - Y_n(\sqrt{\alpha_{nm}} r) J_n'(\sqrt{\alpha_{nm}} a)}{Y_n'(\sqrt{\alpha_{nm}} a)}$

where α_{nm} is the mth solution of the above equation corresponding to the subscript n.   Our formal solution is then

$$\phi(r,\theta,z,t) = 100\,\frac{z}{h}$$

$$+ \sum_{n=0}^{\infty} \sum_{m=1}^{\infty} \sum_{k=1}^{\infty} (a_{kmn} \cos n\theta + b_{kmn} \sin n\theta) \sin \frac{k\pi z}{h}\, U_{mn}(r) e^{-c^2 \lambda_{kmn} t}$$

where
$$\lambda_{kmn} = \frac{k^2 \pi^2}{h^2} + \alpha_{nm}$$

To satisfy the initial condition we must have

$$f(r,\theta,z) - 100\,\frac{z}{h} = \sum_{n=0}^{\infty} \sum_{m=1}^{\infty} \sum_{k=1}^{\infty} (a_{kmn} \cos n\theta + b_{kmn} \sin n\theta) \sin \frac{k\pi z}{h}\, U_{mn}(r)$$

The functions
$$\psi_{kmn}^{(1)} = \cos n\theta \sin (k\pi z/h) U_{mn}(r)$$

and
$$\psi_{kmn}^{(2)} = \sin n\theta \sin (k\pi z/h) U_{nm}(r)$$

form a complete orthogonal set.   If $f(r,\theta,z)$ has continuous second partial derivatives, the series converges uniformly to the required function.

The steady-state solution of the heat-transfer problem just treated satisfies Laplace's equation.   The solution in this case was particularly simple.   To illustrate further the separation-of-variables technique, let us consider the following boundary-value problem: $\nabla^2 \psi = 0$ on the interior of the sphere $r = b$; that is, in spherical coordinates,

$$\frac{\partial^2 \psi}{\partial r^2} + \frac{2}{r}\frac{\partial \psi}{\partial r} + \frac{1}{r^2}\frac{\partial^2 \psi}{\partial \theta^2} + \frac{\cot \theta}{r^2}\frac{\partial \psi}{\partial \theta} + \frac{1}{r^2 \sin^2 \theta}\frac{\partial^2 \psi}{\partial \phi^2} = 0$$

for $0 \leq r < b, 0 \leq \theta \leq \pi, 0 \leq \phi \leq 2\pi$, with $\psi(b,\theta,\phi) = f(\theta,\phi)$ prescribed. ψ may be regarded as the temperature in a steady-state heat-transfer problem, an electrostatic potential, the velocity potential in a homogeneous, incompressible, irrotational, nonviscous fluid, etc.

Let us assume that $\psi(r,\theta,\phi) = U(r)S(\theta,\phi)$.   Then

$$\frac{r^2}{U}\frac{d^2 U}{dr^2} + \frac{2r}{U}\frac{dU}{dr} = -\left(\frac{1}{S}\frac{\partial^2 S}{\partial \theta^2} + \frac{\cot \theta}{S}\frac{\partial S}{\partial \theta} + \frac{1}{S \sin^2 \theta}\frac{\partial^2 S}{\partial \phi^2}\right) = n(n+1)$$

Hence,
$$r^2 \frac{d^2 U}{dr^2} + 2r\frac{dU}{dr} - n(n+1)U = 0$$

$$\sin^2 \theta\, \frac{\partial^2 S}{\partial \theta^2} + \cos \theta \sin \theta\, \frac{\partial S}{\partial \theta} + \frac{\partial^2 S}{\partial \phi^2} + n(n+1)S = 0$$

Therefore, S is a spherical harmonic:

$$S = (A \cos m\phi + B \sin m\phi)P_n^m(\cos \theta)$$

The general solution for U is

$$U(r) = c_1 r^n + c_2 r^{-n-1}$$

However, our solution must be finite at $r = 0$, and we must therefore take $c_2 = 0$. Our formal series solution is then

$$\psi(r,\theta,\phi) = \sum_{n=0}^{\infty} \sum_{m=0}^{n} (A_{mn} \cos m\phi + B_{mn} \sin m\phi) P_n^m(\cos\theta) r^n$$

To satisfy the boundary condition, we must have

$$f(\theta,\phi) = \sum_{n=0}^{\infty} \sum_{m=0}^{n} (A_{mn} \cos m\phi + B_{mn} \sin m\phi) P_n^m(\cos\theta) b^n$$

The coefficients must therefore be

$$A_{0n} = \frac{(2n+1)b^{-n}}{4\pi} \int_0^{2\pi} \int_0^{\pi} P_n(\cos\theta) f(\theta,\phi) \sin\theta \, d\theta \, d\phi$$

$$A_{mn} = \frac{(2n+1)(n-m)! \, b^{-n}}{2(n+m)! \, \pi} \int_0^{2\pi} \int_0^{\pi} P_n^m(\cos\theta) \cos m\phi f(\theta,\phi) \sin\theta \, d\theta \, d\phi$$

$$B_{mn} = \frac{(2n+1)(n-m)! \, b^{-n}}{2(n+m)! \, \pi} \int_0^{2\pi} \int_0^{\pi} P_n^m(\cos\theta) \sin m\phi f(\theta,\phi) \sin\theta \, d\theta \, d\phi$$

In the process of solving this problem we have obtained an expansion for a harmonic function which is finite at $r = 0$ and takes on prescribed values at $r = b$. In some situations which we shall meet in Chap. 5, we shall need expansions of functions not finite at $r = 0$ but decreasing to zero as r approaches infinity. In this case we shall obtain expansions using spherical harmonics and negative powers of r.

Finally, let us consider a problem in quantum mechanics. The hydrogen atom is pictured in terms of a single negatively charged electron orbiting around the positively charged nucleus. With the center of the nucleus as the origin of coordinates, the coulomb potential of the central force field in which the electron is moving is a function of radial distance r only. The steady-state Schrödinger equation is then

$$\nabla^2 \psi + \frac{2m}{\hbar^2} [E - V(r)]\psi = 0$$

where $\psi(r,\theta,\phi)$ is the wave function for the electron.[1] Again using separation of variables, we assume a solution of the form $\psi(r,\theta,\phi) = U(r)S(\theta,\phi)$ in spherical coordinates and obtain

$$\frac{r^2}{U}\frac{d^2U}{dr^2} + \frac{2r}{U}\frac{dU}{dr} + \frac{2mr^2}{\hbar^2}[E - V(r)] = -\frac{1}{S}\left(\frac{\partial^2 S}{\partial\theta^2} + \cot\theta\frac{\partial S}{\partial\theta} + \frac{1}{\sin^2\theta}\frac{\partial^2 S}{\partial\phi^2}\right)$$

[1] Actually, the same equation holds in any central force field problem.

The separation constant must be of the form $n(n + 1)$ with $n = 0, 1, 2, \ldots$ to ensure finite solutions at $\theta = 0$ and π.   The separated equations are then

$$r^2 \frac{d^2 U}{dr^2} + 2r \frac{dU}{dr} + \frac{2mr^2}{\hbar^2} [E - V(r)]U - n(n + 1)U = 0$$

$$\frac{\partial^2 S}{\partial \theta^2} + \cot \theta \frac{\partial S}{\partial \theta} + \frac{1}{\sin^2 \theta} \frac{\partial^2 S}{\partial \phi^2} + n(n + 1)S = 0$$

Solution of the first equation subject to appropriate boundary conditions leads to a determination of the allowable quantum energy values E.   The solution to the second equation leads to solutions in terms of spherical harmonics

$$S_{mn}(\theta, \phi) = A_{mn} \cos m\phi P_n^m(\cos \theta) + B_{mn} \sin m\phi P_n^m(\cos \theta)$$

with $m = 0, 1, 2, 3, \ldots, n$.   The constants m and n are quantum numbers in the following sense.   The allowable values of the z component of angular momentum are $\pm \hbar m$, and the allowable values of orbital angular momentum[1] are $\hbar \sqrt{n(n + 1)}$.   Thus the eigenvalue problem is suggested in the very name of the subject of quantum mechanics.

We have tried to indicate that the method of separation of variables has wide application to the solution of boundary-value problems.   There are other methods, such as the method of conformal mapping for two-dimensional problems in potential theory, which we shall develop in Chap. 7, and finite difference equation methods, which we shall not develop here.   In the next chapter, we shall develop methods based on the construction of Green's functions. In Chap. 6 we shall relate the problem to the solution of integral equations, and in Chap. 8 we shall look at the problem again in terms of integral transform methods.

Exercises 4.4

1. Solve the following problem in the theory of heat conduction:

$$\frac{\partial^2 \phi}{\partial x^2} = \frac{1}{a^2} \frac{\partial \phi}{\partial t} \qquad 0 < x < L \qquad t > 0$$

$$\phi(0,t) = 0 \qquad \phi(L,t) = 100$$

$$\phi(x,0) = f(x)$$

2. Find the steady-state temperature in an infinitely long heat-conducting slab, $0 \leq x < \infty$, $0 \leq y \leq a$, which is insulated on the top and bottom faces so that the flow of heat is two-dimensional.   The faces at $y = 0$ and $y = a$ are kept at temperature zero, while the face at $x = 0$ is heated so that the temperature is $100y(a - y)$.

3. Find the equilibrium displacement of a circular membrane of radius a when the boundary $r = a$ is displaced according to the prescription $\psi(a,\theta) = g(\theta)$.

[1] This does not take into account the spin of the electron.

Find a series solution and show that it can be put in the closed form

$$\psi(r,\theta) = \frac{1}{2\pi} \int_0^{2\pi} \frac{(a^2 - r^2)g(\phi)}{a^2 + r^2 - 2ar\cos(\theta - \phi)}\, d\phi$$

HINT:
$$\sum_{n=0}^{\infty} \frac{2r^n}{a^n}\cos n(\theta - \phi) - 1 = \operatorname{Re}\left(\frac{\zeta + z}{\zeta - z}\right)$$

when $|z| < |\zeta|$ and $z = re^{i\theta}$, $\zeta = ae^{i\phi}$.

4. Find the natural frequencies and normal modes of vibration of a clamped membrane in the shape of a 60° sector of a circle of radius b.

References

Chester, C. R.: "Techniques in Partial Differential Equations," McGraw-Hill Book Company, New York, 1971.

Churchill, Ruel V.: "Fourier Series and Boundary Value Problems," 3d ed., McGraw-Hill Book Company, New York, 1978.

Copson, Edward T.: "Theory of Functions of a Complex Variable," Oxford University Press, Fairlawn, N.J., 1935.

Courant, Richard, and David Hilbert: "Methods of Mathematical Physics," Interscience Publishers (Division of John Wiley & Sons, Inc.), New York, 1953, vol. I.; 1962, vol. II.

Duff, G. F. D., and D. Naylor: "Differential Equations of Applied Mathematics," John Wiley & Sons, Inc., New York, 1966.

Murnaghan, Francis D.: "Introduction to Applied Mathematics," John Wiley & Sons, Inc., New York, 1948.

Page, Chester H.: "Physical Mathematics," D. Van Nostrand Company, Inc., Princeton, N.J., 1955.

Sagan, Hans: "Boundary and Eigenvalue Problems in Mathematical Physics," John Wiley & Sons, Inc., New York, 1961.

Sneddon, Ian N.: "Elements of Partial Differential Equations," McGraw-Hill Book Company, New York, 1957.

Tikhonov, A. N., and A. A. Samarskii: "Equations of Mathematical Physics," Pergamon Press, Oxford, 1963.

Chapter 5. Boundary-value Problems. Green's Functions

5.1 Nonhomogeneous Boundary-value Problems

Up to now we have been concerned mainly with solving the homogeneous partial differential equation

$$\nabla^2 \phi = a\phi + b\frac{\partial \phi}{\partial t} + c\frac{\partial^2 \phi}{\partial t^2}$$

in a bounded three-dimensional region V, where a, b, and c are constants, subject to one of the three homogeneous boundary conditions

$$\phi = 0$$

$$\frac{d\phi}{dn} = 0$$

$$\frac{d\phi}{dn} + \alpha\phi = 0$$

on S, the boundary of V. If we specify the initial conditions

$$\phi(x,y,z,0) = g(x,y,z)$$

$$\left(\frac{\partial \phi}{\partial t}\right)_{t=0} = h(x,y,z)$$

the problem generally has a unique solution. Upon separating the variables, that is, assuming $\phi = \psi(x,y,z)f(t)$, we find we have to solve the homogeneous partial differential equation

$$\nabla^2 \psi_i + \lambda_i \psi_i = 0$$

in V, subject to $\psi_i = 0$, or $d\psi_i/dn = 0$, or $d\psi_i/dn + \alpha\psi_i = 0$ on S, and the

218

linear second-order ordinary differential equation

$$c\ddot{f}_i + b\dot{f}_i + (a + \lambda_i)f_i = 0$$

The latter equation has the solution

$$f_i = e^{-\gamma t}(A_i \cos \omega_i t + B_i \sin \omega_i t)$$

where $\gamma = b/2c$ and $\omega_i = (1/2c)\sqrt{|4c(a + \lambda_i) - b^2|}$.† The problem has a discrete spectrum of eigenvalues λ_i which increases without bound as $i \to \infty$, and a complete set of orthogonal eigenfunctions ψ_i, which we can assume are normalized as follows:

$$\iiint\limits_V \psi_i^2 \, dV = 1$$

The general solution is $$\phi(x,y,z,t) = \sum_{i=1}^{\infty} f_i \psi_i$$

If the initial values g and h are sufficiently regular, g and h can be expanded in convergent series of the eigenfunctions, and the coefficients A_i and B_i are determined as follows:

$$g = \phi(x,y,z,0) = \sum_{i=1}^{\infty} A_i \psi_i$$

$$h = \dot{\phi}(x,y,z,0) = \sum_{i=1}^{\infty} (\omega_i B_i - \gamma A_i)\psi_i$$

where $$A_i = \iiint\limits_V g\psi_i \, dV \quad \text{and} \quad B_i = \frac{1}{\omega_i}\left(\gamma A_i + \iiint\limits_V h\psi_i \, dV\right)$$

The above problem becomes **nonhomogeneous** if the partial differential equation is nonhomogeneous or if the boundary condition is nonhomogeneous, or both. For example, in the membrane problem considered in Sec. 3.6, if we add a distributed force of density $f(x,y,t)$ per unit area, the partial differential equation satisfied by the displacement is

$$\nabla^2 \phi = \frac{\rho}{\sigma}\frac{\partial^2 \phi}{\partial t^2} - \frac{1}{\sigma}f(x,y,t)$$

If no external force is applied to the membrane, the partial differential equation remains homogeneous, but the problem may be nonhomogeneous if the boundary of the membrane is displaced in some prescribed way; that is, $\phi = g(s,t)$ on the boundary curve, where s is a parameter which described the boundary curve. Of course, one could also have both an applied external

† If $4c(a + \lambda_i) - b^2 < 0$, sin and cos become sinh and cosh.

force and a prescribed displacement of the boundary. Then the boundary-value problem would become

$$\nabla^2 \phi = \frac{\rho}{\sigma} \frac{\partial^2 \phi}{\partial t^2} - \frac{1}{\sigma} f(x,y,t)$$

in R, with $\phi = g(s,t)$ on C.

One could enumerate countless other nonhomogeneous boundary-value problems, for example, forced vibrations of an acoustic or electromagnetic cavity; forced vibrations of a string, an elastic rod, or a clamped plate; heat flow or diffusion problems with distributed sources or sources on the boundary; the transmission line with impressed voltages or currents; the Dirichlet, Neumann, or mixed boundary-value problems of potential theory; problems subject to Poisson's equation; etc.   As a rather general case, let us consider the following boundary-value problem:

$$\nabla^2 \phi = a\phi + b \frac{\partial \phi}{\partial t} + c \frac{\partial^2 \phi}{\partial t^2} - f(x,y,z,t)$$

to be satisfied in a region V, subject to

$$\frac{d\phi}{dn} + \alpha\phi = g(s,t)$$

on S, the boundary of V.   We note that we can split the problem up into three distinct problems and then combine the results to give us the desired solution.   Let ϕ_0 be a solution of

$$\nabla^2 \phi_0 = a\phi_0 + b \frac{\partial \phi_0}{\partial t} + c \frac{\partial^2 \phi_0}{\partial t^2}$$

subject to $d\phi_0/dn + \alpha\phi_0 = 0$ on S, and let ϕ_1 be a solution of

$$\nabla^2 \phi_1 = a\phi_1 + b \frac{\partial \phi_1}{\partial t} + c \frac{\partial^2 \phi_1}{\partial t^2} - f(x,y,z,t)$$

subject to $d\phi_1/dn + \alpha\phi_1 = 0$ on S, and let ϕ_2 be a solution of

$$\nabla^2 \phi_2 = a\phi_2 + b \frac{\partial \phi_2}{\partial t} + c \frac{\partial^2 \phi_2}{\partial t^2}$$

subject to $d\phi_2/dn + \alpha\phi_2 = g(s,t)$ on S.   Then $\phi = \phi_0 + \phi_1 + \phi_2$ satisfies

$$\nabla^2 \phi = a\phi + b \frac{\partial \phi}{\partial t} + c \frac{\partial^2 \phi}{\partial t^2} - f(x,y,z,t)$$

subject to $d\phi/dn + \alpha\phi = g(s,t)$ on S.   The problem therefore reduces to solving a homogeneous problem and two nonhomogeneous problems, one of which has a nonhomogeneous differential equation and a homogeneous boundary condition, while the other has a homogeneous differential equation and a nonhomogeneous boundary condition.

Let us first consider the problem defining ϕ_1. Suppose

$$f(x,y,z,t) = u(x,y,z)v(t)$$

In this case, the differential equation is separable, provided

$$\frac{av + b\dot{v} + c\ddot{v}}{v} = -\lambda$$

where λ is a known constant. We assume that $\phi_1 = \psi_1 v$. Then

$$\frac{\nabla^2\psi_1}{\psi_1} + \frac{u}{\psi_1} = \frac{av + b\dot{v} + c\ddot{v}}{v} = -\lambda$$

and the problem is reduced to the solution of the nonhomogeneous boundary-value problem

$$\nabla^2\psi_1 + \lambda\psi_1 = -u \qquad \text{in } V$$

$$\frac{d\psi_1}{dn} + \alpha\psi_1 = 0 \qquad \text{on } S$$

One of the important cases where this situation arises is when $a = b = 0$ and we have a sinusoidal forcing function. Even if the forcing function is only periodic, we may still be able to separate the equation with the help of Fourier series.[1] In this case, $f(x,y,z,t) = u\sum_{i=0}^{\infty} c_i v_i$, where $c\ddot{v}_i = -\lambda_i v_i$ (no summation). In this case we seek a solution of the form $\phi_1 = \sum_{i=0}^{\infty} c_i\psi_i v_i$. Substituting in the differential equation, we have

$$\sum_{i=0}^{\infty} c_i\nabla^2\psi_i v_i = \sum_{i=0}^{\infty} cc_i\psi_i\ddot{v}_i - \sum_{i=0}^{\infty} uc_i v_i$$

or

$$\sum_{i=0}^{\infty} c_i[\nabla^2\psi_i + \lambda_i\psi_i + u]v_i = 0$$

We are thus led again to the nonhomogeneous problem

$$\nabla^2\psi_i + \lambda_i\psi_i = -u \qquad \text{in } V$$

$$\frac{d\psi_i}{dn} + \alpha\psi_i = 0 \qquad \text{on } S$$

For the problem with the nonhomogeneous boundary condition defining ϕ_2 we have similar considerations. Suppose $g(s,t) = u(s)v(t)$; then we seek a

---

[1] If neither case holds, we may be able to resort to integral transform methods which will be discussed in Chap. 8.

solution of the form $\phi_2 = \psi_2 v$. The differential equation becomes

$$\frac{\nabla^2\psi_2}{\psi_2} = \frac{av + b\dot{v} + c\ddot{v}}{v} = -\lambda$$

and we have to solve $\nabla^2\psi_2 + \lambda\psi_2 = 0$. The boundary condition becomes

$$v(t)\left(\frac{d\psi_2}{dn} + \alpha\psi_2\right) = u(s)v(t)$$

to be satisfied for all t. This implies that $d\psi_2/dn + \alpha\psi_2 = u(s)$. This will be possible, in particular, when we have a sinusoidal displacement of the boundary and $a = b = 0$. If g is periodic, we shall expand it in a Fourier series and seek a solution of the form

$$\phi_2 = \sum_{i=0}^{\infty} c_i\psi_i v_i$$

where $c\ddot{v}_i = -\lambda_i v_i$ (no summation). We shall then be led to the boundary-value problem

$$\nabla^2\psi_i + \lambda_i\psi_i = 0 \qquad \text{in } V$$

$$\frac{d\psi_i}{dn} + \alpha\psi_i = u \qquad \text{on } S$$

The boundary-value problem with the nonhomogeneous boundary condition is essentially equivalent to the problem with the nonhomogeneous differential equation, for if there exists a twice differentiable function F defined in V and on S such that

$$\frac{dF}{dn} + \alpha F = u$$

then we let $H = \psi_2 - F$. We then have

$$\frac{dH}{dn} + \alpha H = 0 \qquad \text{on } S$$

$$\nabla^2 H + \lambda H = \nabla^2\psi_2 + \lambda\psi_2 - \nabla^2 F - \lambda F$$

$$\nabla^2 H + \lambda H = -\nabla^2 F - \lambda F \qquad \text{in } V$$

and H satisfies a homogeneous boundary condition and a nonhomogeneous differential equation.

The nonhomogeneous problem

$$\nabla^2\psi + \lambda\psi = -f \qquad \text{in } V$$

$$\frac{d\psi}{dn} + \alpha\psi = 0 \qquad \text{on } S$$

may be solvable in terms of the eigenfunctions of the homogeneous problem

$$\nabla^2 \psi_i + \lambda_i \psi_i = 0 \qquad \text{in } V$$

$$\frac{d\psi_i}{dn} + \alpha \psi_i = 0 \qquad \text{on } S$$

The coefficients in a "best" approximation to ψ by a linear combination $c_i \psi_i$ in the least-mean-square sense are

$$c_i = \iiint_V \psi \psi_i \, dV$$

Let $\qquad \gamma_i = \iiint_V f \psi_i \, dV$

then $\qquad \gamma_i = -\iiint_V (\nabla^2 \psi + \lambda \psi) \psi_i \, dV$

$$= -\iiint_V \psi \nabla^2 \psi_i \, dV + \iint_S \left(\psi_i \frac{d\psi}{dn} - \psi \frac{d\psi_i}{dn} \right) dS - \lambda c_i$$

$$= (\lambda_i - \lambda) c_i$$

Thus $c_i = \gamma_i / (\lambda_i - \lambda)$, provided $\lambda \neq \lambda_i$ for every i. The case where $\lambda = \lambda_i$ for some i is referred to as **resonance**. There may still be a solution in the case of resonance. Suppose that $\lambda = \lambda_k$ but $\gamma_k = \iiint_V f \psi_k \, dV = 0$. Then c_k is arbitrary, and there are an infinite number of solutions. Note the similarities between this discussion and the solution of nonhomogeneous linear equations in Sec. 1.8 and the problem of forced vibrations in Sec. 1.9.

Exercises 5.1

1. Solve for the forced vibrations of a clamped circular membrane of radius 1, if there is a distributed force of density $a(1 - r) \sin \omega t$ (a and ω are known constants), and the initial conditions are $\phi(r, \theta, 0) = g(r, \theta)$ and $\dot\phi(r, \theta, 0) = 0$.

2. Consider the Dirichlet problem, that is,

$$\nabla^2 \phi = 0 \qquad \text{in } V$$

$$\phi = f(x, y, z) \qquad \text{on } S$$

Describe a procedure for obtaining a series solution of the problem in terms of the eigenfunctions of the problem

$$\nabla^2 \psi_i + \lambda_i \psi_i = 0 \qquad \text{in } V$$

$$\psi_i = 0 \qquad \text{on } S$$

3. Prove the uniqueness of the solution of the nonhomogeneous boundary-value problem

$$\nabla^2 \psi + \lambda \psi = -f \qquad \text{in } V$$

$$\frac{d\psi}{dn} + \alpha \psi = g \qquad \text{on } S$$

if $\lambda \neq \lambda_i$, an eigenvalue of the homogeneous problem.   HINT: Consider the difference of two possible different solutions.

5.2   One-dimensional Green's Functions

Another approach to the solution of nonhomogeneous boundary-value problems is by means of the construction of auxiliary functions known as **Green's functions.**   To illustrate this method, let us consider the solution of the following boundary-value problem, which arises in the study of forced vibrations of a string with fixed ends.

$$\frac{d^2\psi}{dx^2} + k^2\psi = -f(x) \qquad 0 \leq x \leq a$$

$$\psi(0) = \psi(a) = 0$$

We shall first solve the problem by the method of variation of parameters. We assume a solution of the form $\psi = A(x) \sin kx + B(x) \cos kx$.   Then

$$\psi' = A' \sin kx + B' \cos kx + kA \cos kx - kB \sin kx$$

We assume $A' \sin kx + B' \cos kx = 0$ and then differentiate again.

$$\psi'' = -k^2 A \sin kx - k^2 B \cos kx + kA' \cos kx - kB' \sin kx$$

Substituting in the differential equation, we have

$$kA' \cos kx - kB' \sin kx = -f(x)$$

Therefore, we have two linear equations in A' and B' with the solution

$$A' = -\frac{1}{k} f(x) \cos kx$$

$$B' = \frac{1}{k} f(x) \sin kx$$

The solution is then

$$\psi = -\frac{\sin kx}{k} \int_{c_1}^{x} f(y) \cos ky \, dy + \frac{\cos kx}{k} \int_{c_2}^{x} f(y) \sin ky \, dy$$

where c_1 and c_2 are constants to be determined by the boundary conditions.

$$\psi(0) = \frac{1}{k} \int_{c_2}^{0} f(y) \sin ky \, dy = 0$$

implies that $c_2 = 0$.

$$\psi(a) = -\frac{\sin ka}{k}\int_{c_1}^{a} f(y)\cos ky\, dy + \frac{\cos ka}{k}\int_{0}^{a} f(y)\sin ky\, dy$$

$$0 = -\frac{\sin ka}{k}\int_{c_1}^{0} f(y)\cos ky\, dy + \frac{1}{k}\int_{0}^{a} f(y)\sin k(y-a)\, dy$$

We can therefore write the solution

$$\psi(x) = \frac{1}{k}\int_{0}^{x} f(y)\sin k(y-x)\, dy - \frac{\sin kx}{k\sin ka}\int_{0}^{a} f(y)\sin k(y-a)\, dy$$

$$= \frac{1}{k}\int_{0}^{x} f(y)\left[\frac{\sin k(y-x)\sin ka - \sin kx \sin k(y-a)}{\sin ka}\right] dy$$

$$+ \frac{1}{k}\int_{x}^{a} f(y)\frac{\sin kx \sin k(a-y)}{\sin ka}\, dy$$

$$= \int_{0}^{x} f(y)\frac{\sin ky \sin k(a-x)}{k\sin ka}\, dy$$

$$+ \int_{x}^{a} f(y)\frac{\sin kx \sin k(a-y)}{k\sin ka}\, dy$$

$$= \int_{0}^{a} f(y)G(x;y)\, dy$$

where
$$G(x;y) = \frac{\sin ky \sin k(a-x)}{k\sin ka} \qquad 0 \le y \le x$$

$$G(x;y) = \frac{\sin kx \sin k(a-y)}{k\sin ka} \qquad x \le y \le a$$

This function is called a **Green's function.** We note that it exists for this problem unless $\sin ka = 0$. In that case k^2 would be an eigenvalue of the homogeneous problem. Thus the Green's function exists unless k^2 is an eigenvalue of the homogeneous problem.

We have reduced the solution of the nonhomogeneous boundary-value problem to a simple formula:

$$\psi(x) = \int_{0}^{a} f(y)G(x;y)\, dy$$

The advantage of this formulation of the problem is that the Green's function is independent of f; that is, it depends only on the form of the differential equation, k, and the boundary conditions. Therefore, the solutions to all possible such problems with different functions f are known, provided the integral $\int_{0}^{a} f(y)G(x;y)\, dy$ exists.

Let us now look at the properties of the above Green's function so that we may use these properties to construct Green's functions for other problems. These properties are:

1. The Green's function satisfies the homogeneous differential equation[1] $G'' + k^2G = 0$ in each of the intervals $0 \leq y < x$ and $x < y \leq a$, but not at $y = x$. This can be easily seen, because in each of these intervals G can be written in the form $C_1 \sin ky + C_2 \cos ky$. The point $y = x$ must be excluded because, as we shall see, the second derivative of G does not exist there.

2. The Green's function is continuous at $y = x$, since

$$\lim_{y \to x-} G(x;y) = \frac{\sin kx \sin k(a - x)}{k \sin ka}$$

$$\lim_{y \to x+} G(x;y) = \frac{\sin kx \sin k(a - x)}{k \sin ka}$$

3. The derivative of the Green's function is discontinuous at $y = x$.

$$G'(x;x-) = \lim_{y \to x-} G'(x;y) = \frac{\cos kx \sin k(a - x)}{\sin ka}$$

$$G'(x;x+) = \lim_{y \to x+} G'(x;y) = -\frac{\sin kx \cos k(a - x)}{\sin ka}$$

and $G'(x;x+) - G'(x;x-) = -1$.   Because the derivative is discontinuous, the second derivative does not exist.

4. The Green's function satisfies the boundary condition of the problem. $G(x;0) = G(x;a) = 0$.

5. The Green's function is symmetric in the two arguments.   If we interchange x and y in the above definition, we do not change the definition; that is,

$$G(x;y) = G(y;x)$$

Let us now try to reconstruct the above Green's function by seeking a function with these properties.   Starting with property 1, we want a function

$$G(x;y) = A \sin ky + B \cos ky \qquad 0 \leq y < x$$
$$G(x;y) = C \sin ky + D \cos ky \qquad x < y \leq a$$

By property 4,

$$G(x;0) = B = 0$$
$$G(x;a) = C \sin ka + D \cos ka = 0$$

By property 2,

$$A \sin kx = C \sin kx + D \cos kx$$

---

[1] The prime refers to differentiation with respect to y.   Throughout this discussion x is treated as a parameter.

By property 3,

$$kC \cos kx - kD \sin kx - kA \cos kx = -1$$

Solving these equations simultaneously, we have

$$A = \frac{\sin k(a - x)}{k \sin ka}$$

$$B = 0$$

$$C = -\frac{1}{k}\frac{\sin kx \cos ka}{\sin ka}$$

$$D = \frac{1}{k}\frac{\sin kx \sin ka}{\sin ka}$$

so that

$$G(x;y) = \frac{\sin ky \sin k(a - x)}{k \sin ka} \qquad 0 \leq y \leq x$$

$$G(x;y) = \frac{\sin kx \sin k(a - y)}{k \sin ka} \qquad x \leq y \leq a$$

Assuming a Green's function with the above properties, we can arrive at the solution of the boundary-value problem directly from the differential equation.

$$\int_0^a f(y)G(x;y)\,dy = -\int_0^a (\psi'' + k^2\psi)G(x;y)\,dy$$

The last integral could be evaluated by integration by parts except for the fact that G'' does not exist at $y = x$. Therefore, we shall write

$$\int_0^a (\psi'' + k^2\psi)G(x;y)\,dy = \lim_{\xi \to x-}\int_0^\xi (\psi'' + k^2\psi)G\,dy + \lim_{\eta \to x+}\int_\eta^a (\psi'' + k^2\psi)G\,dy$$

We can then integrate by parts twice in each interval; that is,

$$\int_0^\xi (\psi'' + k^2\psi)G\,dy = [G\psi' - G'\psi]_0^\xi + \int_0^\xi \psi(G'' + k^2G)\,dy$$

$$= G(x;\xi)\psi'(\xi) - G'(x;\xi)\psi(\xi)$$

$$\int_\eta^a (\psi'' + k^2\psi)G\,dy = [G\psi' - G'\psi]_\eta^a + \int_\eta^a \psi(G'' + k^2G)\,dy$$

$$= G'(x;\eta)\psi(\eta) - G(x;\eta)\psi'(\eta)$$

Then

$$\int_0^a f(y)G(x;y)\,dy = -\psi(x)[G'(x;x+) - G'(x;x-)]$$

$$+ \psi'(x)[G(x;x+) - G(x;x-)] = \psi(x)$$

where we have used properties 1 to 4.   It is now clear how we would modify the properties to obtain the Green's function for the problem

$$\nabla_\psi^2 + k^2\psi = -f(x)$$
$$\psi'(0) = \psi'(a) = 0$$

In this case, $G(x;y)$ would satisfy properties 1 to 3, and property 4 would be changed to $G'(x;0) = G'(x;a) = 0$.

Now let us consider the "function" $G'' + k^2G$.   It is zero for $0 \le y < x$ and zero for $x < y \le a$, and yet if we write

$$-\psi(x) = -\int_0^a f(y)G(x;y)\,dy = \int_0^a (\psi'' + k^2\psi)G\,dy$$

and then formally integrate by parts and use the boundary conditions, we have

$$-\psi(x) = \int_0^a (G'' + k^2G)\psi\,dy$$

We indicate the behavior of $G'' + k^2G$ by writing

$$G'' + k^2G = -\delta(y - x)$$

where $\delta(t)$ is the **Dirac delta function.** It is zero everywhere except at $t = 0$, where it is undefined, and yet

$$\int_{-\infty}^\infty f(t)\,\delta(t)\,dt = f(0)$$

for any continuous function $f(t)$.   In particular, if $f(t) = 1$, then

$$\int_{-\infty}^\infty \delta(t)\,dt = 1$$

The delta function is not a function in the usual sense.   In fact, the integral of a function which is zero everywhere except at one point is necessarily zero. Intuitively we can think of $\delta(t)$ as the "limit" of the function $d_n(t) = n$, $-(1/2n) \le t \le (1/2n)$, $d_n(t) = 0$, elsewhere; for if $f(t)$ is continuous,

$$\lim_{n\to\infty} \int_{-\infty}^\infty f(t)\,d_n(t)\,dt = f(0)$$

Nevertheless, it is not valid to take the limit under the integral sign, so we cannot literally claim that $\delta(t)$ is the limit of $d_n(t)$.   However, there is a theory of **generalized functions,** which we shall outline in the next section, which includes the delta function as well as derivatives of the delta function.

In terms of the equation

$$G'' + k^2G = -\delta(y - x)$$

we have an interesting physical interpretation of the Green's function.

According to this equation, the *Green's function gives the displacement due to a force of density δ applied at the point y = x.* This density would be achieved as the limit of a total force of one unit applied to a region containing the point $y = x$ as the region shrinks to a point keeping the total force equal to one unit. We also have an interpretation of the formula

$$\psi(x) = \int_0^a f(y)G(x;y)\,dy$$

If $G(x;y)$ is the displacement at y due to a unit force at x, then by the symmetry $G(x;y) = G(y;x)$ it also represents the displacement at x due to a unit force at y. Now if we multiply $G(x;y)$ by a "weighting factor" of density $f(y)$ and integrate over all possible points, we get the total displacement at x due to a distributed force of density $f(y)$.

This physical interpretation actually affords a means of computing a certain Green's function. Consider, for example, the equilibrium displacement of a string with fixed ends under a distributed force of density $f(x)$. The differential equation for the displacement is

$$\phi'' = -\frac{f(x)}{\sigma}$$

where σ is the tension in the string. Then the Green's function satisfies the equation

$$G'' = -\delta(y - x) = -\frac{\sigma\,\delta(y - x)}{\sigma}$$

Thus G represents the displacement at y due to a force of strength σ at x.

In equilibrium we have

$$\sigma = \sigma \sin \alpha + \sigma \sin \beta$$
$$= \sigma \tan \alpha + \sigma \tan \beta$$

assuming small displacements. Also, we have

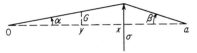

FIGURE 8

$$\tan \alpha = \frac{G}{y}$$

$$\tan \beta = \frac{x}{a - x} \tan \alpha$$

so that

$$G(x;y) = \frac{y(a - x)}{a} \qquad 0 \le y < x$$

By symmetry

$$G(x;y) = \frac{x(a - y)}{a} \qquad x < y \le a$$

Now let us consider the possibility of obtaining a Green's function for the nonhomogeneous Sturm-Liouville problem

$$(p\psi')' - q\psi + \lambda\rho\psi = -f(x) \qquad a \le x \le b$$

We seek a solution of the form

$$\psi(x) = \int_a^b f(y)G(x;y)\,dy$$

$$= -\int_a^b [(p\psi')' - q\psi + \lambda\rho\psi]G\,dy$$

$$= -\lim_{\xi \to x-} \int_a^\xi [(p\psi')' - q\psi + \lambda\rho\psi]G\,dy$$

$$\qquad -\lim_{\eta \to x+} \int_\eta^b [(p\psi')' - q\psi + \lambda\rho\psi]G\,dy$$

Integrating by parts, we have

$$\int_a^\xi [(p\psi')' - q\psi + \lambda\rho\psi]G\,dy = [p(\psi'G - G'\psi)]_a^\xi + \int_a^\xi [(pG')' - qG + \lambda\rho G]\psi\,dy$$

$$\int_\eta^b [(p\psi')' - q\psi + \lambda\rho\psi]G\,dy = [p(\psi'G - G'\psi)]_\eta^b + \int_\eta^b [(pG')' - qG + \lambda\rho G]\psi\,dy$$

If we take as the first property of the Green's function that it satisfies the homogeneous equation

$$(pG')' - qG + \lambda\rho G = 0$$

in each of the intervals $a \le y < x$ and $x < y \le b$, then we can drop the integrals in each of the last two expressions. Furthermore, if we take as another property that G satisfies the same boundary conditions as ψ at $y = a$ and $y = b$, that is, one of the five homogeneous boundary conditions we considered for the homogeneous Sturm-Liouville problem, then $[p(\psi'G - G'\psi)]_a^b = 0$. Taking the limits as $\xi \to x-$ and $\eta \to x+$, we have

$$\psi(x) = -p(x)[G'(x;x+) - G'(x;x-)]\psi(x) + p(x)[G(x;x+) - G(x;x-)]\psi'(x)$$

This equation will be valid if G is continuous at $y = x$ and G' is discontinuous at $y = x$, with a discontinuity of

$$G'(x;x+) - G'(x;x-) = -\frac{1}{p(x)}$$

The four properties just described are sufficient to define the Green's function for the Sturm-Liouville problem, unless λ is an eigenvalue of the homogeneous problem. In that case, properties 1 to 3 determine the eigenfunction corresponding to the eigenvalue λ, which satisfies the homogeneous differential equation and the given boundary conditions and is continuous everywhere. But this solution has a continuous derivative, and therefore property 4 cannot be satisfied. A fifth property, namely, symmetry, that is, $G(x;y) = G(y;x)$, can be proved from the other four. This proof will be left for the exercises. The symmetry results from the fact that the Sturm-Liouville operator is self-adjoint.

As an example, let us find the Green's function associated with the problem

$$(x\psi')' - \frac{n^2}{x}\psi + k^2 x\psi = -f(x) \qquad 0 \le x \le b$$

$$\psi(b) = 0$$

$$\psi(0) \text{ finite}$$

The Green's function should satisfy

$$(yG')' - \frac{n^2}{y}G + k^2 yG = 0$$

in the interval $0 \le y < x$, and be finite at $y = 0$. Hence,

$$G = AJ_n(ky) \qquad 0 \le y < x$$

In the interval $x < y \le b$, G should satisfy the same equation and $G(b) = 0$. Therefore,

$$G = BJ_n(ky) + CY_n(ky) \qquad x < y \le b$$

and

$$BJ_n(kb) + CY_n(kb) = 0$$

G is continuous at $y = x$, so that

$$AJ_n(kx) = BJ_n(kx) + CY_n(kx)$$

G' has a jump discontinuity equal to $-1/x$ at $y = x$, and therefore,

$$BJ'_n(kx) + CY'_n(kx) - AJ'_n(kx) = -\frac{1}{x}$$

Solving simultaneously for A, B, and C, we have

$$A = \frac{1}{x}\frac{J_n(kx)Y_n(kb) - J_n(kb)Y_n(kx)}{J_n(kb)[Y_n(kx)J'_n(kx) - J_n(kx)Y'_n(kx)]}$$

$$B = \frac{1}{x}\frac{J_n(kx)Y_n(kb)}{J_n(kb)[Y_n(kx)J'_n(kx) - J_n(kx)Y'_n(kx)]}$$

$$C = -\frac{1}{x}\frac{J_n(kx)}{Y_n(kx)J'_n(kx) - J_n(kx)Y'_n(kx)}$$

These expressions can be simplified somewhat by use of the Bessel differential equation.   Since

$$J_n'' + \frac{J_n'}{x} - \frac{n^2 J_n}{x^2} + k^2 J_n = 0$$

$$Y_n'' + \frac{Y_n'}{x} - \frac{n^2 Y_n}{x^2} + k^2 Y_n = 0$$

we have $$x(J_n'' Y_n - Y_n'' J_n) + (J_n' Y_n - J_n Y_n') = 0$$

or $$\frac{d}{dx}[x(J_n' Y_n - Y_n' J_n)] = 0$$

$$x(J_n' Y_n - Y_n' J_n) = K$$

where K can be evaluated by considering the limit of the expression as $x \to 0$.† The Green's function is then

$$G(x;y) = \frac{J_n(ky)[J_n(kx) Y_n(kb) - Y_n(kx) J_n(kb)]}{K J_n(kb)}$$

for $a \leq y \leq x$, and

$$G(x;y) = \frac{J_n(kx)[J_n(ky) Y_n(kb) - Y_n(ky) J_n(kb)]}{K J_n(kb)}$$

for $x \leq y \leq b$, provided $J_n(kb) \neq 0$, that is, provided k^2 is not an eigenvalue of the homogeneous equation.

We have seen that we get into difficulty when trying to determine the Green's function by the above procedure, if the constant λ in the nonhomogeneous Sturm-Liouville equation is an eigenvalue of the homogeneous problem.   It is possible nevertheless to obtain a **generalized Green's function** for the nonhomogeneous problem if the function $f(x)$ on the right-hand side of the equation is orthogonal to the eigenfunction corresponding to the eigenvalue λ; that is,

$$(pu')' - qu + \lambda \rho u = 0$$
$$\int_a^b fu\, dx = 0$$

We wish to solve, as before,

$$(p\psi')' - q\psi + \lambda \rho \psi = -f(x)$$

First, we note that $\int_a^b uf\, dx = 0$ is a necessary condition for a solution of the

† $K = -2/\pi.$

nonhomogeneous problem, for if there exists a solution,

$$\int_a^b uf\,dx = -\int_a^b u[L(\psi) + \lambda\rho\psi]\,dx$$

$$= -\int_a^b \psi[L(u) + \lambda\rho u]\,dx$$

$$= 0$$

Next we observe that, without loss of generality, we can seek a solution of the nonhomogeneous problem which is orthogonal to the solution of the homogeneous problem. This follows since for any solution ψ

$$L(\psi - cu) + \lambda\rho(\psi - cu) = L(\psi) + \lambda\rho\psi - c[L(u) + \lambda\rho u]$$

$$= -f$$

where c is arbitrary. Hence c may be chosen so that

$$\int_a^b \rho(\psi - cu)u\,dx = 0$$

that is,

$$c = \int_a^b \rho\psi u\,dx$$

assuming that u has been normalized. We also observe that if there exists a solution of the nonhomogeneous problem, there are actually infinitely many solutions, since any multiple of the solution of the homogeneous problem may be added to the known solution.

In the present case, we seek a Green's function which satisfies

$$L(G) + \lambda\rho G - A\rho u = -\delta(y - x)$$

Then

$$\int_a^b f(y)G(x;y)\,dy = -\int_a^b G[L(\psi) + \lambda\rho\psi]\,dy$$

$$= -\int_a^b \psi[L(G) + \lambda\rho G]\,dy$$

$$= \int_a^b [\delta(y - x)\psi(y) - A\rho u(y)\psi(y)]\,dy$$

$$= \psi(x)$$

The constant A may be determined as follows:

$$A\int_a^b \rho u^2(y)\,dy = \int_a^b \delta(y - x)u(y)\,dy + \int_a^b u[L(G) + \lambda\rho G]\,dy$$

$$A = u(x) + \int_a^b G[L(u) + \lambda\rho u]\,dy$$

$$= u(x)$$

As an example, let us consider the problem

$$\psi'' + \psi = -f(x) \qquad 0 \leq x \leq \pi$$

$$\psi(0) = \psi(\pi) = 0$$

We see that $u'' + u = 0$, $u(0) = u(\pi) = 0$, has the normalized solution $u = \sqrt{2/\pi} \sin x$. Therefore, we seek a Green's function as a solution of

$$G'' + G = -\delta(y - x) + \frac{2}{\pi} \sin x \sin y$$

Hence, $G = A \cos y + B \sin y - \dfrac{1}{\pi} y \sin x \cos y \qquad 0 \leq y < x$

$$G = C \cos y + D \sin y - \frac{1}{\pi} y \sin x \cos y \qquad x < y \leq \pi$$

The boundary conditions imply that

$$G(0) = A = 0$$

$$G(\pi) = -C + \sin x = 0$$

Continuity at $y = x$ implies

$$B \sin x = \sin x \cos x + D \sin x$$

The derivative of the Green's function must have a jump discontinuity of -1 at $y = x$; thus

$$-\sin^2 x + D \cos x - B \cos x = -1$$

The last two equations are satisfied if

$$B - D = \cos x$$

To have symmetry we choose

$$B = \cos x - \frac{x}{\pi} \cos x$$

$$D = - \frac{x}{\pi} \cos x$$

The appropriate Green's function is then

$$G = \cos x \sin y - \frac{x}{\pi} \cos x \sin y - \frac{y}{\pi} \sin x \cos y \qquad 0 \leq y \leq x$$

$$G = \sin x \cos y - \frac{y}{\pi} \sin x \cos y - \frac{x}{\pi} \cos x \sin y \qquad x \leq y \leq \pi$$

and the general solution of the problem is

$$\psi(x) = c \sin x + \int_0^\pi f(y) G(x;y)\, dy$$

where c is arbitrary.

In summary, we have the following: The nonhomogeneous Sturm-Liouville equation

$$(p\psi')' - q\psi + \lambda\rho\psi = -f(x) \qquad a \le x \le b$$

with appropriate homogeneous boundary conditions has a unique solution if λ is not an eigenvalue of the corresponding homogeneous problem. If λ is an eigenvalue of the homogeneous problem with solution u, the nonhomogeneous problem will have a solution if and only if $\int_a^b uf\, dx = 0$, in which case there are infinitely many solutions.

Exercises 5.2

1. Find the Green's function associated with the boundary-value problem $\psi'' + k^2\psi = -f(x)$, $0 \le x \le a$, $\psi'(0) = \psi'(a) = 0$.

2. Prove that the Green's function for the Sturm-Liouville problem is symmetric.

3. Find the Green's function associated with the boundary-value problem $(x\psi')' - (n^2/x)\psi = -f(x)$, $0 \le x \le 1$, $\psi(0)$ finite, $\psi(1) = 0$.

4. Find the generalized Green's function associated with the boundary-value problem

$$\psi'' + \psi = -f(x) \qquad 0 \le x \le \pi \qquad \psi'(0) = \psi'(\pi) = 0$$

if $\int_0^\pi f(x) \cos x\, dx = 0$. Show that the first method outlined in this section fails to yield a Green's function.

5. Consider the nonhomogeneous regular Sturm-Liouville problem; $(py')' - qy + \lambda\rho y = -f(x)$, $y'(a) - \sigma_1 y(a) = 0$, $y'(b) + \sigma_2 y(b) = 0$. Prove that this problem has a unique solution if $f(x)$ is continuous and λ is not an eigenvalue of the homogeneous problem. HINT: Prove the existence of a Green's function using two functions $u(x)$ and $v(x)$ satisfying the homogeneous equation and the initial conditions; $u(a) = 1$, $u'(a) = \sigma_1$, $v(b) = 1$, $v'(b) = -\sigma_2$.

5.3 Generalized Functions

The Dirac delta function was used in physics long before it was made respectable by mathematicians. As we pointed out in the last section, there are no functions which have the properties claimed for the delta function. Nevertheless, the use of the delta function became so firmly entrenched that it become imperative that a mathematical theory be invented which would include the delta function. This was done about 1950 by the mathematician

Schwartz[1] when he formulated the theory of distributions. A better name for this new concept is **generalized function,** which suggests that it includes the old style functions, as it does, and also avoids the confusion with the distributions of probability theory. In this section, we shall outline the theory of generalized functions without being explicit about detailed proofs.

We recall that an important property of the delta function, which we wish to preserve in the theory of generalized functions, is

$$\int_{-\infty}^{\infty} \delta(x)f(x)\,dx = f(0)$$

for every continuous function $f(x)$. In other words, the important thing about the delta function is how it operates on other functions. Therefore, we associate with the delta function an operator $F[\phi] = \phi(0)$. Before we discuss this operator we must decide what its domain shall be. Our first impression is to take $\phi(x)$ in the space of functions continuous on the whole x axis. However, we shall want to include in the theory the concept of derivatives of generalized functions and for this purpose we shall need to have our operator operate on a function space which contains functions with derivatives of all orders. We call this the space of **testing functions.**

Definition. The space of testing functions is the collection of all real-valued functions which have derivatives of all orders and vanish outside of some finite interval of the x axis.

To show that the space of testing functions is not empty, we note that it contains

$$\phi(x) = \begin{cases} A \exp\left[-\dfrac{c}{(x-a)^\alpha}\right] \exp\left[-\dfrac{d}{(-x+b)^\beta}\right] & a < x < b \\ 0 & x \le a \quad \text{or} \quad b \le x \end{cases}$$

where A, a, and b are real numbers and α, β, c, d are any positive real numbers. Note that the interval $a \le x \le b$, outside of which a testing function vanishes, does not have to be the same for all testing functions. Clearly the space of testing functions is a vector space over the real numbers, but it is not a Hilbert space since we have not defined a scalar product. If we used the conventional scalar product it would not be complete.

Returning to the functional $F[\phi] = \phi(0)$ defined for every testing function, we note that it is linear since $F[c_1\phi_1 + c_2\phi_2] = c_1\phi_1(0) + c_2\phi_2(0) = c_1F[\phi_1] + c_2F[\phi_2]$. We shall say that the generalized function, which we call the delta function, is defined by the functional $F[\phi] = \phi(0)$. The integral $\displaystyle\int_{-\infty}^{\infty} \delta(x)\phi(x)\,dx$ is not defined in the ordinary sense. However, we

---

[1] L. Schwartz, "Théorie des distributions," Actualités scientifiques et industrielles, Nos. 1091 and 1122, Hermann & Cie, Paris, 1950–1951.

shall identify the integral with the functional and use the value of the functional to evaluate the integral. In other words, $\int_{-\infty}^{\infty} \delta(x)\phi(x)\,dx$ is symbolic for $\phi(0)$, which gives us the right to state $\int_{-\infty}^{\infty} \delta(x)\phi(x)\,dx = \phi(0)$. The functional does not assign values to the generalized function at each point on the x axis as in the case of ordinary functions. However, by a device which we shall define later we shall be able to say that the delta function is zero except at $x = 0$, where it is undefined.

For a continuous function $f(x)$ the integral $\int_{-\infty}^{\infty} f(x)\phi(x)\,dx$ is defined for all testing functions and it defines a linear functional[1]

$$f[\phi] = \int_{-\infty}^{\infty} f(x)\phi(x)\,dx$$

Furthermore, if $g[\phi] = \int_{-\infty}^{\infty} g(x)\phi(x)\,dx = f[\phi]$ for some continuous function $g(x)$, then $g(x) \equiv f(x)$. To prove this assume that $g(x_0) > f(x_0)$ for some x_0. Then $g(x) > f(x)$ in some interval $a \leq x \leq b$. But there is a testing function $\phi_{ab}(x)$ which is positive in this interval and vanishes outside of it, and

$$0 = \int_{-\infty}^{\infty} [g(x) - f(x)]\phi_{ab}(x)\,dx = \int_{a}^{b} [g(x) - f(x)]\phi_{ab}(x)\,dx > 0$$

which is a contradiction proving that $g(x) \equiv f(x)$. If in this discussion $f(x)$ and $g(x)$ are not continuous but are integrable, then $f[\phi] = g[\phi]$ does not imply that $f(x) \equiv g(x)$, but only that they are equal *almost everywhere*.

In order to obtain the theory of generalized functions, we need to have our functionals more than just linear. We also need to have them continuous in the following sense. A sequence of testing functions $\{\phi_n(x)\}$ converges to a testing function $\phi(x)$ if $\phi_n(x) - \phi(x)$ vanishes outside some finite interval for all n and $\{\phi_n^{(k)}(x) - \phi^{(k)}(x)\}$ converges uniformly to zero for $k = 0, 1, 2, \ldots$. A linear functional F is continuous on the space of testing functions if $\phi_n \to \phi$ implies $F[\phi_n] \to F[\phi]$.

Definition. A generalized function is a continuous linear functional defined on the space of testing functions. Two generalized functions are equal if the functionals defining them are equal for all testing functions. Multiplication by a real scalar c is defined by $cF[\phi] = F[c\phi]$ and addition of two generalized functions is defined by $F_1[\phi] + F_2[\phi] = F[\phi]$.

We can associate values with a generalized function on an open interval by the following device. Let $f(x)$ be a continuous function which defines the

---

[1] From now on we shall use the same letter to denote the functional and the function values, when the function has values.

linear functional

$$f[\phi] = \int_{-\infty}^{\infty} f(x)\phi(x)\,dx$$

Let $F[\phi]$ be a continuous linear functional defining a generalized function. Now we say that $F(x) = f(x)$ on the interval $a < x < b$ if $F[\phi] = f[\phi]$ for all testing functions which vanish outside this interval. This definition allows us to say that $\delta(x) = 0$ except at $x = 0$, because if the interval $a < x < b$ does not include $x = 0$, then $\delta[\phi] = 0$ for all testing functions vanishing outside the interval. As another example consider the continuous linear functional

$$h[\phi] = \int_{0}^{\infty} \phi(x)\,dx$$

Let $0 < a < x < b$ and let $\phi(x)$ vanish outside this interval. Then $h[\phi] = \int_{a}^{b} \phi(x)\,dx = f[\phi]$ for the function $f(x) \equiv 1$. On the other hand if $a < x < b < 0$ and $\phi(x)$ vanishes outside this interval, then $h[\phi] = 0$. This allows us to write $h(x) = 1$, $x > 0$; $h(x) = 0$, $x < 0$; but leaves $h(0)$ undefined. This function is called the **Heaviside function.**

Next we consider the concept of derivative of a generalized function. If $f(x)$ has a continuous derivative $f'(x)$, then both f and f' define continuous linear functions which are related by integration by parts;

$$f'[\phi] = \int_{-\infty}^{\infty} f'(x)\phi(x)\,dx = -\int_{-\infty}^{\infty} f(x)\phi'(x)\,dx = -f[\phi']$$

using the fact that $\phi(x)$ vanishes outside of some finite interval. Now for any continuous linear functional $F[\phi]$ the quantity $-F[\phi']$ defines a continuous linear functional. Therefore, we define

$$F'[\phi] = -F[\phi']$$

and for higher derivatives

$$F^{(k)}[\phi] = (-1)^k F[\phi^{(k)}]$$

This allows us to define all derivatives of the delta function by the formula

$$\delta^{(k)}[\phi] = (-1)^k \phi^{(k)}(0)$$

The reader can show that $h'[\phi] = \delta[\phi]$, where $h(x)$ is the Heaviside function.

Consider a continuous function $f(x)$ with a piecewise continuous derivative. Then

$$f'[\phi] = \int_{-\infty}^{\infty} f'(x)\phi(x)\,dx = -\int_{-\infty}^{\infty} f(x)\phi'(x)\,dx = -f[\phi']$$

Hence, the ordinary derivative is the same as the generalized derivative. On the other hand, consider a function $g(x)$ which is continuous everywhere except at the origin where it has a jump discontinuity of J; that is,

$$g(0+) - g(0-) = J$$

Suppose $g(x)$ is differentiable everywhere except at $x = 0$ where the derivative is undefined. Then

$$\int_{-\infty}^{\infty} g'(x)\phi(x)\, dx = -\int_{-\infty}^{\infty} g(x)\phi'(x)\, dx - J\phi(0)$$

or $-g[\phi'] = \int_{-\infty}^{\infty} g'(x)\phi(x)\, dx + J\delta[\phi]$. In other words, in terms of generalized functions, the generalized derivative is the ordinary derivative plus J times a delta function. This is perfectly consistent with the notion that

$$g(x) = f(x) - Jh(x)$$

where $f(x)$ is continuous and has a piecewise continuous derivative and $h(x)$ is the Heaviside function. Then in terms of generalized functions

$$g'(x) = f'(x) - J\delta(x)$$

and it is clear that, except at the origin where the ordinary derivatives are undefined, $g'(x) = f'(x)$.

Since we have generalized derivatives we can also have generalized antiderivatives, that is, indefinite integrals. We shall say that the generalized function $G[\phi]$ is the indefinite integral of $F[\phi]$ if $G'[\phi] = F[\phi]$ or $F[\phi] = -G[\phi']$. We first prove that if two generalized functions are indefinite integrals of the same generalized function they differ by a constant function $C[\phi] = c \int_{-\infty}^{\infty} \phi(x)\, dx$. Let $G'[\phi] = H'[\phi] = F[\phi]$. We shall prove that $G[\phi] - H[\phi] = C[\phi]$, a constant function. We have $G'[\phi] - H'[\phi] = H[\phi'] - G[\phi'] = 0$. Let ϕ_0 be a testing function such that $\int_{-\infty}^{\infty} \phi_0(x)\, dx = 1$, and let $\phi(x)$ be any testing function. Then we can define

$$\psi(x) = \int_{-\infty}^{x} \phi(\xi)\, d\xi - \int_{-\infty}^{\infty} \phi(\xi)\, d\xi \int_{-\infty}^{x} \phi_0(\xi)\, d\xi$$

It is easy to show that $\psi(x)$ is a testing function and that

$$\psi'(x) = \phi(x) - \phi_0(x)\int_{-\infty}^{\infty} \phi(\xi)\, d\xi$$

Then

$$0 = H[\psi'] - G[\psi'] = H[\phi] - G[\phi] - \left\{ H[\phi_0] - G[\phi_0] \right\}\int_{-\infty}^{\infty} \phi(x)\, dx$$

Therefore,

$$H[\phi] - G[\phi] = c \int_{-\infty}^{\infty} \phi(x)\, dx$$

where $$c = H[\phi_0] - G[\phi_0]$$

Given the generalized function $F[\phi]$, we wish to find an indefinite integral $G[\phi]$ such that $G'[\phi] = F[\phi]$. As above we pick a $\phi_0(x)$ and define a $\psi(x)$, given any $\phi(x)$. Conversely, given a testing function $\psi(x)$ there is a testing function

$$\phi(x) = \psi'(x) + A\phi_0(x)$$

where $$\int_{-\infty}^{\infty} \phi(x)\,dx = \int_{-\infty}^{\infty} \psi'(x)\,dx + A\int_{-\infty}^{\infty} \phi_0(x)\,dx = A$$

We define $G[\phi] = -F[\psi]$. Then $G[\phi_0] = -F[0] = 0$ and

$$G'[\psi] = -G[\psi'] = -G[\phi - A\phi_0] = -G[\phi] = F[\psi]$$

Therefore, $G[\phi]$ is an indefinite integral of $F[\phi]$.

Now let us return to the problem of constructing Green's functions. As an example, let us solve the equation

$$\frac{d^2G}{dy^2} = -\delta(y - x)$$

subject to $G(x;0) = G(x;a) = 0$. Integrating once we have

$$\frac{dG}{dy} = -h(y - x) + f(x)$$

where $f(x)$ is an arbitrary function of x, playing the role of a constant of integration. Integrating again we have

$$G(x;y) = -(y - x)h(y - x) + yf(x) + g(x)$$

where $g(x)$ is an arbitrary function of x. Finally,

$$G(x;0) = g(x) = 0$$
$$G(x;a) = x - a + af(x) = 0$$

must hold for all x. Therefore $g(x) \equiv 0$ and $f(x) = (a - x)/a$, which gives us the result

$$G(x;y) = \begin{cases} \dfrac{(a - x)y}{a} & 0 \leq y < x \\[2ex] \dfrac{(a - y)x}{a} & x < y \leq a \end{cases}$$

which is a Green's function we have met earlier.

As another example, let us solve

$$\frac{d}{dy}\left(p\frac{d\tilde{G}}{dy}\right) = -\delta(y - x)$$

subject to $\tilde{G}(x;0) = \tilde{G}(x;a) = 0$, where $p(y) \neq 0$ for $0 \leq y \leq a$. Integrating once, we have

$$p \frac{d\tilde{G}}{dy} = -h(y - x) + f(x)$$

where $f(x)$ is an arbitrary function of x. Notice that this equation implies immediately that

$$\tilde{G}'(x;x+) - \tilde{G}'(x;x-) = -\frac{1}{p(x)}$$

Integrating again, we have

$$\tilde{G}(x;y) = -h(y - x) \int_x^y \frac{1}{p(\eta)} d\eta + f(x) \int_0^y \frac{1}{p(\eta)} d\eta + g(x)$$

where $g(x)$ is an arbitrary function of x. Now

$$\tilde{G}(x;0) = g(x) \equiv 0$$

$$\tilde{G}(x;a) = \int_a^x \frac{1}{p(\eta)} d\eta + f(x) \int_0^a \frac{1}{p(\eta)} d\eta \equiv 0$$

Therefore,

$$f(x) = A \int_x^a \frac{1}{p(\eta)} d\eta$$

where $A = 1 \Big/ \int_0^a \frac{1}{p(\eta)} d\eta$. Then

$$\tilde{G}(x;y) = \begin{cases} A \int_x^a \frac{1}{p(\eta)} d\eta \int_0^y \frac{1}{p(\eta)} d\eta & 0 \leq y < x \\ \int_y^x \frac{1}{p(\eta)} d\eta + A \int_x^a \frac{1}{p(\eta)} d\eta \int_0^y \frac{1}{p(\eta)} d\eta & x < y \leq a \end{cases}$$

Clearly $\tilde{G}(x;x+) - \tilde{G}(x;x-) = 0$ so that the Green's function is continuous at $y = x$ as expected.

In the general case, for the Sturm-Liouville problem

$$\frac{d}{dy}\left(p \frac{dG}{dy}\right) - qG = -\delta(y - x)$$

subject to appropriate boundary conditions, it is not so easy to see how to solve the differential equation. Of course one could develop the theory of ordinary differential equations for generalized functions and then solve the equation using the boundary conditions. We do not wish to do that here. On the other hand, we could start with the differential equation and from it determine the kinds of conditions enumerated in the last section, which can then be used to find the Green's function. Let us indicate how this can be done.

Let $\tilde{G}$ be the Green's function determined in the above example.   Clearly $\tilde{G}$ is continuous for $0 \leq y \leq a$, satisfies the boundary conditions, and has a jump discontinuity in its derivative at $y = x$ of $-1/p(x)$.   Let G satisfy the differential equation $(pG')' - qG = -\delta(y - x)$ and the same boundary conditions as $\tilde{G}$.   Let $G(x;y) = \tilde{G}(x;y) + H(x;y)$.   Then

$$(pH')' - qH = q\tilde{G}$$

and H satisfies the same boundary conditions as G and $\tilde{G}$.   From the theory of ordinary differential equations, this equation has a unique solution which is continuous and has a continuous derivative and satisfies the boundary conditions, provided the corresponding homogeneous equation has no nontrivial solution.   Therefore, in this case G is continuous, satisfies the boundary conditions, and has a jump discontinuity of $-1/p(x)$ at $y = x$. These are the conditions of the last section which we used to find the Green's functions $G(x;y)$.

Exercises 5.3

1. Prove that the following are continuous linear functionals on the space of testing functions:

a. $\delta[\phi] = \phi(0)$.

b. $f[\phi] = \displaystyle\int_{-\infty}^{\infty} f(x)\phi(x)\, dx, f(x)$ continuous.

c. $f[\phi] = \displaystyle\int_{-\infty}^{\infty} f(x)\phi(x)\, dx, f(x)$ Riemann integrable.

d. $h[\phi] = \displaystyle\int_{-\infty}^{\infty} \phi(x)\, dx$.

e. $\delta'[\phi] = -\phi'(0)$.

2. Prove the following formulas:

a. $\delta[\phi] = h'[\phi]$.

b. $h[\phi] = f'[\phi]$, where $f(x) = x, x \geq 0, f(x) = 0, x < 0$.

c. $(cF)' = cF'$, where c is a constant.

d. $(F_1 + F_2)' = F_1' + F_2'$.

3. Let $f(x)$ have all derivatives.   Show that if $\phi(x)$ is a testing function, then $f(x)\phi(x)$ is a testing function.   Let $F[\phi]$ be an arbitrary continuous linear functional.   Show that $F[f\phi]$ is a continuous linear functional.   Define $f[\phi]F[\phi] = F[f\phi]$ and prove the formula $(fF)' = fF' + f'F$.

4. In ordinary analysis the question of when you can interchange the order of differentiating and taking a limit can be a difficult one.   Prove that in the theory of generalized functions $\lim\limits_{n \to \infty} T'(\phi_n) = T'\left(\lim\limits_{n \to \infty} \phi_n\right)$.

5. Find the Green's function satisfying $G'' = -\delta(y - x)$, $0 \leq y \leq a$, $G(x;0) = G'(x;a) = 0$.

6. Find the Green's function satisfying $(e^{-y}G')' = -\delta(y - x)$, $0 \leq y \leq a$, $G(x;0) = G'(x;a) = 0$.

5.4 Green's Functions in Higher Dimensions

The use of Green's functions to solve boundary-value problems is not restricted to one-dimensional problems. Consider, for example, the two-dimensional boundary-value problem

$$\nabla^2\psi + \lambda\psi = -f(x,y) \qquad \text{in } R$$

subject to $\psi = 0$,

or $\dfrac{d\psi}{dn} = 0$

or $\dfrac{d\psi}{dn} + \alpha\psi = 0 \qquad \text{on } C$

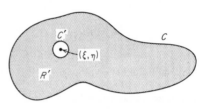

FIGURE 9

the boundary of R. We start by considering the following integral:

$$\iint\limits_{R} f(x,y)G(\xi,\eta;x,y)\, dx\, dy = -\iint\limits_{R} (\nabla^2\psi + \lambda\psi)G\, dx\, dy$$

when G is the Green's function whose properties we wish to determine. We suspect, by analogy with the one-dimensional case, that the Green's function is not well-behaved at $x = \xi$, $y = \eta$. Therefore, we exclude this point from the region by deleting the interior of a small circle $(x - \xi)^2 + (y - \eta)^2 = \rho^2$, with radius ρ and center (ξ,η). We then treat the integral as an improper integral; that is,

$$\iint\limits_{R} (\nabla^2\psi + \lambda\psi)G\, dx\, dy = \lim_{\rho\to 0} \iint\limits_{R'} (\nabla^2\psi + \lambda\psi)G\, dx\, dy$$

where R' is the region between C and C'.

Using Green's theorem, we have

$$\iint\limits_{R'} (\nabla^2\psi + \lambda\psi)G\, dx\, dy = \iint\limits_{R'} (\nabla^2 G + \lambda G)\psi\, dx\, dy$$

$$+ \int_{C}\left(G\frac{d\psi}{dn} - \psi\frac{dG}{dn}\right)ds - \int_{C'}\left(G\frac{\partial\psi}{\partial r} - \psi\frac{\partial G}{\partial r}\right)ds$$

where $r = \sqrt{(x - \xi)^2 + (y - \eta)^2}$. On C', $ds = \rho\, d\theta$, where

$$\theta = \tan^{-1}[(y - \eta)/(x - \xi)]$$

and θ varies from 0 to 2π. Let us now assume that $\nabla^2 G + \lambda G = 0$ in R, except at $x = \xi$, $y = \eta$.   Then $\iint\limits_{R'} (\nabla^2 G + \lambda G)\psi \, dx \, dy = 0$.   If we also

assume that G satisfies the same boundary condition on C as does ψ, then

$\int_C \left(G \dfrac{d\psi}{dn} - \psi \dfrac{dG}{dn} \right) ds = 0$.   Therefore,

$$\iint\limits_{R} f(x,y)G(\xi,\eta;x,y) = \lim_{\rho \to 0} \rho \int_0^{2\pi} \left(G \frac{\partial \psi}{\partial r} - \psi \frac{\partial G}{\partial r} \right)_{r=\rho} d\theta$$

This will yield the value $\psi(\xi,\eta)$ if G behaves in such a way that $\lim\limits_{\rho \to 0} \rho \left(\dfrac{\partial G}{\partial r} \right)_{r=\rho}$
is not zero and yet is finite, while $\lim\limits_{\rho \to 0} \rho G = 0$.   This suggests that G should
behave as $k \log r$ as $r \to 0$.   Therefore, since $Y_0(\sqrt{\lambda} r)$ satisfies the Helmholtz
equation and behaves as $(2/\pi)\log r$ as $r \to 0$, we take

$$G(\xi,\eta;x,y) = -\tfrac{1}{4}Y_0\sqrt{\lambda} r + H(\xi,\eta;x,y)$$

where H and its first and second partial derivatives with respect to x and y are
continuous in R. H will have to be determined so that $\nabla^2 G + \lambda G = 0$ except
at (ξ,η) and G satisfies the same boundary condition on C as ψ.

The same problem can be treated in three dimensions, where this time we
exclude the point (ξ,η,ζ) by deleting the interior of a small sphere with radius
ρ and center (ξ,η,ζ).   Then

$$\iiint\limits_{V} f(x,y,z)G(\xi,\eta,\zeta;x,y,z) \, dx \, dy \, dz = \psi(\xi,\eta,\zeta)$$

$$= -\iiint\limits_{V} (\nabla^2\psi + \lambda\psi)G \, dx \, dy \, dz$$

$$= -\lim_{\rho \to 0} \iiint\limits_{V'} (\nabla^2\psi + \lambda\psi)G \, dx \, dy \, dz$$

$$= -\lim_{\rho \to 0} \iiint\limits_{V'} (\nabla^2 G + \lambda G)\psi \, dx \, dy \, dz$$

$$+ \iint\limits_{S} \left(\psi \frac{dG}{dn} - G \frac{d\psi}{dn} \right) dS$$

$$+ \lim_{\rho \to 0} \rho^2 \int_0^{\pi} \int_0^{2\pi} \left(G \frac{\partial \psi}{\partial r} - \psi \frac{\partial G}{\partial r} \right)_{r=\rho} \sin \theta \, d\theta \, d\phi$$

We wish $\lim\limits_{\rho \to 0} \rho^2 \left(\dfrac{\partial G}{\partial r} \right)_{r=\rho}$ to be nonzero and finite and $\lim \rho^2 G = 0$.   This suggests a behavior like kr^{-1} as $r \to 0$. Therefore, since $r^{-1} \cos \sqrt{\lambda} r$ satisfies the Helmholtz equation and behaves as r^{-1} as $r \to 0$ we take

$$G(\xi,\eta,\zeta;x,y,z) = \frac{\cos \sqrt{\lambda} r}{4\pi r} + H(\xi,\eta,\zeta;x,y,z)$$

where H and its first and second partial derivatives are continuous in V. H is determined so that $\nabla^2 G + \lambda G = 0$ in V except at (ξ,η,ζ) and G satisfies the same boundary conditions as ψ on S.

In each of the above cases, the function H is unique, provided λ is not an eigenvalue of the homogeneous problem. Assume that there are two Green's functions in the three-dimensional case.[1]

$$G_1 = \frac{\cos \sqrt{\lambda} r}{4\pi r} + H_1$$

$$G_2 = \frac{\cos \sqrt{\lambda} r}{4\pi r} + H_2$$

Let $w = G_1 - G_2 = H_1 - H_2$. Then

$$\nabla^2 w + \lambda w = 0$$

except possibly at (ξ,η,ζ). Actually, this equation is satisfied also at (ξ,η,ζ), since w is continuous and has continuous second partial derivatives in V. Furthermore, w satisfies the boundary conditions of the homogeneous boundary-value problem. Therefore, w is a continuous function with continuous first and second partial derivatives, which is a solution of the homogeneous boundary-value problem. But the homogeneous boundary-value problem has no nontrivial solutions except for a certain set of eigenvalues of which λ is not a member. Therefore, $w \equiv 0$, which implies that $H_1 \equiv H_2$.

Now let us see how we would use Green's functions to solve other types of boundary-value problems. Consider the **Dirichlet problem,** that is, to find a function ψ with continuous second partial derivatives in a two-dimensional region R of the xy plane which satisfies Laplace's equation in R and takes on prescribed boundary values f on C, the boundary of R. Let $F(x,y)$ be a function with continuous second partial derivatives in R which takes on the values f on C; then

$$g = \psi - F = 0$$

on C. In R, $$\nabla^2 g = \nabla^2 \psi - \nabla^2 F = -\nabla^2 F$$

Now if we let $G(\xi,\eta;x,y) = -(1/2\pi) \log \sqrt{(x - \xi)^2 + (y - \eta)^2} + H(\xi,\eta;x,y)$

---

[1] The other case is quite similar.

with $\nabla^2 H = 0$ in R and $G = 0$ on C, then[2]

$$g(\xi,\eta) = \iint_R G \nabla^2 F \, dx \, dy$$

$$\psi(\xi,\eta) - F(\xi,\eta) = \iint_R F \nabla^2 G \, dx \, dy + \int_C \left(G \frac{dF}{dn} - F \frac{dG}{dn} \right) ds$$

$$= -\iint_R F \, \delta(x - \xi) \, \delta(y - \eta) \, dx \, dy - \int_C f \frac{dG}{dn} \, ds$$

$$= -F(\xi,\eta) - \int_C f \frac{dG}{dn} \, ds$$

Therefore,
$$\psi(\xi,\eta) = -\int_C f \frac{dG}{dn} \, ds$$

Suppose that R is the interior of a unit circle with center at the origin.   Let the point (ξ,η) be an interior point.   Then the point $[\xi/(\xi^2 + \eta^2), \eta/(\xi^2 + \eta^2)]$ is an exterior point.   The locus of points the ratio of whose distances from these two points is $\sqrt{\xi^2 + \eta^2}$ is the unit circle.   This is easily verified. Let $\rho = \sqrt{\xi^2 + \eta^2}$; then if (x,y) is on the locus,

$$(x - \xi)^2 + (y - \eta)^2 = \rho^2 \left[\left(x - \frac{\xi}{\rho^2} \right)^2 + \left(y - \frac{\eta}{\rho^2} \right)^2 \right]$$

$$\rho^2(x^2 + y^2 - 2x\xi - 2y\eta + \rho^2) = \rho^4(x^2 + y^2) - 2\rho^2 x\xi - 2\rho^2 y\eta + \rho^2$$

$$(x^2 + y^2)(1 - \rho^2) = 1 - \rho^2$$

$$x^2 + y^2 = 1$$

We take

$$G(\xi,\eta;x,y) = -\frac{1}{2\pi} \log \sqrt{(x - \xi)^2 + (y - \eta)^2}$$

$$+ \frac{1}{2\pi} \log \sqrt{\left(x - \frac{\xi}{\rho^2} \right)^2 + \left(y - \frac{\eta}{\rho^2} \right)^2} + \frac{1}{2\pi} \log \sqrt{\xi^2 + \eta^2}$$

This has the desired properties, for

$$H = \frac{1}{2\pi} \log \sqrt{\left(x - \frac{\xi}{\rho^2} \right)^2 + \left(y - \frac{\eta}{\rho^2} \right)^2} + \frac{1}{2\pi} \log \sqrt{\xi^2 + \eta^2}$$

is harmonic inside the unit circle, and when $x^2 + y^2 = 1$,

$$\log \sqrt{(x - \xi)^2 + (y - \eta)^2} - \log \sqrt{\left(x - \frac{\xi}{\rho^2} \right)^2 + \left(y - \frac{\eta}{\rho^2} \right)^2} = \log \sqrt{\xi^2 + \eta^2}$$

---

[2] The delta function is used here to indicate the argument only.   It can be made rigorous by the use of the theory of generalized functions in two dimensions.

On the unit circle $ds = d\theta$ and $dG/dn = \partial G/\partial r$, where $r = \sqrt{x^2 + y^2}$. Let $x = r \cos \theta$, $y = r \sin \theta$, $\xi = \rho \cos \phi$, $\eta = \rho \sin \phi$; then

$$\frac{\partial G}{\partial r} = -\frac{1}{2\pi} \frac{(r \cos \theta - \rho \cos \phi) \cos \theta + (r \sin \theta - \rho \sin \phi) \sin \theta}{(r \cos \theta - \rho \cos \phi)^2 + (r \sin \theta - \rho \sin \phi)^2}$$

$$+ \frac{1}{2\pi} \frac{(r \cos \theta - \cos \phi/\rho) \cos \theta + (r \sin \theta - \sin \phi/\rho) \sin \theta}{(r \cos \theta - \cos \phi/\rho)^2 + (r \sin \theta - \sin \phi/\rho)^2}$$

$$\left(\frac{\partial G}{\partial r}\right)_{r=1} = -\frac{1}{2\pi} \frac{1 - \rho \cos (\theta - \phi)}{1 + \rho^2 - 2\rho \cos (\theta - \phi)} + \frac{1}{2\pi} \frac{\rho^2 - \rho \cos (\theta - \phi)}{1 + \rho^2 - 2\rho \cos (\theta - \phi)}$$

$$= -\frac{1}{2\pi} \frac{1 - \rho^2}{1 + \rho^2 - 2\rho \cos (\theta - \phi)}$$

Therefore,
$$\psi(\xi,\eta) = \frac{1}{2\pi} \int_0^{2\pi} \frac{(1 - \rho^2) f(\theta) \, d\theta}{1 + \rho^2 - 2\rho \cos (\theta - \phi)}$$

This is known as **Poisson's integral formula.**[1] By means of it we can solve the Dirichlet problem for any region which can be mapped conformally onto the interior of the unit circle.[2]

There is actually a closer connection between conformal mapping and Green's functions. Suppose $w = f(z)$ is a conformal mapping which maps the simply connected region R bounded by a simple closed curve C onto the interior of the unit circle so that C maps onto the unit circle and the point $\zeta = \xi + i\eta$ maps into the origin. Then

$$G(\xi,\eta;x,y) = -\frac{1}{2\pi} \log |f(z)|$$

is the Green's function appropriate for the solution of the Dirichlet problem in R at the point (ξ,η). This shall be established in Sec. 7.10.

The **Riemann mapping theorem**[3] implies that any simply connected region bounded by a simple closed curve can be mapped conformally onto the interior of the unit circle so that a given point ζ in R maps into the origin. This ensures the existence of the mapping, which in turn implies the existence of the Green's function, from which one can deduce the existence of the solution of the Dirichlet problem for reasonably well-behaved boundary conditions.

Exercises 5.4

1. Derive the Poisson integral formula for the solution of the Dirichlet problem in the unit sphere

$$\psi(\xi,\eta,\zeta) = \frac{1}{4\pi} \int_0^\pi \int_0^{2\pi} \frac{(\rho^2 - 1) f(\theta,\phi) \sin \theta \, d\theta \, d\phi}{(1 + \rho^2 - 2\rho Q)^{\frac{3}{2}}}$$

[1] Compare with exercise 3, Sec. 4.4, and Sec. 7.10.
[2] See Sec. 7.9.
[3] See Sec. 7.10.

where $Q = \sin \theta \sin \alpha \cos (\phi - \beta) + \cos \theta \cos \alpha$ with

$$x = r \sin \theta \cos \phi$$
$$y = r \sin \theta \sin \phi$$
$$z = r \cos \theta$$
$$\xi = \rho \sin \alpha \cos \beta$$
$$\eta = \rho \sin \alpha \sin \beta$$
$$\zeta = \rho \cos \alpha$$

2. What properties are required for the Green's function associated with the problem

$$\nabla^2 \psi = -f(x,y,z) \qquad \text{in } V$$

$$\frac{d\psi}{dn} = g(x,y,z) \qquad \text{on } S$$

Give a formal expression for the solution in terms of this Green's function.

3. Consider the Dirichlet problem for the upper half plane, $y > 0$. Find a Green's function for the problem. What condition is required on the solution at infinity in order to make it possible to express the solution in terms of the Green's function and the given data on the x axis?

4. Consider the Neumann problem for the region V inside the closed smooth surface S: $\nabla^2 \psi = 0$ in V, $d\psi/dn = g$ on S, where $g(x,y,z)$ is given on S. Does the problem have a unique solution?   Show that a necessary condition for a solution to exist is that the integral of $g(x,y,z)$ over the surface S be zero. State the conditions for a Green's function which will allow you to express a solution in terms of the Green's function and the given data.

5.5   Problems in Unbounded Regions

Up to this point we have considered problems in bounded regions only.   In the homogeneous problem the separation-of-variables technique led to a boundary-value problem with a discrete spectrum of eigenvalues which could be found by finding solutions of a certain homogeneous partial differential equation subject to a homogeneous boundary condition on the surface of a bounded region V.   The nonhomogeneous problem could then be formulated in terms of a series in the solutions of the homogeneous problem or in terms of Green's functions.   When the problem involves an unbounded region, not only do we have to specify conditions on boundaries in the finite region but we must also specify the behavior of the solution at very large distances from the origin in order to ensure a unique solution of the problem.   Part of our investigation will be to determine what these conditions should be.   We shall again be able to formulate the solution of nonhomogeneous problems in terms of Green's functions, provided we impose proper conditions on the Green's function at large distances from the origin.   The homogeneous problem, however, now has only the trivial solution and this guarantees the uniqueness of the solution of the nonhomogeneous problem.

Consider the solution of the following partial differential equation in the unbounded three-dimensional region outside of the bounded volume V

bounded by the surface S:

$$\nabla^2\phi = a\phi + b\frac{\partial\phi}{\partial t} + c\frac{\partial^2\phi}{\partial t^2} - f(x,y,z,t)$$

where $f \equiv 0$ outside of a bounded region and a, b, and c are real constants. We shall assume a boundary condition of the form

$$\frac{d\phi}{dn} + \alpha\phi = g(x,y,z,t)$$

on S, and leave open for the moment the condition to be satisfied at large distances from V. The problem is linear, so that we may write the solution as $\phi = \phi_0 + \phi_1 + \phi_2$, where ϕ_0 is a solution of the homogeneous problem $\nabla^2\phi_0 = a\phi_0 + b(\partial\phi_0/\partial t) + c(\partial^2\phi_0/\partial t^2)$ outside V, subject to $d\phi_0/dn + \alpha\phi_0 = 0$ on S; ϕ_1 in a solution of the nonhomogeneous problem $\nabla^2\phi_1 = a\phi_1 + b(\partial\phi_1/\partial t) + c(\partial^2\phi_1/\partial t^2) - f$, subject to $d\phi_1/dn + \alpha\phi_1 = 0$ on S; and ϕ_2 is a solution of $\nabla^2\phi_2 = a\phi_2 + b(\partial\phi_2/\partial t) + c(\partial^2\phi_2/\partial t^2)$ subject to $d\phi_2/dn + \alpha\phi_2 = g$ on S.

Let us first consider the homogeneous problem. As in the problem for the bounded region, we attempt a separation of variables:

$$\phi_0 = \psi_0(x,y,z)F(t)$$

Then
$$\frac{\nabla^2\psi_0}{\psi_0} = \frac{aF + b\dot{F} + c\ddot{F}}{F} = -\lambda$$

and therefore the equation separates into

$$c\ddot{F} + b\dot{F} + (a + \lambda)F = 0$$
$$\nabla^2\psi_0 + \lambda\psi_0 = 0$$

In the bounded problem we next showed that $\lambda \geq 0$ using Green's identity.

$$\lambda\iiint\limits_{V'}\psi_0^2\,dV = -\iiint\limits_{V'}\psi_0\nabla^2\psi_0\,dV$$

$$= \iiint\limits_{V'}\nabla\psi_0\cdot\nabla\psi_0\,dV + \iint\limits_{S}\alpha\psi_0^2\,dS$$

$$- \lim_{R\to\infty}\iint\limits_{S'}\psi_0\frac{d\psi_0}{dn}\,dS$$

where V' is the region outside V and S' is a large sphere of radius R. The proof fails this time because we have not specified the behavior of ψ_0 at large distances from V.

Before proceeding further, let us consider some elementary solutions of

$$\nabla^2\psi + k^2\psi = 0$$

and by so doing we may gain some insight into the nature of the conditions we must specify on our solution at large distances from the origin.   Let us first look for solutions with spherical symmetry, that is, solutions which depend on r only in the usual spherical coordinate system.   In spherical coordinates the equation becomes

$$\frac{1}{r^2}\frac{d}{dr}\left(r^2\frac{d\psi}{dr}\right) + k^2\psi = 0$$

Let $\psi(r) = h(r)/r$, then $\psi' = h'/r - h/r^2$, $r^2\psi' = rh' - h$, and the equation becomes

$$h'' + k^2h = 0$$

This equation has the following solutions: $\sin kr$, $\cos kr$, e^{ikr}, e^{-ikr}.   If in the original problem $a = b = 0$, then we can write $F = e^{-i\omega t}$, where $\omega = k/\sqrt{c}$. We would then have the wave equation with a periodic time dependence. This is the case which will be of primary interest to us.   It will be convenient to write the time dependence as $e^{-i\omega t}$ with the agreement that the final solution will be the real or imaginary part of the complex solution.   Combining this time dependence with the above spherically symmetric solution, we have solutions of the form $(\sin kr/r)\cos \omega t$, $(\sin kr/r)\sin \omega t$, $(\cos kr/r)\cos \omega t$, $(\cos kr/r)\sin \omega t$, $[\cos (kr + \omega t)]/r$, $[\sin (kr + \omega t)]/r$, $[\cos (kr - \omega t)]/r$, $[\sin (kr - \omega t)]/r$.   The first four of these represent standing waves, the next two inward-traveling waves, and the last two outward-traveling waves.[1]   We shall take it as a basic assumption that a disturbance in the finite region should produce only outward-traveling waves.   The choice of a spherically symmetric solution satisfying this requirement would have to be $\psi = e^{ikr}/r$. It is easily shown that for this solution

$$\lim_{r \to \infty} r\left(\frac{\partial\psi}{\partial r} - ik\psi\right) = 0$$

This is called the **radiation condition.**   We shall take this as the required behavior of the solution of the boundary-value problem for large values of r. We shall see that this implies that the solution of the homogeneous boundary-value problem is identically zero, which in turn implies that the solution of the nonhomogeneous problem is unique.

The basic theorem of this section is the following: Let ψ be a complex-valued function with continuous second partial derivatives outside of a

---

[1] This can be seen by considering the location of the nodes as determined by the equation $\psi(r)e^{-i\omega t} = 0$.

bounded volume V bounded by S, which satisfies $\nabla^2\psi + k^2\psi = 0$ outside V, $d\psi/dn + \alpha\psi = 0$ on S, with k and α both real, and let $\lim\limits_{r\to\infty} r\left(\dfrac{\partial\psi}{\partial r} - ik\psi\right) = 0$; then $\psi \equiv 0$ outside V. We begin the proof by considering the solution expanded in a series of spherical harmonics; that is,

$$\psi(r,\theta,\phi) = \sum_{n=0}^{\infty} v_n(r) S_n(\theta,\phi)$$

where
$$S_n(\theta,\phi) = \sum_{m=0}^{n} A_m P_n^m(\cos\theta) \cos m\phi + B_m P_n^m(\cos\theta) \sin m\phi$$

and
$$\int_0^\pi \int_0^{2\pi} S_n S_p \sin\theta \, d\theta \, d\phi = 0$$

for $n \neq p$. We shall assume that these functions have been normalized; that is,

$$\int_0^\pi \int_0^{2\pi} S_n^2 \sin\theta \, d\theta \, d\phi = 1$$

By Bessel's inequality we have

$$\sum_{n=0}^{\infty} |v_n|^2 \leq \int_0^\pi \int_0^{2\pi} |\psi|^2 \sin\theta \, d\theta \, d\phi$$

Either $v_n \equiv 0$ for all n, in which case $\psi \equiv 0$, or $v_N \not\equiv 0$ for some N. In the latter case,

$$|v_N|^2 \leq \sum_{n=0}^{\infty} |v_n|^2 \leq \int_0^\pi \int_0^{2\pi} |\psi|^2 \sin\theta \, d\theta \, d\phi$$

Now we know that $\psi_N = v_N(r) S_N(\theta,\phi)$ is a solution of the differential equation. Hence,

$$\nabla^2\psi_N + k^2\psi_N = 0$$

or
$$\frac{r^2 v_N'' + 2r v_N'}{v_N} + k^2 r^2 = -\frac{1}{S_N}\left[\frac{1}{\sin\theta}\frac{\partial}{\partial\theta}\left(\sin\theta\frac{\partial S_N}{\partial\theta}\right) + \frac{1}{\sin^2\theta}\frac{\partial^2 S_N}{\partial\phi^2}\right]$$

$$= N(N+1)$$

Therefore, v_N satisfies the equation

$$r^2 v_N'' + 2r v_N' + [k^2 r^2 - N(N+1)]v_N = 0$$

To solve this we first make the substitution $v_N = e^{ikr}h(r)$. Then

$$v_N' = e^{ikr}h' + ike^{ikr}h$$
$$v_N'' = e^{ikr}h'' + 2ike^{ikr}h' - k^2 e^{ikr}h$$

Substituting in the equation for v_N, we have

$$r^2 h'' + (2ikr + 2)rh' + [2ikr - N(N+1)]h = 0$$

When we assume a solution of the form

$$h = r^\beta \sum_{j=0}^\infty c_j r^j$$

the indicial equation is $(\beta - N)(\beta + N + 1) = 0$. If we take the root $\beta_2 = N$, the solution will contain positive powers of r which are not consistent with the radiation condition. By use of $\beta_1 = -N - 1$ the equation becomes

$$\sum_{j=0}^\infty j(j - 2N - 1)c_j r^{j-N-1} + 2ik \sum_{j=0}^\infty (j - N)c_j r^{j-N} = 0$$

from which we get the recurrence relation

$$c_j = \frac{2ik(N + 1 - j)}{j(j - 2N - 1)} c_{j-1}$$

Thus we see that $c_{N+1} = c_{N+2} = \cdots = 0$. Therefore,

$$v_N = \frac{e^{ikr}}{r} \sum_{j=0}^N c_{N-j} r^{-j}$$

Now

$$\int_{R_0}^R |v_N|^2 r^2 \, dr \geq KR$$

where K is a constant, and

$$\int_{R_0}^R \int_0^\pi \int_0^{2\pi} |\psi|^2 r^2 \sin\theta \, dr \, d\theta \, d\phi \geq \int_{R_0}^R |v_N|^2 r^2 \, dr \geq KR$$

Therefore, either $\psi \equiv 0$ or $\int_{R_0}^R \int_0^\pi \int_0^{2\pi} |\psi|^2 r^2 \sin\theta \, dr \, d\theta \, d\phi$ becomes large at least as fast as KR as $R \to \infty$.

Applying Green's identity, we have

$$0 = \iiint_{V'} (\psi \nabla^2 \psi^* - \psi^* \nabla^2 \psi) \, dV = \iint_S \left(\psi \frac{d\psi^*}{dn} - \psi^* \frac{d\psi}{dn} \right) dS$$

$$+ \iint_{S'} \left(\psi \frac{d\psi^*}{dn} - \psi^* \frac{d\psi}{dn} \right) dS$$

where S' is an arbitrary surface completely surrounding S. But the surface integral over S vanishes by the boundary condition, and therefore the integral over S' must also vanish.

Using the radiation condition, we have

$$0 = \lim_{r \to \infty} \int_0^\pi \int_0^{2\pi} r^2 \left(\frac{\partial \psi}{\partial r} - ik\psi \right) \left(\frac{\partial \psi^*}{\partial r} + ik\psi^* \right) \sin \theta \, d\theta \, d\phi$$

$$= \lim_{r \to \infty} \int_0^\pi \int_0^{2\pi} r^2 \left\{ \left| \frac{\partial \psi}{\partial r} \right|^2 + k^2 |\psi|^2 + ik \left(\psi^* \frac{\partial \psi}{\partial r} - \psi \frac{\partial \psi^*}{\partial r} \right) \right\} \sin \theta \, d\theta \, d\phi$$

$$= \lim_{r \to \infty} \int_0^\pi \int_0^{2\pi} r^2 \left\{ \left| \frac{\partial \psi}{\partial r} \right|^2 + k^2 |\psi|^2 \right\} \sin \theta \, d\theta \, d\phi$$

But this implies that

$$\lim_{r \to \infty} \int_0^\pi \int_0^{2\pi} r^2 |\psi|^2 \sin \theta \, d\theta \, d\phi = 0$$

However, this is impossible unless $\psi \equiv 0$; for if it is not, then

$$\int_{R_0}^R \int_0^\pi \int_0^{2\pi} r^2 |\psi|^2 \sin \theta \, d\theta \, d\phi \geq KR$$

But $\lim_{r \to \infty} F(r) = 0$ implies that $\lim_{R \to \infty} \frac{1}{R} \int_{R_0}^R F(r) \, dr = 0$. To show this, let ρ be sufficiently large that $|F(r)| < \epsilon$ for all r such that $R_0 < \rho < r < R$. Then

$$\left| \frac{1}{R} \int_{R_0}^R F(r) \, dr \right| \leq \frac{1}{R} \int_{R_0}^\rho |F(r)| \, dr + \frac{1}{R} \int_\rho^R \epsilon \, dr$$

$$\leq \frac{M}{R} (\rho - R_0) + \frac{\epsilon}{R} (R - \rho)$$

$$\leq 2\epsilon$$

by taking R sufficiently large that $M(\rho - R_0)/R < \epsilon$. Letting

$$F(r) = \int_0^\pi \int_0^{2\pi} r^2 |\psi|^2 \sin \theta \, d\theta \, d\phi$$

we see clearly that we cannot have at the same time

$$\frac{1}{R} \int_{R_0}^R \int_0^\pi \int_0^{2\pi} r^2 |\psi|^2 \sin \theta \, d\theta \, d\phi \geq K$$

and

$$\lim_{r \to \infty} \int_0^\pi \int_0^{2\pi} r^2 |\psi|^2 \sin \theta \, d\theta \, d\phi = 0$$

This completes the proof.

Now let us consider one of the nonhomogeneous problems. Assuming that $a = b = 0$ and that $f(x,y,z,t) = \text{Re} \{ F(x,y,z)e^{-i\omega t} \}$, we take

$$\phi_1 = \psi_1 e^{-i\omega t}$$

Then we write $\nabla^2 \psi_1 e^{-i\omega t} = -\omega^2 c \psi_1 e^{-i\omega t} - F(x,y,z) e^{-i\omega t}$

and we have the separated equation

$$\nabla^2 \psi_1 + k^2 \psi_1 = -F(x,y,z)$$

with $k^2 = \omega^2 c$.   The boundary condition for ψ_1 is

$$\frac{d\psi_1}{dn} + \alpha \psi_1 = 0$$

on S.   This problem has a unique solution, for if

$$w = \psi_1 - \bar{\psi}_1$$

the difference of two possibly different solutions, then

$$\nabla^2 w + k^2 w = 0 \qquad \text{outside } V$$

$$\frac{dw}{dn} + \alpha w = 0 \qquad \text{on } S$$

and w satisfies the radiation condition

$$\lim_{r \to \infty} r \left(\frac{\partial w}{\partial r} - ikw \right) = 0$$

But by the theorem just proved $w \equiv 0$, and therefore $\psi_1 \equiv \bar{\psi}_1$.

The other nonhomogeneous problem is

$$\nabla^2 \psi_2 + k^2 \psi_2 = 0 \qquad \qquad \text{outside } V$$

$$\frac{d\psi_2}{dn} + \alpha \psi_2 = g(x,y,z) \qquad \text{on } S$$

This also has a unique solution by the theorem proved above.

Now let us turn to the representation of the solution of these nonhomogeneous problems in terms of Green's functions.   For the first problem we would want to construct a Green's function with the following properties[1]:

1. $\nabla^2 G + k^2 G = 0$ outside V except at (ξ,η,ζ).
2. $G = (\cos kR)/4\pi R + H(\xi,\eta,\zeta;x,y,z)$, where H has continuous second partial derivatives at (ξ,η,ζ).
3. $dG/dn + \alpha G = 0$ on S.
4. $\lim_{r \to \infty} r \left(\frac{\partial G}{\partial r} - ikG \right) = 0$ uniformly in θ and ϕ.

We see how properties 1, 2, and 3 carry over from the corresponding problem for bounded regions.   Property 4, the radiation condition, is appended on the grounds that the Green's function should physically have the interpretation

---

[1] $R = \sqrt{(x - \xi)^2 + (y - \eta)^2 + (z - \zeta)^2}$

of outward-traveling spherical waves originating from the point (ξ,η,ζ). Also we find that this condition is needed to assure uniqueness of the Green's function.

In this case it is somewhat more convenient to let

$$G(\xi,\eta,\zeta;x,y,z) = \frac{e^{ikR}}{4\pi R} + H(\xi,\eta,\zeta;x,y,z)$$

Then since $e^{ikR}/4\pi R$ satisfies the differential equation except at (ξ,η,ζ) and behaves properly as $R \to 0$, we can take H as a solution of the differential equation regular everywhere outside V. H satisfies the nonhomogeneous boundary condition

$$\frac{dH}{dn} + \alpha H = -\left(\frac{d}{dn} + \alpha\right)\frac{1}{4\pi}\frac{e^{ikR}}{R}$$

on S. $e^{ikR}/4\pi R$ satisfies the radiation condition, since

$$R = \sqrt{(x - \xi)^2 + (y - \eta)^2 + (z - \zeta)^2}$$
$$= r\sqrt{1 + \frac{\rho^2}{r^2} - \frac{2\rho}{r}Q}$$

where $Q = \sin\theta \sin\gamma \cos(\phi - \beta) + \cos\theta \cos\gamma$, with

$$x = r\sin\theta\cos\phi$$
$$y = r\sin\theta\cos\phi$$
$$z = r\cos\theta$$
$$\xi = \rho\sin\gamma\cos\beta$$
$$\eta = \rho\sin\gamma\sin\beta$$
$$\zeta = \rho\cos\gamma$$

Then
$$\frac{e^{ikR}}{4\pi R} = \frac{\exp\{ikr[1 - (2\rho/r)Q + (\rho^2/r^2)]^{\frac{1}{2}}\}}{4\pi r}\left(1 - \frac{2\rho Q}{r} + \frac{\rho^2}{r^2}\right)^{-\frac{1}{2}}$$
$$= \frac{e^{ikr}}{4\pi r}h(\theta,\phi) + O\left(\frac{1}{r^2}\right)^{\dagger}$$

Since the Green's function satisfies the radiation condition and $e^{ikR}/4\pi R$ also does, then H must as well. Therefore, H is a solution of a nonhomogeneous boundary-value problem and satisfies the radiation condition and hence is unique.

---

† $f(r) = O(1/r^2)$ as $r \to \infty$ implies that $|r^2 f(r)|$ is bounded as $r \to \infty$.

As for the solution of the original nonhomogeneous boundary-value problem, this can be formulated as follows:

$$-\iiint\limits_{V} FG \, dV = \iiint\limits_{V'} [G(\nabla^2\psi_1 + k^2\psi_1) - \psi_1(\nabla^2 G + k^2 G)] \, dV$$

$$= \iint\limits_{S} \left[G\left(\frac{d\psi_1}{dn} + \alpha\psi_1\right) - \psi_1\left(\frac{dG}{dn} + \alpha G\right) \right] \, dS$$

$$-\iint\limits_{\Sigma} \left(G \frac{\partial\psi_1}{\partial R} - \psi_1 \frac{\partial G}{\partial R}\right) \, dS + \iint\limits_{\Sigma'} \left(G \frac{\partial\psi_1}{\partial r} - \psi_1 \frac{\partial G}{\partial r}\right) \, dS$$

where V' is a volume bounded by S; Σ, a small sphere of radius δ about (ξ,η,ζ); and Σ', a large sphere of radius D with center at the origin.   The nature of the singularity of G at (ξ,η,ζ) ensures that the surface integral on Σ will approach $-\psi(\xi,\eta,\zeta)$ as $\delta \to 0$.   The surface integral on S will vanish by the boundary conditions on ψ_1 and G, while the surface integral on Σ' approaches zero as $D \to \infty$ by the radiation conditions on G and ψ_1.   Indeed

$$\psi_1 = \frac{e^{ikr}}{r} f(\theta,\phi) + O\left(\frac{1}{r^2}\right)$$

$$G = \frac{1}{4\pi} \frac{e^{ikr}}{r} h(\theta,\phi) + O\left(\frac{1}{r^2}\right)$$

as $r \to \infty$.   Therefore,

$$\lim_{D\to\infty} \iint\limits_{\Sigma'} \left(G \frac{\partial\psi_1}{\partial r} - \psi_1 \frac{\partial G}{\partial r}\right)_{r=D} D^2 \sin\theta \, d\theta \, d\phi$$

$$= \lim_{D\to\infty} \iint\limits_{\Sigma'} \frac{1}{4\pi} \left\{ \frac{e^{ikr}}{r} h\left[\frac{ike^{ikr}}{r} f + O\left(\frac{1}{r^2}\right)\right] \right.$$

$$\left. - \frac{e^{ikr}}{r} f\left[\frac{ik}{4\pi} \frac{e^{ikr}}{r} h + O\left(\frac{1}{r^2}\right)\right] \right\}_{r=D} D^2 \sin\theta \, d\theta \, d\phi$$

$$= 0$$

Therefore, we have an integral representation for the solution

$$\psi_1(\xi,\eta,\zeta) = \iiint\limits_{\bar{V}} F(x,y,z) G(\xi,\eta,\zeta;x,y,z) \, dx \, dy \, dz$$

where $\bar{V}$ is that part of the volume outside V where $F \not\equiv 0$.

The same Green's function can be used to formulate the solution of the other nonhomogeneous problem.   In this case, the solution becomes

$$\psi_2(\xi,\eta,\zeta) = \iint_S g(x,y,z)G(\xi,\eta,\zeta;x,y,z)\, dS$$

One could argue that the Green's function formulation of the problem just substitutes for the original problem the equally difficult problem of finding the Green's function.   In a few cases, the Green's function can be found without too much difficulty, and then we have an explicit representation of the solution. Another approach is to pick a Green's function which does not satisfy all the conditions and is thus simpler to find.   However, then we lose the explicit representation of the solution.   On the other hand, we get an **integral equation** which may be solvable by other techniques.   For example, in the first nonhomogeneous boundary-value problem above, if we take what is called the **free-space Green's function**

$$G = \frac{1}{4\pi}\frac{e^{ikR}}{R}$$

the Green's function no longer satisfies the boundary condition.   We then have

$$\psi_1(\xi,\eta,\zeta) = \iiint_V FG\, dV - \iint_S \psi_1\left(\frac{dG}{dn} + \alpha G\right) dS$$

If (ξ,η,ζ) is a point on S, then this becomes an integral equation for the boundary values of ψ_1 on S.   Having determined the values of ψ_1 on S, we can then substitute back into the above equation to obtain values of ψ_1 at points not on S.   We shall explore this avenue of approach in the next chapter.

Now let us turn to the solution of some particular problems in terms of Green's functions.   In electrostatics we are sometimes interested in the potential due to a continuous distribution of charge over a finite volume $\bar{V}$. In this case, we must solve **Poisson's equation**[1] $\nabla^2\psi = -f(x,y,z)$ in $\bar{V}$ and Laplace's equation outside $\bar{V}$.   This is the first nonhomogeneous boundary-value problem treated above if we take $k = 0$ and let S shrink to a point. Then we want the Green's function to satisfy the following properties:

1. $\nabla^2 G = 0$ except at (ξ,η,ζ).
2. $G = 1/4\pi R + H(\xi,\eta,\zeta;x,y,z)$, where H has continuous second partial derivatives at (ξ,η,ζ).
3. $G = O(1/r)$ as $r \to \infty$.

In this case we can find G explicitly as $1/4\pi R$.   Then the solution is

$$\psi(\xi,\eta,\zeta) = \frac{1}{4\pi}\iiint_V \frac{f(x,y,z)}{\sqrt{(x-\xi)^2 + (y-\eta)^2 + (z-\zeta)^2}}\, dx\, dy\, dz$$

[1] Here f is proportional to the charge density.

It is a problem in potential theory to determine general conditions on f which will assure that ψ is a solution of the problem.[1]

Next, let us consider the Dirichlet problem for unbounded regions; that is,

$$\nabla^2\psi = 0 \qquad \text{outside } V$$

$$\psi = g(x,y,z) \qquad \text{on } S$$

The Green's function should satisfy

1. $\nabla^2 G = 0$ except at (ξ,η,ζ).
2. $G = 1/4\pi R + H(\xi,\eta,\zeta;x,y,z)$, where H has continuous second partial derivatives at (ξ,η,ζ).
3. $G = O(1/r)$ as $r \to \infty$.
4. $G = 0$ on S.

The solution is then

$$\psi(\xi,\eta,\zeta) = -\iint\limits_S g(x,y,z)\frac{dG}{dn}\,dS$$

For example, suppose we wish to solve Laplace's equation in the upper half space, $z > 0$, with $\psi(x,y,0) = g(x,y)$ on the xy plane. The Green's function can be written explicitly as

$$G = \frac{1}{4\pi}\frac{1}{\sqrt{(x-\xi)^2 + (y-\eta)^2 + (z-\zeta)^2}}$$

$$- \frac{1}{4\pi}\frac{1}{\sqrt{(x-\xi)^2 + (y-\eta)^2 + (z+\zeta)^2}}$$

The first term has the proper singularity at (ξ,η,ζ) and satisfies Laplace's equation except at that point. The second term is obtained by reflecting the singularity in the xy plane. It therefore satisfies Laplace's equation everywhere in the upper half space. When $z = 0$, $G = 0$, so that the boundary condition is satisfied.

$$\left(\frac{dG}{dn}\right)_{z=0} = -\left(\frac{\partial G}{\partial z}\right)_{z=0}$$

$$= -\frac{1}{2\pi}\frac{\zeta}{[(x-\xi)^2 + (y-\eta)^2 + \zeta^2]^{\frac{3}{2}}}$$

so that

$$\psi(\xi,\eta,\zeta) = \frac{1}{2\pi}\int_{-\infty}^{\infty}\int_{-\infty}^{\infty}g(x,y)\frac{\zeta}{[(x-\xi)^2 + (y-\eta)^2 + \zeta^2]^{\frac{3}{2}}}\,dx\,dy$$

[1] See O. D. Kellogg, "Foundations of Potential Theory," Dover Publications, Inc., New York, 1953.

In this case, since the surface S is unbounded, to verify the Green's function representation of the solution, we have to integrate over a large circle of radius D in the xy plane and over the surface of a large hemisphere of radius D in the upper half space, take the limit at $D \to \infty$, and show that the integral over the surface of the hemisphere approaches zero as $D \to \infty$. This is not hard to do, since $G = O(1/r)$ and $\psi = O(1/r)$ as $r \to \infty$, and therefore

$$r^2 \left(G \frac{d\psi}{\partial r} - \psi \frac{\partial G}{\partial r} \right) = O\left(\frac{1}{r} \right)$$

as $r \to \infty$.

Exercises 5.5

1. Find the Green's function associated with the problem

$$\psi''(x) - k^2 \psi(x) = -f(x) \qquad 0 \le x < \infty$$
$$\psi(0) = 0$$
$$\psi \to 0 \quad \text{as} \quad x \to \infty$$

which will give the integral representation of the solution

$$\psi(x) = \int_0^\infty G(x;\xi) f(\xi) \, d\xi$$

2. Solve the exterior Dirichlet problem for the unit sphere; that is, $\nabla^2 \psi = 0$ for $x^2 + y^2 + z^2 > 1$, with $\psi = g(1,\theta,\phi)$ on S: $x^2 + y^2 + z^2 = 1$. Find an explicit integral representation.

3. Find an integral representation for the solution of the following problem: $\nabla^2 \psi = 0$ for $z > 0$, $y > 0$, with $\psi(x,y,0) = g(x,y)$, $\psi(x,0,z) = h(x,z)$, $\psi = O(1/r)$ as $r = \sqrt{x^2 + y^2 + z^2} \to \infty$.

4. A square membrane is vibrating in its fundamental mode. Formulate an integral equation whose solution will give approximately the acoustic field produced by the membrane. List the assumptions which you make to make this formulation possible.

5.6   A Problem in Diffraction Theory

In some problems the source of excitation is at a large distance from the volume V. For example, in diffraction theory we are often interested in the interaction between a passive surface S and a plane wave. In this case, the source of excitation of the field is a plane wave. The plane wave itself does not satisfy the boundary condition on S. The effect of the surface is then to produce a scattered field which together with the plane wave will satisfy the boundary condition. Consider a plane wave

$$\phi_p = e^{i(ku - \omega t)}$$

where $u = x \cos \alpha + y \cos \beta + z \cos \gamma$, a coordinate with axis in a direction making angles α, β, and γ with the x, y, and z coordinate axes. It is easily shown that ϕ_p satisfies the wave equation

$$\nabla^2 \phi_p = \frac{k^2}{\omega^2} \frac{\partial^2 \phi_p}{\partial t^2}$$

and is a plane wave proceeding in the direction of the u-coordinate axis.   Now we let the whole field be

$$\phi = \phi_p + \phi_s$$

where ϕ_s represents the scattered field. ϕ must satisfy the wave equation

$$\nabla^2 \phi = \frac{1}{a^2} \frac{\partial^2 \phi}{\partial t^2}$$

Then letting $\phi = \psi e^{-i\omega t}$, we have the result that ψ satisfies

$$\nabla^2 \psi + k^2 \psi = 0$$

with $k^2 = \omega^2/a^2$.   We can write

$$\psi = \psi_p + \psi_s$$
$$= e^{iku} + \psi_s$$

In the acoustic case, if S is a rigid boundary, there cannot be a normal component of velocity across it.   Therefore, $d\phi/dn = 0$ and $d\psi/dn = 0$ on S. We see that ψ solves a homogeneous boundary-value problem.   This does not, however, imply that $\psi \equiv 0$, for it contains a plane wave which does not satisfy the radiation condition.   We will require, however, that ψ_s satisfy the radiation condition. ψ_s will then solve the following nonhomogeneous boundary-value problem:

$$\nabla^2 \psi_s + k^2 \psi_s = 0 \qquad \text{outside } V$$

$$\frac{d\psi_s}{dn} = -\frac{d\psi_p}{dn} \quad \text{on } S$$

$$\lim_{r \to \infty} r\left(\frac{\partial \psi_s}{\partial r} - ik\psi_s\right) = 0$$

ψ_s is therefore unique, and the problem can be formulated in terms of Green's functions.

To be a little more specific, let us consider a classical diffraction theory problem, the diffraction of a plane wave by a rigid semi-infinite plane sheet. We shall assume a plane wave with perpendicular incidence, striking the edge (z axis) of the rigid plane sheet (xz plane, $x \leq 0$).   From the geometry we see that the solution will not depend on the z coordinate.   The velocity potential $\phi(x,y,t)$ will satisfy the wave equation except on the xz plane, $x \leq 0$, where it must satisfy the boundary condition $d\phi/dn = 0$. We write

$$\phi(x,y,t) = \phi_p + \phi_s$$
$$= [e^{ikr\cos(\theta-\alpha)} + \psi_s(x,y)]e^{-i\omega t}$$

In this case, geometrical optics tells us to expect a shadow (no plane wave) in region 1 and a reflected plane wave in region 2. Therefore, we write

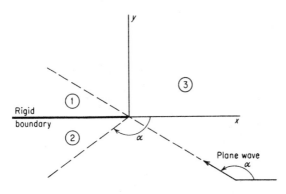

FIGURE 10

$\psi_s = \bar{\psi}_s$ for $-\alpha < \theta < \alpha$, $\psi_s = \bar{\psi}_s - e^{ikr\cos(\theta-\alpha)}$ for $\alpha < \theta < \pi$, and $\psi_s = \bar{\psi}_s + e^{ikr\cos(\theta+\alpha)}$ for $\pi < \theta < 2\pi - \alpha$. In this problem we have reflection and diffraction, and the diffraction effect is entirely contained in the function $\bar{\psi}_s$. The reflected plane wave does not satisfy the radiation condition, but we shall require that $\bar{\psi}_s$ does so.

Our procedure will be to introduce a Green's function for the purpose of formulating an integral equation whose solution will allow us to find the acoustic field. To determine the appropriate Green's function we shall consider axially symmetric solutions of the two-dimensional Helmholtz equation. These solutions must satisfy the zeroth-order Bessel equation

$$\psi'' + \frac{1}{r}\psi' + k^2\psi = 0$$

which has solutions

$$J_0(kr) \qquad Y_0(kr) \qquad H_0^{(1)}(kr) = J_0(kr) + iY_0(kr)$$

and
$$H_0^{(2)}(kr) = J_0(kr) - iY_0(kr)$$

It can be shown that these functions behave as

$$\frac{\cos kr}{\sqrt{r}} \qquad \frac{\sin kr}{\sqrt{r}} \qquad \frac{e^{ikr}}{\sqrt{r}} \quad \text{and} \quad \frac{e^{-ikr}}{\sqrt{r}}$$

respectively, as $r \to \infty$. Therefore, we take the solution $H_0^{(1)}(kr)$ as the one representing outward traveling cylindrical waves. To have the proper behavior as (x,y) approaches (ξ,η), the Green's function must be asymptotic to $(i/4)H_0^{(1)}(kR)$ where $R = \sqrt{(x - \xi)^2 + (y - \eta)^2}$. That the constant must be $i/4$ follows from the fact that $J_0(kR)$ is bounded and $Y_0(kR)$ behaves as $(2/\pi) \log R$ as $R \to 0$. In the two-dimensional problem the radiation condition takes the form

$$\lim_{r \to \infty} \sqrt{r}\left(\frac{\partial \psi}{\partial r} - ik\psi\right) = 0$$

Getting back to the diffraction problem, for the purpose of formulating the integral equation, we shall use the following Green's function:

$$G(\xi,\eta;x,y) = \frac{i}{4}\,[H_0^{(1)}(k\sqrt{(x - \xi)^2 + (y - \eta)^2}) + H_0^{(1)}(k\sqrt{(x - \xi)^2 + (y + \eta)^2})]$$

This function satisfies the conditions $(\partial G/\partial y)_{y=0} = 0$, and

$$G = \frac{e^{ikr}}{\sqrt{r}}\,g(\theta) + O(r^{-\frac{3}{2}})$$

as $r \to \infty$.

Consider the regions R_1 and R_2 bounded by the contours C_1 and C_2 (see Fig. 11). We apply the Green's identity to $\psi^{(1)} = \psi_s^{(1)} + e^{ikr \cos (\theta-\alpha)}$, which is the whole field in the upper half plane, in the region R_1; that is,

$$\psi^{(1)}(\xi,\eta) = \int_{C_1}\left(G\,\frac{d\psi^{(1)}}{dn} - \psi^{(1)}\,\frac{dG}{dn}\right)\,ds$$

$$= -\int_{-\infty}^{\infty} G\left(\frac{\partial \psi}{\partial y}\right)_{y=0+}\,dx$$

provided we can show that the contributions of the small and large semicircles approach zero as the radii approach zero and infinity, respectively. We also apply the Green's identity to $\psi^{(2)} - e^{ikr \cos (\theta-\alpha)} - e^{ikr \cos (\theta+\alpha)}$, which is the whole field with the plane wave and reflected wave subtracted out, in the region R_2; that is,

$$\psi^{(2)}(\xi,\eta) - e^{ik(\xi \cos \alpha + \eta \sin \alpha)} - e^{ik(\xi \cos \alpha - \eta \sin \alpha)} = \int_{-\infty}^{\infty} G\left(\frac{\partial \psi}{\partial y}\right)_{y=0-}\,dx$$

provided again that we can show that the contributions of the small and large semicircles approach zero. Here we have also made use of the fact that

$$\frac{\partial}{\partial y}\,[e^{ikr \cos (\theta-\alpha)} + e^{ikr \cos (\theta+\alpha)}]_{y=0} = 0$$

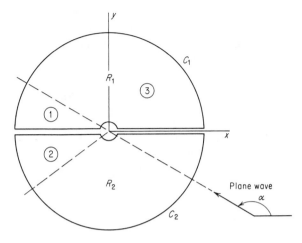

FIGURE 11

To complete the derivation of the integral equation we proceed as follows. Noting that for $x > 0$, $y = 0$, ψ and $\partial\psi/\partial y$ are continuous, we have

$$-2e^{ik\xi \cos \alpha} = 2\int_{-\infty}^{\infty} G\left(\frac{\partial\psi}{\partial y}\right)_{y=0} dx$$

$$-2e^{ik\xi \cos \alpha} = i\int_{-\infty}^{\infty} H_0^{(1)}(k\,|x-\xi|)I(x)\,dx$$

where $I(x) = (\partial\psi/\partial y)_{y=0}$. For $\xi < 0$

$$\psi(\xi,0-) - \psi(\xi,0+) - 2e^{ik\xi \cos \alpha} = i\int_{-\infty}^{\infty} H_0^{(1)}(k\,|x-\xi|)I(x)\,dx$$

Hence, we have

$$i\int_{-\infty}^{\infty} H_0^{(1)}(k\,|x-\xi|)I(x)\,dx = g(\xi)$$

where $g(\xi) = -2e^{ik\xi \cos \alpha} \qquad \xi > 0$

$$g(\xi) = \psi(\xi,0-) - \psi(\xi,0+) - 2e^{ik\xi \cos \alpha} \qquad \xi < 0$$

The integral equation we have just derived is known as the **Wiener-Hopf integral equation.** It also occurs in the theory of stochastic processes. The function $g(\xi)$ is not completely known, but this does not turn out to be serious. We shall find that the equation can be solved by the **Wiener-Hopf technique,** which will be discussed in Chap. 8 in connection with the study of Fourier transforms.

To complete the discussion we must show that the integrals over the small and large semicircles approach zero as their radii approach zero and infinity, respectively.   Let us take the radius of the small semicircle as δ.   If the edge of the rigid sheet ($r = 0$) were a source point, then $\psi = O(\log r)$ as $r \to 0$.   In this case

$$\int_0^{2\pi} \left(G \frac{\partial \psi}{\partial r} - \psi \frac{\partial G}{\partial r} \right)_{r=\delta} \delta \, d\theta = O(1)$$

as $\delta \to 0$, and the integrals over the small semicircles would not vanish as $\delta \to 0$.   Therefore, we must assume that $\psi = o(\log r)$ as $r \to 0$,† which means physically that $r = 0$ is not a source point.   Under this assumption,

$$\lim_{\delta=0} \int_0^{\pi} \left(G \frac{\partial \psi^{(1)}}{\partial r} - \psi^{(1)} \frac{\partial G}{\partial r} \right)_{r=\delta} \delta \, d\theta = 0$$

$$\lim_{\delta \to 0} \int_{\pi}^{2\pi} \left(G \frac{\partial \psi^{(2)}}{\partial r} - \psi^{(2)} \frac{\partial G}{\partial r} \right)_{r=\delta} \delta \, d\theta = 0$$

As far as the large semicircles are concerned, we note that for $\alpha < \theta < 2\pi - \alpha$ we have, in effect, applied Green's identity to $\bar{\psi}_s$, which satisfies the radiation condition, and hence

$$\bar{\psi}_s = \frac{e^{ikr}}{\sqrt{r}} h(\theta) + O(r^{-\frac{3}{2}})$$

Clearly
$$\int_0^{2\pi} \left(G \frac{\partial \bar{\psi}_s}{\partial r} - \bar{\psi}_s \frac{\partial G}{\partial r} \right)_{r=D} D \, d\theta = O\left(\frac{1}{D}\right)$$

as $D \to \infty$, where D is the radius of the large semicircle.   Therefore,

$$\lim_{D \to \infty} \int_0^{2\pi} \left(G \frac{\partial \bar{\psi}_s}{\partial r} - \bar{\psi}_s \frac{\partial G}{\partial r} \right)_{r=D} D \, d\theta = 0$$

It remains to show that

$$\int_0^{\alpha} \left(G \frac{\partial}{\partial r} e^{ikr \cos(\theta - \alpha)} - e^{ikr \cos(\theta-\alpha)} \frac{\partial G}{\partial r} \right)_{r=D} D \, d\theta = O\left(\frac{1}{\sqrt{D}}\right)$$

$$\int_{2\pi-\alpha}^{2\pi} \left(G \frac{\partial}{\partial r} e^{ikr \cos(\theta+\alpha)} - e^{ikr \cos(\theta+\alpha)} \frac{\partial G}{\partial r} \right)_{r=D} D \, d\theta = O\left(\frac{1}{\sqrt{D}}\right)$$

as $D \to \infty$. We consider the first of these integrals only. The second is very similar.

$$\int_0^\alpha \left(G \frac{\partial}{\partial r} e^{ikr \cos(\theta - \alpha)} - e^{ikr \cos(\theta - \alpha)} \frac{\partial G}{\partial r} \right)_{r=D} D \, d\theta$$

$$= \int_0^\alpha ik \sqrt{D} g(\theta) [\cos(\theta - \alpha) - 1] e^{ikD[\cos(\theta - \alpha) - 1]} \, d\theta + O\left(\frac{1}{\sqrt{D}}\right)$$

$$= -\int_0^\alpha 2ik \sqrt{D} g(\theta) \sin^2\left(\frac{\theta - \alpha}{2}\right) e^{2ikD \cos^2(\theta - \alpha)/2} \, d\theta + O\left(\frac{1}{\sqrt{D}}\right)$$

$$= \frac{1}{\sqrt{D}} \int_0^\alpha g(\theta) \tan\left(\frac{\theta - \alpha}{2}\right) e^{2ikD \cos^2(\theta - \alpha)/2} \, d[e^{ikD \cos^2(\theta - \alpha)/2}] + O\left(\frac{1}{\sqrt{D}}\right)$$

$$= \left[\frac{g(\theta)}{\sqrt{D}} \tan\left(\frac{\theta - \alpha}{2}\right) e^{2ikD \cos^2(\theta - \alpha)/2} \right]_0^\alpha$$

$$\quad - \frac{1}{\sqrt{D}} \int_0^\alpha e^{2ikD \cos^2(\theta - \alpha)/2} \frac{d}{d\theta}\left[g(\theta) \tan\left(\frac{\theta - \alpha}{2}\right) \right] d\theta + O\left(\frac{1}{\sqrt{D}}\right)$$

$$= O\left(\frac{1}{\sqrt{D}}\right)$$

The integration by parts is valid since $(\theta - \alpha)/2 \neq \pi/2$ in the range of integration. The discussion of the other integral is identical except for the change of the limits of integration and the change from $\theta - \alpha$ to $\theta + \alpha$ in the integrand. This completes the derivation of the Wiener-Hopf integral equation for the half-plane problem. We shall return to the problem in Sec. 8.4.

Exercises 5.6

Derive the appropriate Wiener-Hopf integral equation for the half-plane diffraction problem with the boundary condition $\psi = 0$ on the sheet. HINT: Use the Green's function

$$G(x,y;\xi,\eta) = \frac{i}{4} H_0^{(1)}[k \sqrt{(x - \xi)^2 + (y - \eta)^2}] - \frac{i}{4} H_0^{(1)} k[\sqrt{(x - \xi)^2 + (y + \eta)^2}]$$

References

Courant, Richard, and David Hilbert: "Methods of Mathematical Physics," Interscience Publishers (Division of John Wiley & Sons, Inc.), New York, 1953, vol. I; 1962, vol. II.

Erdélyi, Arthur: "Operational Calculus and Generalized Functions," Holt, Rinehart and Winston, Inc., New York, 1962.

Friedman, Bernard: "Principles and Techniques of Applied Mathematics," John Wiley & Sons, Inc., New York, 1956.

Hildebrand, F. B.: "Methods of Applied Mathematics," Prentice-Hall, Inc., Englewood Cliffs, N.J., 1952.

Lighthill, M. J.: "Introduction to Fourier Analysis and Generalized Functions," Cambridge University Press, London, 1959.

Murnaghan, Francis D.: "Introduction to Applied Mathematics," John Wiley & Sons, Inc., New York, 1948.

Noble, Benjamin: "The Wiener-Hopf Technique," Pergamon Press, Inc., New York, 1958.

Page, Chester H.: "Physical Mathematics," D. Van Nostrand Company, Inc., Princeton, N.J., 1955.

Sagan, Hans: "Boundary and Eigenvalue Problems in Mathematical Physics," John Wiley & Sons, Inc., New York, 1961.

Sneddon, Ian N.: "Elements of Partial Differential Equations," McGraw-Hill Book Company, New York, 1957.

Sommerfeld, Arnold J.: "Partial Differential Equations in Physics," Academic Press Inc., New York, 1949.

Chapter 6. Integral Equations

6.1 Integral-equation Formulation of Boundary-value Problems

As we have already indicated in the previous chapter, certain nonhomogeneous boundary-value problems can be formulated in terms of integral equations by use of Green's functions. The procedure is much more general than these earlier remarks would indicate and applies to homogeneous problems as well as to nonhomogeneous problems. Consider, for example, the homogeneous boundary-value problem

$$\nabla^2 \psi + \lambda \psi = 0 \qquad \text{in } V$$

$$\frac{d\psi}{dn} + \alpha \psi = 0 \qquad \text{on } S$$

where V is a bounded volume bounded by the surface S. Let G be a Green's function with the following properties:

1. $\nabla^2 G = 0$ except at (ξ, η, ζ) in V.
2. $dG/dn + \alpha G = 0$ on S.
3. $G = \dfrac{1}{4\pi} \dfrac{1}{R} + H(\xi, \eta, \zeta; x, y, z)$

where $R = \sqrt{(x - \xi)^2 + (y - \eta)^2 + (z - \zeta)^2}$, and H is a solution of Laplace's equation everywhere in V. Then

$$\iiint_V \left(G\nabla^2\psi - \psi\nabla^2 G \right) dV = \iint_S \left[G\left(\frac{d\psi}{dn} + \alpha\psi\right) - \psi\left(\frac{dG}{dn} + \alpha G\right) \right] dS = 0$$

or

$$\psi(\xi, \eta, \zeta) = \lambda \iiint_V \psi G \, dV$$

The last equation is a **homogeneous integral equation of the second**

267

kind.[1]   Using the same Green's function we can also formulate the solution of the nonhomogeneous boundary-value problem

$$\nabla^2\psi + \lambda\psi = -f \qquad \text{in } V$$

$$\frac{d\psi}{dn} + \alpha\psi = 0 \qquad \text{on } S$$

In this case, the integral equation is

$$\psi(\xi,\eta,\zeta) = F(\xi,\eta,\zeta) + \lambda\iiint_V \psi G \, dV$$

where $F(\xi,\eta,\zeta) = \iiint_V fG \, dV$. This time we have a **nonhomogeneous integral equation of the second kind.** In this case, λ is a known constant, whereas in the homogeneous integral equation above λ is an eigenvalue to be determined in the course of solving the integral equation.

In the one-dimensional case, we have integral equations corresponding to Sturm-Liouville problems. Here the general homogeneous problem is

$$(py')' - qy + \lambda\rho y = 0 \qquad a \leq x \leq b$$

subject to one of five homogeneous boundary conditions at a and b:

1. $y = 0$
2. $y' = 0$
3. $y' + \alpha y = 0$
4. y and y' finite when $p = 0$
5. $y(a) = y(b)$ and $p(a)y'(a) = p(b)y'(b)$

Let $L(y) = (py')' - qy$. Then the differential equation can be written $L(y) + \lambda\rho y = 0$. We take the Green's function with the following properties:

1. $L(G) = 0$ except at $x = \xi$
2. G continuous at $x = \xi$
3. $G'(\xi;\xi+) - G'(\xi;\xi-) = -1/p(\xi)$
4. G satisfies the same boundary conditions as y

Then

$$\int_a^b [GL(y) - yL(G)] \, dx = 0$$

and

$$y(\xi) = \lambda\int_a^b G(\xi;x)\rho(x)y(x) \, dy$$

---

[1] An integral equation of the form $f(x) = \int_a^b K(\xi;x)\psi(\xi) \, d\xi$ is known as an **integral equation of the first kind.** Here the unknown ψ appears only under the integral sign and f and K are known functions, whereas in the integral equation of the second kind the unknown appears both under the integral and outside.

We introduce the new unknown $\psi(x) = \sqrt{\rho(x)}\, y(x)$.   Then

$$\sqrt{\rho(\xi)}\, y(\xi) = \lambda \int_a^b \sqrt{\rho(\xi)\rho(x)}\, G(\xi;x)\sqrt{\rho(x)}\, y(x)\, dx$$

or
$$\psi(\xi) = \lambda \int_a^b K(\xi;x)\psi(x)\, dx$$

where $K(\xi;x) = \sqrt{\rho(\xi)\rho(x)}\, G(\xi;x)$.   We again have a homogeneous integral equation of the second kind. K is called the **kernel** of the integral equation. The same Green's function leads to the equation

$$y(\xi) = \int_a^b f(x)G(\xi;x)\, dx + \lambda \int_a^b G(\xi;x)\rho(x)y(x)\, dx$$

for the solution of the nonhomogeneous Sturm-Liouville equation

$$(py')' - qy + \lambda\rho y = -f$$

Then we have the nonhomogeneous integral equation of the second kind

$$\psi(\xi) = F(\xi) + \lambda \int_a^b K(\xi;x)\psi(x)\, dx$$

where $\psi = \sqrt{\rho}\, y$, $F = \int_a^b \sqrt{\rho}\, fG\, dx$, and $K = \sqrt{\rho(\xi)\rho(x)}\, G(\xi;x)$.

As an example of an integral equation, consider the problem corresponding to the Sturm-Liouville problem $y'' + \lambda y$, $0 \le x \le 1$, with $y(0) = y(1) = 0$. Then $G'' = 0$ except at $x = \xi$.   Therefore,

$$G = ax + b \qquad 0 \le x < \xi$$
$$G = cx + d \qquad \xi < x \le 1$$

$G(0) = 0$ implies that $b = 0$. $G(1) = 0$ implies that $d = -c$. G is continuous at $x = \xi$.   Hence, $a\xi = c\xi - c$.   Finally, $G'(\xi+) - G'(\xi-) = -1$ implies that $c - a = -1$.   Therefore

$$G = (1 - \xi)x \qquad 0 \le x < \xi$$
$$G = \xi(1 - x) \qquad \xi < x \le 1$$

The integral equation is then

$$y(\xi) = \lambda \int_0^\xi (1 - \xi)xy(x)\, dx + \lambda \int_\xi^1 \xi(1 - x)y(x)\, dx$$

It may not always be possible to define the Green's function from the four properties listed above.   Consider, for example, the same equation $y'' + \lambda y = 0$, $0 \le x \le 1$, but with the boundary condition $y'(0) = y'(1) = 0$. If we proceed as above, we have $G'' = 0$, implying

$$G = ax + b \qquad 0 \le x < \xi$$
$$G = cx + d \qquad \xi < x \le 1$$

But the other conditions give $a = c = 0$ and $b = d$, and hence we have no Green's function. The difficulty here is that $\lambda = 0$ is an eigenvalue of the differential equation. In fact, the properties listed above will never yield a Green's function for the Sturm-Liouville problem in the case where $\lambda = 0$ is an eigenvalue of the differential equation. Let u be the solution of the differential equation when $\lambda = 0$, satisfying a set of homogeneous boundary conditions. Then

$$L(u) = 0$$

Let v be a solution of the differential equation for $\lambda \neq 0$, satisfying the same boundary conditions. Then

$$L(v) + \lambda \rho v = 0$$

Therefore, $$\lambda \int_a^b \rho uv \, dx = \int_a^b [vL(u) - uL(v)] \, dx = 0$$

and u and v are orthogonal. Because of this orthogonality we can define a Green's function satisfying the differential equation

$$L(G) - A\rho u = 0$$

except at $x = \xi$, leaving the other three properties unchanged.

$$\int_a^b \{GL(v) - v[L(G) - A\rho u]\} \, dx = 0$$

$$v(\xi) = \lambda \int_a^b \rho vG \, dx$$

If u is normalized, that is,

$$\int_a^b \rho u^2 \, dx = 1$$

then we can determine A, for

$$0 = \int_a^b [GL(u) - uL(G)] \, dx = u(\xi) - A \int_a^b \rho u^2 \, dx$$

Hence, $A = u(\xi)$.

Getting back to our example, we wish to solve $G'' = 1$. Therefore,

$$G = \tfrac{1}{2}x^2 + ax + b \qquad 0 \leq x < \xi$$
$$G = \tfrac{1}{2}x^2 + cx + d \qquad \xi < x \leq 1$$

$G'(0) = 0$ implies $a = 0$. $G'(1) = 0$ implies $c = -1$. $G(\xi+) = G(\xi-)$ implies $b = d - \xi$. Then

$$G = \tfrac{1}{2}x^2 + d - \xi \qquad 0 \leq x < \xi$$
$$G = \tfrac{1}{2}x^2 + d - x \qquad \xi < x \leq 1$$

In order to make G symmetric, we pick $d = \frac{1}{2}\xi^2$. The integral equation is then

$$y(\xi) = \lambda \int_0^\xi [\tfrac{1}{2}(x^2 + \xi^2) - \xi]y(x)\,dx + \lambda \int_\xi^1 [\tfrac{1}{2}(x^2 + \xi^2) - x]y(x)\,dx$$

Exercises 6.1

1. Formulate an integral equation for the solution of the Sturm-Liouville problem $(xy')' + \lambda xy = 0$, $1 \le x \le e$, $y(1) = y(e) = 0$.

2. Formulate an integral equation for the solution of the Sturm-Liouville problem $[(1 - x^2)y']' + \lambda y = 0$, $-1 \le x \le 1$, $y(-1)$ and $y(1)$ finite.

6.2 Hilbert-Schmidt Theory

We saw, in the previous section, that the integral equations associated with the homogeneous and nonhomogeneous Sturm-Liouville problems are the **homogeneous Fredholm equation**

$$\psi(\xi) = \lambda \int_a^b K(\xi;x)\psi(x)\,dx$$

and the **nonhomogeneous Fredholm equation**

$$\psi(\xi) = F(\xi) + \lambda \int_a^b K(\xi;x)\psi(x)\,dx$$

In this case, the kernel K is real, symmetric, and continuous over the rectangle $a \le x \le b$, $a \le \xi \le b$. Among other things, this implies that

$$\int_a^b \int_a^b K^2(\xi;x)\,d\xi\,dx$$

exists. If we define the linear operator A by

$$Af = \int_a^b K(\xi;x)f(x)\,dx$$

then according to Sec. 2.8 this is a completely continuous self-adjoint operator and the theorems of that section apply. We could end the discussion there, but this would gloss over many important details which a more explicit treatment can bring to light. For this reason, we shall develop the Hilbert-Schmidt theory along lines parallel to the general treatment of completely continuous self-adjoint operators. The reader should make the obvious comparisons.

In the homogeneous equation λ is an eigenvalue to be determined in the course of solving the equation. In the nonhomogeneous equation λ is a known constant, and we shall find that, in general, the equation will have a solution unless perhaps λ is an eigenvalue of the homogeneous equation. If λ is an eigenvalue of the homogeneous equation, a necessary condition for the existence of a solution of the nonhomogeneous equation is that the function $F(x)$ be orthogonal to the eigenfunctions corresponding to that eigenvalue; that is, if $\lambda = \lambda_i$, where

$$\psi_i(\xi) = \lambda_i \int_a^b K(\xi;x)\psi_i(x)\,dx$$

then
$$\int_a^b \psi(\xi)\psi_i(\xi)\,d\xi = \int_a^b F(\xi)\psi_i(\xi)\,d\xi + \lambda_i \int_a^b \int_a^b K(\xi;x)\psi(x)\psi_i(\xi)\,dx\,d\xi$$

$$= \int_a^b F(\xi)\psi_i(\xi)\,d\xi + \int_a^b \psi(x)\psi_i(x)\,dx$$

Hence, $\int_a^b F(\xi)\psi_i(\xi)\,d\xi = 0$. In this case, if a solution ψ of the nonhomogeneous equation exists, it is not unique, for

$$\lambda_i \int_a^b K(\xi;x)[\psi(x) + c\psi_i(x)]\,dx + F(\xi) = F(\xi) + \lambda_i \int_a^b K(\xi;x)\psi(x)\,dx + c\psi_i(\xi)$$

$$= \psi(\xi) + c\psi_i(\xi)$$

implying that $\psi(\xi) + c\psi_i(\xi)$ is also a solution for arbitrary c.

Let ψ_i and ψ_j be solutions of the homogeneous equation corresponding to different eigenvalues λ_i and λ_j. Then

$$\psi_i(\xi) = \lambda_i \int_a^b K(\xi;x)\psi_i(x)\,dx$$

$$\psi_j(\xi) = \lambda_j \int_a^b K(\xi;x)\psi_j(x)\,dx$$

and
$$(\lambda_i - \lambda_j)\int_a^b \psi_i(\xi)\psi_j(\xi)\,d\xi = \lambda_i\lambda_j \int_a^b \int_a^b K(\xi;x)\psi_i(x)\psi_j(\xi)\,dx\,d\xi$$

$$- \lambda_i\lambda_j \int_a^b \int_a^b K(\xi;x)\psi_j(x)\psi_i(\xi)\,dx\,d\xi$$

$$= \lambda_i\lambda_j \int_a^b \int_a^b K(\xi;x)\psi_i(x)\psi_j(\xi)\,dx\,d\xi$$

$$- \lambda_i\lambda_j \int_a^b \int_a^b K(x;\xi)\psi_i(x)\psi_j(\xi)\,d\xi\,dx$$

Interchanging the order of integration and making use of the symmetry of K, we have

$$(\lambda_i - \lambda_j)\int_a^b \psi_i(\xi)\psi_j(\xi)\,d\xi = 0$$

But $\lambda_i \neq \lambda_j$; therefore, $\int_a^b \psi_i(\xi)\psi_j(\xi)\,d\xi = 0$. Hence, solutions corresponding to different eigenvalues are orthogonal. Also, since the equation is homogeneous, we can assume that the solutions are normalized; that is,

$$\int_a^b |\psi|^2 \, d\xi = 1$$

Next we show that the eigenvalues are real. Let ψ be a solution with eigenvalue λ. Then

$$\psi(\xi) = \lambda \int_a^b K(\xi;x)\psi(x) \, dx$$

$$\psi^*(\xi) = \lambda^* \int_a^b K(\xi;x)\psi^*(x) \, dx$$

If $\lambda \neq \lambda^*$, then

$$\int_a^b \psi\psi^* \, d\xi = \int_a^b |\psi|^2 \, d\xi = 0$$

which implies that $\psi \equiv 0$. Therefore, either λ is real or ψ is the trivial solution.

If there are an infinite number of solutions of the homogeneous equation, then there can be only a finite number of linearly independent solutions corresponding to the same eigenvalue. We shall show this as follows. Let $\psi_1, \psi_2, \ldots$ be an infinite set of linearly independent solutions. Since they are linearly independent, we can assume that they are orthonormal; that is,

$$\int_a^b \psi_i\psi_j \, dx = \delta_{ij}$$

Let us consider the problem of approximating in the least-mean-square sense the kernel by a series of a finite number of these functions. In other words, we wish to minimize

$$\int_a^b \left| K(\xi;x) - \sum_{i=1}^n c_i\psi_i(x) \right|^2 dx$$

We know from Chap. 2 that to do this we must choose

$$c_i = \int_a^b K(\xi;x)\psi_i(x) \, dx = \frac{\psi_i(\xi)}{\lambda_i}$$

In this case,

$$\min \int_a^b \left| K(\xi;x) - \sum_{i=1}^n c_i\psi_i(x) \right|^2 dx = \int_a^b K^2(\xi;x) \, dx - \sum_{i=1}^n c_i^2 \geq 0$$

therefore,

$$\sum_{i=1}^n \frac{\psi_i^2(\xi)}{\lambda_i^2} \leq \int_a^b K^2(\xi;x) \, dx$$

and

$$\sum_{i=1}^n \frac{\int_a^b \psi_i^2(\xi) \, d\xi}{\lambda_i^2} = \sum_{i=1}^n \frac{1}{\lambda_i^2} \leq \int_a^b \int_a^b K^2(\xi;x) \, dx \, d\xi < \infty$$

But this is true for any n.   Therefore,

$$\sum_{i=1}^{\infty} \frac{1}{\lambda_i^2}$$

converges.   This implies that

$$\lim_{i \to \infty} \frac{1}{\lambda_i^2} = 0$$

and that there can be at most a finite number of linearly independent solutions corresponding to the same eigenvalue.   The problem can be degenerate; that is, more than one linearly independent solution can correspond to the same eigenvalue, but the degeneracy must be finite.

Let us assume that there are an infinite number of linearly independent solutions ψ_i of the homogeneous equation.   We can assume that these functions are orthonormal.   Consider the nonhomogeneous integral equation

$$\psi(\xi) = F(\xi) + \lambda \int_a^b K(\xi;x)\psi(x)\, dx$$

or

$$g(\xi) = \psi(\xi) - F(\xi) = \lambda \int_a^b K(\xi;x)\psi(x)\, dx$$

We wish to approximate g by a series of solutions of the homogeneous equation. The best approximation in the least-mean-square sense would have coefficients.

$$\begin{aligned}
c_i &= \int_a^b g(\xi)\psi_i(\xi)\, d\xi \\
&= \lambda \int_a^b \int_a^b K(\xi;x)\psi(x)\psi_i(\xi)\, dx\, d\xi \\
&= \frac{\lambda}{\lambda_i} \int_a^b \psi_i(x)\psi(x)\, dx \\
&= \frac{\lambda}{\lambda_i} \int_a^b \psi_i(x)[g(x) + F(x)]\, dx \\
&= \frac{\lambda}{\lambda_i}(c_i + \gamma_i) \qquad \text{(no summation on } i\text{)}
\end{aligned}$$

where

$$\gamma_i = \int_a^b F(x)\psi_i(x)\, dx$$

Solving for c_i, we have

$$c_i = \frac{\lambda \gamma_i}{\lambda_i - \lambda}$$

Therefore, the c's can be determined provided $\lambda \neq \lambda_i$ for any i. We shall be able to show that

$$g(\xi) = \sum_{i=1}^{\infty} c_i \psi_i(\xi)$$

and the series converges uniformly. Hence, the solution of the nonhomogeneous equation exists and is unique if $\lambda \neq \lambda_i$. If $\lambda = \lambda_i$ for some i, the solution will still exist if $\gamma_i = 0$. But in this case c_i is arbitrary, and so the solution is not unique. We have thus shown that we can concentrate on the solution of the homogeneous equation.

As a first case, let us consider the homogeneous integral equation with a **degenerate kernel.** The kernel is said to be degenerate if it can be expressed as

$$K(\xi;x) = \sum_{i=1}^{n} \alpha_i(\xi)\beta_i(x)$$

Consider the set of $2n$ functions $\alpha_1(x)$, $\alpha_2(x)$, ..., $\alpha_n(x)$, $\beta_1(x)$, $\beta_2(x)$, ..., $\beta_n(x)$. From this set we can construct a set of linearly independent functions $\gamma_1(x)$, $\gamma_2(x)$, ..., $\gamma_p(x)$, with $p \leq 2n$. We can assume that this set is orthonormal; that is,

$$\int_a^b \gamma_i(x)\gamma_j(x)\,dx = \delta_{ij}$$

Then
$$\alpha_i(\xi) = a_{ij}\gamma_j(\xi) \qquad i = 1, 2, 3, \ldots, n$$
$$\beta_i(x) = b_{ij}\gamma_j(x) \qquad j = 1, 2, 3, \ldots, p$$

and
$$K(\xi;x) = a_{ij}b_{ik}\gamma_j(\xi)\gamma_k(x)$$
$$= c_{jk}\gamma_j(\xi)\gamma_k(x)$$

By the symmetry of K, we have

$$K(\xi;x) - K(x;\xi) = c_{jk}\gamma_j(\xi)\gamma_k(x) - c_{jk}\gamma_j(x)\gamma_k(\xi)$$
$$= (c_{jk} - c_{kj})\gamma_j(\xi)\gamma_k(x) = 0$$

Therefore, $c_{jk} = c_{kj}$, so the matrix C is symmetric. The integral equation is then

$$\psi(\xi) = \lambda \int_a^b c_{jk}\gamma_j(\xi)\gamma_k(x)\psi(x)\,dx$$
$$= \lambda c_{jk}\gamma_j(\xi)y_k$$

where
$$y_k = \int_a^b \gamma_k(x)\psi(x)\,dx$$

Multiplying by $\gamma_i(\xi)$ and integrating,

$$y_i = \lambda c_{jk}\delta_{ij}y_k$$
$$= \lambda c_{ik}y_k$$

or $[C - (1/\lambda)I]Y = O$. This is a system of p homogeneous linear equations in p unknowns. The characteristic equation is

$$\left| C - \frac{1}{\lambda} I \right| = 0$$

This equation has p roots, and since C is real and symmetric, there are p real roots, $\mu_i = 1/\lambda_i$. Corresponding to each root μ_i there is an eigenvector $\mathbf{Y}_i$ and a solution of the integral equation

$$\psi_i(\xi) = \lambda_i c_{jk} \gamma_j(\xi) y_{ik}$$

(no summation on i). Here y_{ik} is the kth component of the ith eigenvector. One can show directly that the solutions are orthonormal, for

$$\int_a^b \psi_i \psi_j \, d\xi = \lambda_i \lambda_j c_{mn} y_{in} c_{pq} y_{jq} \int_a^b \gamma_m(\xi) \gamma_p(\xi) \, d\xi$$

$$= y_{im} y_{jp} \delta_{mp}$$

$$= (\mathbf{Y}_i, \mathbf{Y}_j) = \delta_{ij}$$

We can approach the general problem from the point of view of the calculus of variations. We first define the quadratic functional

$$Q[f] = \int_a^b \int_a^b K(\xi;x) f(\xi) f(x) \, d\xi \, dx$$

where f is a piecewise continuous function. $Q[f]$ is bounded if f is normalized, since Schwarz's inequality implies

$$(Q[f])^2 \leq \int_a^b f^2(x) \, dx \int_a^b \left[\int_a^b K(\xi;x) f(\xi) \, d\xi \right]^2 dx$$

$$\leq \int_a^b f^2(x) \, dx \int_a^b f^2(\xi) \, d\xi \int_a^b \int_a^b K^2(\xi;x) \, d\xi \, dx$$

$$\leq \int_a^b \int_a^b K^2(\xi;x) \, d\xi \, dx < \infty$$

Furthermore, $Q[f]$ cannot be zero for all f unless $K \equiv 0$, for then

$$Q[f,g] = \tfrac{1}{2}\{Q[f+g] - Q[f] - Q[g]\}$$

would also be zero for all f and g. Then letting

$$g(x) = \int_a^b K(\xi;x) f(\xi) \, d\xi$$

we have
$$Q[f,g] = \int_a^b \left[\int_a^b K(\xi;x) f(\xi) \, d\xi \right]^2 dx = 0$$

This implies that

$$\int_a^b K(\xi;x)f(\xi)\,d\xi = 0$$

for arbitrary f. Let $f(\xi) = K(\xi;x)$. Then

$$\int_a^b K^2(\xi;x)\,d\xi = 0$$

which implies that $K \equiv 0$.

Thus $Q[f]$ is greater than zero or less than zero for some f. We also know that it is bounded. Let us assume that $Q[f] > 0$ for some f, and furthermore that it takes on a maximum value $1/\lambda_1$ for some function ψ_1.† Letting $f = \psi_1 + \epsilon\eta$, we have

$$Q[\psi_1 + \epsilon\eta] \leq \frac{1}{\lambda_1} N[\psi_1 + \epsilon\eta]$$

for an arbitrary function η and an arbitrary constant ϵ. Here $N[f] = \int_a^b f^2\,dx$. A necessary condition for this inequality to hold is

$$Q[\psi_1,\eta] - \frac{1}{\lambda_1} N[\psi_1,\eta] = 0$$

or

$$\int_a^b \left[\int_a^b K(\xi;x)\psi_1(\xi)\,d\xi - \frac{1}{\lambda_1}\psi_1(x)\right]\eta(x)\,dx = 0$$

But since η is arbitrary,

$$\lambda_1 \int_a^b K(\xi;x)\psi_1(\xi)\,d\xi - \psi_1(x) = 0$$

and this is the integral equation. Therefore, a necessary condition for the maximizing function is that it satisfy the integral equation.

To get the next eigenvalue and eigenfunction we maximize $Q[f]$ subject to $N[f] = 1$ and $N[f,\psi_1] = 0$. This can be accomplished by introducing a new symmetric kernel

$$K_1(\xi;x) = K(\xi;x) - \frac{\psi_1(\xi)\psi_1(x)}{\lambda_1}$$

and a new functional

$$Q_1[f] = \int_a^b \int_a^b K_1(\xi;x)f(\xi)f(x)\,d\xi\,dx$$

We maximize $Q_1[f]$ subject to $N[f] = 1$. If the maximum is $1/\lambda_2$ and the

† The existence of ψ_1 is guaranteed by the second theorem of Sec. 2.8, which also tells us that $1/\lambda_1 = \|A\|$.

maximizing function is ψ_2, then

$$\psi_2(x) = \lambda_2 \int_a^b K_1(\xi;x)\psi_2(\xi)\, d\xi$$

$$= \lambda_2 \int_a^b K(\xi;x)\psi_2(\xi)\, d\xi - \frac{\lambda_2}{\lambda_1} \int_a^b \psi_1(x)\psi_1(\xi)\psi_2(\xi)\, d\xi$$

and

$$\int_a^b \psi_1(x)\psi_2(x)\, dx = \lambda_2 \int_a^b \psi_2(\xi)\left[\int_a^b K(\xi;x)\psi_1(x)\, dx\right] d\xi$$

$$- \frac{\lambda_2}{\lambda_1}\int_a^b \psi_1(\xi)\psi_2(\xi)\, d\xi \int_a^b \psi_1^2(x)\, dx$$

$$= \frac{\lambda_2}{\lambda_1}\int_a^b \psi_1(\xi)\psi_2(\xi)\, d\xi - \frac{\lambda_2}{\lambda_1}\int_a^b \psi_1(\xi)\psi_2(\xi)\, d\xi$$

$$= 0$$

Therefore, the integral equation is in reality

$$\psi_2(x) = \lambda_2 \int_a^b K(\xi;x)\psi_2(\xi)\, d\xi$$

Also

$$\frac{1}{\lambda_2} = Q_1[\psi_2] = \int_a^b \int_a^b K(\xi;x)\psi_2(\xi)\psi_2(x)\, d\xi\, dx - \frac{1}{\lambda_1}\left[\int_a^b \psi_1(x)\psi_2(x)\, dx\right]^2$$

$$= \int_a^b \int_a^b K(\xi;x)\psi_2(\xi)\psi_2(x)\, d\xi\, dx = Q[\psi_2] \le Q[\psi_1] = \frac{1}{\lambda_1}$$

Therefore, $\lambda_1 \le \lambda_2$. We continue in this way generating a sequence of eigen-functions and eigenvalues

$$\lambda_1 \le \lambda_2 \le \lambda_3 \le \cdots$$

keeping in mind that we can get only a finite number of solutions corresponding to the same eigenvalue. This process will continue indefinitely, unless at some stage

$$Q_n[f] = \int_a^b \int_a^b \left[K(\xi;x) - \sum_{i=1}^n \frac{\psi_i(\xi)\psi_i(x)}{\lambda_i}\right] f(\xi)f(x)\, d\xi\, dx$$

is zero for all f. But this would imply that

$$K(\xi;x) = \sum_{i=1}^n \frac{\psi_i(\xi)\psi_i(x)}{\lambda_i}$$

and the kernel would be degenerate.

It may happen that $Q[f]$ is negative for some f. We then would minimize $Q[f]$ subject to $N[f] = 1$. If the minimum is $1/\lambda_{-1}$, and the minimizing function is ψ_{-1}, then

$$Q[\psi_{-1} + \epsilon\eta] \ge \frac{1}{\lambda_{-1}} N[\psi_{-1} + \epsilon\eta]$$

for arbitrary ϵ and η. A necessary condition for ψ_{-1} is

$$Q[\psi_{-1},\eta] - \frac{1}{\lambda_{-1}} N[\psi_{-1},\eta] = 0$$

and we have the integral equation

$$\psi_{-1}(x) = \lambda_{-1} \int_a^b K(\xi;x)\psi_{-1}(\xi)\, d\xi$$

ψ_{-1} is orthogonal to every solution with a positive eigenvalue, since λ_{-1} is negative. We next minimize

$$Q_{-1}[f] = \int_a^b \int_a^b \left[K(\xi;x) - \frac{\psi_{-1}(\xi)\psi_{-1}(x)}{\lambda_{-1}} \right] f(\xi)f(x)\, d\xi\, dx$$

subject to $N[f] = 1$. If the minimum is $1/\lambda_{-2}$ and the minimizing function is ψ_{-2}, then we arrive at the integral equation

$$\psi_{-2}(x) = \lambda_{-2} \int_a^b K(\xi;x)\psi_{-2}(\xi)\, d\xi$$

with $\int_a^b \psi_{-1}(x)\psi_{-2}(x)\, dx = 0$ and $\lambda_{-2} \leq \lambda_{-1}$. Continuing, we generate a sequence of eigenvalues

$$\lambda_{-1} \geq \lambda_{-2} \geq \lambda_{-3} \geq \cdots$$

Let us renumber the eigenvalues and eigenfunctions according to increasing absolute value of the eigenvalue; that is,

$$|\lambda_1| \leq |\lambda_2| \leq |\lambda_3| \leq \cdots$$

We then have the following expansion theorem. Every continuous function $g(x)$ which is an integral transform of $K(\xi;x)$, that is,

$$g(x) = \int_a^b K(\xi;x)h(\xi)\, d\xi$$

where h is piecewise continuous, can be expanded in a uniformly convergent series in the eigenfunctions of the integral equation

$$\psi(x) = \lambda \int_a^b K(\xi;x)\psi(\xi)\, d\xi$$

that is,

$$g(x) = \sum_{i=1}^{\infty} c_i \psi_i(x)$$

where

$$c_i = \int_a^b g(x)\psi_i(x)\, dx$$

We prove this as follows.

$$c_i = \int_a^b g(x)\psi_i(x)\,dx = \int_a^b \int_a^b K(\xi;x)h(\xi)\psi_i(x)\,d\xi\,dx$$

$$= \frac{b_i}{\lambda_i} \quad \text{(no summation on } i\text{)}$$

where $b_i = \int_a^b h(x)\psi_i(x)\,dx$. By Bessel's inequality, we have

$$\sum_{i=1}^\infty b_i^2 \le \int_a^b h^2(x)\,dx$$

$$\sum_{i=1}^\infty \frac{\psi_i^2(x)}{\lambda_i^2} \le \int_a^b K^2(\xi;x)\,d\xi \le M^2(b-a)$$

where M is the maximum of $|K(\xi;x)|$ in the square $a \le \xi \le b$, $a \le x \le b$. Now consider the series

$$\sum_{i=1}^\infty c_i\psi_i(x) = \sum_{i=1}^\infty b_i\frac{\psi_i(x)}{\lambda_i}$$

The remainder after n terms in this series is

$$R_n = \frac{b_{n+1}\psi_{n+1}(x)}{\lambda_{n+1}} + \frac{b_{n+2}\psi_{n+1}(x)}{\lambda_{n+2}} + \cdots$$

By Schwarz's inequality we have

$$R_n^2 \le (b_{n+1}^2 + b_{n+2}^2 + \cdots)\left[\frac{\psi_{n+1}^2(x)}{\lambda_{n+1}^2} + \frac{\psi_{n+2}^2(x)}{\lambda_{n+2}^2} + \cdots\right]$$

Since $\sum_{i=1}^\infty b_i^2$ converges, $\sum_{i=n+1}^\infty b_i^2$ can be made arbitrarily small by making n sufficiently large. Also $\sum_{i=n+1}^\infty \psi_i^2(x)/\lambda_i^2$ is bounded uniformly in x. Therefore, $R_n \to 0$ as $n \to \infty$, and the convergence is uniform in x. Therefore, the series converges to a continuous function $\gamma(x)$. It remains to show that $\gamma(x)$ and $g(x)$ are the same. Let

$$K_n(\xi;x) = K(\xi;x) - \sum_{i=1}^n \frac{\psi_i(\xi)\psi_i(x)}{\lambda_i}$$

Then

$$\int_a^b K_n(\xi;x)h(\xi)\,d\xi = g(x) - \gamma_n(x)$$

where

$$\gamma_n(x) = \sum_{i=1}^n \frac{b_i\psi_i(x)}{\lambda_i}$$

and we know that

$$\lim_{n\to\infty} \gamma_n(x) = \gamma(x)$$

Next we have

$$\int_a^b [g(x) - \gamma_n(x)]\eta(x)\,dx = \int_a^b \int_a^b K_n(\xi;x)h(\xi)\eta(x)\,dx$$
$$= Q_n[h,\eta]$$

where $\eta(x)$ is an arbitrary continuous function. From the variational problem we have

$$|Q_n[h]| \le \frac{1}{|\lambda_{n+1}|}\,N[h] \to 0 \qquad \text{as } n \to \infty$$

$$|Q_n[\eta]| \le \frac{1}{|\lambda_{n+1}|}\,N[\eta] \to 0 \qquad \text{as } n \to \infty$$

$$|Q_n[h+\eta]| \le \frac{1}{|\lambda_{n+1}|}\,N[h+\eta] \to 0 \qquad \text{as } n \to \infty$$

Therefore, $|Q_n[h,\eta]| \le \tfrac{1}{2}\{|Q_n[h+\eta]| + |Q_n[h]| + |Q_n[\eta]|\}$

and $Q_n[h,\eta] = \int_a^b [g(x) - \gamma_n(x)]\eta(x)\,dx \to 0 \qquad \text{as } n \to \infty$

Then, by the uniform convergence,

$$\lim_{n \to \infty} \int_a^b [g(x) - \gamma_n(x)]\eta(x)\,dx = \int_a^b \left[g(x) - \lim_{n \to \infty} \gamma_n(x)\right]\eta(x)\,dx$$
$$= \int_a^b [g(x) - \gamma(x)]\eta(x)\,dx = 0$$

But $\eta(x)$ is arbitrary, implying that $g(x) \equiv \gamma(x)$. This completes the proof of the expansion theorem.

Next we let $h(\xi) = K(\xi;y)$. Then

$$\bar{K}(x;y) = \int_a^b K(\xi;x)K(\xi;y)\,d\xi$$

is an integral transform. Therefore, K has an expansion

$$\bar{K}(x;y) = \sum_{i=1}^{\infty} \frac{\psi_i(x)}{\lambda_i} \int_a^b K(\xi;y)\psi_i(\xi)\,d\xi$$
$$= \sum_{i=1}^{\infty} \frac{\psi_i(x)\psi_i(y)}{\lambda_i^2}$$

which is uniformly convergent in x. Also

$$\bar{K}(x;x) = \sum_{i=1}^{\infty} \frac{\psi_i^2(x)}{\lambda_i^2} = \int_a^b K^2(\xi;x)\,d\xi$$

Then $\displaystyle\int_a^b \left[K(\xi;x) - \sum_{i=1}^n \frac{\psi_i(\xi)\psi_i(x)}{\lambda_i} \right]^2 d\xi$

$$= \bar{K}(x;x) + \sum_{i=1}^n \frac{\psi_i^2(x)}{\lambda_i^2} - 2\sum_{i=1}^n \frac{\psi_i(x)}{\lambda_i} \int_a^b K(\xi;x)\psi_i(\xi)\, d\xi$$

$$= \bar{K}(x;x) - \sum_{i=1}^n \frac{\psi_i^2(x)}{\lambda_i^2}$$

Therefore, $\displaystyle\lim_{n\to\infty} \int_a^b \left[K(\xi;x) - \sum_{i=1}^n \frac{\psi_i(\xi)\psi_i(x)}{\lambda_i} \right]^2 d\xi = 0$

and $\displaystyle\sum_{i=1}^\infty \frac{\psi_i(\xi)\psi_i(x)}{\lambda_i}$

converges in mean to the kernel $K(\xi;x)$.

If the nonhomogeneous Fredholm equation has a solution, then

$$g(\xi) = \psi(\xi) - F(\xi) = \lambda \int_a^b K(\xi;x)\psi(x)\, dx$$

is an integral transform, and hence

$$g(x) = \sum_{i=1}^\infty c_i \psi_i(x)$$

where
$$c_i = \int_a^b g(x)\psi_i(x)\, dx$$

$$= \frac{\lambda}{\lambda_i - \lambda} \int_a^b F(x)\psi_i(x)\, dx$$

converges uniformly. Hence,

$$\psi(x) = F(x) + \sum_{i=1}^\infty \frac{\lambda \psi_i(x)}{\lambda_i - \lambda} \int_a^b F(x)\psi_i(x)\, dx$$

Alternatively, consider the series $\displaystyle\sum_{i=1}^\infty [\lambda\gamma_i/(\lambda_i - \lambda)]\psi_i(x)$, where $\gamma_i = \displaystyle\int_a^b F(x)\psi_i(x)\, dx$. Let R_n be the remainder after n terms of the series. Then by Schwarz's inequality

$$R_n^2 \le \sum_{i=n+1}^\infty \left[\frac{\lambda\lambda_i\gamma_i}{\lambda_i - \lambda} \right]^2 \sum_{i=n+1}^\infty \frac{[\psi_i(x)]^2}{\lambda_i^2}$$

For n sufficiently large $[\lambda\lambda_i/(\lambda_i - \lambda)]^2 < 4\lambda^2$ for all $i \ge n + 1$. Thus,

$$\sum_{i=n+1}^\infty \left[\frac{\lambda\lambda_i\gamma_i}{\lambda_i - \lambda} \right]^2 < 4\lambda^2 \sum_{i=n+1}^\infty \gamma_i^2 \to 0$$

as $n \to \infty$. $\sum\limits_{i=n+1}^{\infty} [\psi_i(x)]^2/\lambda_i^2$ is uniformly bounded. Therefore, $R_n \to 0$ as $n \to \infty$ uniformly in x, and hence

$$\sum_{i=1}^{\infty} \frac{\lambda \gamma_i}{\lambda_i - \lambda} \psi_i(x)$$

converges uniformly. Substituting in the integral equation, we have

$$\sum_{i=1}^{\infty} \frac{\lambda \gamma_i}{\lambda_i - \lambda} \psi_i(x) = \lambda \int_a^b K(\xi;x)F(\xi)\,d\xi + \lambda^2 \sum_{i=1}^{\infty} \frac{\gamma_i}{\lambda_i - \lambda} \int_a^b K(\xi;x)\psi_i(\xi)\,d\xi$$

$$= \lambda \int_a^b K(\xi;x)F(\xi)\,d\xi + \sum_{i=1}^{\infty} \frac{\lambda \gamma_i}{\lambda_i - \lambda} \psi_i(x) - \lambda \sum_{i=1}^{\infty} \frac{\gamma_i}{\lambda_i} \psi_i(x)$$

Now if $F(x)$ is a piecewise continuous function, then

$$\int_a^b K(\xi;x)F(\xi)\,d\xi = \sum_{i=1}^{\infty} \frac{\gamma_i}{\lambda_i} \psi_i(x)$$

where the convergence is uniform. Thus we see that the integral equation is satisfied by

$$\psi(x) = F(x) + \sum_{i=1}^{\infty} \frac{\lambda \gamma_i}{\lambda_i - \lambda} \psi_i(x)$$

We have thus established both existence and uniqueness of the solution of the nonhomogeneous integral equation in the case where λ is not an eigenvalue of the homogeneous equation. We have already seen that no solution can exist if $\lambda = \lambda_i$ for some i and $\int_a^b F(x)\psi_i(x)\,dx \neq 0$. However, if

$$\gamma_i = \int_a^b F(x)\psi_i(x)\,dx = 0$$

then the troublesome term in the series does not appear, and the above argument still holds, showing the existence of a solution. In this case, as we have seen, the solution is not unique, for any multiple of $\psi_i(x)$ can be added to the solution.

Recall that in solving boundary-value problems in Chap. 4, we had to obtain series expansions for the initial value of the solution in terms of eigenfunctions of the appropriate linear differential operator. For example, if we were to solve

$$\frac{\partial}{\partial x}\left(p\,\frac{\partial \phi}{\partial x}\right) - q\phi = \rho(x)\,\frac{\partial^2 \phi}{\partial t^2} \qquad a \leq x \leq b \qquad 0 \leq t$$

$$\phi(a,t) = \phi(b,t) = 0$$

$$\phi(x,0) = g(x)$$

$$\phi_t(x,0) = h(x)$$

we would attempt to find a solution in the form

$$\phi(x,t) = \sum_{i=1}^{\infty} (A_i \cos \sqrt{\lambda_i} t + B_i \sin \sqrt{\lambda_i} t) y_i(x)$$

where
$$(py_i')' - qy_i + \lambda_i \rho y_i = 0 \qquad a \le x \le b$$
$$y_i(a) = y_i(b) = 0$$

For the method to be valid it must be possible to obtain series expansions for the initial values; that is,

$$g(x) = \sum_{i=1}^{\infty} A_i y_i(x)$$

$$h(x) = \sum_{i=1}^{\infty} \sqrt{\lambda_i} B_i y_i(x)$$

Using the results of the present section, we can find conditions on $g(x)$ and $h(x)$ which will guarantee that such expansions can be found.

Let $g(x)$ be an arbitrary continuous function with a continuous first derivative and a piecewise continuous second derivative, which satisfies the same homogeneous boundary conditions as $y(x)$ in the Sturm-Liouville problem

$$(py')' - qy + \lambda \rho y = 0$$

Then
$$(pg')' - qg = -f(x)$$

where $f(x)$ is a piecewise continuous function. Let $G(\xi;x)$ be the Green's function satisfying

$$(pG')' - qG = -\delta(x - \xi)$$

Then
$$g(x) = -\int_a^b gL[G]\, d\xi = -\int_a^b GL[g]\, d\xi = \int_a^b G(\xi;x)f(\xi)\, d\xi$$

$$\sqrt{\rho(x)}g(x) = \int_a^b \sqrt{\rho(x)\rho(\xi)}\, G(\xi;x)\frac{f(\xi)}{\sqrt{\rho(\xi)}}\, d\xi$$

$$= \int_a^b K(\xi;x)\frac{f(\xi)}{\sqrt{\rho(\xi)}}\, d\xi$$

Therefore, $\sqrt{\rho(x)}g(x)$ is an integral transform of a piecewise continuous function,[1] and as such has a uniformly convergent series expansion

$$\sqrt{\rho(x)}g(x) = \sum_{i=1}^{\infty} c_i \psi_i$$

in terms of the normalized eigenfunctions of the integral equation

$$\psi(x) = \lambda \int_a^b K(x;\xi)\psi(\xi)\, d\xi$$

---

[1] Here we are assuming that $\rho(x) > 0$.

with

$$c_i = \int_a^b \sqrt{\rho(x)} g(x) \psi_i(x) \, dx$$

$$= \int_a^b \rho(x) g(x) y_i(x) \, dx$$

since $\psi_i(x) = \sqrt{\rho(x)} y_i(x)$.

Finally, we shall generalize the entire discussion to the corresponding problems in two and three dimensions. The corresponding nonhomogeneous integral equations associated with Helmholtz's equation are, in two dimensions,

$$\psi(\xi,\eta) = F(\xi,\eta) + \lambda \iint_R K(\xi,\eta;x,y) \psi(x,y) \, dx \, dy$$

where

$$K(\xi,\eta;x,y) = -\frac{1}{2\pi} \log \sqrt{(x-\xi)^2 + (y-\eta)^2} + H(\xi,\eta;x,y)$$

where H is a harmonic function with appropriate boundary values so that $dK/dn + \alpha K = 0$ on the boundary of R; and, in three dimensions,

$$\psi(\xi,\eta,\zeta) = F(\xi,\eta,\zeta) + \lambda \iiint_V K(\xi,\eta,\zeta;x,y,z) \psi(x,y,z) \, dx \, dy \, dz$$

where

$$K(\xi,\eta,\zeta;x,y,z) = \frac{1}{4\pi\sqrt{(x-\xi)^2 + (y-\eta)^2 + (z-\zeta)^2}} + H(\xi,\eta,\zeta;x,y,z)$$

and H is a harmonic function with appropriate boundary values so that $dK/dn + \alpha K = 0$ on the boundary of V.

In these cases, the kernel is not continuous, but the integrals

$$\iint_R [K(\xi,\eta;x,y)]^2 \, dx \, dy$$

$$\iiint_V [K(\xi,\eta,\zeta;x,y,z)]^2 \, dx \, dy \, dz$$

exist and are continuous functions of the parameters in each case. The demonstration of this will be left for the exercises. The reader should reread the above development, making appropriate changes where necessary to obtain the same results in these more general cases.

In these cases, we again have expansion theorems for the initial conditions in the initial value problem. For example, in the two-dimensional case, let $g(x,y)$ be a function defined in R and on C, the boundary of R, where it is continuous, with continuous first partial derivatives and piecewise continuous second partial derivatives, and satisfying the boundary condition $dg/dn + \alpha g = 0$ on C. Then $\nabla^2 g = -f(x,y)$ in R, where f is piecewise continuous and

$$g(x,y) = \iint\limits_{R} K(x,y;\xi,\eta)f(\xi,\eta)\,d\xi\,d\eta$$

is the integral transform of a piecewise continuous function. Hence,

$$g(x,y) = \sum_{i=1}^{\infty} c_i \psi_i(x,y)$$

where

$$c_i = \iint\limits_{R} g(x,y)\psi_i(x,y)\,dx\,dy$$

and the convergence is uniform. A similar theorem holds in the three-dimensional case.

Exercises 6.2

1. Let $K(\xi;x)$ be a continuous complex-valued "hermitian kernel" defined on the square $a \leq \xi \leq b$, $a \leq x \leq b$ such that $K(\xi;x) = K^*(x;\xi)$. Prove that the eigenvalues of the homogeneous integral equation

$$\psi(\xi) = \lambda \int_a^b K(\xi;x)\psi(x)\,dx$$

are real. Also show that eigenfunctions corresponding to different eigenvalues are orthogonal in the sense

$$\int_a^b \psi_i(x)\psi_j^*(x)\,dx = 0$$

if $\lambda_i \neq \lambda_j$. Note that the theory of Fredholm integral equations with hermitian kernels runs parallel to the theory for real symmetric kernels, just as the theory of hermitian forms parallels that for quadratic forms.[1] How then would one expect to solve the nonhomogeneous equation

$$\psi(\xi) = F(\xi) + \lambda \int_a^b K(\xi;x)\psi(x)\,dx$$

2. Describe a Rayleigh-Ritz procedure for obtaining sequences of upper bounds for the eigenvalues of the homogeneous Fredholm equation with real, symmetric, continuous kernel. In the notation of Sec. 3.9, prove that the nth approximation to the kth eigenvalue converges to λ_k as n approaches infinity, provided the Rayleigh-Ritz procedure is based on a complete set of functions. HINT: If not, $|\mu_k^{(n)}| > |\lambda_k| + \delta$ for some fixed δ and all n, and this leads to a contradiction to the fact that there exists a sequence of functions converging in mean to ψ_k.

3. Solve the integral equation $(\psi)\xi = \xi + \int_0^1 (1 + \xi x)\psi(x)\,dx$.

[1] See Sec. 1.8.

4. Solve the integral equation

$$\psi(\xi) = \xi + \int_0^\xi \left(\frac{x}{\xi} - x\xi\right)\psi(x)\,dx + \int_\xi^1 \left(\frac{\xi}{x} - x\xi\right)\psi(x)\,dx.$$

HINT: Find the boundary-value problem equivalent to the integral equation.

5. Find a series representation of the kernel

$$K(\xi;x) = x \qquad 0 \le x \le \xi$$
$$K(\xi;x) = \xi \qquad \xi \le x \le 1$$

Show that the series converges uniformly to the kernel.

6. Show that $\int_0^\rho (\log r)^2 r\,dr$, $\int_0^\rho r \log r\,dr$ exist. Use these results to show that

$$\iint\limits_R [K(\xi,\eta;x,y)]^2\,dx\,dy$$

exists and is a continuous function of ξ and η in R.

*7. Show that the homogeneous Fredholm equation can have no nontrivial continuous solution for $|\lambda| < [M(b - a)]^{-1}$, where $|K(\xi;x)| \le M$.

6.3   Fredholm Theory

In the last section we obtained solutions of the nonhomogeneous Fredholm equation

$$\psi(\xi) = F(\xi) + \lambda \int_a^b K(\xi;x)\psi(x)\,dx$$

under the assumption that the kernel $K(\xi;x)$ was symmetric. In this section, we shall obtain series solutions of the equation, assuming only that the kernel is continuous in the square $a \le \xi \le b$, $a \le x \le b$. The kernel need not be symmetric.

To begin with, let $\psi_0(x)$ be any piecewise continuous function. Then

$$\lambda \int_a^b K(\xi;x)\psi_0(x)\,dx$$

is a continuous function of ξ, and if $F(\xi)$ is piecewise continuous,

$$\psi_1(\xi) = F(\xi) + \lambda \int_a^b K(\xi;t_1)\psi_0(t_1)\,dt_1$$

is a piecewise continuous first approximation to the solution of the integral equation. A second approximation is given by

$$\psi_2(\xi) = F(\xi) + \lambda \int_a^b K(\xi;t_1)\psi_1(t_1)\,dt_1$$

$$= F(\xi) + \lambda \int_a^b K(\xi;t_1)F(t_1)\,dt_1 + \lambda^2 \int_a^b \int_a^b K(\xi;t_1)K(t_1;t_2)\psi_0(t_2)\,dt_1\,dt_2$$

Continuing by successive substitution we obtain an nth approximation as follows:

$$\psi_n(\xi) = F(\xi) + \lambda \int_a^b K(\xi;t_1)F(t_1) \, dt_1 + \lambda^2 \int_a^b \int_a^b K(\xi;t_1)K(t_1;t_2)F(t_2) \, dt_1 \, dt_2$$

$$+ \cdots + \lambda^n \int_a^b \cdots \int_a^b K(\xi;t_1)K(t_1;t_2) \cdots K(t_{n-1};t_n)\psi_0(t_n) \, dt_1 \cdots dt_n$$

Without loss of generality, we can assume that $|\psi_0(x)| \leq 1$; hence

$$\left| \lambda^n \int_a^b \cdots \int_a^b K(\xi;t_1)K(t_1;t_2) \cdots K(t_{n-1};t_n)\psi_0(t_n) \, dt_1 \cdots dt_n \right|$$

$$\leq |\lambda|^n M^n (b-a)^n$$

where $|K(\xi;x)| \leq M$. Therefore, if $|\lambda| < [M(b-a)]^{-1}$,

$$\lim_{n \to \infty} \psi_n(\xi) = F(\xi) + \lambda \int_a^b K(\xi;t_1)F(t_1) \, dt_1$$

$$+ \lambda^2 \int_a^b \int_a^b K(\xi;t_1)K(t_1;t_2)F(t_2) \, dt_1 \, dt_2 + \cdots$$

$$+ \lambda^n \int_a^b \cdots \int_a^b K(\xi;t_1)K(t_1;t_2) \cdots K(t_{n-1};t_n)F(t_n) \, dt_1 \cdots dt_n + \cdots$$

exists, and the convergence of the series is uniform in ξ. It remains to show that this is actually a solution of the integral equation. Multiplying the right-hand side of the last equation by $\lambda K(x;\xi)$ and integrating term by term, using the uniform convergence, we obtain the original series back again except for the term $F(x)$. Hence, by adding $F(x)$, we obtain again

$$\psi(x) = \lim_{n \to \infty} \psi_n(x)$$

which we have now shown is a solution of the integral equation provided $|\lambda| < [M(b-a)]^{-1}$.

Notice that the result is independent of the choice of ψ_0. This fact allows us to prove the uniqueness of the solution in the present case. If a second solution $\bar{\psi}$ exists, let $\psi_0 = \bar{\psi}$. But then $\psi_n = \bar{\psi}$ for all n, and $\psi \equiv \bar{\psi}$. This of course implies that there can be no nontrivial solution of the homogeneous integral equation for $|\lambda| < [M(b-a)]^{-1}$.

We notice that we can write the solution of the integral equation as follows:

$$\psi(\xi) = F(\xi) + \lambda \int_a^b k(\xi,x,\lambda)F(x) \, dx$$

where $k(\xi,x,\lambda) = K(\xi;x) + \lambda \int_a^b K(\xi;t_1)K(t_1;x) \, dt_1 + \cdots$

$$+ \lambda^n \int_a^b \cdots \int_a^b K(\xi;t_1)K(t_1;t_2) \cdots K(t_n;x) \, dt_1 \cdots dt_n + \cdots$$

and the series converges uniformly if $|\lambda| < [M(b-a)]^{-1}$. $k(\xi,x,\lambda)$ is called the **reciprocal kernel.** We see from the definition of the reciprocal kernel that

$$k(\xi,x,\lambda) = K(\xi;x) + \lambda \int_a^b K(\xi;t)k(t,x,\lambda)\,dt$$

$$k(\xi,x,\lambda) = K(\xi;x) + \lambda \int_a^b k(\xi,t,\lambda)K(t;x)\,dt$$

Conversely, if we find a reciprocal kernel which satisfies this integral equation, we can construct a solution of the nonhomogeneous Fredholm equation as

$$\psi(\xi) = F(\xi) + \lambda \int_a^b k(\xi,x,\lambda)F(x)\,dx$$

for then
$$\lambda \int_a^b K(\xi;x)\psi(x)\,dx = \lambda \int_a^b F(x)K(\xi;x)\,dx$$
$$+ \lambda^2 \int_a^b \int_a^b k(x,t,\lambda)F(t)K(\xi;x)\,dt\,dx$$
$$= \lambda \int_a^b K(\xi;x)F(x)\,dx$$
$$+ \lambda \int_a^b [k(\xi,t,\lambda) - K(\xi;t)]F(t)\,dt$$
$$= \lambda \int_a^b k(\xi,t,\lambda)F(t)\,dt$$
$$= \psi(\xi) - F(\xi)$$

The restriction $|\lambda| < [M(b-a)]^{-1}$ on the above series solution of the integral equation is much too severe. Fredholm showed that it is possible to obtain series solutions for almost all values of λ. To illustrate his approach let us consider a special case where we can get an explicit solution. We shall consider the degenerate kernel.

$$K(\xi;x) = \alpha_1(\xi)\beta_1(x) + \alpha_2(\xi)\beta_2(x) + \alpha_3(\xi)\beta_3(x)$$

then
$$\psi(\xi) = F(\xi) + \lambda \int_a^b \alpha_i(\xi)\beta_i(x)\psi(x)\,dx$$
$$= F(\xi) + \lambda\alpha_i y_i \qquad i = 1, 2, 3$$

where
$$y_i = \int_a^b \beta_i(x)\psi(x)\,dx$$

Therefore,
$$y_j = \int_a^b \beta_j(\xi)\psi(\xi)\,d\xi = c_j + \lambda a_{ji} y_i$$

where
$$c_j = \int_a^b \beta_j(\xi)F(\xi)\,d\xi \qquad \text{and} \qquad a_{ji} = \int_a^b \beta_j(\xi)\alpha_i(\xi)\,d\xi$$

We thus have a system of nonhomogeneous linear equations which can be written in matrix notation,

$$Y = C + \lambda A Y$$

$$(I - \lambda A)Y = C$$

If $|I - \lambda A| \neq 0$, then by Cramer's rule we have

$$y_i = \frac{D_i}{|I - \lambda A|}$$

where D_i is the determinant of the matrix formed from $I - \lambda A$ by replacing the ith column by the vector C. Then

$$\psi(\xi) = F(\xi) + \frac{\lambda \alpha_i D_i}{|I - \lambda A|}$$

The denominator of the last expression can be expanded as follows:

$$|I - \lambda A| = \begin{vmatrix} 1 - \lambda a_{11} & -\lambda a_{12} & -\lambda a_{13} \\ -\lambda a_{21} & 1 - \lambda a_{22} & -\lambda a_{23} \\ -\lambda a_{31} & -\lambda a_{32} & 1 - \lambda a_{33} \end{vmatrix}$$

$$= 1 - \lambda(a_{11} + a_{22} + a_{33})$$

$$+ \lambda^2(a_{11}a_{22} - a_{11}a_{33} + a_{22}a_{33} - a_{12}a_{21} - a_{13}a_{31} - a_{23}a_{32})$$

$$- \lambda^3(a_{11}a_{22}a_{33} + a_{12}a_{23}a_{31} + a_{13}a_{21}a_{32}$$

$$- a_{11}a_{23}a_{32} - a_{13}a_{22}a_{31} - a_{12}a_{21}a_{33})$$

$$= 1 - \lambda \int_a^b K(t;t)\, dt + \frac{\lambda^2}{2!} \int_a^b \int_a^b \begin{vmatrix} K(t_1;t_1) & K(t_1;t_2) \\ K(t_2;t_1) & K(t_2;t_2) \end{vmatrix} dt_1\, dt_2$$

$$- \frac{\lambda^3}{3!} \int_a^b \int_a^b \int_a^b \begin{vmatrix} K(t_1;t_1) & K(t_1;t_2) & K(t_1;t_3) \\ K(t_2;t_1) & K(t_2;t_2) & K(t_2;t_3) \\ K(t_3;t_1) & K(t_3;t_2) & K(t_3;t_3) \end{vmatrix} dt_1\, dt_2\, dt_3$$

We see that these are the first four terms of a series

$$D(\lambda) = \sum_{k=0}^\infty \frac{(-1)^k \lambda^k}{k!} \gamma_k$$

where $\quad \gamma_k = \int_a^b \cdots \int_a^b \begin{vmatrix} K(t_1;t_1) & K(t_1;t_2) & \cdots & K(t_1;t_k) \\ K(t_2;t_1) & K(t_2;t_2) & \cdots & K(t_2;t_k) \\ \cdots\cdots\cdots\cdots\cdots\cdots\cdots\cdots\cdots \\ K(t_k;t_1) & K(t_k;t_2) & \cdots & K(t_k;t_k) \end{vmatrix} dt_1 \cdots dt_k$

We shall show that this series converges for all values of λ for any continuous kernel. However, before returning to the general case, let us consider the numerator in our above explicit solution. Expanding the determinants, we have

$$\lambda\alpha_i(\xi)D_i = \lambda\alpha_1(\xi)\begin{vmatrix} c_1 & -\lambda a_{12} & -\lambda a_{13} \\ c_2 & 1-\lambda a_{22} & -\lambda a_{23} \\ c_3 & -\lambda a_{32} & 1-\lambda a_{33} \end{vmatrix}$$

$$+ \lambda\alpha_2(\xi)\begin{vmatrix} 1-\lambda a_{11} & c_1 & -\lambda a_{13} \\ -\lambda a_{21} & c_2 & -\lambda a_{23} \\ -\lambda a_{31} & c_3 & 1-\lambda a_{33} \end{vmatrix}$$

$$+ \lambda\alpha_3(\xi)\begin{vmatrix} 1-\lambda a_{11} & -\lambda a_{12} & c_1 \\ -\lambda a_{21} & 1-\lambda a_{22} & c_2 \\ -\lambda a_{31} & -\lambda a_{32} & c_3 \end{vmatrix}$$

$$= \lambda\int_a^b K(\xi;x)F(x)\,dx - \lambda^2\int_a^b\int_a^b \begin{vmatrix} K(\xi;x) & (\xi;t_1) \\ K(t_1;x) & K(t_1;t_1) \end{vmatrix}F(x)\,dt_1\,dx$$

$$+ \frac{\lambda^3}{2!}\int_a^b\int_a^b\int_a^b \begin{vmatrix} K(\xi;x) & K(\xi;t_1) & K(\xi;t_2) \\ K(t_1;x) & K(t_1;t_1) & K(t_1;t_2) \\ K(t_2;x) & K(t_2;t_1) & K(t_2;t_2) \end{vmatrix}F(x)\,dt_1\,dt_2\,dx$$

We see that we can write our solution in terms of the reciprocal kernel

$$k(\xi,x,\lambda) = \frac{1}{D(\lambda)}\left[K(\xi;x) - \lambda\int_a^b \begin{vmatrix} K(\xi;x) & K(\xi;t_1) \\ K(t_1;x) & K(t_1;t_1) \end{vmatrix}dt_1 \right.$$

$$\left. + \frac{\lambda^2}{2!}\int_a^b\int_a^b \begin{vmatrix} K(\xi;x) & K(\xi;t_1) & K(\xi;t_2) \\ K(t_1;x) & K(t_1;t_1) & K(t_1;t_2) \\ K(t_2;x) & K(t_2;t_1) & K(t_2;t_2) \end{vmatrix}dt_1\,dt_2 \right]$$

if λ is not a zero of $D(\lambda)$. Then

$$\psi(\xi) = F(\xi) + \lambda\int_a^b k(\xi,x,\lambda)F(x)\,dx$$

Taking our cue from this special case, we attempt to find a solution of the nonhomogeneous Fredholm equation in the general case along the same lines. We begin by defining two infinite series which we shall show converge for all values of λ. The first is **Fredholm's determinant**

$$D(\lambda) = \sum_{k=0}^{\infty} \frac{(-1)^k \lambda^k}{k!} \gamma_k$$

where

$$\gamma_k = \int_a^b \cdots \int_a^b \begin{vmatrix} K(t_1;t_1) & K(t_1;t_2) & \cdots & K(t_1;t_k) \\ K(t_2;t_1) & K(t_2;t_2) & \cdots & K(t_2;t_k) \\ \hdotsfor{4} \\ K(t_k;t_1) & K(t_k;t_2) & \cdots & K(t_k;t_k) \end{vmatrix} dt_1 \cdots dt_k$$

$$\gamma_0 = 1$$

The second is **Fredholm's first minor**

$$D_1(\xi,x,\lambda) = \sum_{k=0}^{\infty} \frac{(-1)^k \lambda^k}{k!} \delta_k$$

where

$$\delta_k = \int_a^b \cdots \int_a^b \begin{vmatrix} K(\xi;x) & K(\xi;t_1) & \cdots & K(\xi;t_k) \\ K(t_1;x) & K(t_1;t_1) & \cdots & K(t_1;t_k) \\ \hdotsfor{4} \\ K(t_k;x) & K(t_k;t_1) & \cdots & K(t_k;t_k) \end{vmatrix} dt_1 \cdots dt_k$$

$$\delta_0 = K(\xi;x)$$

To show the convergence of these two series we need **Hadamard's inequality** for determinants: If A is an nth-order determinant with elements a_{ij} satisfying the inequality $|a_{ij}| \le M$ for all i and j, then

$$|A| \le M^n n^{n/2}$$

We prove this as follows. If any row contains all zeros, $A = 0$, and the inequality follows immediately. Otherwise we can write

$$A = a_1 a_2 \cdots a_n B$$

where

$$a_i = \left(\sum_{j=1}^n |a_{ij}|^2 \right)^{\frac{1}{2}}$$

and B is a determinant with normalized rows. Obviously

$$a_i \le M n^{\frac{1}{2}}$$

Hence, it remains to show that $|B| \le 1$. Now

$$B = b_{\alpha j} B_{\alpha j} \qquad j = 1, 2, 3, \ldots, n$$

where $B_{\alpha j}$ is the cofactor of $b_{\alpha j}$. By Schwarz's inequality,

$$|B|^2 \le \sum_{j=1}^n |b_{\alpha j}|^2 \sum_{j=1}^n |B_{\alpha j}|^2$$

Equality holds if and only if $b_{\alpha j} = \mu_\alpha B_{\alpha j}^*$.[†] Now $|B|^2$ is equal to the

† See exercise 3, Sec. 1.6.

determinant of the product of the matrix with elements b_{ij} with its own conjugate transposed matrix. But when b_{ij} is proportional to the conjugate of its cofactor, this product is the identity matrix, and hence the maximum value of $|B|$ is 1. Therefore,

$$|A| \leq a_1 a_2 \cdots a_n \leq M^n n^{n/2}$$

Now consider $|\gamma_k|$ and $|\delta_k|$. By Hadamard's inequality,

$$|\gamma_k| \leq (b-a)^k M^k k^{k/2}$$

$$|\delta_k| \leq (b-a)^k M^{k+1}(k+1)^{(k+1)/2}$$

Then

$$\lim_{k \to \infty} \frac{|\lambda|\,(b-a)M(k+1)^{(k+1)/2}}{(k+1)k^{k/2}} = |\lambda|\,(b-a)M \lim_{k \to \infty} \left(1+\frac{1}{k}\right)^{k/2} \frac{1}{(k+1)^{\frac{1}{2}}} = 0$$

$$\lim_{k \to \infty} \frac{|\lambda|\,(b-a)M(k+2)^{(k+2)/2}}{(k+1)(k+1)^{(k+1)/2}}$$
$$= |\lambda|\,(b-a)M \lim_{k \to \infty} \left(1+\frac{1}{k+1}\right)^{(k+1)/2}\left[\frac{k+2}{(k+1)^2}\right]^{\frac{1}{2}} = 0$$

Therefore, both series converge absolutely for all values of λ, and the series for $D_1(\xi,x,\lambda)$ converges uniformly in ξ and x.

Finally, we define

$$k(\xi,x,\lambda) = \frac{D_1(\xi,x,\lambda)}{D(\lambda)}$$

provided $D(\lambda) \neq 0$, and show that this is a reciprocal kernel and hence that

$$\psi(\xi) = F(\xi) + \lambda \int_a^b \frac{D_1(\xi,x,\lambda)}{D(\lambda)}\,F(x)\,dx$$

is the solution of the integral equation. To this end we expand δ_k, using the elements of the first column:

$$\delta_k(\xi,x) = \int_a^b \cdots \int_a^b K(\xi;x) \begin{vmatrix} K(t_1;t_1) & K(t_1;t_2) & \cdots & K(t_1;t_k) \\ K(t_2;t_1) & K(t_2;t_2) & \cdots & K(t_2;t_k) \\ \cdots\cdots\cdots\cdots\cdots\cdots\cdots\cdots \\ K(t_k;t_1) & K(t_k;t_2) & \cdots & K(t_k;t_k) \end{vmatrix} dt_1 \cdots dt_k$$

$$+ \sum_{i=1}^k (-1)^i \int_a^b \cdots \int_a^b K(t_i;x) \begin{vmatrix} K(\xi;t_1) & K(\xi;t_2) & \cdots & K(\xi;t_k) \\ \cdots\cdots\cdots\cdots\cdots\cdots\cdots\cdots \\ K(t_{i-1};t_1) & K(t_{i-1};t_2) & \cdots & K(t_{i-1};t_k) \\ K(t_{i+1};t_1) & K(t_{i+1};t_2) & \cdots & K(t_{i+1};t_k) \\ \cdots\cdots\cdots\cdots\cdots\cdots\cdots\cdots \\ K(t_k;t_1) & K(t_k;t_2) & \cdots & K(t_k;t_k) \end{vmatrix} dt_1 \cdots dt_k$$

By replacing t_i by t, t_{i+1} by t_i, t_{i+2} by t_{i+1}, etc., and moving the ith column to the first position, we can write

$$\delta_k(\xi;x) = K(\xi;x)\gamma_k + \sum_{i=1}^{k} (-1)^{2i-1} \int_a^b \cdots \int_a^b K(t;x)$$

$$\times \begin{vmatrix} K(\xi;t) & K(\xi;t_1) & \cdots & K(\xi;t_{k-1}) \\ K(t_1;t) & K(t_1;t_1) & \cdots & K(t_1;t_{k-1}) \\ \cdots\cdots\cdots\cdots\cdots\cdots\cdots\cdots\cdots\cdots\cdots \\ K(t_{k-1};t) & K(t_{k-1};t_1) & \cdots & K(t_{k-1};t_{k-1}) \end{vmatrix} dt_1 \cdots dt_{k-1}\, dt$$

$$= K(\xi;x)\gamma_k - k \int_a^b K(t;x)\, \delta_{k-1}(\xi,t)\, dt$$

Then

$$D_1(\xi,x,\lambda) = \sum_{k=0}^{\infty} \frac{(-1)^k \lambda^k \delta_k}{k!}$$

$$= K(\xi;x) \sum_{k=0}^{\infty} \frac{(-1)^k \lambda^k \gamma_k}{k!} + \lambda \int_a^b \sum_{k=1}^{\infty} \frac{(-1)^{k-1}\lambda^{k-1}\, \delta_{k-1}(\xi,t)}{(k-1)!} K(t;x)\, dt$$

$$= K(\xi;x)D(\lambda) + \lambda \int_a^b D_1(\xi,t,\lambda)K(t;x)\, dt$$

where we have used the uniform convergence to interchange the summation and the integration. If $D(\lambda) \neq 0$, we have

$$k(\xi,x,\lambda) = \frac{D_1(\xi,x,\lambda)}{D(\lambda)} = K(\xi;x) + \lambda \int_a^b \frac{D_1(\xi,t,\lambda)}{D(\lambda)} K(t;x)\, dt$$

$$= K(\xi;x) + \lambda \int_a^b k(\xi,t,\lambda)K(t;x)\, dt$$

Similarly, if we expand δ_k by the first row rather than the first column, we find

$$\delta_k(\xi,x) = K(\xi;x)\gamma_k - k \int_a^b K(\xi;t)\, \delta_{k-1}(t,x)\, dt$$

and it follows that

$$D_1(\xi,x,\lambda) = K(\xi;x)D(\lambda) + \lambda \int_a^b K(\xi;t)D_1(t,x,\lambda)\, dt$$

and

$$k(\xi,x,\lambda) = K(\xi;x) + \lambda \int_a^b K(\xi;t)k(t,x,\lambda)\, dt$$

Hence, we have a unique solution if $D(\lambda) \neq 0$, given by

$$\psi(\xi) = F(\xi) + \lambda \int_a^b k(\xi,x,\lambda)F(x)\, dx$$

Suppose $D(\lambda_1) = 0$ for some λ_1. We know that $\lambda_1 \neq 0$, since $D(0) = 1$. Then we have

$$D_1(\xi, x_1, \lambda_1) = \lambda_1 \int_a^b K(\xi; t) D_1(t, x_1, \lambda_1)\, dt$$

Thus we see that $D_1(\xi, x_1, \lambda_1)$ is a nontrivial solution of the homogeneous Fredholm equation for some x_1, unless of course it is identically zero for all values of x. We shall show, however, that this cannot be the case if $D'(\lambda_1) \neq 0$. Differentiating $D(\lambda)$, we have

$$-D'(\lambda) = \sum_{k=0}^{\infty} \frac{(-1)^k \lambda^k}{k!}\, \gamma_{k+1}$$

Now

$$\gamma_{k+1} = \int_a^b \cdots \int_a^b \begin{vmatrix} K(t_1; t_1) & K(t_1; t_2) & \cdots & K(t_1; t_{k+1}) \\ K(t_2; t_1) & K(t_2; t_2) & \cdots & K(t_2; t_{k+1}) \\ \hdotsfor{4} \\ K(t_{k+1}; t_1) & K(t_{k+1}; t_2) & \cdots & K(t_{k+1}; t_{k+1}) \end{vmatrix} dt_1 \cdots dt_{k+1}$$

$$= \int_a^b \left\{ \int_a^b \cdots \int_a^b \begin{vmatrix} K(\xi; \xi) & K(\xi; t_1) & \cdots & K(\xi; t_k) \\ K(t_1; \xi) & K(t_1; t_1) & \cdots & K(t_1; t_k) \\ \hdotsfor{4} \\ K(t_k; \xi) & K(t_k; t_1) & \cdots & K(t_k; t_k) \end{vmatrix} dt_1 \cdots dt_k \right\} d\xi$$

$$= \int_a^b \delta_k(\xi; \xi)\, d\xi$$

Hence,

$$-D'(\lambda) = \sum_{k=0}^{\infty} \frac{(-1)^k \lambda^k}{k!} \int_a^b \delta_k(\xi, \xi)\, d\xi$$

$$= \int_a^b \left[\sum_{k=0}^{\infty} \frac{(-1)^k \lambda^k}{k!}\, \delta_k(\xi, \xi) \right] d\xi$$

$$= \int_a^b D_1(\xi, \xi, \lambda)\, d\xi$$

This shows that if $D'(\lambda_1) \neq 0$, then $D_1(\xi, \xi, \lambda_1) \not\equiv 0$, and an x_1 exists such that $D_1(\xi, x_1, \lambda_1) \not\equiv 0$. We now have a solution of the homogeneous Fredholm equation corresponding to the eigenvalue λ_1. We shall show that in this case, that is, $D(\lambda_1) = 0$, $D'(\lambda_1) \neq 0$, any other solution of the homogeneous equation is a multiple of this solution, and hence there is a one-parameter family of nontrivial solutions, and therefore λ_1 is not a degenerate eigenvalue.

To show this we introduce **Fredholm's second minor,**

$$D_2(\xi, \eta, x, y, \lambda) = \sum_{k=0}^{\infty} \frac{(-1)^k \lambda^k}{k!}\, \Gamma_k(\xi, \eta, x, y)$$

where

$$\Gamma_k(\xi,\eta,x,y) = \int_a^b \cdots \int_a^b \begin{vmatrix} K(\xi;\eta) & K(\xi;y) & K(\xi;t_1) & \cdots & K(\xi;t_k) \\ K(x;\eta) & K(x;y) & K(x;t_1) & \cdots & K(x;t_k) \\ K(t_1;\eta) & K(t_1;y) & K(t_1;t_1) & \cdots & K(t_1;t_k) \\ \cdots\cdots\cdots\cdots\cdots\cdots\cdots\cdots\cdots\cdots\cdots \\ K(t_k;\eta) & K(t_k;y) & K(t_k;t_1) & \cdots & K(t_k;t_k) \end{vmatrix} dt_1 \cdots dt_k$$

$$\Gamma_0(\xi,\eta,x,y) = \begin{vmatrix} K(\xi;\eta) & K(\xi;y) \\ K(x;\eta) & K(x;y) \end{vmatrix}$$

By use of Hadamard's inequality it can be shown that the series converges absolutely for all λ and uniformly in ξ, η, x, and y. Expanding $\Gamma_k(\xi,\eta,x,y)$ using the elements in the first column, we have

$$\Gamma_k(\xi,\eta,x,y) = K(\xi;\eta)\delta_k(x,y) - K(x;\eta)\delta_k(\xi,y)$$

$$-k\int_a^b K(t;\eta)\Gamma_{k-1}(\xi,t,x,y)\,dt$$

$$\Gamma_0(\xi,\eta,x,y) = K(\xi;\eta)\delta_0(x,y) - K(x;\eta)\delta_0(\xi,y)$$

Hence,
$$D_2(\xi,\eta,x,y,\lambda) = K(\xi;\eta)D_1(x,y,\lambda) - K(x;\eta)D_1(\xi,y,\lambda)$$

$$+ \lambda\int_a^b K(t;\eta)D_2(\xi,t,x,y,\lambda)\,dt$$

Let $\psi(\xi)$ be a solution of the homogeneous Fredholm equation corresponding to the eigenvalue λ_1; that is,

$$\psi(\xi) = \lambda_1\int_a^b K(\xi;\eta)\psi(\eta)\,d\eta$$

Then
$$\psi(\xi) = \lambda_1\int_a^b K(\xi;\eta)\psi(\eta)\,d\eta$$

$$= \lambda_1\int_a^b \frac{D_2(\xi,\eta,x_0,x_1,\lambda_1)}{D_1(x_0,x_1,\lambda_1)}\,\psi(\eta)\,d\eta$$

$$+ \lambda_1\int_a^b K(x_0;\eta)\frac{D_1(\xi,x_1,\lambda_1)}{D_1(x_0,x_1,\lambda_1)}\,\psi(\eta)\,d\eta$$

$$- \lambda_1^2\int_a^b\int_a^b K(t;\eta)\frac{D_2(\xi,t,x_0,x_1,\lambda_1)}{D_1(x_0,x_1,\lambda_1)}\,\psi(\eta)\,dt\,d\eta$$

$$= \left[\frac{\lambda_1}{D_1(x_0,x_1,\lambda_1)}\int_a^b K(x_0;\eta)\psi(\eta)\,d\eta\right]D_1(\xi,x_1,\lambda_1)$$

Here we have used the fact that $D'(\lambda_1) \neq 0$ to ensure us that there exist numbers x_0 and x_1 such that $D_1(x_0,x_1,\lambda_1) \neq 0$.

This discussion can be generalized to the case where

$$D(\lambda_1) = D'(\lambda_1) = \cdots = D^{(p-1)}(\lambda_1) = 0 \qquad D^{(p)}(\lambda_1) \neq 0$$

In this case[1] there are p linearly independent solutions of the homogeneous Fredholm equation corresponding to the eigenvalue λ_1, and hence there is a p-fold infinity of solutions. Each eigenvalue must have a finite degeneracy, however, or $D(\lambda)$ would vanish identically.

Finally, let us consider the solution of the nonhomogeneous Fredholm equation in the case where λ is an eigenvalue of the homogeneous equation. To do this we must consider the **associated integral equation**

$$\bar{\psi}(\xi) = F(\xi) + \lambda \int_a^b K(x;\xi)\bar{\psi}(x)\,dx$$

We see immediately that $\bar{D}(\lambda) = D(\lambda)$, since interchanging the rows and columns will not change the values of the determinants involved. On the other hand, $\bar{D}_1(\xi,x,\lambda) = D_1(x,\xi,\lambda)$. Therefore, a solution of the associated equation is

$$\bar{\psi}(\xi) = F(\xi) + \lambda \int_a^b k(x,\xi,\lambda)F(x)\,dx$$

provided that $D(\lambda) \neq 0$. We also see that the eigenvalues of the associated homogeneous equation are the same as for the original equation, and the degeneracy, if there is any, is the same.

Let $D(\lambda_1) = D'(\lambda_1) = \cdots = D^{(p-1)}(\lambda_1) = 0$ and $D^{(p)}(\lambda_1) \neq 0$. Then there are p linearly independent solutions of the associated homogeneous equation, $\bar{\psi}_1, \bar{\psi}_2, \ldots, \bar{\psi}_p$. A necessary condition for the existence of a solution of the nonhomogeneous equation

$$\psi(\xi) = F(\xi) + \lambda_1 \int_a^b K(\xi;x)\psi(x)\,dx$$

is that $F(\xi)$ be orthogonal to $\bar{\psi}_1, \bar{\psi}_2, \ldots, \bar{\psi}_p$. We show this as follows:

$$\int_a^b F(\xi)\bar{\psi}_j(\xi)\,d\xi = \int_a^b \psi(\xi)\bar{\psi}_j(\xi)\,d\xi - \lambda_1 \int_a^b \int_a^b K(\xi;x)\psi(x)\psi_j(\xi)\,dx\,d\xi$$

$$= \int_a^b \psi(\xi)\bar{\psi}_j(\xi)\,d\xi - \int_a^b \psi(x)\bar{\psi}_j(x)\,dx$$

$$= 0 \qquad j = 1, 2, \ldots, p$$

This condition is also sufficient. We shall show this in the case where λ_1 is a nondegenerate eigenvalue. The proof involves Fredholm's second minor. The extension of the proof to degenerate eigenvalues is straightforward, but it involves the definition of Fredholm's third minor, fourth minor, etc., and we shall not carry it out here.[2]

---

[1] See William V. Lovitt, "Linear Integral Equations," Dover Publications, Inc., New York, 1950, pp. 46ff.

[2] *Ibid.*, pp. 64ff.

If λ_1 is nondegenerate, there exists just one nontrivial solution $\bar{\psi}(x)$ of the associated homogeneous integral equation satisfying

$$\bar{\psi}(\xi) = \lambda_1 \int_a^b K(x;\xi)\bar{\psi}(x)\, dx$$

We are assuming that $F(\xi)$ is orthogonal to $\bar{\psi}(\xi)$; that is,

$$\int_a^b F(\xi)\bar{\psi}(\xi)\, d\xi = 0$$

By previous considerations we can take

$$\bar{\psi}(\xi) = D_1(x_1,\xi,\lambda_1)$$

and we know by the condition $D'(\lambda_1) \neq 0$ that there exists an x_1 such that $D_1(x_1,\xi,\lambda_1) \neq 0$. From the definition of Fredholm's second minor, if we expand $\Gamma_k(\xi,\eta,x,y)$ using the elements in the first row, we obtain the relation

$$D_2(\xi,\eta,x,y,\lambda) = K(\xi;\eta)D_1(x,y,\lambda) - K(\xi;y)D_1(x,\eta,\lambda)$$
$$+ \lambda \int_a^b K(\xi;t)D_2(t,\eta,x,y,\lambda)\, dt$$

Now $\quad 0 = \lambda_1 \int_a^b K(\xi;\eta)D_1(x_1,y,\lambda_1)F(y)\, dy$

$$= \lambda_1 \int_a^b D_2(\xi,\eta,x_1,y,\lambda_1)F(y)\, dy + \lambda_1 \int_a^b K(\xi;y)D_1(x_1,\eta,\lambda_1)F(y)\, dy$$
$$- \lambda_1^2 \int_a^b \int_a^b K(\xi;t)D_2(t,\eta,x_1,y,\lambda_1)F(y)\, dt\, dy$$

There exists an η_0 such that $D_1(x_1,\eta_0,\lambda_1) \neq 0$. Therefore,

$$0 = \lambda_1 \int_a^b \frac{D_2(\xi,\eta_0,x_1,y,\lambda_1)}{D_1(x_1,\eta_0,\lambda_1)} F(y)\, dy + \lambda_1 \int_a^b K(\xi;y)F(y)\, dy$$
$$- \lambda_1^2 \int_a^b \int_a^b K(\xi;y)\frac{D_2(y,\eta_0,x_1,t,\lambda_1)}{D_1(x_1,\eta_0,\lambda_1)} F(t)\, dt\, dy$$

Hence,

$$F(\xi) - \lambda_1 \int_a^b \frac{D_2(\xi,\eta_0,x_1,y,\lambda_1)}{D_1(x_1,\eta_0,\lambda_1)} F(y)\, dy$$
$$= F(\xi) + \lambda_1 \int_a^b K(\xi;y)\left[F(y) - \lambda_1 \int_a^b \frac{D_2(y,\eta_0,x_1,t,\lambda_1)}{D_1(x_1,\eta_0,\lambda_1)} F(t)\, dt \right] dy$$

Therefore, $\qquad \psi(\xi) = F(\xi) - \lambda_1 \int_a^b \frac{D_2(\xi,\eta_0,x_1,y,\lambda_1)}{D_1(x_1,\eta_0,\lambda_1)} F(y)\, dy$

is a solution of the nonhomogeneous integral equation.

If λ_1 is a degenerate eigenvalue of the homogeneous integral equation corresponding to p linearly independent solutions $\psi_1, \psi_2, \ldots, \psi_p$, then there exist solutions of the nonhomogeneous equation, provided $F(\xi)$ is orthogonal to each of the solutions of the associated homogeneous equation. However, the solution is not unique, for if

$$u(\xi) = F(\xi) + \lambda_1 \int_a^b K(\xi;x)u(x)\, dx$$

$$v(\xi) = F(\xi) + \lambda_1 \int_a^b K(\xi;x)v(x)\, dx$$

then
$$u(\xi) - v(\xi) = \lambda_1 \int_a^b K(\xi;x)[u(x) - v(x)]\, dx$$

and
$$u(\xi) - v(\xi) = c_i \psi_i(\xi) \qquad i = 1, 2, \ldots, p$$

Exercises 6.3

1. If the kernel in the homogeneous Fredholm integral equation is nonsymmetric, the equation may have no eigenfunctions. Show that there are no nontrivial solutions of $\psi(\xi) = \lambda \int_0^\pi \sin \xi \cos x \, \psi(x) \, dx$.

2. If $K(\xi;x)$ is symmetric, prove that the reciprocal kernel takes the form

$$k(\xi,x,\lambda) = K(\xi;x) + \lambda \sum_{i=1}^\infty \frac{\psi_i(\xi)\psi_i(x)}{\lambda_i(\lambda_i - \lambda)}$$

where $\psi_i(\xi) = \lambda_i \int_a^b K(\xi;x)\psi_i(x)\, dx$. Show that the series converges for all values of $\lambda \neq \lambda_i$.

3. Show the following: If $\psi(\xi)$ is a nontrivial solution of

$$\psi(\xi) = \lambda_1 \int_a^b K(\xi;x)\psi(x)\, dx$$

and $\bar{\psi}(\xi)$ is a nontrivial solution of the associated equation

$$\bar{\psi}(\xi) = \lambda_2 \int K(x;\xi)\bar{\psi}(x)\, dx$$

with $\lambda_1 \neq \lambda_2$, then $\int_a^b \psi(x)\bar{\psi}(x)\, dx = 0$.

4. Solve the following Fredholm equations:

a. $\psi(\xi) = 3 \int_0^1 \xi x \psi(x)\, dx$

b. $\psi(\xi) = \sin \xi + \int_0^\pi \cos \xi \sin x \, \psi(x)\, dx$

c. $\psi(\xi) = 2\xi - 1 + 2 \int_0^1 \xi \psi(x)\, dx$

5. The integral equation

$$\psi(\xi) = F(\xi) + \lambda \int_a^\xi K(\xi;x)\psi(x)\,dx$$

where $K(\xi;x)$ is continuous for $a \le x \le b$, $a \le \xi \le b$, is known as the **Volterra integral equation.** Construct a series solution by successive approximations and show that it converges for all values of λ.

HINT:
$$\left| \int_a^\xi K(\xi;x)\psi_0(x)\,dx \right| \le M \int_a^\xi dx = M(\xi - a)$$

6. Show that the solution of the Volterra equation $\psi(\xi) = 1 + \int_0^\xi (x - \xi)\psi(x)\,dx$ satisfies the differential equation $\psi'' + \psi = 0$ and the boundary conditions

$$\psi(0) = 1 \qquad \psi'(0) = 0$$

The solution is evidently $\psi(\xi) = \cos \xi$, but find the series representation by the method or exercise 5, for illustrative purposes.

7. Let $AX - \lambda X = C$ be a system of n linear equations in n unknowns. The matrix A is nonsymmetric. Prove the following:

a. The system has a unique solution if λ is not an eigenvalue of A.

b. If λ is an eigenvalue of A, the system has a solution if and only if the vector C is orthogonal to all the eigenvectors of the associated equation $A'Z - \lambda Z = O$.

Relate this algebraic problem to the Fredholm integral equation with degenerate kernel.

6.4 Integral Equations of the First Kind

We shall conclude this chapter with a brief discussion of integral equations of the first kind. The general linear integral equation of the first kind can be written in the one-dimensional case as

$$f(x) = \int_a^b K(x;\xi)\psi(\xi)\,d\xi$$

where $f(x)$ is a known function, $K(x;\xi)$ is a given kernel, and $\psi(\xi)$ is the unknown to be determined. We have already seen an example of this type in the **Wiener-Hopf equation**

$$f(x) = \int_{-\infty}^\infty K(|x - \xi|)\psi(\xi)\,d\xi$$

which appeared in Sec. 5.6. This equation comes up quite frequently in diffraction theory and also in the theory of smoothing and prediction.[1]

Historically, one of the first integral equations to be studied was **Abel's equation**

$$f(x) = \int_0^x \frac{\psi(\xi)}{\sqrt{x - \xi}}\,d\xi$$

---

[1] See Norbert Wiener, "Extrapolation, Interpolation, and Smoothing of Stationary Time Series," Technology Press, M.I.T., Cambridge, Mass., and John Wiley & Sons, Inc., New York, 1949.

This equation arises in the following physical problem (see Fig. 12). A particle slides down a smooth curve under the influence of gravity, starting from rest. The total time of descent is

$$T = -\int_0^x \frac{ds}{\sqrt{2g(x-\xi)}} = \int_0^x \frac{\frac{1}{\sqrt{2g}}\frac{ds}{d\xi}}{\sqrt{x-\xi}}\,d\xi$$

$$= \int_0^x \frac{\psi(\xi)\,d\xi}{\sqrt{x-\xi}}$$

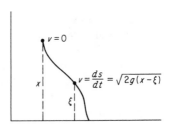

FIGURE 12

where $\psi(\xi) = (1/\sqrt{2g})(ds/d\xi)$. If the path is known, the time of descent can be computed. The problem of finding the path which leads to a given time as a function of the initial height x leads to Abel's integral equation.[1]

In Chap. 8 we shall discuss integral transform methods. Let $\psi(\xi)$ be a function for which the integral transform

$$f(x) = T[\psi] = \int_a^b K(x;\xi)\psi(\xi)\,d\xi$$

exists. Then the problem of finding the inverse transform, that is,

$$\psi(\xi) = T^{-1}[f]$$

reduces to solving an integral equation of the first kind. Some examples of these are

$$f(x) = \frac{1}{\sqrt{2\pi}}\int_{-\infty}^{\infty} \psi(\xi)e^{-ix\xi}\,d\xi$$

$$f(x) = \sqrt{\frac{2}{\pi}}\int_0^{\infty} \psi(\xi)\cos{(x\xi)}\,d\xi$$

$$f(x) = \sqrt{\frac{2}{\pi}}\int_0^{\infty} \psi(\xi)\sin{(x\xi)}\,d\xi$$

$$f(x) = \int_0^{\infty} \psi(\xi)e^{-x\xi}\,d\xi$$

$$f(x) = \int_0^{\infty} \xi J_v(\xi x)\psi(\xi)\,d\xi$$

$$f(x) = \int_0^{\infty} \xi^{x-1}\psi(\xi)\,d\xi$$

[1] In Sec. 3.2 we considered the problem of finding the curve which minimizes the time of descent between two points. This is a problem in the calculus of variations.

These are respectively the **Fourier, Fourier cosine, Fourier sine, Laplace, Hankel,** and **Mellin transforms.** In each of these cases the inverse transformation problem has been solved for large classes of functions. Hence, in each case the corresponding integral equations have been solved.

As a further example of an integral equation of the first kind, consider the diffraction of a plane acoustic wave by a slit with perpendicular incidence (see Fig. 13). The xz plane is a plane rigid barrier except where it is slit between $x = \pm 1$. The plane wave is incident perpendicular to the slit so that there is no z variation, making the problem essentially two-dimensional. The problem is to find an acoustic field $\phi(x,y,t)$ satisfying the two-dimensional wave equation

$$\nabla^2 \phi = \frac{1}{a^2} \frac{\partial^2 \phi}{\partial t^2}$$

throughout the xy plane except on the x axis for $x \leq -1$ and $1 \leq x$. On the rigid boundary ($x \leq -1$, $1 \leq x$) the boundary condition is $\partial \phi / \partial y = 0$. We expect a reflected wave in the lower half plane, so we write

$$\phi = \phi_i + \phi_r + \phi_s$$

where ϕ_i is the incident plane wave, ϕ_r is a reflected plane wave, and ϕ_s is the diffracted wave. The diffraction effects are contained in ϕ_s, and it is on this part of the solution that we shall concentrate our attention. Now

$$\phi_i = e^{i(ku - \omega t)}$$
$$= e^{i(xk_x + yk_y)} e^{-i\omega t}$$
$$= e^{i(r \cos\theta \cos\alpha + r \sin\theta \sin\alpha)} e^{-i\omega t}$$
$$= e^{ikr \cos(\theta - \alpha)} e^{-i\omega t}$$

where $k_x^2 + k_y^2 = k^2 = \omega^2/a^2$. Then ϕ_i satisfies the wave equation. We take

$$\phi_r = e^{ikr \cos(\theta + \alpha)} e^{-i\omega t}$$

If the slit were closed, then we would have simply the incident plane wave plus the reflected wave in the lower half plane as an unperturbed solution which we designate as ϕ_0; that is,

$$\phi_0 = [e^{ikr \cos(\theta - \alpha)} + e^{ikr \cos(\theta + \alpha)}] e^{-i\omega t}$$
$$= 2e^{ikx \cos\alpha} \cos(k_y y) e^{-i\omega t}$$

ϕ_0 obviously satisfies the wave equation and the boundary condition $\partial \phi_0 / \partial y = 0$ at $y = 0$, which is the boundary condition for the unperturbed solution. We, of course, take $\phi_0 \equiv 0$ in the upper half plane. Next we break down the solution as follows:

$$\phi = \phi_0 + \phi_1 \qquad y < 0$$
$$\phi = \phi_2 \qquad y > 0$$

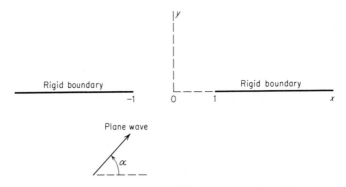

FIGURE 13

Now ϕ_0, ϕ_1, ϕ_2 must all satisfy the wave equation.   In addition the boundary conditions are $\partial\phi_1/\partial y = \partial\phi_2/\partial y = 0$ on the rigid boundary.   In the slit we must have continuity of ϕ and its first partial derivatives.   Hence,

$$\left(\frac{\partial\phi_0}{\partial y}\right)_{y=0} + \left(\frac{\partial\phi_1}{\partial y}\right)_{y=0} = \left(\frac{\partial\phi_2}{\partial y}\right)_{y=0}$$

$$\left(\frac{\partial\phi_1}{\partial y}\right)_{y=0} = \left(\frac{\partial\phi_2}{\partial y}\right)_{y=0}$$

since $(\partial\phi_0/\partial y)_{y=0} = 0$.   Also

$$\phi_0(x,0) + \phi_1(x,0) = \phi_2(x,0)$$

for $-1 < x < 1$.   We note that trivially $(\partial\phi_1/\partial y)_{y=0} = (\partial\phi_2/\partial y)_{y=0}$ outside the slit and therefore this boundary condition holds everywhere on the x axis. This suggests that the scattered field is skew-symmetric in y,[†] that is,

$$\phi_2(x,y) = -\phi_1(x,-y)$$

It follows that $\phi_2 - \phi_1 = 2\phi_2 = \phi_0$ in the slit; that is,

$$\phi_2(x,0) = e^{ikx\,\cos\alpha}e^{-i\omega t} \quad -1 < x < 1$$

As a Green's function we pick

$$G(x,y;\xi,\eta) = \frac{i}{4}H_0^{(1)}(k\sqrt{(x-\xi)^2 + (y-\eta)^2})$$

which satisfies $\nabla^2 G + k^2 G = -\delta(x-\xi)\delta(y-\eta)$

[†] Another attack on the problem is to break up ϕ into symmetric and skew-symmetric parts; that is, $\phi(x,y) = \frac{1}{2}[\phi(x,y) + \phi(x,-y)] + \frac{1}{2}[\phi(x,y) - \phi(x,-y)] = \phi_+ + \phi_-$.   Then by continuity $\partial\phi_+/\partial y = 0$ on the x axis.   Therefore, $\phi_+ = \phi_0$ for $y < 0$ and $\phi_+ = 0$ for $y > 0$ is the solution with the slit closed.   It follows that $\phi_- = \phi_1$, $y < 0$, $\phi_- = \phi_2$, $y > 0$.

Let $\phi_1 = \psi_1(x,y)e^{-i\omega t}$ and $\phi_2 = \psi_2(x,y)e^{-i\omega t}$.   Then

$$\nabla^2\psi_1 + k^2\psi_1 = 0 \qquad \nabla^2\psi_2 + k^2\psi_2 = 0$$

away from the rigid boundary.   From the usual theory of Green's functions we have

$$\psi_2(\xi,\eta) = -\int_{-\infty}^{\infty} \psi_2 \frac{\partial G}{\partial y}\,dx + \int_{-1}^{1} G \frac{\partial \psi_2}{\partial y}\,dx$$

$$-\psi_1(\xi,\eta) = -\int_{-\infty}^{\infty} \psi_1 \frac{\partial G}{\partial y}\,dx + \int_{-1}^{1} G \frac{\partial \psi_1}{\partial y}\,dx$$

We obtain the first equation by integrating along the x axis and then along a large semicircle in the upper half plane.   The second equation is obtained by integrating along the x axis and along a large semicircle in the lower half plane.   The contributions of the semicircular arcs can be shown to go to zero as their radii go to infinity using the radiation condition on ψ_1 and ψ_2. Adding and letting $\eta \to 0$, we have for $-1 < \xi < 1$

$$2e^{ik\xi \cos\alpha} = \psi_2(\xi,0) - \psi_1(\xi,0) = \int_{-1}^{1} G\left(\frac{\partial\psi_1}{\partial y} + \frac{\partial\psi_2}{\partial y}\right)dx - \int_{-\infty}^{\infty}(\psi_1 + \psi_2)\frac{\partial G}{\partial y}\,dx$$

But by the skew-symmetry, $\psi_1(x,0-) + \psi_2(x,0+) = 0$ and

$$\left(\frac{\partial\psi_1}{\partial y}\right)_{y=0} = \left(\frac{\partial\psi_2}{\partial y}\right)_{y=0}$$

Therefore $$e^{ik\xi \cos\alpha} = \frac{i}{4}\int_{-1}^{1} H_0^{(1)}(k\,|x - \xi|)\left(\frac{\partial\psi_1}{\partial y}\right)_{y=0} dx$$

which is again an integral equation of the first kind.[1]   Having solved this equation for $(\partial\psi_1/\partial y)_{y=0} = (\partial\psi_2/\partial y)_{y=0}$ for $-1 < x < 1$, we obtain

$$2\psi_2(\xi,\eta) = \psi_2(\xi,\eta) - \psi_1(\xi,-\eta)$$

$$= \frac{i}{4}\int_{-1}^{1} H_0^{(1)}(k\sqrt{(x-\xi)^2 + \eta^2})\left[\left(\frac{\partial\psi_1}{\partial y}\right)_{y=0} + \left(\frac{\partial\psi_2}{\partial y}\right)_{y=0}\right]dx$$

giving us the solution throughout the plane.

There is one case where the integral equation of the first kind has a particularly easy solution.   That is the case where $K(x;\xi)$ is a continuous symmetric kernel and $f(x)$ has a series representation $\sum_{i=1}^{\infty} c_i\psi_i(x)$ in terms of the eigenfunctions of the kernel, such that $\sum_{i=1}^{\infty} \lambda_i c_i\psi_i(x)$ converges uniformly.   If

---

[1] Integral equations of this type have been studied by G. Latta, The Solution of a Class of Integral Equations. *Journal of Rational Mechanics and Analysis*, vol. 5, pp. 821–834, 1956.

this is the case, this gives us a solution, since the uniform convergence allows us to integrate term by term giving

$$\int_a^b \left[K(\xi;x) \sum_{i=1}^{\infty} \lambda_i c_i \psi_i(x) \right] dx = \sum_{i=1}^{\infty} c_i \lambda_i \int_a^b K(\xi;x) \psi_i(x) \, dx$$

$$= \sum_{i=1}^{\infty} c_i \psi_i(\xi)$$

$$= f(\xi)$$

If a piecewise continuous solution to the integral equation exists, $\sum_{i=1}^{\infty} c_i \psi_i(x)$ with $c_i = \int_a^b f(x) \psi_i(x) \, dx$ is uniformly convergent. This follows from the work of Sec. 6.2, since in this case $f(x)$ is an integral transform. However, here the restriction on $f(x)$ is more severe, since $\sum_{i=1}^{\infty} \lambda_i c_i \psi_i(x)$ must converge uniformly and the λ_i are increasing as $i \to \infty$.

We shall see in Chap. 8 that integral transform methods are sometimes quite effective in solving integral equations of the first kind.

Exercises 6.4

1. Solve Abel's equation by the following device. Multiply both sides of the equation by $(t - x)^{-\frac{1}{2}}$ and integrate from 0 to t.

HINT:
$$\int_\xi^t \frac{dx}{\sqrt{(x - \xi)(t - x)}} = \int_0^1 \frac{du}{\sqrt{u(1 - u)}} = \pi.$$

2. Discuss the diffraction of a plane acoustic wave by a circular hole in an infinite plane rigid barrier. Set up the integral equation whose solution leads to the scattered field. What is the appropriate Green's function?

3. Solve the integral equation

$$x(1 - x) = \int_0^1 K(x;\xi)\psi(\xi) \, d\xi$$

where
$$K(x;\xi) = \xi(1 - x) \qquad 0 \le \xi \le x$$
$$K(x;\xi) = x(1 - \xi) \qquad x \le \xi \le 1$$

References

Courant, Richard, and David Hilbert: "Methods of Mathematical Physics," Interscience Publishers (Division of John Wiley & Sons, Inc.), New York, 1953, vol. I; 1962, vol. II.

Friedman, Bernard: "Principles and Techniques of Applied Mathematics," John Wiley & Sons, Inc., New York, 1956.

Hildebrand, F. B.: "Methods of Applied Mathematics," Prentice-Hall, Inc., Englewood Cliffs, N.J., 1952.

Hochstadt, Harry: "Integral Equations," Interscience Publishers (Division of John Wiley & Sons, Inc.), New York, 1973.

Lovitt, William V.: "Linear Integral Equations," Dover Publications, Inc., New York, 1950.

Mikhlin, Solomon G.: "Integral Equations," Pergamon Press, Inc., New York, 1957.

Murnaghan, Francis D.: "Introduction to Applied Mathematics," John Wiley & Sons, Inc., New York, 1948.

Petrovskii, Ivan G.: "Integral Equations," Graylock Press, Rochester, N.Y., 1957.

Tricomi: "Integral Equations," Interscience Publishers (Division of John Wiley & Sons, Inc.), New York, 1957.

Chapter 7. Analytic Function Theory

7.1 Introduction

Up to this point we have been able to avoid, for the most part, using functions of a complex variable although we have made extensive use of complex-valued functions of a real variable. However, for the discussion of integral transforms in Chap. 8 we shall need some of the basic properties of analytic functions of one complex variable. Also, in developing this theory we shall discover a strong connection between analytic functions and real-valued functions of two real variables which satisfy Laplace's equation. This will allow us to introduce the method of conformal mapping for solving many boundary-value problems in potential theory.

We have been assuming that the reader is familiar with complex numbers and the elementary arithmetic associated with them. However, in order to establish our notation and give a quick review, we collect together in this section some of the basic facts about complex numbers.

Complex numbers are of the form $z = x + iy$, where x and y are real. We call x the real part and y the imaginary part; that is, $x = \text{Re}\,(z)$, $y = \text{Im}\,(z)$. Addition, which is defined as follows:

$$z_1 + z_2 = x_1 + x_2 + i(y_1 + y_2)$$

is commutative and associative. Multiplication, which is defined as follows:

$$z_1 z_2 = x_1 x_2 - y_1 y_2 + i(x_1 y_2 + x_2 y_1)$$

is also commutative and associative. The complex zero is $0 = 0 + i0$, and the identity is $1 = 1 + i0$. The negative of $z = x + iy$ is $-z = -x + i(-y)$. With these definitions we have the following obvious properties:[1]

$$z + 0 = z \qquad z + (-z) = 0 \qquad 1z = z$$

[1] Two complex numbers are equal if and only if their real parts are equal and their imaginary parts are equal.

Subtraction is easily defined in terms of addition of the negative; that is,

$$z_1 - z_2 = z_1 + (-z_2) = x_1 + iy_1 + (-x_2 - iy_2) = x_1 - x_2 + i(y_1 - y_2)$$

Reciprocals are defined for every complex number other than zero by the following:

$$\frac{1}{z} = \frac{x}{x^2 + y^2} - i\frac{y}{x^2 + y^2} \qquad x^2 + y^2 \neq 0$$

and then it follows that $(1/z)z = 1$. Division by any complex number other than zero is defined by multiplication by the reciprocal; that is,

$$\frac{z_1}{z_2} = z_1\left(\frac{1}{z_2}\right) = (x_1 + iy_1)\left(\frac{x_2}{x_2^2 + y_2^2} - i\frac{y_2}{x_2^2 + y_2^2}\right)$$

$$= \frac{x_1 x_2 + y_1 y_2}{x_2^2 + y_2^2} + i\frac{x_2 y_1 - x_1 y_2}{x_2^2 + y_2^2}$$

The distributive law is easy to verify; that is,

$$z_1(z_2 + z_3) = z_1 z_2 + z_1 z_3$$

What we have said so far can be easily summarized by the following statement: *the complex numbers are a field under the operations of addition and multiplication.* The reader will recognize that so far the algebra of the complex numbers is just like that of the real numbers. In other words, the complex numbers and the real numbers are fields. However, the real numbers are an ordered field whereas the complex numbers are not. Recall that for any two real numbers a and b we have precisely one of the following satisfied: (1) $a < b$, (2) $a = b$, or (3) $a > b$. Also there are positive real numbers ($a > 0$) and negative real numbers ($a < 0$), and the square of any nonzero real number is positive. If we were to have the same properties for the complex numbers, we would be led to the following contradiction: $(-1)^2 = 1$, therefore 1 is positive; but $i^2 = -1$ and therefore -1 is positive. Since the complex numbers are not ordered we shall not write inequalities between complex numbers.

The real numbers are all of the form $a = a + i0$ and therefore the real numbers are included in the complex numbers. We say that the real numbers are a subfield of the complex number field. That they are a proper subfield is indicated by the statement $i^2 = -1$, and we know that the square of every nonzero real number is positive.

Multiplication of a complex number z by a real number a is obviously defined; that is,

$$az = (a + i0)(x + iy) = ax + iay$$

This is reminiscent of the vector-space operation of multiplication by a real scalar. In fact, the reader has already shown (exercise 1a, Sec. 1.5) that the collection of complex numbers is a two-dimensional vector space over the reals. This suggests an obvious geometrical interpretation of complex numbers in terms of two-dimensional analytic geometry. There is, in fact, a one-to-one correspondence between the complex numbers and the points in the euclidean plane (see Fig. 14). For each complex number $z = x + iy$ there is a point (x,y), and for each point there is a complex number. We shall henceforth refer to this plane as the complex plane.

Using polar coordinates we have $x = r \cos \theta$ and $y = r \sin \theta$ and hence $z = r(\cos \theta + i \sin \theta)$, where

$$r = \sqrt{x^2 + y^2}$$

$$\theta = \tan^{-1} \frac{y}{x}$$

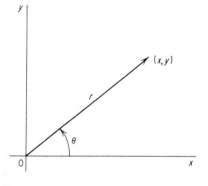

The magnitude of the vector corresponding to z is r, which is usually called the modulus of z and denoted by $|z|$. The angle θ is called the argument of z and is denoted by arg z. The modulus is single-valued but the argument is multivalued because of the periodicity of $\cos \theta$ and $\sin \theta$; that is, there are infinitely many

FIGURE 14

permissible values for the argument, differing by multiples of 2π.

Addition of two complex numbers has the usual geometric interpretation of the addition of two two-dimensional vectors. Multiplication is a little more complicated but is easily interpreted in terms of the polar form of the complex numbers; that is,

$$\begin{aligned}
z_1 z_2 &= r_1 r_2 (\cos \theta_1 + i \sin \theta_1)(\cos \theta_2 + i \sin \theta_2) \\
&= r_1 r_2 [\cos \theta_1 \cos \theta_2 - \sin \theta_1 \sin \theta_2 + i(\sin \theta_1 \cos \theta_2 + \cos \theta_1 \sin \theta_2)] \\
&= r_1 r_2 [\cos(\theta_1 + \theta_2) + i \sin(\theta_1 + \theta_2)]
\end{aligned}$$

In other words, $|z_1 z_2| = |z_1|\,|z_2|$ and arg $z_1 z_2 =$ arg $z_1 +$ arg z_2. It is also easy to establish the following properties:

$$|z_1/z_2| = |z_1|/|z_2| \qquad \text{and} \qquad \arg (z_1/z_2) = \arg z_1 - \arg z_2$$

The operation of conjugation is important. If $z = x + iy$, then the conjugate of z is $z^* = x - iy$. It is easy to verify the following properties:

$$z^{**} = z \qquad (z_1^* \pm z_2^*) = z_1^* \pm z_2^* \qquad (-z)^* = -z^* \qquad (z_1 z_2)^* = z_1^* z_2^*$$

$$\left(\frac{z_1}{z_2}\right)^* = \frac{z_1^*}{z_2^*} \qquad |z^*| = |z| \qquad \arg z^* = -\arg z \qquad |z|^2 = zz^*$$

There are certain important inequalities which are extremely useful. Since $z_1 + z_2$ is represented by the diagonal of the parallelogram determined by z_1 and z_2, the inequality

$$|z_1 + z_2| \leq |z_1| + |z_2|$$

expresses the fact that the sum of two sides of a triangle is greater than the third side. This inequality can be extended by induction to

$$|z_1 + z_2 + \cdots + z_n| \leq |z_1| + |z_2| + \cdots + |z_n|$$

for $n = 3, 4, 5, \ldots$. It is easy to establish the related inequality

$$|z_1 - z_2| \geq \big||z_1| - |z_2|\big|$$

The distance between two points z_1 and z_2 in the complex plane is

$$\sqrt{(x_1 - x_2)^2 + (y_1 - y_2)^2} = |z_1 - z_2|$$

It is obvious that $|z_1 - z_2| = |z_2 - z_1|$, $|z_1 - z_2| \geq 0$, $|z_1 - z_2| = 0$ if and only if $z_1 = z_2$, and $|z_1 - z_2| \leq |z_1 - z_3| + |z_2 - z_3|$. Therefore, the complex plane is a metric space and we can use the conventional topology of the euclidean plane to study convergence, etc., in complex variable theory.

It should be clear what we mean by a set of points in the complex plane. We also use the conventional definitions of union, intersection, and complements of sets. We shall use the usual notation to denote sets; that is,

$$S = \{z \mid \text{defining property of the set}\}$$

An ϵ-neighborhood of a point z_0 is

$$N_\epsilon(z_0) = \{z \mid |z - z_0| < \epsilon\}$$

which is the set of points within a circle of radius ϵ centered at z_0. The point z_0 of a set S is an interior point of S if S contains an ϵ-neighborhood of z_0. A set is open if all of its points are interior points. A set is closed if it is the complement of an open set. The point z_0 is a limit point of a set S if every ϵ-neighborhood of z_0 contains an infinite number of points of S. The closure of a set S, denoted $\bar{S}$, is the union of S with the set of limit points of S. The interior of a set S is the collection of interior points of S. The exterior of a set S is the collection of interior points of the complement of S. The boundary of a set S is the closure $\bar{S}$ with the interior of S deleted.

A set of points S in the complex plane is said to be compact if every infinite subset of S has at least one limit point belonging to S. Finite sets are compact because there are no infinite subsets and therefore the definition is vacuously satisfied. The finite sets are closed and bounded.[1] As a matter of fact, the compact sets of the complex plane are the closed bounded

---

[1] A set is bounded if some ϵ-neighborhood of one of its points contains the whole set.

sets. The reader can show that the compact sets are closed and bounded. That the closed bounded sets are compact follows from the Bolzano-Weierstrass theorem: *every bounded infinite set of points in the complex plane has at least one limit point.*

A collection of sets is a covering of a set S if S is contained in the union of the collection. A subcovering of a covering is a subcollection of sets from the covering which also covers the set S. We will make use of the Heine-Borel property: *a set S is compact if and only if every covering of S by open sets has a subcovering consisting of a finite number of open sets.*

Exercises 7.1

1. Let $z_1 = 1 + 3i$, $z_2 = 2 - i$. Compute $z_1 + z_2$, $z_1 - z_2$, $z_1 z_2$, and z_1/z_2.
2. Prove that there is only one complex number 0 such that $z + 0 = z$ for all z. Prove that there is only one complex number 1 such that $1z = z$ for all z.
3. Let $z \neq 0$. Prove that there is a unique complex number w such that $zw = 1$. Prove that $az = bz$ implies that $a = b$.
4. Let n be an integer. Prove that $z^n = |z|^n [\cos(n \arg z) + i \sin(n \arg z)]$.
5. How many distinct complex numbers are there satisfying $z^4 = 1$? Find them.
6. Find the roots of $az^2 + bz + c = 0$, where a, b, and c are complex.
7. Give an algebraic proof for $|z_1 + z_2| \leq |z_1| + |z_2|$.
8. Prove the inequality $|z_1 - z_2| \geq \big| |z_1| - |z_2| \big|$.
9. Prove the formulas: Re $(z) = \frac{1}{2}(z + z^*)$, Im $(z) = (i/2)(z^* - z)$.
10. Prove that a finite point set is closed.
11. Prove that a set is closed if it contains all of its limit points or has no limit points.
12. Prove that the closure of any set is closed.
13. Prove that the boundary of a set S is the intersection of $\overline{S}$ and the closure of the complement of S.
14. Prove that a compact set is closed and bounded.
15. Show that a bounded set which is not closed is not compact. Show that a closed set which is not bounded is not compact.

7.2 Analytic Functions

A complex-valued function of a complex variable assigns to each point of some subset of the complex z plane a definite complex number. The set of points on which the function is defined is called the domain of the function and the set of values of the function is called the range. If the range is some subset of the set S, then we say that the function is into S. If the range is S we say the function is onto S. If no two points map into the same point, we say that the function is one-to-one. A one-to-one function has an inverse function, that is, a function which assigns to each point of the range a definite value in the domain of the original, namely, the point which mapped into that point of the range under the original function.

Although it is considered bad practice, we shall use the notation $f(z)$ to denote a function of a complex variable z.† Since $f(z)$ is complex it can be written as $u + iv$. Hence,

$$w = u + iv = f(z) = u(x,y) + iv(x,y)$$

In other words, defining $f(z)$ is equivalent to defining two real-valued functions of two real variables.

Let D be the domain of the function $f(z)$ and let z_0 be in D. Then $f(z)$ is continuous at z_0 if for every ϵ-neighborhood of $f(z_0)$ there exists a δ-neighborhood of z_0 such that $f(z)$ is in $N_\epsilon(f(z_0))$ whenever z is in D and z is in $N_\delta(z_0)$. In other words, given any $\epsilon > 0$ there exists a δ such that $|f(z) - f(z_0)| < \epsilon$ whenever z is in D and $|z - z_0| < \delta$. As an example of this concept, consider $f(z) = z^2$, which is defined for all z. The domain D is the whole plane. Let z_0 be any complex number. Then $|f(z) - f(z_0)| = |z^2 - z_0^2| = |z - z_0|\,|z + z_0|$. Let $|z - z_0| < \delta$. Then $|z + z_0| = |z - z_0 + 2z_0| \leq 2\,|z_0| + |z - z_0| < 2\,|z_0| + \delta$ and $|f(z) - f(z_0)| < 2\,|z_0|\,\delta + \delta^2$. Therefore, given any $\epsilon > 0$, $|f(z) - f(z_0)| < \epsilon$ provided $|z - z_0| < \delta$ where $\epsilon = 2\,|z_0|\,\delta + \delta^2$. Solving for δ we have $\delta = \sqrt{|z_0|^2 + \epsilon} - |z_0|$. Therefore, $f(z) = z^2$ is continuous everywhere in its domain.

In the above example, we found that δ depended on z_0 as well as on ϵ. This is true in general. However, suppose we restrict the domain of $f(z) = z^2$ to $S = \{z \mid |z| \leq M\}$. Then $|f(z) - f(z_0)| < 2M\delta + \delta^2 = \epsilon$ for $\delta = \sqrt{M^2 + \epsilon} - M$, and so a δ can be found for each ϵ, which is independent of z_0 in S. This is an example of what we call uniform continuity. A function $f(z)$, defined on S, is uniformly continuous on S, if given any $\epsilon > 0$ there exists a $\delta > 0$ such that $|f(z_1) - f(z_2)| < \epsilon$ whenever z_1 and z_2 are in S and $|z_1 - z_2| < \delta$.

Theorem 1. If $f(z)$ is continuous on a compact set S, then it is uniformly continuous on S.

Proof. Given $\epsilon > 0$, we can construct a covering of S by neighborhoods $N_{\frac12\delta}(\zeta)$, ζ in S, using the δ's furnished by the continuity at each ζ of S. This is a covering by open sets and, because S is compact, it has a finite subcovering. In other words, each z in S is in

$$N_{\frac12\delta_1}(\zeta_1) \cup N_{\frac12\delta_2}(\zeta_2) \cup \cdots \cup N_{\frac12\delta_n}(\zeta_n)$$

for some finite n. Let $2\rho = \min\,[\delta_1, \delta_2, \ldots, \delta_n]$ and let z_1 and z_2 be in S and $|z_1 - z_2| < \rho \leq \delta_k/2$, for $k = 1, 2, 3, \ldots, n$. There is a ζ_j such that $|z_1 - \zeta_j| < \delta_j/2$. Then

$$|z_2 - \zeta_j| = |z_2 - z_1 + z_1 - \zeta_j| \leq |z_1 - z_2| + |z_1 - \zeta_j| < \delta_j$$

† Most people use f to denote the function and $f(z)$ to denote the value of f at the point z. However, for our purposes no serious confusion will arise if we use $f(z)$ to denote both the function and the function value.

and $\qquad |f(z_1) - f(z_2)| = |f(z_1) - f(\zeta_j) + f(\zeta_j) - f(z_2)|$

$$\leq |f(z_1) - f(\zeta_j)| + |f(z_2) - f(\zeta_j)| < 2\epsilon$$

Therefore, $f(z)$ is uniformly continuous.

Next we define derivative. Let $f(z)$ be defined in some neighborhood of z_0. Then for z sufficiently close to z_0 we can define the difference quotient

$$F(z) = \frac{f(z) - f(z_0)}{z - z_0}$$

This is a function of z which is not defined at z_0. However, if there exists a complex number $f'(z_0)$ such that by defining $F(z_0) = f'(z_0)$ we make $F(z)$ continuous at z_0, then $f(z)$ is said to be **differentiable** at z_0 and $f'(z_0)$ is the derivative at z_0. In other words, given any $\epsilon > 0$ there exists a $\delta > 0$ such that

$$\left| \frac{f(z) - f(z_0)}{z - z_0} - f'(z_0) \right| < \epsilon$$

when $0 < |z - z_0| < \delta$. Another way of putting it is that

$$\lim_{z \to z_0} \frac{f(z) - f(z_0)}{z - z_0} = f'(z_0)$$

As an example, let us consider $f(z) = z^2$. Then

$$\frac{f(z) - f(z_0)}{z - z_0} = \frac{z^2 - z_0^2}{z - z_0} = z + z_0 \qquad z \neq z_0$$

and, if we take $f'(z_0) = 2z_0$, we have $|z + z_0 - 2z_0| = |z - z_0| < \epsilon$ for $\delta = \epsilon$. We see that this function is differentiable everywhere.

As another example, consider $f(z) = |z|^2 = zz^*$. Then

$$\frac{f(z) - f(z_0)}{z - z_0} = \frac{zz^* - z_0 z_0^*}{z - z_0} = z_0 \frac{z^* - z_0^*}{z - z_0} + z^*$$

If $z_0 = 0$, then the difference quotient is z^* which approaches 0 and therefore $f'(0) = 0$. On the other hand, if $z_0 \neq 0$ then the difference quotient has no limit since $(z - z_0)^*/(z - z_0)$ does not approach a unique complex number as z approaches z_0. The function $f(z) = |z|^2$ is therefore differentiable only at the origin.

As a third example, consider $f(z) = \text{Re}(z) = (z + z^*)/2$. The difference quotient is

$$\frac{\text{Re}(z) - \text{Re}(z_0)}{z - z_0} = \frac{1}{2} + \frac{1}{2} \frac{(z - z_0)^*}{z - z_0}$$

which has no limit, and hence the function is nowhere differentiable.

Theorem 2. If $f(z)$ is differentiable at z_0, then $f(z)$ is continuous at z_0.

Proof. The domain of $f(z)$ must contain some neighborhood of z_0, and in this neighborhood, with z_0 deleted

$$f(z) - f(z_0) = \frac{f(z) - f(z_0)}{z - z_0}(z - z_0)$$

Then
$$\lim_{z \to z_0}[f(z) - f(z_0)] = f'(z_0)\lim_{z \to z_0}(z - z_0) = 0$$

which shows that $\lim_{z \to z_0} f(z) = f(z_0)$. This proof is based on some simple properties of limits, which the reader should verify.

The reader should have no trouble showing, following the usual procedures of the calculus, that if $f(z)$ and $g(z)$ are both differentiable at z, then

$$[f(z) \pm g(z)]' = f'(z) \pm g'(z)$$

$$[f(z)g(z)]' = f(z)g'(z) + f'(z)g(z)$$

$$\left[\frac{f(z)}{g(z)}\right]' = \frac{g(z)f'(z) - f(z)g'(z)}{[g(z)]^2} \qquad \text{provided } g(z) \neq 0$$

The proof of the last of these makes use of the fact that if $g(z_0) \neq 0$, then there is a neighborhood of z_0 throughout which $g(z) \neq 0$. This follows from the continuity of $g(z)$ at z_0. In fact, if $\epsilon = \frac{1}{2}|g(z_0)|$, then $|g(z) - g(z_0)| < \frac{1}{2}|g(z_0)|$ and

$$|g(z)| = |g(z) - g(z_0) + g(z_0)| \geq ||g(z_0)| - |g(z) - g(z_0)||$$
$$> |g(z_0)| - \tfrac{1}{2}|g(z_0)| = \tfrac{1}{2}|g(z_0)|$$

in some neighborhood of z_0.

Another simple property, which is easy to prove, is that if c is constant and $f(z)$ is differentiable at z, then $[cf(z)]' = cf'(z)$. Also if $f(z)$ is constant $f'(z) = 0$. This is so because the difference quotient is identically zero. If $f(z) = z$, then the difference quotient is identically 1. Therefore, the derivative of z is 1. Now let us prove by induction that[1]

$$\frac{d}{dz}z^n = nz^{n-1}$$

We have already established it for $n = 0, 1, 2$. Now if

$$\frac{d}{dz}z^k = kz^{k-1}$$

then
$$\frac{d}{dz}z^{k+1} = \frac{d}{dz}(zz^k) = z^k + kz^k = (k+1)z^k$$

---

[1] $(d/dz)f(z)$ is another notation for $f'(z)$.

which completes the induction. We now have enough properties to establish the differentiability of any polynomial. Also any rational function, that is, the ratio of two polynomials, will be differentiable everywhere except where the denominator vanishes. We shall later show that the polynomial in the denominator can vanish at only a finite number of points.

Now let us consider the problem of differentiating composite functions. Suppose $w = f(z)$ is a function differentiable at z_0. Let $w_0 = f(z_0)$ and suppose $\zeta = g(w)$ is differentiable at w_0. Then $F(z) = g(f(z))$ is differentiable at z_0. To prove this, we form the difference quotient

$$\frac{F(z) - F(z_0)}{z - z_0} = \frac{g(w) - g(w_0)}{z - z_0} = \frac{g(w) - g(w_0)}{w - w_0}\frac{w - w_0}{z - z_0}$$

By the continuity of $f(z)$, as z approaches z_0, w approaches w_0 and

$$F'(z_0) = \lim_{z \to z_0}\frac{F(z) - F(z_0)}{z - z_0} = \lim_{w \to w_0}\frac{g(w) - g(w_0)}{w - w_0}\lim_{z \to z_0}\frac{f(z) - f(z_0)}{z - z_0}$$

$$= g'(w_0)f'(z_0)$$

This is the chain rule for differentiating a composite function.

Next we define **analyticity**. *A function is analytic at z_0 if it has a continuous derivative in some neighborhood of z_0.* Most people require only the existence of the derivative in some neighborhood of z_0, and then prove the continuity of the derivative. However, we shall proceed this way in order to simplify the proof of Cauchy's theorem, which is quite lengthy without the assumption of continuity of the derivative. For the other approach see the author's "Applied Complex Variables," The Macmillan Company, New York, 1965.

Polynomials are differentiable everywhere and hence are analytic everywhere. Rational functions are differentiable except at a finite number of points which are isolated, and therefore are analytic everywhere except at a finite number of points. The function $f(z) = |z|^2$ is nowhere analytic in spite of its being differentiable at the origin. The functions $|z|$, z^*, Re (z), Im (z), and arg z are nowhere differentiable and hence nowhere analytic. We shall consider some other elementary functions in the next section.

Next we obtain necessary and sufficient conditions for a function to be analytic at an interior point of its domain. These are known as the Cauchy-Riemann conditions.

Theorem 3. Let z_0 be an interior point of the domain of $f(z) = u(x,y) + iv(x,y)$. Then $f(z)$ is analytic at z_0 if and only if u and v have continuous first partial derivatives satisfying $u_x = v_y$ and $u_y = -v_x$ in some neighborhood of z_0.

Proof. If $f(z)$ is analytic at z_0, then it has a continuous derivative in some neighborhood of z_0. Then if ξ and η are real

$$f'(z) = \lim_{\xi \to 0} \frac{f(z + \xi) - f(z)}{\xi}$$

$$= \lim_{\xi \to 0} \frac{u(x + \xi, y) - u(x,y)}{\xi} + i \lim_{\xi \to 0} \frac{v(x + \xi, y) - v(x,y)}{\xi}$$

$$= u_x + iv_x$$

$$f'(z) = \lim_{\eta \to 0} \frac{f(z + i\eta) - f(z)}{i\eta}$$

$$= -i \lim_{\eta \to 0} \frac{u(x, y + \eta) - u(x,y)}{\eta} + \lim_{\eta \to 0} \frac{v(x, y + \eta) - v(x,y)}{\eta}$$

$$= v_y - iu_y$$

This proves that the first partial derivatives exist and are continuous and that the Cauchy-Riemann equations $u_x = v_y$ and $u_x = -v_y$ are satisfied.

Conversely, let us assume that the first partial derivatives are continuous and satisfy the Cauchy-Riemann equations. Then

$$\frac{f(z + \xi + i\eta) - f(z)}{\xi + i\eta} = \frac{u(x + \xi, y + \eta) - u(x,y)}{\xi + i\eta} + i\frac{v(x + \xi, y + \eta) - v(x,y)}{\xi + i\eta}$$

Using the continuity of the partial derivatives and the mean value theorem, we have

$$u(x + \xi, y + \eta) - u(x,y) = \xi u_x(x,y) + \eta u_y(x,y) + \theta_1 \xi + \theta_2 \eta$$

$$v(x + \xi, y + \eta) - v(x,y) = \xi v_x(x,y) + \eta v_y(x,y) + \theta_3 \xi + \theta_4 \eta$$

where θ_1, θ_2, θ_3, and θ_4 all approach zero as $|\xi + i\eta| \to 0$. Using the Cauchy-Riemann equations we have

$$\lim_{|\xi + i\eta| \to 0} \frac{f(z + \xi + i\eta) - f(z)}{\xi + i\eta} = u_x(x,y) + iv_x(x,y)$$

$$= v_y(x,y) - iu_y(x,y)$$

Hence, $f'(z)$ exists and is continuous in some neighborhood of z_0. This completes the proof.

As an example, let us consider the function

$$f(z) = e^x \cos y + ie^x \sin y$$

In this case, $u = e^x \cos y$ and $v = e^x \sin y$ and $u_x = e^x \cos y = v_y$, $u_y = -e^x \sin y = -v_x$. Since u_x, u_y, v_x, and v_y are continuous everywhere and satisfy the Cauchy-Riemann equations, $f(z)$ is analytic everywhere and

$$f'(z) = e^x \cos y + ie^x \sin y = f(z)$$

In the next section, we shall define this to be the exponential function e^z.

Let $f(z)$ be analytic at z_0. Then the Cauchy-Riemann conditions are satisfied in some neighborhood of z_0. Let us also assume that the second partial derivatives of u and v are continuous[1] in the same neighborhood. Differentiating the Cauchy-Riemann equations, we have

$$u_{xx} = v_{yx} \qquad u_{yy} = -v_{xy}$$
$$v_{xx} = -u_{yx} \qquad v_{yy} = u_{xy}$$

and by the equality of the mixed derivatives, we have

$$u_{xx} + u_{yy} = 0 \qquad v_{xx} + v_{yy} = 0$$

so both Re $[f(z)]$ and Im $[f(z)]$ satisfy Laplace's equation. We say that a function of two variables is harmonic at z_0 if it has continuous second partial derivatives and satisfies Laplace's equation in some neighborhood of z_0. Hence u and v are harmonic at z_0.

Exercises 7.2

1. Prove that $f(z) = u(x,y) + iv(x,y)$ is continuous at $z_0 = x_0 + iy_0$ if and only if u and v are continuous at (x_0,y_0).

2. Prove that $|z|$, z^*, Re (z), and Im (z) are continuous everywhere.

3. Prove that $|z|$, z^*, Im (z), arg z are nowhere differentiable. Restrict arg z to an interval of 2π.

4. Consider $z = |z|^{\frac{1}{2}} (\cos \frac{1}{2} \arg z + i \sin \frac{1}{2} \arg z)$, $0 \leq \arg z < 2\pi$. Prove that this function is continuous except on the positive real axis. Prove that it is differentiable everywhere except on the positive real axis and at the origin.

5. Let $f(z) = u(r \cos \theta, r \sin \theta) + iv(r \cos \theta, r \sin \theta) = U(r,\theta) + iV(r,\theta)$ be analytic. Prove that $rU_r = V_\theta$ and $rV_r = -U_\theta$.

6. Prove that $f(z) = 1/z$ is continuous on $S = \{z \mid 0 < |z| \leq 1\}$ but not uniformly continuous on S.

7. Show that $f(z) = \sin x \cosh y + i \cos x \sinh y$ is analytic everywhere.

8. Show that $f(z) = \frac{1}{2} \ln (x^2 + y^2) + i \arg z$, $0 \leq \arg z < 2\pi$ is analytic everywhere except on the positive real axis and at the origin.

7.3 Elementary Functions

In this section, we shall define some of the elementary functions of a complex variable and examine some of their properties. After polynomials and rational functions, the next simplest functions are roots. Let $w = \sqrt[n]{z}$, where n is a positive integer greater than or equal to 2. If $w = \rho(\cos \phi + i \sin \phi)$ and $z = r(\cos \theta + i \sin \theta)$, then

$$\rho^n(\cos n\phi + i \sin n\phi) = r(\cos \theta + i \sin \theta)$$

which implies that $\rho = r^{1/n}$ and $\phi = \theta/n$. However, since the various arg z

---

[1] We shall later show that the analyticity implies the existence and continuity of all the partial derivatives of u and v.

differ by multiples of 2π, there are n distinct values of ϕ differing by multiples of $2\pi/n$.   To define a single-valued function $w = \sqrt[n]{z}$, we must restrict θ to an interval of 2π, for example, $0 \leq \theta < 2\pi$.   With this choice of arg z we have a single-valued function which is continuous everywhere except on the positive real axis.   That it is discontinuous on the positive real axis is indicated by the fact that arg w experiences a jump of $2\pi/n$ there.   Except on the positive real axis and at the origin, this function is analytic.   To see this, let $w = w(z)$, $w_0 = w(z_0)$, and then

$$\frac{w - w_0}{z - z_0} = \frac{w - w_0}{w^n - w_0^n} = \frac{1}{w^{n-1} + w^{n-2}w_0 + \cdots + ww_0^{n-2} + w_0^{n-1}}$$

If $w_0 \neq 0$ and w_0 is not on the positive real axis, $\lim\limits_{z-z_0} w(z) = w_0 \neq 0$, and hence

$$w'(z_0) = \lim_{z \to z_0} \frac{w - w_0}{z - z_0} = \frac{1}{nw_0^{n-1}} = \frac{1}{n} z_0^{(1/n)-1}$$

The exponential function[1] is defined as follows:

$$\exp z = e^z = e^x(\cos y + i \sin y)$$

By the Cauchy-Riemann conditions we can easily show that this function is everywhere analytic and

$$\frac{d}{dz} e^z = e^z$$

The familiar property $e^{z_1+z_2} = e^{z_1}e^{z_2}$ can be proved as follows:

$$
\begin{aligned}
e^{z_1+z_2} &= e^{x_1+x_2}[\cos (y_1 + y_2) + i \sin (y_1 + y_2)] \\
&= e^{x_1+x_2}[\cos y_1 \cos y_2 - \sin y_1 \sin y_2 + i(\sin y_1 \cos y_2 + \cos y_1 \sin y_2)] \\
&= e^{x_1}(\cos y_1 + i \sin y_2)e^{x_2}(\cos y_2 + i \sin y_2) \\
&= e^{z_1}e^{z_2}
\end{aligned}
$$

It is easy to show that $1/e^z = e^{-z}$ and therefore that

$$e^{z_1-z_2} = \frac{e^{z_1}}{e^{z_2}}$$

When $x = 0$ we have $e^{iy} = \cos y + i \sin y$ from which it follows that

$$\cos y = \frac{e^{iy} + e^{-iy}}{2} \qquad \sin y = \frac{e^{iy} - e^{-iy}}{2i}$$

We use these formulas to generalize to

$$\cos z = \frac{e^{iz} + e^{-iz}}{2} \qquad \sin z = \frac{e^{iz} - e^{-iz}}{2i}$$

[1] Clearly, if $y = 0$, $e^z = e^x$, the real exponential. We shall later show that e^z is the only function analytic in the whole complex plane which is equal to e^x on the real axis.

Using the chain rule, we have

$$\frac{d}{dz} \cos z = \frac{ie^{iz} - ie^{-iz}}{2} = -\sin z$$

$$\frac{d}{dz} \sin z = \frac{e^{iz} + e^{-iz}}{2} = \cos z$$

Therefore, $\cos z$ and $\sin z$ are analytic everywhere. The remaining trigonometric functions are defined by analogy

$$\tan z = \frac{\sin z}{\cos z} \qquad \cot z = \frac{\cos z}{\sin z}$$

$$\sec z = \frac{1}{\cos z} \qquad \csc z = \frac{1}{\sin z}$$

These functions will be analytic except where their denominators vanish. For example,

$$\sin z = \frac{e^{-y+ix} - e^{y-ix}}{2i} = \sin x \cosh y + i \cos x \sinh y$$

is zero when $\sin x \cosh y = 0$ and $\cos x \sinh y = 0$. Since $\cosh y \neq 0$, $\sin x = 0$ implies that $x = k\pi$, $k = 0, \pm 1, \pm 2, \ldots$. However, $\cos k\pi \neq 0$, and therefore $\sinh y = 0$ implies that $y = 0$. Hence, $\sin z = 0$ if and only if $z = k\pi$, $k = 0, \pm 1, \pm 2, \ldots$. Similarly, $\cos z = 0$, if and only if $z = (2k + 1)\pi/2$, $k = 0, \pm 1, \pm 2, \ldots$. Except where $\cos z = 0$,

$$\frac{d}{dz} \tan z = \frac{\cos^2 z + \sin^2 z}{\cos^2 z} = \sec^2 z$$

$$\frac{d}{dz} \sec z = \frac{\sin z}{\cos^2 z} = \tan z \sec z$$

Except where $\sin z = 0$,

$$\frac{d}{dz} \cot z = \frac{-\sin^2 z - \cos^2 z}{\sin^2 z} = -\csc^2 z$$

$$\frac{d}{dz} \csc z = \frac{-\cos z}{\sin^2 z} = -\cot z \csc z$$

The hyperbolic functions are defined as follows:

$$\sinh z = \frac{e^z - e^{-z}}{2} \qquad \cosh z = \frac{e^z + e^{-z}}{2}$$

$$\tanh z = \frac{\sinh z}{\cosh z} \qquad \coth z = \frac{\cosh z}{\sinh z}$$

$$\operatorname{sech} z = \frac{1}{\cosh z} \qquad \operatorname{csch} z = \frac{1}{\sinh z}$$

Clearly, sinh z and cosh z are analytic everywhere, and

$$\frac{d}{dz} \sinh z = \cosh z \qquad \frac{d}{dz} \cosh z = \sinh z$$

The others are analytic everywhere except where their denominators vanish. Since $\sinh iz = i \sin z$ and $\cosh iz = \cos z$, $\sinh z = 0$ if and only if $z = ik\pi$, $k = 0, \pm 1, \pm 2, \ldots$ and $\cosh z = 0$ if and only if $z = (2k + 1)i\pi/2$, $k = 0, \pm 1, \pm 2, \ldots$.   Except where $\cosh z = 0$

$$\frac{d}{dz} \tanh z = \frac{\cosh^2 z - \sinh^2 z}{\cosh^2 z} = \operatorname{sech}^2 z$$

$$\frac{d}{dz} \operatorname{sech} z = \frac{-\sinh z}{\cosh^2 z} = -\tanh z \operatorname{sech} z$$

Except where $\sinh z = 0$

$$\frac{d}{dz} \coth z = \frac{\sinh^2 z - \cosh^2 z}{\sinh^2 z} = -\operatorname{csch}^2 z$$

$$\frac{d}{dz} \operatorname{csch} z = \frac{-\cosh z}{\sinh^2 z} = -\coth z \operatorname{csch} z$$

The logarithmic function we define as the inverse of the exponential function.   However, since

$$\exp (z + 2k\pi i) = \exp z$$

$k = 0, \pm 1, \pm 2, \ldots$, we shall be able to define an inverse function only for restricted values of the imaginary part of the function.   Let

$$z = e^w = e^u(\cos v + i \sin v)$$

then $|z| = e^u$, $u = \ln |z|$, $\arg z = v$.   Therefore,

$$w = \log z = \ln |z| + i \arg z$$

where $\arg z$ is restricted to some interval of 2π.   Suppose, for example, that $0 \leq \arg z < 2\pi$.   Then there are discontinuities of the function on the positive real axis and at the origin ($\ln 0$ is undefined).   Elsewhere the function is analytic and

$$\frac{d}{dz} \log z = \frac{\partial}{\partial x} \frac{1}{2} \ln (x^2 + y^2) + i \frac{\partial}{\partial x} \tan^{-1} \frac{y}{x}$$

$$= \frac{x - iy}{x^2 + y^2} = \frac{z^*}{zz^*} = \frac{1}{z}$$

Other values of $\log z$ differ from this by multiples of $2\pi i$ and hence have the same derivative.

The function z^a where a is real or complex is defined by

$$z^a = e^{a \log z}$$

where a definite determination of log z has been chosen. This function is analytic where log z is and

$$\frac{d}{dz} z^a = \frac{a}{z} e^{a \log z} = az^{a-1}$$

Exercises 7.3

1. Let $w = \sqrt[n]{z} = |z|^{1/n}[\cos(\theta/n) + i \sin(\theta/n)]$, $0 \le \theta < 2\pi$, where n is a positive integer greater than or equal to 2. Prove that w is continuous except on the positive real axis.

2. Find all the distinct solutions of $z^4 = 1 + i$.

3. Let p and q be relatively prime integers. We define $z^{p/q} = \exp[(p/q) \log z]$. How many distinct values of $z^{p/q}$ are there? If we restrict $0 \le \arg z < 2\pi$, where is the function continuous, analytic?

4. Prove that $1/e^z = e^{-z}$ and $(e^z)^* = e^{z^*}$.

5. Prove that e^z has no zeros.

6. Prove that $2\pi i$ is the minimum period of e^z.

7. Prove that $\cos^2 z + \sin^2 z = 1$.

8. Prove that $\sin(z_1 + z_2) = \sin z_1 \cos z_2 + \cos z_1 \sin z_2$ and that

$$\cos(z_1 + z_2) = \cos z_1 \cos z_2 - \sin z_1 \sin z_2$$

9. Prove that $|\sin z|$ and $|\cos z|$ are not bounded.

10. Prove that $(\sin z)^* = \sin z^*$ and $(\cos z)^* = \cos z^*$.

11. Find all the distinct solutions of $\sin z = i$.

12. Prove that $\cosh^2 z - \sinh^2 z = 1$.

13. Prove that $\sinh(z_1 + z_2) = \sinh z_1 \cosh z_2 + \cosh z_1 \sinh z_2$ and that

$$\cosh(z_1 + z_2) = \cosh z_1 \cosh z_2 + \sinh z_1 \sinh z_2$$

14. Prove that $(\sinh z)^* = \sinh z^*$ and $(\cosh z)^* = \cosh z^*$.

15. Find all the distinct solutions of $\cosh z = i$.

16. Find all the distinct solutions of $\log z = i$.

17. Find all the distinct values of i^i.

18. Criticize the following "proof" of $\pi = 0$: $0 = \log 1 = \log(-1)(-1) = \log(-1) + \log(-1) = \pi i + \pi i = 2\pi i$.

19. We define $a^z = \exp(z \log a)$, $a \ne 0$. How many values of a^z are there? Where is a particular determination of the function analytic?

7.4 Complex Integration

We shall define the integral of a function of a complex variable in terms of a pair of line integrals. First, however, we must define certain kinds of curves and restrict our attention to certain special curves. The concept of curve in the complex plane is really too broad to allow us to consider integrals over arbitrary curves and still end up with a reasonably simple theory.

By an **arc** we shall mean a set of points determined by a parametric equation $z(t) = x(t) + iy(t)$, $a \le t \le b$, where $x(t)$ and $y(t)$ are continuous and $t_1 \ne t_2$ implies $z(t_1) \ne z(t_2)$. By a **smooth arc** we shall mean an arc determined by a parametric equation $z(t) = x(t) + iy(t)$, $a \leq t \leq b$, where $\dot{x} = dx/dt$ and $\dot{y} = dy/dt$ are continuous and $\dot{x}^2 + \dot{y}^2 \neq 0$. In effect, a smooth arc is an arc which has a continuous tangent vector everywhere. The initial point of an arc is $z(a)$ and the final point is $z(b)$. A **curve** is the union of a finite number of arcs joined so that the final point of the first arc is the initial point of the second, the final point of the second is the initial point of the third, etc. A **piecewise smooth curve** is a curve made up of smooth arcs. A curve can obviously have multiple points, where an arc in the chain meets another arc other than at the end points as already specified. If there are no such multiple points, then the curve is said to be simple. If the only multiple point is the common initial point of the first arc and final point of the last arc, then the curve is called a simple closed curve. For brevity we shall refer to a simple piecewise smooth curve as a **contour** and to a simple closed piecewise smooth curve as a **closed contour**.

We shall need a result known as the **Jordan curve theorem**, which states the following: A simple closed curve in the complex plane divides the plane into two disjoint open sets, one bounded (the interior) and the other unbounded (the exterior). We shall not prove this theorem but shall use it when needed.

A smooth arc has length equal to $\int_a^b \sqrt{\dot{x}^2 + \dot{y}^2}\, dt$ and therefore a piecewise smooth curve has length equal to the finite sum of the lengths of its component arcs.

We now define the integral of a continuous function of a complex variable on a smooth arc. Let the arc Γ have a parametric representation $z(t) = x(t) + iy(t)$, $a \leq t \leq b$, and let $f(z)$ be continuous on Γ. Then $f(z) = u(x,y) + iv(x,y)$, where u and v are continuous on Γ. The line integrals

$$\int_\Gamma u\, dx - v\, dy$$

$$\int_\Gamma v\, dx + u\, dy$$

exist and we define[1]

$$\int_\Gamma f(z)\, dz = \int_\Gamma u\, dx - v\, dy + i\int_\Gamma v\, dx + u\, dy$$

The definition of $\int f(z)\, dz$ for a piecewise smooth curve is the same except that the line integrals are carried out over the finite number of smooth arcs, and the definition represents the finite sum of line integrals over these smooth arcs.

We can evaluate the integral over a smooth arc in terms of a parametric

---

[1] Note that the definition gives the result expected from writing $f(z) = u + iv$, $dz = dx + i\, dy$ and formally expanding $\int (u + iv)(dx + i\, dy)$ into real and imaginary parts.

representation of the arc. In fact,

$$\int_\Gamma f(z)\,dz = \int_a^b [U(t)\dot{x} - V(t)\dot{y}]\,dt + i\int_a^b [V(t)\dot{x} + U(t)\dot{y}]\,dt$$

where $U(t) = u[x(t),y(t)]$, $V(t) = v[x(t),y(t)]$.

We can infer many properties of the integral directly from well-known properties of line integrals. Some of these are as follows:

1. $\displaystyle\int kf(z)\,dz = \int ku\,dx - kv\,dy + i\int kv\,dx + ku\,dy$

$$= k\left(\int u\,dx - v\,dy + i\int v\,dx + u\,dy\right) = k\int f(z)\,dz$$

where k is a constant.

2. $\displaystyle\int [f_1(z) + f_2(z)]\,dz = \int (u_1 + u_2)\,dx - (v_1 + v_2)\,dy$

$$+ i\int (v_1 + v_2)\,dx + (u_1 + u_2)\,dy$$

$$= \int u_1\,dx - v_1\,dy + i\int v_1\,dx + u_1\,dy$$

$$+ \int u_2\,dx - v_2\,dy + i\int v_2\,dx + u_2\,dy$$

$$= \int f_1(z)\,dz + \int f_2(z)\,dz$$

3. $\displaystyle\int_{-\Gamma} f(z)\,dz = \int_{-\Gamma} u\,dx - v\,dy + i\int_{-\Gamma} v\,dx + v\,dy$

$$= -\int_\Gamma u\,dx - v\,dy - i\int_\Gamma v\,dx + v\,dy$$

$$= -\int_\Gamma f(z)\,dz$$

where $-\Gamma$ refers to the same curve as Γ but with the initial and final points interchanged.[1]

4. $\displaystyle\int_{\Gamma_1+\Gamma_2} f(z)\,dz = \int_{\Gamma_1+\Gamma_2} u\,dx - v\,dy + i\int_{\Gamma_1+\Gamma_2} v\,dx + u\,dy$

$$= \int_{\Gamma_1} u\,dx - v\,dy + i\int_{\Gamma_1} v\,dx + u\,dy$$

$$+ \int_{\Gamma_2} u\,dx - v\,dy + i\int_{\Gamma_2} v\,dx + u\,dy$$

$$= \int_{\Gamma_1} f(z)\,dz + \int_{\Gamma_2} f(z)\,dz$$

[1] If Γ is given by $z(t)$, $a \le t \le b$, then $-\Gamma$ is given by $z(-\tau + a + b)$, $a \le \tau \le b$.

where $\Gamma_1 + \Gamma_2$ is the curve whose initial point of Γ_2 is the final point of Γ_1.

5. $\left| \int_\Gamma f(z)\, dz \right| \leq 4 \int_\Gamma |f(z)|\, ds \leq 4ML(\Gamma)$, where $M = \max |f(z)|$ on Γ and $L(\Gamma)$ is the length of Γ.

Property 5 can be proved as follows.   Recalling that $ds = \sqrt{(dx)^2 + (dy)^2}$, we have

$$\left| \int_\Gamma u\, dx \right| \leq \int_\Gamma |u|\, ds \qquad \left| \int_\Gamma u\, dy \right| \leq \int_\Gamma |u|\, ds$$

$$\left| \int_\Gamma v\, dx \right| \leq \int_\Gamma |v|\, ds \qquad \left| \int_\Gamma v\, dy \right| \leq \int_\Gamma |v|\, ds$$

Also $|u| \leq |f| \leq M$ and $|v| \leq |f| \leq M$ on Γ.   Hence,

$$\left| \int_\Gamma f(z)\, dz \right| \leq \left| \int_\Gamma u\, dx - v\, dy \right| + \left| \int_\Gamma v\, dx + u\, dy \right|$$

$$\leq 2 \int_\Gamma |u|\, ds + 2 \int_\Gamma |v|\, ds$$

$$\leq 4 \int_\Gamma |f|\, ds \leq 4\, M \int_\Gamma ds = 4ML(\Gamma)$$

It should be noted that by making more refined estimates one can actually prove that $|\int f(z)\, dz| \leq \int |f(z)|\, ds$, without the factor 4.   However, for the proofs which appear in this book our property 5 will be sufficient.

As an example, consider $\int_0^{1+i} z^*\, dz$ along two different arcs, (1) a straight-line segment, and (2) a parabola $y = x^2$.   In the first case, $y = x$, $dy = dx$, and

$$\int_0^{1+i} z^*\, dz = \int_0^1 (x - ix)(dx + i\, dx) = 2 \int_0^1 x\, dx = 1$$

In the second case, $y = x^2$, $dy = 2x\, dx$, and

$$\int_0^{1+i} z^*\, dz = \int_0^1 (x - ix^2)(dx + 2ix\, dx)$$

$$= \int_0^1 (x + 2x^3)\, dx + i \int_0^1 x^2\, dx = 1 + \tfrac{1}{3}i$$

This example illustrates that, in general, the integral depends on the curve used in joining the two end points.

As another example, consider $\int_1^{-1} (1/z)\, dz$ along two different arcs, (1) the unit semicircle in the upper half plane, (2) the unit semicircle in the lower half plane.   In the first case, $z = e^{i\theta}$, $0 \leq \theta \leq \pi$, and $dz = ie^{i\theta}\, d\theta$, and

$$\int_1^{-1} (1/z)\, dz = i \int_0^\pi d\theta = \pi i$$

In the second case, $z = e^{i\theta}$, $-\pi \leq \theta \leq 0$, and $dz = ie^{i\theta}\, d\theta$, and

$$\int_1^{-1} (1/z)\, dz = i \int_0^{-\pi} d\theta = -\pi i$$

Again we see that the integral depends on the curve of integration. However, as we shall soon see, the reason for the difference in this case is that the origin, where $1/z$ is not analytic, lies between the two arcs, whereas in the first example, z^* is not analytic on any of the curves joining the two points. The situations in the two examples are quite different.

We now consider necessary and sufficient conditions for the integral to be independent of the contour joining the two end points.

Theorem 1. Let $f(z)$ be continuous in a region[1] R. Then $\int_\Gamma f(z)\, dz$ depends only on the end points of Γ in R, if and only if $f(z)$ is the derivative of a function analytic in R.

Proof. If there exists a function $F(z) = U(x,y) + iV(x,y)$ such that $F'(z) = f(z)$ in R, then $U_x = V_y = u$ and $-U_y = V_x = v$, and

$$\int_\Gamma f(z)\, dz = \int_\Gamma U_x\, dx + U_y\, dy + i \int_\Gamma V_x\, dx + V_y\, dy$$

$$= \int_\Gamma dU + i \int_\Gamma dV = F(\beta) - F(\alpha)$$

where α and β are, respectively, initial and final points of Γ. Hence the integral only depends on the end points, not on the particular curve Γ in R joining the end points.

On the other hand, suppose $z_0 = x_0 + iy_0$ is the initial point and $z = x + iy$ is the final point, where z_0 is fixed, and the integral is independent of the curve Γ in R. Then, if $\zeta = \xi + i\eta$

$$F(z) = \int_{z_0}^z f(\zeta)\, d\zeta = \int_{z_0}^z u\, d\xi - v\, d\eta + i \int_{z_0}^z v\, d\xi + u\, d\eta$$

$$= U(x,y) + iV(x,y)$$

is a function defined in R. By virtue of well-known properties of line integrals $U_x = u$, $U_y = -v$, $V_x = v$, and $V_y = u$. Hence $F(z)$ has a continuous derivative $F'(z) = u + iv = f(z)$. This completes the proof.

The effect of this theorem is that if we can find an analytic function $F(z)$ such that $F'(z) = f(z)$ in a region containing the curve Γ, then

$$\int_\Gamma f(z)\, dz = F(\beta) - F(\alpha)$$

---

[1] A region is a nonempty connected open set. An open set is connected if it is not the union of two disjoint open sets.

where α and β are, respectively, initial and final points of Γ. For example,

$$\int_{-i}^{i} e^{z} \, dz = e^{i} - e^{-i}$$

We now come to one of the key theorems in the study of analytic functions, **Cauchy's theorem.**

Theorem 2. Let $f(z)$ be analytic in a region R. Then $\int_{\Gamma} f(z) \, dz = 0$ for every closed contour Γ which lies in R and has its interior in R.

Proof. Since $f'(z)$ is continuous in R, so are u_x, u_y, v_x, v_y and $u_x = v_y$, $-u_y = v_x$. Therefore, Green's theorem holds and

$$\int_{\Gamma} f(z) \, dz = \int_{\Gamma} u \, dx - v \, dy + i \int_{\Gamma} v \, dx + u \, dy$$

$$= -\iint_{S} (v_x + u_y) \, dx \, dy + i \iint_{S} (u_x - v_y) \, dx \, dy = 0$$

where S is the interior of Γ.

It should be noted that the continuity of the partial derivatives of u and v within and on Γ is needed in order to use Green's theorem. This is the reason we assumed the continuity of the derivative in the definition of analyticity. Cauchy's theorem can be proved on the basis of the existence of $f'(z)$ in R, without assuming its continuity, but the proof is very much longer. Later when the existence of the second derivative is established in R, the continuity of $f'(z)$ is inferred. Hence, we do not assume a smaller class of analytic functions by our more restricted definition.

If R is simply connected,[1] then $\int f(z) \, dz = 0$ for every closed contour implies that $\int_{\alpha}^{\beta} f(z) \, dz$ is independent of the contour joining α and β in R. This implies the existence of a function $F(z)$ analytic in R such that $F'(z) = f(z)$ and

$$\int_{\alpha}^{\beta} f(z) \, dz = F(\beta) - F(\alpha)$$

Also the fact that the integral is independent of the contour makes it possible to use whatever contour in R is convenient for the evaluation of the line integrals, in case it is not apparent what $F(z)$ is. For example, suppose we are asked to evaluate

$$\int_{1}^{i/2} \frac{1}{1 + z^2} \, dz$$

along the line $2y + x = 1$. The integrand is analytic except at $\pm i$. Therefore, we may replace the path of integration by the broken line from 1 to 0

[1] This means that every closed contour in R has its interior in R.

and then from 0 to $i/2$. Consequently,

$$\int_1^{i/2} \frac{1}{1+z^2}\,dz = \int_1^0 \frac{dx}{1+x^2} + i\int_0^{1/2} \frac{dy}{1-y^2} = -\frac{\pi}{4} + \frac{i}{2}\ln 3$$

Exercises 7.4

1. Consider the following definition of $\int_\Gamma f(z)\,dz$ of $f(z)$ continuous on a smooth arc Γ with parametric representation $z(t) = x(t) + iy(t)$, $0 \le t \le 1$. Let $0 = t_0 < t_1 < t_2 < \cdots < t_n = 1$ be an arbitrary partition of the interval $[0,1]$. Let $t_{k-1} \le \tau_k \le t_k$. We define

$$\int_\Gamma f(z)\,dz = \lim_{\substack{n\to\infty \\ \max |t_k - t_{k-1}| \to 0}} \sum_{k=1}^n f[z(\tau_k)][z(t_k) - z(t_{k-1})]$$

Show that this limit exists and yields the same result as the definition given in this section.

2. Prove properties 1–4 of the integral using the definition of exercise 1.

3. Prove the improved version of property 5,

$$\left| \int_\Gamma f(z)\,dz \right| \le \int_\Gamma |f(z)|\,ds \le ML$$

using the definition of exercise 1.

4. Evaluate the integrals of Re (z), Im (z), $|z|$, $|z|^2$ along the two paths $y = x$ and $y = x^2$ joining the points $(0,0)$ and $(1,1)$.

5. Evaluate the integrals of z, z^2, e^z, $\cos z$, $\sinh z$ from $(0,0)$ to $(1,1)$ along arbitrary contours joining the two points.

6. Evaluate the integral of $1/(1 + z^2)$ from $(0,0)$ to $(0,2)$ along two different circular arcs centered on $(0,1)$, one lying to the right of i and the other to the left. Explain the difference in values of the two integrals.

7. Show that the integral around the unit circle $x^2 + y^2 = 1$ of z^n is zero for all integers n except $n = -1$.

8. Prove that $f(z) = u(x,y) + iv(x,y)$, where u and v have continuous first partial derivatives in a region R, is analytic if $\int_\Gamma f(z)\,dz = 0$ for any closed contour in R. HINT: Using Green's theorem prove that u and v satisfy the Cauchy-Riemann equations.

7.5 Integral Representations

Let $f(z)$ be analytic within and on a closed contour C. Then since each point of C must be an interior point of some region of analyticity of $f(z)$, the function is actually analytic in a simply connected region containing C and hence $\int_C f(z)\,dz = 0$. Now let z_0 be inside C. Then $f(z)/(z - z_0)$ is not analytic everywhere inside C and

$$\frac{1}{2\pi i} \int_C \frac{f(z)}{z - z_0}\,dz$$

is not necessarily zero.   In fact, we shall show that

$$\frac{1}{2\pi i} \int_{C+} \frac{f(z)}{z - z_0} \, dz = f(z_0)$$

where $C+$ denotes the positive direction on C, which is the direction an observer must take in order to keep the interior of C on his left and the exterior of C on his right.   For a circle the positive direction is obviously the counterclockwise direction.   We prove the formula first for a small circle inside C, which has z_0 as its center.   Let

$$C_\rho = \{z \mid |z - z_0| = \rho\}$$

where ρ is sufficiently small so that C_ρ is inside C.   Then

$$\frac{1}{2\pi i} \int_{C_\rho+} \frac{f(z_0)}{z - z_0} \, dz = \frac{f(z_0)}{2\pi i} \int_{C_\rho+} \frac{1}{z - z_0} \, dz = f(z_0)$$

Now

$$\frac{1}{2\pi i} \int_{C_\rho+} \frac{f(z) - f(z_0)}{z - z_0} \, dz = \frac{1}{2\pi i} \int_{C_\delta+} \frac{f(z) - f(z_0)}{z - z_0} \, dz$$

for $0 < \delta < \rho$, and

$$\left| \frac{1}{2\pi i} \int_{C_\delta+} \frac{f(z) - f(z_0)}{z - z_0} \, dz \right| \leq 4\epsilon(\delta)$$

where $\epsilon(\delta) = \max|f(z) - f(z_0)|$ on C_δ.   Clearly, by the continuity of $f(z)$ at z_0, ϵ can be made arbitrarily small by picking δ sufficiently small.   This proves that

$$\frac{1}{2\pi i} \int_{C_\rho+} \frac{f(z)}{z - z_0} \, dz = f(z_0)$$

Finally, we show that

$$\frac{1}{2\pi i} \int_{C+} \frac{f(z)}{z - z_0} \, dz$$

$$= \frac{1}{2\pi i} \int_{C_\rho+} \frac{f(z)}{z - z_0} \, dz$$

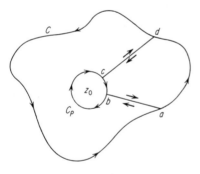

FIGURE 15

Refer to Fig. 15. Starting at d, we integrate along C in the positive direction to a, then to b along the straight-line segment, then to c in the clockwise direction along C_ρ, then to d along the straight-line segment.   Since $f(z)/(z - z_0)$ is analytic within and on this closed contour, the integral is zero.   Again starting at d, we integrate along the straight-line segment to c, then to b along C_ρ in the clockwise direction, then to a along the straight-line segment, and then along

C to d in the positive direction. The result is again zero. Adding the two integrals, the integrals along the line segments cancel, and we have

$$\frac{1}{2\pi i}\int_{C+}\frac{f(z)}{z-z_0}\,dz + \frac{1}{2\pi i}\int_{C_\rho-}\frac{f(z)}{z-z_0}\,dz = 0$$

where $C_\rho-$ is the negative or clockwise direction. The result follows by reversing the direction on C_ρ which changes the sign of the integral. This completes the proof of the **Cauchy integral formula**:

$$f(z_0) = \frac{1}{2\pi i}\int_{C+}\frac{f(z)}{z-z_0}\,dz$$

where $f(z)$ is analytic within and on C.

Under the same hypotheses, we can get an integral representation of $f'(z_0)$. In fact,

$$f'(z_0) = \lim_{h\to 0}\frac{f(z_0+h)-f(z_0)}{h} = \lim_{h\to 0}\frac{1}{2\pi i h}\int_{C+}\left[\frac{f(z)}{z-z_0-h} - \frac{f(z)}{z-z_0}\right]dz$$

$$= \lim_{h\to 0}\frac{1}{2\pi i}\int_{C+}\frac{f(z)}{(z-z_0)(z-z_0-h)}\,dz$$

$$= \frac{1}{2\pi i}\int_{C+}\frac{f(z)}{(z-z_0)^2}\,dz$$

To verify that we can take the limit under the integral sign, we must show that

$$\lim_{h\to 0}\int_{C+}f(z)\left[\frac{1}{(z-z_0)(z-z_0-h)} - \frac{1}{(z-z_0)^2}\right]dz = 0$$

We have

$$\left|\int_{C+}f(z)\left[\frac{1}{(z-z_0)(z-z_0-h)} - \frac{1}{(z-z_0)^2}\right]dz\right|$$

$$= \left|h\int_{C+}\frac{f(z)}{(z-z_0)^2(z-z_0-h)}\,dz\right|$$

Let d = minimum distance from C to z_0, L = length of C, $M = \max|f(z)|$ on C, and let $|h| < \delta$. Then

$$\left|h\int_{C+}\frac{f(z)}{(z-z_0)^2(z-z_0-h)}\,dz\right| \le \frac{4\delta ML}{d^2(d-\delta)}$$

Clearly, this can be made arbitrarily small by making δ sufficiently small, which establishes the limit.

So far we have obtained integral representations of $f(z_0)$ and $f'(z_0)$, but we have not established the existence of any new properties of $f(z)$. However, now using the representation of $f'(z)$ we can prove that the second derivative exists inside C as follows:

$$f''(z_0) = \lim_{h \to 0} \frac{f'(z_0 + h) - f'(z_0)}{h}$$

$$= \lim_{h \to 0} \frac{1}{2\pi i h} \int_{C+} \left[\frac{f(z)}{(z - z_0 - h)^2} - \frac{f(z)}{(z - z_0)^2} \right] dz$$

$$= \lim_{h \to 0} \frac{1}{\pi i} \int_{C+} \frac{f(z)}{(z - z_0)(z - z_0 - h)^2} dz$$

$$= \frac{1}{\pi i} \int_{C+} \frac{f(z)}{(z - z_0)^3} dz$$

The reader should prove that it is permissible to take the limit under the integral sign, by an argument similar to the one above for the first-derivative representation.

There is no reason to stop with the second derivative. In fact, one can show by induction that

$$f^{(n)}(z_0) = \frac{n!}{2\pi i} \int_{C+} \frac{f(z)}{(z - z_0)^{n+1}} dz$$

This will be left as an exercise for the reader. We now have the following important theorem:

Theorem 1. Let $f(z)$ be analytic in a region R. Then $f(z)$ has all derivatives in R. The real and imaginary parts of $f(z)$ are harmonic in R.

Proof. Since R is open, if z_0 is in R, there is a small circle C with center at z_0 such that $f(z)$ is analytic within and on C. Then the Cauchy formula holds for $f(z_0)$, using $C+$ for the integration. It then follows, by the above, that $f^{(n)}(z_0)$ exists for $n = 1, 2, 3, \ldots$. But z_0 is any point in R, and therefore $f(z)$ has all derivatives in R. The existence of $f'''(z)$ implies the continuity of the second partial derivatives of u and v and this completes the proof, indicated in Sec. 7.2, that u and v are harmonic.

A couple of important theorems immediately follow. One of these is a converse for Cauchy's theorem known as **Morera's theorem.**

Theorem 2. Let $f(z)$ be continuous in a region R and $\int_C f(z)\, dz = 0$ for every closed contour C in R. Then $f(z)$ is analytic in R.

Proof. By the condition $\int_C f(z)\, dz = 0$ we have that

$$\int_a^z f(\zeta)\, d\zeta$$

is independent of the path of integration connecting a and z in R. Therefore, there is a function

$$F(z) = \int_a^z f(\zeta)\, d\zeta$$

defined in R such that $F'(z) = f(z)$. But now we know that $F''(z) = f'(z)$ exists and is continuous in R. Therefore, $f(z)$ is analytic in R.

Another result, which depends on the integral representation of the derivative, is known as **Liouville's theorem.**

Theorem 3. If $f(z)$ is analytic everywhere in the complex plane and bounded, then $f(z)$ is constant.

Proof. Let z_0 be any point in the complex plane. Then

$$f'(z_0) = \frac{1}{2\pi i} \int_{C_{\rho+}} \frac{f(z)}{(z - z_0)^2} \, dz$$

where $C_\rho = \{z \mid |z - z_0| = \rho\}$. Then

$$f'(z_0) = \frac{1}{2\pi} \left| \int_{C_{\rho+}} \frac{f(z)}{(z - z_0)^2} \, dz \right| \leq \frac{4M}{\rho}$$

where $|f(z)| \leq M$ in the plane, since $f(z)$ is bounded. Now ρ can be made arbitrarily large and hence $4M/\rho$ can be made arbitrarily small. This proves that $f'(z_0) = 0$, but since z_0 is any point in the plane, $f'(z) \equiv 0$. This proves that $u_x = u_y = v_x = v_y = 0$ everywhere. Let z_1 and z_2 be any two distinct points. Then

$$f(z_1) - f(z_2) = u(x_1, y_1) - u(x_2, y_2) + i[v(x_1, y_1) - v(x_2, y_2)]$$
$$= u_x(x_1 - x_2) + u_y(y_1 - y_2) + iv_x(x_1 - x_2) + iv_y(y_1 - y_2)$$

by the mean-value theorem, where the partial derivatives are evaluated between z_1 and z_2. Our conclusion is that $f(z_1) = f(z_2)$, which proves that $f(z)$ is constant.

The integral

$$\int_C \frac{f(\zeta)}{\zeta - z} \, d\zeta$$

is called a Cauchy-type integral. The fact that it is a differentiable function of z does not depend on $f(\zeta)$ being analytic within and on C, or for that matter, that C be a closed contour. We can prove that

$$F(z) = \int_C \frac{f(\zeta)}{\zeta - z} \, d\zeta$$

which is defined for $f(\zeta)$ continuous on a piecewise smooth curve C as long as z is not on C, is analytic. In fact

$$F'(z) = \lim_{h \to 0} \frac{F(z + h) - F(z)}{h} = \lim_{h \to 0} \frac{1}{h} \int_C \left(\frac{f(\zeta)}{\zeta - z - h} - \frac{f(\zeta)}{\zeta - z} \right) d\zeta$$

$$= \lim_{h \to 0} \int_C \frac{f(\zeta)}{(\zeta - z - h)(\zeta - z)} \, d\zeta = \int_C \frac{f(\zeta)}{(\zeta - z)^2} \, d\zeta$$

where the justification for taking the limit under the integral sign is the same as we used in obtaining the integral representation for the derivative of a function analytic within and on a closed contour. This result is a special case of a more general theorem:

Theorem 4. Let $f(\zeta,z)$ be defined for each ζ on a piecewise smooth curve C and each z in some region R. Let $f(\zeta,z)$ be continuous on C for each z in R and let $f(\zeta,z)$ be analytic in R for each ζ on C. Then

$$F(z) = \int_C f(\zeta,z)\, d\zeta$$

is analytic in R where

$$F'(z) = \int_C \frac{\partial}{\partial z} f(\zeta,z)\, d\zeta$$

Proof. Let $\zeta(t) = \xi(t) + i\eta(t)$, $a \le t \le b$, be a parametric representation of C, and let

$$f(\zeta(t),z) = u(t,x,y) + iv(t,x,y)$$

for z in R. Then u_x, u_y, v_x, v_y are continuous in R and $u_x = v_y$, $-u_y = v_x$. Now

$$F(z) = \int_a^b (u\dot{\xi} - v\dot{\eta})\, dt + i \int_a^b (v\dot{\xi} + u\dot{\eta})\, dt$$
$$= U(x,y) + iV(x,y)$$

U and V have continuous first partial derivatives in R and

$$U_x = \int_a^b (u_x\dot{\xi} - v_x\dot{\eta})\, dt = \int_a^b (v_y\dot{\xi} + u_y\dot{\eta})\, dt = V_y$$

$$U_y = \int_a^b (u_y\dot{\xi} - v_y\dot{\eta})\, dt = -\int_a^b (v_x\dot{\xi} + u_x\dot{\eta})\, dt = -V_x$$

Therefore, $F(z)$ is analytic in R and

$$F'(z) = U_x + iV_x = \int_a^b (u_x\dot{\xi} - v_x\dot{\eta})\, dt + i \int_a^b (v_x\dot{\xi} + u_x\dot{\eta})\, dt$$

$$= \int_a^b \frac{\partial}{\partial z} f(\zeta,z)\, d\zeta$$

This completes the proof.

Exercises 7.5

1. Using the improved estimate of exercise 3, Sec. 7.4, prove Cauchy's inequalities:

$$|f^{(n)}(z_0)| \le \frac{n!\, M}{\rho^n}$$

where $f(z)$ is analytic within and on the circle $C = \{z \mid |z - z_0| = \rho\}$ and $M = $ max $|f(z)|$ on C.

2. Prove the Fundamental Theorem of Algebra: Every polynomial $P(z) = a_0 + a_1 z + a_2 z^2 + \cdots + a_n z^n$ has at least one root if $n \geq 1$. HINT: Assume that $P(z)$ has no root and apply Liouville's theorem to $Q(z) = 1/P(z)$.

3. Let $f(z)$ be analytic everywhere in the complex plane and $|f(re^{i\theta})| \leq Mr^\alpha$, where M and α are real constants, $\alpha > 0$. Prove that $f(z)$ is a polynomial of degree at most α.

4. Evaluate $\displaystyle\int_{C+} (\sin z/z^n)\, dz$, $n = $ integer, where C is the unit circle.

5. Evaluate $\displaystyle\int_{C+} dz/(z^3 - 1)$, where $C = \{z \mid |z - 1| = 1\}$.

7.6   Sequences and Series

In the next section we shall obtain series representations of analytic functions, but first we must introduce the concept of convergence in the complex plane.

Let $w_n = f(n)$ be a complex-valued function defined on the positive integers $n = 1, 2, 3, \ldots$. We say that we have a **sequence** denoted by $\{w_n\}$. We say that $\{w_n\}$ has a **limit** w if and only if given any $\epsilon > 0$ there is an N such that $|w_n - w| < \epsilon$ for all $n \geq N$. If the sequence has a limit, we say that it **converges**. Otherwise we say that it diverges.

The limit of a sequence $\{w_n\}$ is unique. This can be shown as follows. Let w and w' be limits where $|w - w'| = \delta \neq 0$. There exists an N such that $|w_n - w| < \delta/2$ and $|w_n - w'| < \delta/2$ for all $n \geq N$. But then $|w - w'| = |w - w_n + w_n - w'| \leq |w - w_n| + |w' - w_n| < \delta$ for all $n \geq N$. This is a contradiction, and so $w = w'$.

If $w_n = u_n + iv_n$, then $\{w_n\}$ defines two real sequences $\{u_n\}$ and $\{v_n\}$. It is easily shown (exercise 1, Sec. 7.6) that $\{w_n\}$ has the limit $w = u + iv$ if and only if $\{u_n\}$ converges to u and $\{v_n\}$ converges to v. The Cauchy criterion for convergence of a real sequence is: the sequence $\{u_n\}$ has a limit if and only if given any $\epsilon > 0$ there exists an N such that $|u_n - u_m| < \epsilon$ for all $n \geq N$ and all $m \geq N$. The importance of this is clear since it settles the question of convergence of a sequence without the explicit knowledge of the limit. The Cauchy criterion for complex sequences $\{w_n\}$ is the following: the sequence $\{w_n\}$ has a limit if and only if given any $\epsilon > 0$ there exists an N such that $|w_n - w_m| < \epsilon$ for all $n \geq N$ and all $m \geq N$. This can be proved easily using the inequalities $|u_n - u_m| \leq |w_n - w_m|$, $|v_n - v_m| \leq |w_n - w_m|$, and $|w_n - w_m| \leq |u_n - u_m| + |v_n - v_m|$. The proof will be left for the exercises (exercise 2, Sec. 7.6).

The customary notation used to indicate that $\{w_n\}$ has a limit w is $\lim_{n \to \infty} w_n = w$. Using this notation the following properties hold: if $\lim_{n \to \infty} w_n = w$ and $\lim_{n \to \infty} w'_n = w'$, then

1. $\lim\limits_{n \to \infty} (w_n + w'_n) = w + w'$

2. $\lim\limits_{n \to \infty} (w_n - w'_n) = w - w'$

3. $\lim\limits_{n \to \infty} w_n w'_n = ww'$

4. $\lim\limits_{n \to \infty} \dfrac{w_n}{w'_n} = \dfrac{w}{w'}$ provided $w' \neq 0$

The proofs of these properties will be left for the reader.

We can also consider $\{w_n(z)\}$, where each element of the sequence is a function of the complex variable z. We can have convergence for certain values of z and divergence for other values. If each $w_n(z)$ is defined on some set S, then we say that we have **pointwise convergence** on S if given any $\epsilon > 0$ there exists an N (depending possibly on z) such that $|w_n(z) - w_m(z)| < \epsilon$ for all $n \geq N$ and all $m \geq N$. If we have pointwise convergence on S, then the limit $w(z)$, which generally depends on z, is a function of z on S. If it happens in the definition of pointwise convergence that for every $\epsilon > 0$ we can find an N which is independent of z in S, then we say that $\{w_n(z)\}$ **converges uniformly** to $w(z)$ on S. To illustrate this concept, we consider the following example. Let

$$w_n(z) = 1 + z + z^2 + \cdots + z^n = \frac{1 - z^{n+1}}{1 - z}$$

and let $S = \{z \mid |z| < 1\}$. Then

$$\left| w_n(z) - \frac{1}{1 - z} \right| = \frac{|z|^{n+1}}{|1 - z|}$$

where the denominator is not zero since $|1 - z| \geq 1 - |z| \neq 0$. Clearly

$$\lim_{n \to \infty} \frac{|z|^{n+1}}{|1 - z|} = 0$$

for all z in S, so that $\{w_n\}$ converges pointwise on S. In fact, we may take

$$N = \left[\frac{\ln \epsilon \, |1 - z| - \ln |z|}{\ln |z|} \right] + 1$$

where $[\alpha]$ denotes the greatest integer in α. Obviously N depends on z and it is impossible to find an N which works for all z in S for a given ϵ. This is because, having fixed ϵ, our assumption that an N exists satisfying the convergence requirement independent of z in S leads to a contradiction, because there are always z close to 1 in S requiring a larger N. Therefore, we do not have uniform convergence in S. However, if $S_r = \{z \mid |z| \leq r < 1\}$, we have uniform convergence in S_r because

$$N_r = \left[\frac{\ln \epsilon \, |1 - r| - \ln r}{\ln r} \right] + 1$$

will work for all z in S_r.

Uniform convergence is important for the reason that it implies basic properties of the limit function. For example, we have the following theorem:

Theorem 1. A uniformly convergent sequence of continuous functions converges to a continuous function.

Proof. Let $\{w_n(z)\}$ be a sequence of functions all continuous on S. Let $\lim_{n \to \infty} w_n(z) = w(z)$ uniformly on S. We shall prove that $w(z)$ is continuous at z_0, where z_0 is an arbitrary point in S. Consider $|w(z) - w(z_0)|$. We have

$$|w(z) - w(z_0)| = |w(z) - w_n(z) + w_n(z) - w_n(z_0) + w_n(z_0) - w(z_0)|$$
$$\leq |w(z) - w_n(z)| + |w_n(z) - w_n(z_0)| + |w_n(z_0) - w(z_0)|$$

Given $\epsilon > 0$ there exists an N, independent of z in S, such that $|w(z) - w_n(z)| < \epsilon/3$ and $|w(z_0) - w_n(z_0)| < \epsilon/3$ for $n \geq N$. Let us fix $n \geq N$; then $|w_n(z) - w_n(z_0)| < \epsilon/3$ provided z is sufficiently close to z_0, because $w_n(z)$ is continuous at z_0. Consequently $|w(z) - w(z_0)| < \epsilon$ provided z is sufficiently close to z_0. This completes the proof.

Uniform convergence also gives us an important result about taking a limit under an integral sign. We have the following theorem:

Theorem 2. Let $w_n(z)$ be continuous on a contour C for $n = 1, 2, 3, \ldots$ and converge uniformly to $w(z)$ on C. Then

$$\lim_{n \to \infty} \int_C w_n(z)\, dz = \int_C \lim_{n \to \infty} w_n(z)\, dz = \int_C w(z)\, dz$$

Proof. Given $\epsilon > 0$ there exists an N, independent of z on C, such that $|w_n(z) - w(z)| < \epsilon/4L$ for all $n \geq N$, where L is the length of C. Then

$$\left| \int_C w(z)\, dz - \int_C w_n(z)\, dz \right| = \left| \int_C [w(z) - w_n(z)]\, dz \right|$$
$$< \frac{\epsilon}{L} \cdot L = \epsilon$$

for all $n \geq N$. This proves that

$$\int_C w(z)\, dz = \lim_{n \to \infty} \int_C w_n(z)\, dz$$

We now can prove an important theorem about uniformly convergent sequences of analytic functions.

Theorem 3. Let $w_n(z)$ be an analytic function on a simply connected region S for $n = 1, 2, 3, \ldots$, and let $\lim_{n \to \infty} w_n(z) = w(z)$ uniformly on S. Then $w(z)$ is analytic on S.

Proof. Let C be any simple closed contour in S. Then by Cauchy's theorem $\int_C w_n(z)\, dz = 0$. By the previous theorem

$$\int_C w(z)\, dz = \lim_{n \to \infty} \int_C w_n(z)\, dz = 0$$

By the continuity of $w(z)$ and Morera's theorem, $w(z)$ is analytic on S.

We now turn our attention to series of complex numbers. We denote an infinite series of complex numbers by

$$\sum_{k=1}^{\infty} w_k$$

where $\{w_k\}$ is a sequence of complex numbers. The nth partial sum of the series is

$$S_n = \sum_{k=1}^{n} w_k$$

We define the convergence of the series in terms of the convergence of the sequence of partial sums $\{S_n\}$; that is, the series $\sum_{k=1}^{\infty} w_k$ converges to the sum S if and only if $\lim_{n \to \infty} S_n = S$. If the sequence $\{S_n\}$ diverges, then the series diverges.

A necessary condition for convergence is that

$$\lim_{n \to \infty} w_n = 0$$

This is so because

$$\lim_{n \to \infty} w_n = \lim_{n \to \infty} (S_n - S_{n-1}) = S - S = 0$$

This condition is not sufficient, however, as indicated by the example $\sum_{k=1}^{\infty} \dfrac{1}{k}$ which diverges although $\lim_{n \to \infty} \dfrac{1}{n} = 0$.

The Cauchy criterion for convergence of the partial sums becomes: for every $\epsilon > 0$ there exists an N such that

$$|S_n - S_m| = |w_n + w_{n-1} + \cdots + w_{m+1}| < \epsilon$$

for all $n \geq N$ and all $m \geq N$. A real series closely related to $\sum_{k=1}^{\infty} w_k$ is $\sum_{k=1}^{\infty} |w_k|$. We say that $\sum_{k=1}^{\infty} w_k$ **converges absolutely** if $\sum_{k=1}^{\infty} |w_k|$ converges. It then follows easily from the Cauchy criterion that absolute convergence implies convergence.   We prove this as follows: given $\epsilon > 0$ there exists N such that

$$|w_n| + |w_{n-1}| + \cdots + |w_{m+1}| < \epsilon$$

for $n \geq N$ and $m \geq N$.   Now convergence follows from

$$|w_n + w_{n-1} + \cdots + w_{m+1}| \leq |w_n| + |w_{n-1}| + \cdots + |w_{m+1}| < \epsilon$$

for $n \geq N$ and $m \geq N$.

The fact that absolute convergence implies convergence means that any of the tests developed for proving convergence of real series with nonnegative terms can be used to prove convergence of the corresponding complex series. We review a couple of these for reference.

Ratio Test.   Let $L = \lim\limits_{n \to \infty} |w_{n+1}|/|w_n|$, if it exists.   Then the series $\sum\limits_{k=1}^{\infty} w_k$ (1) converges if $L < 1$, (2) diverges if $L > 1$, and (3) the test gives no information if $L = 1$.

Root Test.   Let $L = \lim\limits_{n \to \infty} |w_n|^{1/n}$, if it exists.   Then the series $\sum\limits_{k=1}^{\infty} w_k$ (1) converges if $L < 1$, (2) diverges if $L > 1$, and (3) the test gives no information if $L = 1$.

If $w_k(z)$ is a function defined on some set T for $k = 1, 2, 3, \ldots$ and we have convergence of $\sum\limits_{k=1}^{\infty} w_k(z)$ for each z in T, then we say we have pointwise convergence in T.   If an N in the convergence criterion (or Cauchy criterion) can be found for each $\epsilon > 0$, independent of z in T, then we say that we have uniform convergence in T.   Based on our work on uniformly convergent sequences we have the following theorems for series of complex functions:

Theorem 4.   A uniformly convergent series of continuous functions converges to a continuous function.

Theorem 5.   If T is a region,[1] $w_k(z)$ is analytic in T for $k = 1, 2, 3, \ldots$, and $\sum\limits_{k=1}^{\infty} w_k(z)$ converges uniformly in T, then the series converges to a function analytic in T.

Theorem 6.   If $w_k(z)$ is continuous on a contour C for $k = 1, 2, 3, \ldots$ and $\sum\limits_{k=1}^{\infty} w_k(z)$ converges uniformly on C, then

$$\int_C \sum_{k=1}^{\infty} w_k(z) \, dz = \sum_{k=1}^{\infty} \int_C w_k(z) \, dz$$

[1] The corresponding theorem for sequences was stated for simply connected regions. However, the present theorem can be proved by considering simply connected subregions of T.

Theorem 7. If T is a region, $w_k(z)$ is analytic in T for $k = 1, 2, 3, \ldots$, and $\sum\limits_{k=1}^{\infty} w_k(z)$ converges uniformly in T, then

$$\frac{d}{dz} \sum_{k=1}^{\infty} w_k(z) = \sum_{k=1}^{\infty} w_k'(z)$$

in T.

Proof. Let z_0 be any point in T and let $S(z) = \sum\limits_{k=1}^{\infty} w_k(z)$. There exists a circle $C = \{z \mid |z - z_0| = \rho\}$ such that $w_k(z)$ is analytic within and on C for all k. Therefore,

$$S'(z_0) = \frac{1}{2\pi i} \int_{C+} \frac{\sum\limits_{k=1}^{\infty} w_k(z)}{(z - z_0)^2} \, dz$$

By the uniform convergence on C we can interchange the integration and the summation. Hence,

$$S'(z_0) = \sum_{k=1}^{\infty} \frac{1}{2\pi i} \int_{C+} \frac{w_k(z)}{(z - z_0)^2} \, dz = \sum_{k=1}^{\infty} w_k'(z_0)$$

This completes the proof.

There is a simple test for uniform convergence of a series which is discussed next.

Weierstrass M-test. Let $w_k(z)$ be defined in T where $|w_k(z)| \leq M_k$, $k = 1, 2, 3, \ldots$, where M_k is independent of z in T. Let $\sum\limits_{k=1}^{\infty} M_k$ converge. Then $\sum\limits_{k=1}^{\infty} w_k(z)$ converges uniformly in T.

Proof. Given $\epsilon > 0$ there exists N, independent of z in T, such that

$$M_n + M_{n-1} + M_{n-2} + \cdots + M_{m+1} < \epsilon$$

for $n \geq N$ and $m \geq N$. Then

$$|w_n(z) + w_{n-1}(z) + \cdots + w_{m+1}(z)| \leq |w_n(z)| + |w_{n-1}(z)| + \cdots + |w_{m+1}(z)|$$
$$\leq M_n + M_{n-1} + \cdots + M_{m+1} < \epsilon$$

which proves uniform convergence in T.

A very important type of series is the **power series,** of the form $\sum\limits_{k=0}^{\infty} a_k z^k$, where $\{a_k\}$, $k = 0, 1, 2, 3, \ldots$ is a sequence of constants. Obviously, all power series converge for $z = 0$. If $\sum\limits_{k=0}^{\infty} a_k z^k$ converges for $z = z_0 \neq 0$, then it converges absolutely for $|z| < |z_0|$. This is so because $\lim\limits_{n \to \infty} a_n z_0^n = 0$, which implies that $|a_n z_0^n|$ is bounded for all n; that is, there is an M such that

$|a_n|\,|z_0|^n \leq M.$ Therefore

$$|a_n z^n| = |a_n z_0^n| \left| \frac{z}{z_0} \right|^n \leq M\rho^n$$

where $\rho < 1$ if $|z| < |z_0|$. The series $\sum\limits_{k=0}^{\infty} M\rho^k$ converges and so by comparison $\sum\limits_{k=0}^{\infty} a_k z^k$ converges absolutely for $|z| < |z_0|$. If the power series diverges for $z = z_0$, then it must diverge for all z such that $|z| > |z_0|$, because if it converged for some z satisfying this inequality it would have to converge for $z = z_0$, which is a contradiction. Therefore, unless the power series converges only for $z = 0$ or converges for all z, there must be a circle $|z| = R \neq 0$ such that the series converges for $|z| < R$ and diverges for $|z| > R$. R is called the radius of convergence. In case the power series converges only for $z = 0$, we say that $R = 0$. If the series converges for all z, we say $R = \infty$. On the circle[1] of radius R anything can happen (see exercise 13, Sec. 7.6). The radius of convergence can usually be determined by the ratio test or the root test.

If R is the radius of convergence of a power series, then it converges uniformly for $|z| \leq \rho < R$. This is proved by the M-test; that is,

$$|a_k z^k| = |a_k|\,|z|^k \leq |a_k|\,\rho^k = M_k$$

The series $\sum\limits_{k=0}^{\infty} M_k$ converges by the absolute convergence of the power series for $|z| = \rho < R$.

The uniform convergence of a power series for $|z| \leq \rho < R$ immediately implies several important properties:

1. A power series is a continuous function within its circle of convergence.

2. A power series represents an analytic function within its circle of convergence.

3. A power series may be differentiated term by term within its circle of convergence.

4. A power series may be integrated term by term along any contour lying within its circle of convergence.

5. A power series may be differentiated term by term as many times as we wish within its circle of convergence.

The last property can be proved in the same manner as was employed in the above theorem on term-by-term differentiation of a uniformly convergent series of analytic functions, only now one makes use of the Cauchy formula for the nth derivative. If

$$f(z) = \sum\limits_{k=0}^{\infty} a_k z^k$$

[1] This is called the **circle of convergence**.

then
$$f^{(n)}(z) = \sum_{k=n}^{\infty} k(k-1)\cdots(k-n+1)a_k z^{k-n}$$

It is easy to show that the radius of convergence of the last series is R for all n. Now if $z = 0$

$$f^{(n)}(0) = n!\, a_n$$

so that
$$f(z) = \sum_{k=0}^{\infty} \frac{f^{(k)}(0)}{k!} z^k$$

which is called the **Maclaurin series** representation of $f(z)$. If the origin is shifted to the point z_0, then we have the corresponding results

$$f(z) = \sum_{k=0}^{\infty} a_k (z - z_0)^k$$

and
$$f(z) = \sum_{k=0}^{\infty} \frac{f^{(k)}(z_0)}{k!} (z - z_0)^k$$

The latter series is called the **Taylor series** representation of $f(z)$. We shall show in the next section that any function $f(z)$ analytic at z_0 has a Taylor series representation at z_0 with a positive radius of convergence. Hence, power series representations are very important in analytic function theory.

Exercises 7.6

1. Let $\{w_n\} = \{u_n + iv_n\}$ be a complex sequence with u_n and v_n real. Prove that $\{w_n\}$ has the limit $w = u + iv$ if and only if u_n converges to u and v_n converges to v.

2. Prove that the sequence w_n has a limit if and only if given any $\epsilon > 0$ there exists an N such that $|w_n - w_m| < \epsilon$ for all $n \geq N$ and all $m \geq N$.

3. Prove that $\lim_{n\to\infty} w_n = w \neq 0$ implies that there is a $\delta > 0$ such that $|w_n| \geq \delta$ for all sufficiently large n.

4. Prove the limit properties for sums, differences, products, and quotients.

5. Prove that $\lim_{n\to\infty} w_n = w$ implies that $\lim_{n\to\infty} |w_n| = |w|$. Is the converse true? Is the corresponding property true for arg w_n?

6. Prove that z^n, $n = 1, 2, 3, \ldots$ converges for $|z| < 1$ but not uniformly.

7. Prove that n^z, $n = 1, 2, 3, \ldots$ converges in $S = \{z \mid \text{Re}\,(z) < 0 \text{ or } z = 0\}$ but nor uniformly in S. Prove that $\{n^z\}$ converges uniformly in

$$S_\rho = \{z \mid \text{Re}\,(z) \leq \rho < 0\}$$

8. Prove that the series $\sum_{k=1}^{\infty} w_k$, with $w_k = u_k + iv_k$, u_k and v_k real, converges to $S = U + iV$ if and only if $\sum_{k=1}^{\infty} u_k$ converges to U and $\sum_{k=1}^{\infty} v_k$ converges to V.

9. Give an example of a series which converges conditionally, that is, converges but does not converge absolutely.

10. Prove the ratio test. HINT: (1) implies absolute convergence, (2) implies that $\lim_{n \to \infty} |w_n| = \infty$, and (3) can be handled by counterexamples.

11. Prove the root test. See hint in exercise 10.

12. Let $w_k(z)$ be analytic in a region T, and let $\sum_{k=1}^{\infty} w_k(z) = w(z)$, converging uniformly in T. Prove that $w^{(n)}(z) = \sum_{k=1}^{\infty} w_k^{(n)}(z)$, $n = 1, 2, 3, \ldots$. HINT: Use the Cauchy formula for the nth derivative.

13. Find the radius of convergence of the series $\sum_{k=0}^{\infty} z^k$, $\sum_{k=1}^{\infty} z^k/k$, and $\sum_{k=1}^{\infty} z^k/k^2$. Find the points on the circles of convergence where the series converge and diverge.

14. Find the radius of convergence of the series $\sum_{k=1}^{\infty} kz^k$, $\sum_{k=0}^{\infty} k! \, z^k$, and $\sum_{k=0}^{\infty} z^k/k!$.

15. Find the sum of the series $\sum_{k=1}^{\infty} k^2 z^k$. HINT: Consider derivatives of the geometric series $\sum_{k=0}^{\infty} z^k = 1/(1 - z)$, $|z| < 1$.

16. If $f(z) = \sum_{k=0}^{\infty} z^k/k!$, what is $f'(z)$? Is $f(z) = e^z$?

17. Show that the root test works on the series with terms $w_{2k} = (2i/3)^{2k}$, $w_{2k+1} = (i/3)^{2k+1}$, $k = 1, 2, 3, \ldots$, although the ratio test fails.

18. What is the region of uniform convergence of $\sum_{k=0}^{\infty} (z + 1)^k/(z - 1)^k$?

19. Prove that $\sum_{k=0}^{\infty} a_k z^k = 0$ for $|z| < R$ implies that $a_k = 0$ for all k. What does this say about the uniqueness of power series representations of functions?

7.7 Series Representations of Analytic Functions

In the last section, we found that a power series represented an analytic function within its circle of convergence. We now show that the converse is also true, that is, that if $f(z)$ is analytic at z_0, then it has a power series representation in some circle centered at z_0.

Theorem 1. Let $f(z)$ be analytic at z_0. Then

$$f(z) = \sum_{k=0}^{\infty} \frac{f^{(k)}(z_0)}{k!} (z - z_0)^k$$

where the power series has a positive radius of convergence.

Proof. Since $f(z)$ is analytic at z_0, there exists a neighborhood of z_0 throughout which it is analytic. Let C be a circle centered at z_0 lying in this neighborhood. Then if z lies inside C

$$f(z) = \frac{1}{2\pi i} \int_{C+} \frac{f(\zeta)}{\zeta - z} \, d\zeta$$

Now

$$\frac{1}{\zeta - z} = \frac{1}{\zeta - z_0 - (z - z_0)} = \frac{1}{\zeta - z_0}\left[\frac{1}{1 - (z - z_0)/(\zeta - z_0)}\right] = \sum_{k=0}^{\infty}\frac{(z - z_0)^k}{(\zeta - z_0)^{k+1}}$$

and this series converges uniformly with respect to ζ on C provided z is inside C. Therefore, we can substitute the series in the integral and integrate term by term. Hence,

$$\begin{aligned}
f(z) &= \frac{1}{2\pi i}\int_{C+}\sum_{k=0}^{\infty}\frac{f(\zeta)(z - z_0)^k}{(\zeta - z_0)^{k+1}}\,d\zeta \\
&= \sum_{k=0}^{\infty}\frac{(z - z_0)^k}{2\pi i}\int_{C+}\frac{f(\zeta)}{(\zeta - z_0)^{k+1}}\,d\zeta \\
&= \sum_{k=0}^{\infty}\frac{f^{(k)}(z_0)}{k!}(z - z_0)^k
\end{aligned}$$

The series has a positive radius of convergence since the way it was derived guarantees that it converge for some $z \neq z_0$. This completes the proof.

In order to apply the last theorem, $f(z)$ must be defined in a region D_1, which contains an ϵ-neighborhood of z_0. The interior of the circle of convergence of the Taylor series representation of $f(z)$ at z_0 is a region D_2. Clearly, if D_2 is contained in D_1, then the Taylor series represents $f(z)$ everywhere in D_2. On the other hand, what if D_2 is not contained in D_1? In this case, D_1 and D_2 intersect in an open set where $f(z)$ agrees with the Taylor series. Now the union of D_1 and D_2 is a region $D_1 \cup D_2$. Suppose we define an analytic function $F(z)$ in $D_1 \cup D_2$ as follows:

$$F(z) = \begin{cases} f(z) & \text{in } D_1 \\ \displaystyle\sum_{k=0}^{\infty}\frac{f^{(k)}(z_0)}{k!}(z - z_0)^k & \text{in } D_2 \end{cases}$$

It turns out that this is the only possible analytic function defined in $D_1 \cup D_2$ which takes on the common values of $f(z)$ and its Taylor series representation. This follows from the following theorem:

Theorem 2. Let $f(z)$ and $g(z)$ both be analytic in a region D, where they agree on a set of points having a limit point z_0 in D. Then $f(z) \equiv g(z)$ in D.

Proof. Let $h(z) = f(z) - g(z)$. Then $h(z)$ is analytic in D and is zero on a set of points with a limit point z_0 in D. We shall prove that $h(z) \equiv 0$ in D. By continuity $h(z_0) = 0$. Therefore, $h(z)$ has a Taylor series

$$h(z) = \sum_{k=1}^{\infty}\frac{h^{(k)}(z_0)}{k!}(z - z_0)^k = (z - z_0)h_1(z)$$

which converges in some ϵ-neighborhood of z_0. Then

$$h_1(z) = \frac{h(z)}{z - z_0} = \frac{h(z) - h(z_0)}{z - z_0}$$

Let $\{z_n\}$ be a sequence of distinct points with limit z_0 such that $h(z_n) = 0$. Then

$$h'(z_0) = \lim_{n \to \infty} h_1(z_n) = \lim_{n \to \infty} \frac{h(z_n) - h(z_0)}{z_n - z_0} = 0$$

This proves that $h'(z_0) = 0$, and a simple induction shows that $h^{(n)}(z_0) = 0$ for all n. This shows that $h(z) \equiv 0$ in some ϵ-neighborhood of z_0. Since D is connected, an arbitrary point ζ in D and z_0 can be joined by a polygonal path, which by the Heine-Borel property can be covered by a finite number of ϵ-neighborhoods lying in D. These open neighborhoods form a finite chain of overlapping open sets reaching from z_0 to ζ. Now we have already shown that $h(z) \equiv 0$ in the first. Again using the Taylor series representation of $h(z)$, we can show that $h(z) \equiv 0$ in the second, and so on until, after a finite number of steps, that $h(\zeta) = 0$. This proves the theorem.

The method we have described above for extending the definition of an analytic function to a larger domain by a power series representation is called **analytic continuation**. More generally, we can describe the process as follows:

Let $f(z)$ be analytic in a region D_1 and let $g(z)$ be analytic in a region D_2, and let D_1 intersect D_2, where $f(z) \equiv g(z)$. Then $f(z)$ is called the analytic continuation of $g(z)$ into D_1 and $g(z)$ is called the analytic continuation of $f(z)$ into D_2. The theorem we have just proved implies that the continuation of an analytic function into an overlapping region is unique.

If we start with a given analytic function $f(z)$ defined on a region D, not the whole plane, then either we can or cannot continue $f(z)$ to a larger region. If we cannot, then the boundary of D is called a natural boundary (see exercise 4, Sec. 7.7). If we can, then we say that $f(z)$ is continuable. Now we can ask the same question relative to the continued analytic function, etc. The global definition of an analytic function will be the result of starting with a given analytic function defined on a region D, taking all possible continuations of it, all possible continuations of continuations, etc. We must be prepared for the possibility of a globally defined analytic function being multivalued. For example,

$$f(z) = \sum_{k=1}^{\infty} \frac{(-1)^{k+1}(z - 1)^k}{k} \qquad |z - 1| < 1$$

is a power series representation of a part of the branch of $\log z$ which takes on the value zero at $z = 1$. If this is continued once around the origin in the counterclockwise direction using power series representations with centers on the unit circle, we arrive back at the point $z = 1$ with the value $2\pi i$. This is the value of log z on another branch. In fact, starting with the given function defined for $|z - 1| < 1$, the global function defined from it by all possible analytic continuations is precisely the multivalued function log z. The proper way to reconcile the global definition of an analytic function with the requirement that a function be single-valued is to define analytic functions on **Riemann surfaces** in which we reproduce the complex plane as many times as we have different values of the global function at the same complex number.[1]

Considering all possible continuations of an analytic function, there may still be points at which the function will not be analytic in any continuation process. Such a point is called a singularity of the analytic function if it is the limit point of at least one of the regions on which the function is continued. An example of this is the origin with respect to the function log z, discussed above. Also every point on a natural boundary of a function is a singularity of that function. It is difficult to know where the singularities of a function are unless one has the global definition conveniently stated. However, one thing that can be said is that there is at least one singularity of an analytic function on the circle of convergence of every one of its power series representations. To prove this, consider the circle of convergence $C = \{z \mid |z - z_0| = R\}$ covered by open sets consisting of the interiors of circles of convergence of individual power series continuations, assuming no singularities on C. This is an open covering of C, which is a compact set. Therefore, by the Heine-Borel property, it has a finite subcovering. The union of this finite subcovering with the interior of C is an open set which contains a circle $C_\delta = \{z \mid |z - z_0| = R + \delta, \delta > 0\}$ within and on which $f(z)$ is analytic. But then C_δ could be used to prove that the Taylor series representation at z_0 actually converges outside C, contradicting the fact that R is the radius of convergence of the Taylor series at z_0.

There is a more general type of expansion valid for a function analytic in an annulus. This is known as the **Laurent expansion**.

Theorem 3. Let $f(z)$ be analytic in the annulus $R_1 < |z - z_0| < R_2$. Then $f(z) = \sum\limits_{k=-\infty}^{\infty} a_k(z - z_0)^k$ in the annulus, where

$$a_k = \frac{1}{2\pi i} \int_{C+} \frac{f(z)}{(z - z_0)^{k+1}} \, dz$$

where C is any closed contour lying in the annulus with z_0 in its interior.

Proof. Let z be in the annulus and $\rho = |z - z_0|$. There are circles C_1 and C_2 with radii r_1 and r_2, respectively, such that $R_1 < r_1 < \rho < r_2 < R_2$.

---

[1] See John W. Dettman, "Applied Complex Variables," The Macmillan Company, New York, 1965.

Using the generalized Cauchy formula for a multiply connected domain, we have

$$f(z) = -\frac{1}{2\pi i}\int_{C_1+}\frac{f(\zeta)}{\zeta-z}\,d\zeta + \frac{1}{2\pi i}\int_{C_2+}\frac{f(\zeta)}{\zeta-z}\,d\zeta$$

On C_1

$$\frac{1}{\zeta-z} = \frac{1}{\zeta-z_0-(z-z_0)}$$

$$= \frac{-1}{(z-z_0)\,[1-(\zeta-z_0)/(z-z_0)]} = -\sum_{k=-\infty}^{-1}\frac{(z-z_0)^k}{(\zeta-z_0)^{k+1}}$$

On C_2

$$\frac{1}{\zeta-z} = \frac{1}{\zeta-z_0-(z-z_0)}$$

$$= \frac{1}{(\zeta-z_0)[(1-(z-z_0)/(\zeta-z_0)]} = \sum_{k=0}^{\infty}\frac{(z-z_0)^k}{(\zeta-z_0)^{k+1}}$$

The series converge uniformly with respect to ζ on their respective circles. We may therefore insert them in the integrals and integrate term by term. This gives us

$$f(z) = \sum_{k=-\infty}^{-1}\frac{(z-z_0)^k}{2\pi i}\int_{C_1+}\frac{f(\zeta)}{(\zeta-z_0)^{k+1}}\,d\zeta + \sum_{k=0}^{\infty}\frac{(z-z_0)^k}{2\pi i}\int_{C_2+}\frac{f(\zeta)}{(\zeta-z_0)^{k+1}}\,d\zeta$$

Noting that $f(\zeta)/(\zeta-z_0)^{k+1}$ is analytic in the annulus for all k, we can replace the circles C_1 and C_2 by an arbitrary closed contour C lying in the annulus provided z_0 is in the interior of C. This completes the proof.

A case which is especially important is the case where z_0 is an isolated singularity of $f(z)$; that is, there exists a deleted neighborhood of z_0, $0 < |z-z_0| < R$ in which $f(z)$ is analytic. Then $f(z)$ has a Laurent expansion at z_0, valid for $0 < |z-z_0| < R$. It is customary to break this into two parts, as follows:

$$f(z) = \sum_{k=0}^{\infty}a_k(z-z_0)^k + \sum_{k=1}^{\infty}\frac{b_k}{(z-z_0)^k}$$

The part $\sum_{k=1}^{\infty} b_k(z-z_0)^{-k}$ is called the **principal part** of $f(z)$ at z_0. There are three cases, which can be classified as follows:

1. If the principal part is zero we say that the function has a **removable singularity** at z_0. In this case, by defining $f(z_0)$ as a_0 we make the function analytic at z_0.

2. If $b_m \neq 0$ and $b_k = 0$ for all $k > m$, we say that $f(z)$ has a **pole of order** m at z_0. In this case $(z-z_0)^m f(z)$ has a removable singularity at z_0.

3. If the principal part has an infinite number of nonzero terms in it, we say that $f(z)$ has an **essential singularity** at z_0.

The coefficient b_1 is especially important. This is because

$$b_1 = \frac{1}{2\pi i} \int_{C+} f(z)\, dz$$

and so b_1 gives the value of the integral of $f(z)$ taken around a closed contour lying in a region of analyticity with the only possible singularity inside C at z_0. We shall take advantage of this fact in the next section to evaluate integrals of various types. The coefficient b_1 is called the **residue** of $f(z)$ at z_0.

It will be necessary to calculate the residue of a function at a pole of order m. This is accomplished as follows. The function

$$g(z) = \begin{cases} (z - z_0)^m f(z) & z \neq z_0 \\ b_m & z = z_0 \end{cases}$$

is analytic at z_0 and has the Taylor expansion

$$g(z) = b_m + b_{m-1}(z - z_0) + \cdots + b_1(z - z_0)^{m-1} + \sum_{k=0}^{\infty} a_k(z - z_0)^{k+m}$$

Therefore,

$$b_1 = \frac{g^{(m-1)}(z_0)}{(m-1)!} = \lim_{z \to z_0} \frac{1}{(m-1)!} \frac{d^{m-1}}{dz^{m-1}} (z - z_0)^m f(z)$$

Exercises 7.7

1. Find the Taylor series representation of e^z about z_0. Where does it converge?

2. Find the Taylor series representation of the branch of $\log z$ which takes the value zero at $z = 1$, about $z = 1$. Where does it converge?

3. The series $\sum_{k=0}^{\infty} z^k$ converges only for $|z| < 1$. Where is it analytic? What is the global definition of this function? Is it single-valued? multivalued? What are the singularities?

4. Consider the series $f(z) = \sum_{k=0}^{\infty} z^{k!}$. Where does it converge? Show that for p and q any positive integers

$$\lim_{r \to 1^-} f(re^{2\pi pi/q}) = \infty$$

This shows that $f(z)$ cannot be continuous at any point on the unit circle of the form $e^{2\pi pi/q}$, and hence there cannot be analytic continuations through any of these points. This shows that $f(z)$ has a **natural boundary** on the unit circle.

5. Let $f(x) = \sum_{k=0}^{\infty} a_k(x - x_0)^k$, $|x - x_0| < R$, a_k and x_0 real. Show that $f(z) = \sum_{k=0}^{\infty} a_k(z - x_0)^k$ is analytic for $|z - x_0| < R$ and equals $f(x)$ when z is real. Show that $f(z)$ is the only function analytic for $|z - x_0| < R$ with this property.

6. Prove that the Laurent expansion is unique. HINT: Given a series of the Laurent form multiply by $(z - z_0)^n$ and integrate term by term.

7. Show that $\sin z/z$ has a removable singularity at $z = 0$.

8. Show that $\tan z$ has a pole of order one at each of the zeros of $\cos z$. What are the residues?

9. Show that $e^{1/z}$ has an essential singularity at $z = 0$.

10. Show that if $f(z) = 1/g(z)$ has a pole of order one at z_0, then $g(z_0) = 0$, $g'(z_0) \neq 0$. Show that the residue of $f(z)$ at z_0 is $1/g'(z_0)$. Extend this result to the case of a pole of order two.

11. Let $f(z)$ have a pole at z_0. Show that $\lim\limits_{z \to z_0} |f(z)| = \infty$.

12. A function with an essential singularity at z_0 has the property that it takes on every value, with one possible exception, in every neighborhood of z_0. Show that this is the case with $e^{1/z}$ where the exceptional value is zero.

13. A function may have several Laurent expansions. Find expansions of $f(z) = 1/z(z - 1)$ valid in (a) $0 < |z| < 1$, (b) $0 < |z - 1| < 1$, (c) $|z| > 1$, and (d) $|z - 1| > 1$.

14. A zero of an analytic function is a point z_0 where $f(z)$ is analytic and $f(z_0) = 0$. Prove that, unless $f(z) \equiv 0$, a zero is isolated.

15. Let $f(z)$ be analytic everywhere in the complex plane except for certain poles. Prove that it has but a finite number of poles in any bounded region.

7.8 Contour Integration

The remark of the last section about the use of the residue at an isolated singularity of an analytic function to compute the integral of the function around a closed contour surrounding the singularity is very important in the calculation of integrals around closed contours inside which there are only a finite number of isolated singularities. In most cases of this kind, it is easy to show that the integral around the closed contour can be expressed as a finite sum of integrals around circles inside each of which there is but one singularity. Consider for example the situation illustrated in Fig. 16. Let $f(z)$ be analytic within and on the closed contour C except at n points z_1, $z_2, \ldots, z_n$ inside C. If we start at β and integrate to α along the upper part of C and then return to β along an obvious path made up of line segments and

arcs of circles, the integral will have the value zero by Cauchy's theorem. On the other hand, starting at α we can integrate along the lower part of C to β and then return to α along a path consisting of line segments and arcs of circles so that no singularities of $f(z)$ are inside this closed path. The result will again be zero. Adding these two integrals, the part along the line segments will cancel, and we will have shown that the

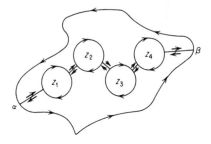

FIGURE 16

integral around C in the positive direction plus the sum of the integrals around the circles in the negative direction equals zero. Taking the integrals around the circles to the other side of the equation and reversing the direction of integration shows that

$$\int_{C+} f(z)\, dz = 2\pi i \sum_{j=1}^{n} K_j$$

where K_j is the residue of $f(z)$ at z_j. This is essentially the **residue theorem.**

Theorem 1. If $f(z)$ is analytic within and on a closed contour C except for a finite number of isolated singularities inside C, then $\int_{C+} f(z)\, dz$ is equal to $2\pi i$ times the sum of the residues at the isolated singularities inside C.

To illustrate the use of this theorem, consider $\int_{C+} [1/(z^2 + 1)^2]\, dz$, where $C = \{z \mid |z - i| = 1\}$. The singularities of the integrand are at $\pm i$, the only one of which inside C is i. The integrand can be written as

$$\frac{1}{(z^2 + 1)^2} = \frac{1}{(z - i)^2 (z + i)^2}$$

and hence there is a pole of order 2 at $z = i$. To calculate the residue at i we multiply by $(z - i)^2$ and differentiate:

$$K = \lim_{z \to i} \frac{d}{dz} \frac{1}{(z + i)^2} = \frac{1}{4i}$$

Therefore,

$$\int_{C+} \frac{1}{(z^2 + 1)^2}\, dz = \frac{\pi}{2}$$

One of the applications of contour integration is in the evaluation of certain real integrals. Consider for example $\int_0^{2\pi} (a + b \cos \theta)^{-1}\, d\theta$, where $|a| > |b|$, a and b real. Let $z = e^{i\theta}$; then $\cos \theta = \frac{1}{2}(e^{i\theta} + e^{-i\theta}) = \frac{1}{2}(z + z^{-1})$, $d\theta = dz/iz$, and

$$\int_0^{2\pi} \frac{1}{a + b \cos \theta}\, d\theta = -2i \int_{C+} \frac{dz}{bz^2 + 2az + b}$$

where C is the unit circle. The only singularities of the integrand are at $-(a/b) \pm \sqrt{a^2 - b^2}/b$. If $0 < b < a$, then $-(a/b) + \sqrt{a^2 - b^2}/b$ is inside C and the other is outside. In this case, by the residue theorem,

$$\int_0^{2\pi} \frac{1}{a + b \cos \theta}\, d\theta = \frac{-2i}{b} \int_C \frac{dz}{z^2 + 2(a/b)z + 1} = 2\pi i \left(\frac{-i}{\sqrt{a^2 - b^2}} \right) = \frac{2\pi}{\sqrt{a^2 - b^2}}$$

If $a < b < 0$, then the value of the integral is $-2\pi/\sqrt{a^2 - b^2}$. If $b < 0 < a$, then the pole at $-(a/b) - \sqrt{a^2 - b^2}/b$ is inside C while the other is outside.

In this case, using the residue theorem again we obtain the value $2\pi/\sqrt{a^2 - b^2}$. If $a < 0 < b$, the value is $-2\pi/\sqrt{a^2 - b^2}$.

The residue theorem is also useful in evaluating certain real improper integrals. Consider for example $\int_{-\infty}^{\infty} (1 + x^4)^{-1} \, dx$. It is easy to show that this improper integral exists and is equal to

$$\lim_{R \to \infty} \int_{-R}^{R} \frac{1}{1 + x^4} \, dx$$

We evaluate this limit using a contour integral which goes from $-R$ to R on the real axis and then returns to $-R$ along the arc of the semicircle $z = Re^{i\theta}$, $0 \leq \theta \leq \pi$. We call this closed contour C, and by the residue theorem the value of the integral around $C+$ (if $R > 1$) is $2\pi i$ times the sum of the residues at $(1/\sqrt{2})(\pm 1 + i)$. Therefore,

$$\int_{C+} \frac{1}{1 + z^4} \, dz = \int_{-R}^{R} \frac{1}{1 + x^4} \, dx + \int_{0}^{\pi} \frac{iRe^{i\theta} \, d\theta}{1 + R^4 e^{4i\theta}}$$

$$= 2\pi i \left[\frac{\sqrt{2}}{2i(2 + 2i)} + \frac{\sqrt{2}}{2i(2 - 2i)} \right] = \frac{\pi}{\sqrt{2}}$$

Letting R approach infinity, the first integral is the improper integral we wish to evaluate, while

$$\left| \int_{0}^{\pi} \frac{iRe^{i\theta} \, d\theta}{1 + R^4 e^{4i\theta}} \right| \leq \frac{4\pi R}{R^4 - 1} \to 0$$

as $R \to \infty$. Therefore, we have

$$\int_{-\infty}^{\infty} \frac{1}{1 + x^4} \, dx = \frac{\pi}{\sqrt{2}}$$

In order to consider certain integral transforms in the next section, we must have some information about functions defined as improper integrals along infinite contours. We say that C is an **infinite contour** if it is a curve with a parametric representation $z(t) = x(t) + iy(t)$, $a \leq t < b$, such that for each T, $a < T < b$, the part of the curve for $a \leq t \leq T$ is a contour with length $L(T)$ and $\lim_{T \to b} L(T) = \infty$. We shall consider functions of the form

$$\int_{C} f(\zeta, z) \, d\zeta$$

where ζ is on C and z is in some region D. If $f(\zeta, z)$ is continuous on C for each z in D and analytic in D for each ζ on C, then the function

$$F_n(z) = \int_{C_n} f[\zeta(t), z] \, d\zeta(t)$$

$a \leq t \leq t_n < b$ is analytic in D and

$$F'_n(z) = \int_{C_n} \frac{\partial f(\zeta,z)}{\partial z}\, d\zeta$$

Let $\{t_n\}$ be a sequence of real numbers approaching b. If $\{F_n(z)\}$ converges uniformly in D, then $F(z) = \lim_{n \to \infty} F_n(z)$ is analytic in D and

$$F'(z) = \int_C \frac{\partial f}{\partial z}\, d\zeta$$

We need a test for uniform convergence of $\{F_n(z)\}$. Let $\zeta(t) = \xi(t) + i\eta(t)$ be a parameterization of C. Let

$$|f(\zeta,z)|\, |\dot{\xi}(t) + i\dot{\eta}(t)| \leq M(t)$$

for all z in D, and let $\int_a^b M(t)\, dt$ exist. Then $\{F_n(z)\}$ converges uniformly in D. To prove this, let $\{t_n\}$ be a sequence approaching b from below. Then

$$I_n = \int_a^{t_n} M(t)\, dt$$

converges to $\int_a^b M(t)\, dt$. Therefore, given $\epsilon > 0$, there is an N such that $|I_n - I_m| = \int_{t_m}^{t_n} M(t)\, dt < \epsilon$ for $n \geq N$ and $m \geq N$. Then

$$|F_n(z) - F_m(z)| = \left| \int_{t_m}^{t_n} f[\zeta(t),z](\dot{\xi} + i\dot{\eta})\, dt \right|$$

$$\leq \int_{t_m}^{t_n} M(t)\, dt < \epsilon$$

for $n \geq N$ and $m \geq N$. This completes the proof.

As an example, let $g(t)$ be continuous for $0 \leq t < \infty$ and $|g(t)| \leq K e^{\alpha t}$ for all positive t. We consider the Laplace transform of $g(t)$; that is,

$$\int_0^\infty e^{-zt} g(t)\, dt$$

Let z be in the half plane Re $(z) \geq \rho > \alpha$. e^{-zt} is analytic in this half plane for each positive t. Now

$$|e^{-zt} g(t)| \leq K e^{(\alpha - x)t} \leq K e^{(\alpha - \rho)t}$$

for all z satisfying Re $(z) \geq \rho > \alpha$. Finally,

$$\int_0^\infty K e^{(\alpha - \rho)t}\, dt = \frac{K}{\rho - \alpha}$$

Therefore, since ρ is any number greater than α, we have proved analyticity of the Laplace transform for Re $(z) > \alpha$.

As another example, consider the Fourier transform of a function $g(t)$ which is continuous for $-\infty < t < \infty$ and such that $|g(t)| \leq Ke^{-\alpha t}$ for $t \geq 0$ and $|g(t)| \leq Me^{\beta t}$ for $t \leq 0$. The Fourier transform of $g(t)$ is

$$\frac{1}{\sqrt{2\pi}} \int_{-\infty}^{\infty} e^{-izt} g(t)\, dt$$

Let $-\beta < \rho_1 \leq \operatorname{Im}(z) \leq \rho_2 < \alpha$. The function e^{-izt} is analytic for z in this strip for every t. Also

$$|g(t)e^{-izt}| \leq Ke^{(y-\alpha)t} \leq Ke^{(\rho_2-\alpha)t} \qquad t \geq 0$$
$$|g(t)e^{-izt}| \leq Me^{(y+\beta)t} \leq Me^{(\rho_1+\beta)t} \qquad t \leq 0$$

Therefore, the Fourier transform is analytic in the strip $-\beta < \operatorname{Im}(z) < \alpha$, since

$$\int_0^{\infty} Ke^{(\rho_2-\alpha)t}\, dt = \frac{K}{\alpha - \rho_2}$$

$$\int_{-\infty}^0 Me^{(\rho_1+\beta)t}\, dt = \frac{M}{\beta + \rho_1}$$

The problem of inverting a Laplace or Fourier transform can involve contour integration. For example, it is shown in the next chapter that, under rather general circumstances, if $F(z)$ is a Laplace transform of $f(t)$, analytic in a half plane $\operatorname{Re}(z) > \alpha$, then for $\gamma > \alpha$, $t > 0$,

$$f(t) = \frac{1}{2\pi i} \int_{\gamma-i\infty}^{\gamma+i\infty} F(z)e^{zt}\, dz$$

Let us consider the function $F(z) = 1/(z^2 + 1)$ for $\operatorname{Re}(z) > 0$. We take $\gamma > 0$ and evaluate

$$f(t) = \frac{1}{2\pi i} \int_{\gamma-i\infty}^{\gamma+i\infty} \frac{e^{zt}}{z^2 + 1}\, dz$$

by integrating along a line from $\gamma - iR$ to $\gamma + iR$ and then along a semi-circular arc $z = \gamma + Re^{i\theta}$, $\pi/2 \leq \theta \leq 3\pi/2$. If R is sufficiently large, the value of this integral, by the residue theorem, is $2\pi i$ times the sum of the residues at $\pm i$. Therefore,

$$\frac{1}{2\pi i} \int_{\gamma-iR}^{\gamma+iR} \frac{e^{zt}}{z^2 + 1}\, dz + \frac{1}{2\pi i} \int_{\pi/2}^{3\pi/2} \frac{e^{t(\gamma+Re^{i\theta})} Rie^{i\theta}}{(\gamma + Re^{i\theta})^2 + 1}\, d\theta = \frac{e^{it} - e^{-it}}{2i} = \sin t$$

As $R \to \infty$ the first integral becomes $f(t)$, which is the inverse Laplace transform of $F(z)$. The other integral can be shown to approach zero as $R \to \infty$; that is,

$$\left| \int_{\pi/2}^{3\pi/2} \frac{e^{t(\gamma+Re^{i\theta})} Rie^{i\theta}}{(\gamma + Re^{i\theta})^2 + 1}\, d\theta \right| \leq \frac{4\pi e^{t\gamma} R}{R^2 - 2\gamma R - \gamma^2 - 1} \to 0$$

as $R \to \infty$.

Exercises 7.8

1. Evaluate $\displaystyle\int_{C+} (\tan z)/(1 - e^z)\, dz$, where C is the circle with radius 2 and center at the origin.

2. Evaluate $\displaystyle\int_{C+} z/(1 - e^z)^2\, dz$, where C is the circle with radius 7 and center at the origin.

3. Let $f(z) = P(z)/Q(z)$ be a rational function with the degree of $P(z)$ two or more less than the degree of $Q(z)$. Prove that the sum of the residues at the zeros of $Q(z)$ is zero. HINT: Prove that the integral of $f(z)$ around a very large circle is zero.

4. Evaluate $\displaystyle\int_{C+} ze^{1/z}\, dz$, where C is the unit circle.

5. Evaluate $\displaystyle\int_0^{2\pi} 1/(a + b\sin\theta)\, d\theta$, $|a| > |b|$.

6. Evaluate $\displaystyle\int_{-\infty}^{\infty} 1/(1 + x^2)^2\, dx$.

7. Evaluate $\displaystyle\int_{-\infty}^{\infty} (\cos x)/(1 + x^2)\, dx$. HINT: Integrate $e^{iz}/(1 + z^2)$ around an appropriate contour.

8. Evaluate $\displaystyle\int_0^{\infty} (\sin x)/x\, dx$. HINT: Integrate e^{iz}/z around an appropriate contour indented at the origin.

9. Evaluate $\displaystyle\int_0^{\infty} e^{-x^2}\cos 2mx\, dx$ by integrating around a rectangular contour with corners at $-R$, R, $R + im$, and $-R + im$. HINT: $\displaystyle\int_0^{\infty} e^{-x^2}\, dx = \sqrt{\pi}/2$.

10. The gamma function $\Gamma(z)$ is defined by the improper integral $\displaystyle\int_0^{\infty} t^{z-1}e^{-t}\, dt$ for Re $(z) > 0$. Show that it is analytic as a function of z. Show that $\Gamma(z + 1) = z\Gamma(z)$. HINT: Show that for x real, $\Gamma(x + 1) = x\Gamma(x)$, and obtain the general formula by analytic continuation.

11. Let $f(t)$ be continuous for $0 \le t < \infty$ and let $\displaystyle\int_0^{\infty} |f(t)|\, dt$ exist. Prove that the Mellin transform of $f(t)$, $\displaystyle\int_0^{\infty} t^{z-1}f(t)\, dt$, is analytic in a strip $0 < $ Re $(z) < \beta$ for some β.

12. Use the inversion integral to compute the inverse Laplace transform of $1/z^2$.

13. Use the inversion integral to compute the inverse Laplace transform of $z/(1 + z^2)$.

7.9 Conformal Mapping

Whenever we have a function $w = f(z)$ defined on some region D of the z plane, it defines a mapping of D onto some set of the w plane, the range of f.

The geometric properties of this mapping will be especially interesting and useful if $f(z)$ is analytic and $f'(z) \neq 0$ in D.

Theorem 1. If $w = f(z)$ is analytic at z_0 and $f'(z_0) \neq 0$, then there exists a neighborhood of $w_0 = f(z_0)$ where $f(z)$ has a local inverse which is analytic.

Proof. Let $f(z) = u(x,y) + iv(x,y)$. Then the equations $u = u(x,y)$, $v = v(x,y)$ define a coordinate transformation from the z plane to the w plane. The functions u and v are continuous and differentiable at $z_0 = x_0 + iy_0$. The Jacobian of the transformation is

$$J(x,y) = \begin{vmatrix} u_x(x,y) & u_y(x,y) \\ v_x(x,y) & v_y(x,y) \end{vmatrix} = u_x^2(x,y) + u_y^2(x,y) = |f'(z)|^2$$

using the Cauchy-Riemann equations. Now $J(x_0,y_0) = |f'(z_0)|^2 \neq 0$, and by a well-known theorem the coordinate transformation has a local inverse $x = x(u,v)$, $y = y(u,v)$ defined and continuous in some neighborhood of z_0. Let

$$z = x(u,v) + iy(u,v) = g(w)$$

Then
$$g'(w_0) = \lim_{w \to w_0} \frac{g(w) - g(w_0)}{w - w_0}$$

$$= \lim_{z \to z_0} \frac{z - z_0}{f(z) - f(z_0)} = \frac{1}{f'(z_0)}$$

In fact, in an entirely similar manner we can prove that

$$g'(w) = \frac{1}{f'(z)}$$

for all points in some neighborhood of w_0. This proves that $g(w)$ is analytic at w_0.

It should be noted that this result is only a local result. For example, if $w = z^2$ and $|z_0| = \delta \neq 0$, then the theorem applies at z_0; in other words, there is an analytic inverse $z = \sqrt{w}$ defined in some neighborhood of w_0. However, this neighborhood cannot contain $w = 0$. Since δ may be very small we cannot be sure that the theorem holds for anything but a small distance from w_0.

Theorem 2. Let $w = f(z)$ be analytic at z_0, where $f'(z_0) \neq 0$. Then the mapping preserves angles at z_0 both in magnitude and sense.

Proof. Consider two smooth curves C_1 and C_2 passing through z_0 such that their tangent vectors meet at an angle α (see Fig. 17). Now

$$f'(z_0) = \lim_{z \to z_0} \frac{w - w_0}{z - z_0}$$

Let z approach z_0 on C_1 as w approaches w_0 on Γ_1. Then

$$\theta_1 = \lim_{z \to z_0} \arg (z - z_0)$$

$$\phi_1 = \lim_{w \to w_0} \arg (w - w_0)$$

$$\arg f'(z_0) = -\lim_{z \to z_0} \arg (z - z_0) + \lim_{w \to w_0} \arg (w - w_0)$$

$$= -\theta_1 + \phi_1$$

Similarly for C_2 and Γ_2, we have $\arg f'(z_0) = -\theta_2 + \phi_2$. Therefore, $\phi_2 - \phi_1 = \theta_2 - \theta_1 = \alpha$, which proves the theorem.

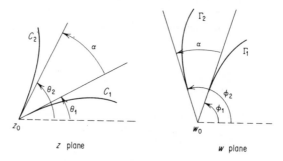

z plane   w plane

FIGURE 17

A mapping which preserves angles both in magnitude and sense is said to be **conformal**. Hence, $w = f(z)$ is conformal at a point z_0, where it is analytic and $f'(z_0) \neq 0$.

Theorem 3. Let $\phi(x,y)$ be harmonic at $z_0 = x_0 + iy_0$. Let $w = f(z)$ be analytic at z_0, where $f'(z_0) \neq 0$. Let $z = g(w) = x(u,v) + iy(u,v)$ be the inverse of $f(z)$ at $w_0 = f(z_0)$. Then

$$\Phi(u,v) = \phi[x(u,v), y(u,v)]$$

is harmonic at w_0.

Proof. There exists a neighborhood N of z_0 and an analytic function $F(z)$ such that $\phi(x,y) = \text{Re}\,[F(z)]$ in N. We have just proved that $z = g(w)$ is analytic at w_0. Therefore, $F[g(w)]$ is analytic at w_0. Now $\Phi(u,v) = \text{Re}\,F[g(w)]$ is the real part of a function analytic at w_0. Therefore, $\Phi(u,v)$ is harmonic at w_0.

This last theorem is the basis for the conformal mapping method of solving boundary-value problems in two-dimensional potential theory. We seek a function harmonic in a given region D_1 such that the function or its normal derivative take on certain prescribed values on the boundary of D_1 (see Sec. 3.6). Problems of this type come up in electrostatics, heat conduction, fluid mechanics, elasticity, and many other places. The conformal mapping method depends on finding an analytic function $f(z)$ which maps D_1 in the z plane onto D_2 in the w plane, where a solution of the boundary-value problem is known. The inverse image of the solution under the mapping is then the solution of the given problem. Of course, to formulate a new problem after the mapping, we have to know how the boundary values map. Clearly, if z_0 is a boundary point of D_1, then $w_0 = f(z_0)$ will be a boundary point of D_2. If $\Phi(u,v)$ is harmonic in D_2 and approaches the value a as $w \to w_0$ from the interior of D_2, then $\phi(x,y) = \Phi[u(x,y), v(x,y)]$ will approach a as $z \to z_0$ from the interior of D_1. Hence, boundary values should correspond. On the other hand, if we are specifying values of the normal derivative, it can easily be shown that

$$\frac{d\phi}{dn_z} = |f'(z_0)| \frac{d\Phi}{dn_w}$$

where $d\phi/dn_z$ is the normal derivative at z_0 and $d\Phi/dn_w$ is the normal derivative at w_0. Therefore, in either case, as soon as the mapping is known, corresponding boundary data can be determined.

An extensive literature[1] has been developed which catalogs conformal mappings, giving the two regions and the mapping function. It would be impossible in this brief treatment to give even a modest list. However, we can give a few examples to give some idea of the scope of the method.

Consider $w = z^2$. This mapping is conformal everywhere except at $z = 0$. There is no harm in having the origin on the boundary of the region, provided care is taken that the mapping is one-to-one. For example, if D_1 is the interior of the first quadrant, $|z| > 0$, $0 < \arg z < \pi/2$, then D_2 is the upper half plane, $\text{Im}(w) > 0$.

The mapping

$$w = \frac{az + b}{cz + d}$$

with $ad - bc \neq 0$, is called the **linear fractional transformation**. The derivative

$$\frac{dw}{dz} = \frac{ad - bc}{(cz + d)^2}$$

is never zero and therefore the mapping is conformal wherever it is defined. The mapping is defined everywhere except at $z = -d/c$ and is one-to-one;

---

[1] See, for example, H. Kober, "Dictionary of Conformal Representations," Dover Publications, Inc., New York, 1952.

that is, if $w_1 = w_2$, then

$$\frac{az_1 + b}{cz_1 + d} = \frac{az_2 + b}{cz_2 + d}$$

$$(ad - bc)(z_1 - z_2) = 0$$

which implies that $z_1 = z_2$.

The linear fractional transformation can be written as

$$w = \frac{a}{c} + \frac{bc - ad}{c} \frac{1}{cz + d}$$

which means that it is the composition of transformations of the following types:

1. $w = z + b = x + \text{Re}(b) + i[y + \text{Im}(b)]$. This is obviously a simple translation of the plane by the complex vector b.

2. $w = az = |a|\, re^{i(\theta + \arg a)}$. This is a rotation of the plane through the angle $\arg a$, followed by an expansion ($|a| > 1$) or a contraction ($|a| < 1$).

3. $w = 1/z$. Let $z = re^{i\theta}$ and $w = \rho e^{i\phi}$. Then $\rho = 1/r$ and $\phi = -\theta$. This mapping takes the interior of the unit circle to the exterior and vice versa. At the same time it reflects points in the real axis.

It is obvious that cases 1 and 2 map circles into circles and straight lines into straight lines. It is not so obvious what happens to circles and straight lines in case 3. Let

$$\alpha(x^2 + y^2) + \beta x + \gamma y + \delta = 0$$

be the equation of a circle ($\alpha \neq 0$) or straight line ($\alpha = 0$) in the z plane. This can be written as

$$\alpha\,|z|^2 + \frac{\beta}{2}(z + z^*) + \frac{\gamma}{2i}(z - z^*) + \delta = 0$$

Under the transformation $w = 1/z$, the circle or straight line maps into

$$\delta\,|w|^2 + \frac{\beta}{2}(w + w^*) - \frac{\gamma}{2i}(w - w^*) + \alpha = 0$$

This is the equation of a circle ($\delta \neq 0$) or straight line ($\delta = 0$). Therefore, circles and straight lines map into circles or straight lines. It is now obvious that the linear fractional transformation, which is the composition of a finite number of mappings of the types 1, 2, or 3, maps straight lines and circles into straight lines or circles.

There are really only three independent parameters in the linear fractional transformation. For example, one of the constants must be different from zero. Suppose it is d. Then

$$w = \frac{(a/d)z + (b/d)}{(c/d)z + 1}$$

Therefore, specifying three points z_1, z_2, and z_3 with definite images w_1, w_2, w_3 will determine the transformation. If z_1, z_2, z_3 are collinear, then this line maps onto the circle or straight line determined by w_1, w_2, w_3. If z_1, z_2, z_3 are noncollinear then they determine a unique circle whose image is the circle or straight line determined by w_1, w_2, w_3. Therefore, the linear fractional transformation is an obvious choice if we wish to map a half plane onto the interior (or exterior) of a circle. For example, suppose we wish to map the upper half plane onto the interior of the unit circle. If we set up the correspondence $-1 \to -i$, $0 \to 1$, and $1 \to i$ we shall have the real axis mapping onto the unit circle. Solving the equations $-i = (-a + b)/(-c + d)$, $1 = b/d$, and $i = (a + b)/(c + d)$, we have

$$w = \frac{iz + 1}{-iz + 1}$$

The inverse of this transformation is

$$z = \frac{w - 1}{iw + i}$$

If we let $w = \rho e^{i\phi}$, with $\rho < 1$, then

$$\text{Im}\,(z) = \frac{1 - \rho^2}{1 + \rho^2 + 2\rho \cos \phi}$$

Obviously $\text{Im}\,(z) > 0$, which shows that the upper half plane maps onto the interior of the unit circle.

Consider the function $w = e^z$. Then $u = e^x \cos y$ and $v = e^x \sin y$. Consider the strip in the z plane, $0 \leq y \leq \pi$. Clearly, this maps into the upper half plane $v \geq 0$. To show that this is onto, we have $y = \tan^{-1}(v/u)$. There is always a unique y satisfying $0 < y < \pi$ for every u and v, $v > 0$. The boundary points map as follows:

$$y = 0 \qquad x \geq 0 \to v = 0 \qquad u \geq 1$$
$$y = 0 \qquad x \leq 0 \to v = 0 \qquad 0 \leq u \leq 1$$
$$y = \pi \qquad x \geq 0 \to v = 0 \qquad u \leq -1$$
$$y = \pi \qquad x \leq 0 \to v = 0 \qquad -1 \leq u \leq 0$$

Consider the function $w = \sin z$. Then $u = \sin x \cosh y$ and $v = \cos x \sinh y$. Consider the half strip in the z plane, $-\pi/2 \leq x \leq \pi/2$, $y \geq 0$. Clearly this maps into the upper half plane, $v \geq 0$. To show that it is onto, we have

$$\frac{u^2}{\cosh^2 y} + \frac{v^2}{\sinh^2 y} = 1$$

$$\frac{u^2}{\sin^2 x} - \frac{v^2}{\cos^2 x} = 1$$

These equations represent a family of ellipses and hyperbolas.   For any given u and v in the upper half plane ($v > 0$) there is an ellipse and hyperbola from these families intersecting at this point.   These determine a point from the given half strip which maps into the point with coordinates (u,v).   The boundary points map as follows:

$$y = 0 \qquad -\frac{\pi}{2} \leq x \leq \frac{\pi}{2} \rightarrow v = 0 \qquad -1 \leq u \leq 1$$

$$y \geq 0 \qquad x = \frac{\pi}{2} \rightarrow v = 0 \qquad 1 \leq u$$

$$y \geq 0 \qquad x = -\frac{\pi}{2} \rightarrow v = 0 \qquad u \leq -1$$

By composing various mappings we can obtain other mappings which take care of more complicated situations.   For example,

$$w = \frac{iz^2 + 1}{-iz^2 + 1}$$

maps the first quadrant onto the interior of the unit circle,

$$w = \frac{ie^z + 1}{-ie^z + 1}$$

maps the strip $0 < y < \pi$ onto the interior of the unit circle, and

$$w = \frac{i \sin z + 1}{-i \sin z + 1}$$

maps the half strip $0 < y$, $-\pi/2 < x < \pi/2$ onto the interior of the unit circle.

We conclude this section with an example of a boundary-value problem which can be solved by the method of conformal mapping.   We wish to find a function which is harmonic inside the unit circle and which takes on the value zero on the upper unit semicircle and the value 1 on the lower unit semicircle.   This function represents the potential in a cylindrical electrostatic capacitor with the voltage zero on one side and 1 on the other.   We first map the interior of the unit circle onto the upper half plane using

$$w = -i \frac{z - 1}{z + 1}$$

This mapping takes the upper semicircle onto the positive real axis and the lower semicircle onto the negative real axis.   The problem now asks for a function harmonic in the upper half plane which takes on the value zero on the positive real axis and the value 1 on the negative real axis. Next we map the upper half plane onto the strip $0 < \mathrm{Im}\,(\zeta) < \pi$ using the mapping

$$\zeta = \log w$$

This mapping takes the positive real axis onto the real axis and the negative real axis onto the line $\mathrm{Im}\,(\zeta) = \pi$. The problem now asks for a function which is harmonic in the strip $0 < \mathrm{Im}\,(\zeta) < \pi$ and which takes the value zero on the lower side and the value 1 on the upper side. This function is obviously $\phi = (1/\pi)\mathrm{Im}\,(\zeta)$. Let $w = u + iv$. Then

$$\phi = \frac{1}{\pi} \tan^{-1} \frac{v}{u}$$

However,

$$u + iv = \frac{2y - i(x^2 + y^2 - 1)}{(x + 1)^2 + y^2}$$

and therefore,

$$\phi = \frac{1}{\pi} \tan^{-1} \frac{1 - x^2 - y^2}{2y}$$

where the values of the inverse tangent are taken between zero and π.

Exercises 7.9

1. Show that $w = z^2$ does not preserve angles at the origin, but that it doubles angles and preserves the sense.

2. Let $w = f(z)$ be analytic at z_0 and $f'(z)$ have a zero of order n at z_0. Show that angles are multiplied by a factor of $n + 1$ at z_0. HINT: Consider the Taylor expansion.

3. Show that small distances at z_0, where $f(z)$ is conformal, are multiplied by a factor $|f'(z_0)|$.

4. Show that if C is a smooth curve passing through z_0, where $f(z)$ is conformal, C' is the image of C passing through $w_0 = f(z_0)$, and $\phi(x,y)$ is a differentiable function defined in a neighborhood of z_0, then the mapping $w = f(z)$ multiplies the normal derivative at z_0 by a factor of $1/|f'(z_0)|$.

5. Find a linear fractional transformation which maps the right half plane $\mathrm{Re}\,(z) > 0$ onto the exterior of the unit circle.

6. Find the most general linear fractional transformation which maps the upper half plane $\mathrm{Im}\,(z) > 0$ onto the interior of the unit circle.

7. Prove that under a linear fractional transformation the cross ratio $[(z_1 - z_4)(z_2 - z_3)]/[(z_3 - z_4)(z_2 - z_1)]$ is invariant.

8. Show that, unless it is the identity, the linear fractional transformation can have at most two fixed points. Can it have one fixed point? no fixed points?

9. Find a mapping which maps the half strip $-\pi/2 < x < \pi/2$, $y > 0$ onto the full strip $0 < v < \pi$.

10. Solve the boundary-value problem: $\nabla^2 \phi = 0$ inside the unit circle, while on the circle $z = e^{i\theta}$, $0 \leq \theta \leq 2\pi$, ϕ satisfies $\phi = 0$ for $0 < \theta < \pi$, $\phi = 1$ for $\pi < \theta < 3\pi/2$, $\partial\phi/\partial r = 0$ for $3\pi/2 < \theta < 2\pi$. HINT: Map the interior of the unit circle onto the upper half plane so that $-1 \to \infty$, $1 \to \pi/2$, and $-i \to -\pi/2$; then map the upper half plane onto a half strip.

7.10   Potential Theory

The connections between analytic function theory and two-dimensional potential theory, the study of harmonic functions, have already been pointed out. In this section, we shall examine in more detail the solutions of two very important boundary-value problems in potential theory, the Dirichlet problem and the Neumann problem.

The Dirichlet problem is the problem of finding a function ψ harmonic in a bounded region R, bounded by a finite number of closed contours, such that $\nabla^2 \psi = 0$ in R, ψ is continuous in the closure of R, and ψ takes on prescribed values on the boundary of R. If the problem has a solution it is unique.[1] We shall establish existence of the solution if R is simply connected.   Consider first the case where R is the interior of the circle $C = \{z \mid |z| = \rho\}$. We wish to find a function harmonic inside C, continuous within and on C, which takes on prescribed boundary values $g(\phi)$, $0 \le \phi \le 2\pi$, on C. For these conditions to be met it is necessary that $g(\phi)$ be continuous on C and therefore $g(0) = g(2\pi)$.

Suppose, for the moment, we assume that $g(\phi)$ is the real part on C of a function $f(z)$ analytic within and on C. Then obviously $\psi = \operatorname{Re} f(z)$ is the solution of the Dirichlet problem for R, since the real part of an analytic function is harmonic. We wish to obtain an integral representation of this solution.   By the Cauchy formula, if $|z| < \rho$

$$
\begin{aligned}
f(z) &= \frac{1}{2\pi i} \int_{C+} \frac{f(\zeta)}{\zeta - z} \, d\zeta \\
&= \frac{1}{2\pi} \int_0^{2\pi} \frac{f(\rho e^{i\phi}) \rho e^{i\phi}}{\rho e^{i\phi} - r e^{i\theta}} \, d\phi
\end{aligned}
$$

where $z = re^{i\theta}$ and $\zeta = \rho e^{i\phi}$.   The point $z' = \rho^2 e^{i\theta}/r$ is outside C and hence

$$
\begin{aligned}
0 &= \frac{1}{2\pi i} \int_{C+} \frac{f(\zeta)}{\zeta - z'} \, d\zeta \\
&= \frac{1}{2\pi} \int_0^{2\pi} \frac{f(\rho e^{i\phi}) \rho e^{i\phi}}{\rho e^{i\phi} - (\rho^2/r)e^{i\theta}} \, d\phi \\
&= \frac{1}{2\pi} \int_0^{2\pi} \frac{f(\rho e^{i\phi}) r e^{i\phi}}{r e^{i\phi} - \rho e^{i\theta}} \, d\phi
\end{aligned}
$$

---

[1] For a proof of uniqueness see J. W. Dettman, "Applied Complex Variables," The Macmillan Company, New York, 1965, p. 237.

Since $|\rho e^{i\phi} - re^{i\theta}|^2 = |re^{i\theta} - \rho e^{i\theta}|^2 = \rho^2 + r^2 - 2\rho r \cos(\theta - \phi)$, we have

$$f(re^{i\theta}) = \frac{1}{2\pi} \int_0^{2\pi} \frac{f(\rho e^{i\phi})(\rho^2 - \rho r e^{i(\phi - \theta)})}{\rho^2 + r^2 - 2\rho r \cos(\theta - \phi)} \, d\phi$$

$$0 = \frac{1}{2\pi} \int_0^{2\pi} \frac{f(\rho e^{i\phi})(r^2 - \rho r e^{i(\phi - \theta)})}{\rho^2 + r^2 - 2\rho r \cos(\theta - \phi)} \, d\phi$$

Subtracting we obtain

$$f(re^{i\theta}) = \frac{1}{2\pi} \int_0^{2\pi} \frac{f(\rho e^{i\phi})(\rho^2 - r^2)}{\rho^2 + r^2 - 2\rho r \cos(\theta - \phi)} \, d\phi$$

Taking real parts, we have

$$\psi(r,\theta) = \operatorname{Re} f(re^{i\theta}) = \frac{1}{2\pi} \int_0^{2\pi} \frac{g(\phi)(\rho^2 - r^2)}{\rho^2 + r^2 - 2\rho r \cos(\theta - \phi)} \, d\phi$$

This is the Poisson integral formula (compare with exercise 3, Sec. 4.4, and Sec. 5.4).

Next consider the function

$$\psi(r,\theta) = \frac{1}{2\pi} \int_0^{2\pi} \frac{g(\phi)(\rho^2 - r^2)}{\rho^2 + r^2 - 2\rho r \cos(\theta - \phi)} \, d\phi$$

where $g(\phi)$ is continuous and $g(0) = g(2\pi)$. The integral clearly exists if $r < \rho$. Also for $r < \rho$ the conditions are met for differentiating under the integral sign with respect to r and θ. We show that $\psi(r,\theta)$ is harmonic inside C by taking its Laplacian in polar coordinates, where on the right-hand side we operate under the integral sign and use the fact that

$$\nabla^2 \left[\frac{\rho^2 - r^2}{\rho^2 + r^2 - 2\rho r \cos(\theta - \phi)} \right] = 0$$

The reader should verify this. Next we show that[1]

$$\lim_{r \to \rho-} \psi(r,\theta) = g(\theta)$$

By taking $f(z) \equiv 1$ in the above derivation of the Poisson integral formula, we obtain

$$1 = \frac{1}{2\pi} \int_0^{2\pi} \frac{\rho^2 - r^2}{\rho^2 + r^2 - 2\rho r \cos(\theta - \phi)} \, d\phi$$

Hence, $$\psi(r,\theta) - g(\theta) = \frac{1}{2\pi} \int_0^{2\pi} \frac{[g(\phi) - g(\theta)](\rho^2 - r^2)}{\rho^2 + r^2 - 2\rho r \cos(\theta - \phi)} \, d\phi$$

[1] For a proof that $\psi(r,\theta)$ approaches $g(\phi)$ as $z \to \rho e^{i\phi}$ from the inside of C in an arbitrary manner see L. V. Ahlfors, "Complex Analysis," 2d ed., McGraw-Hill Book Company, New York, 1966, pp. 167, 168.

We wish to show that $\lim_{r \to \rho^-} [\psi(r,\theta) - g(\theta)] = 0$. We have

$$\frac{1}{2\pi} \int_0^{2\pi} \frac{[g(\phi) - g(\theta)](\rho^2 - r^2)}{\rho^2 + r^2 - 2\rho r \cos(\theta - \phi)} \, d\phi = \frac{1}{2\pi} \int_0^{\theta-\delta} \frac{[g(\phi) - g(\theta)](\rho^2 - r^2)}{\rho^2 + r^2 - 2\rho r \cos(\theta - \phi)} \, d\phi$$

$$+ \frac{1}{2\pi} \int_{\theta-\delta}^{\theta+\delta} \frac{[g(\phi) - g(\theta)](\rho^2 - r^2)}{\rho^2 + r^2 - 2\rho r \cos(\theta - \phi)} \, d\phi$$

$$+ \frac{1}{2\pi} \int_{\theta+\delta}^{2\pi} \frac{[g(\phi) - g(\theta)](\rho^2 - r^2)}{\rho^2 + r^2 - 2\rho r \cos(\theta - \phi)} \, d\phi$$

$$= I_1 + I_2 + I_3$$

For δ sufficiently small $|g(\phi) - g(\theta)| < \epsilon$ for $|\phi - \theta| < \delta$. Then

$$|I_2| \leq \frac{\epsilon}{2\pi} \int_0^{2\pi} \frac{\rho^2 - r^2}{\rho^2 + r^2 - 2\rho r \cos(\theta - \phi)} \, d\phi = \epsilon$$

Also if $|g(\phi)| \leq M$

$$|I_1| \leq \frac{2M}{2\pi} \int_0^{\theta-\delta} \frac{\rho^2 - r^2}{\rho^2 + r^2 - 2\rho r \cos(\theta - \phi)} \, d\phi$$

$$\leq \frac{M}{\pi} \int_0^{\theta-\delta} \frac{\rho^2 - r^2}{\rho^2 + r^2 - 2\rho r \cos \delta} \, d\phi$$

$$\leq 2M \frac{\rho^2 - r^2}{(\rho - r \cos \delta)^2} \leq 2M \frac{\rho^2 - r^2}{\rho^2(1 - \cos \delta)^2} < \epsilon$$

provided $|\rho - r|$ is sufficiently small. Similarly, $|I_3| < \epsilon$. Therefore, for $|\rho - r|$ sufficiently small

$$\left| \frac{1}{2\pi} \int_0^{2\pi} \frac{[g(\phi) - g(\theta)](\rho^2 - r^2)}{\rho^2 + r^2 - 2\rho r \cos(\theta - \phi)} \, d\phi \right| \leq |I_1| + |I_2| + |I_3| < 3\epsilon$$

which proves the desired result since ϵ is arbitrary.

We are now in a position to solve the Dirichlet problem for any simply connected region R, bounded by a closed contour, which can be conformally mapped onto the interior of the circle C such that the boundary of R maps onto C. The existence of such a mapping is assured by the **Riemann mapping theorem,**[1] which states that any simply connected region bounded by a closed contour can be mapped conformally onto the interior of the circle C such that the boundary of R maps onto C and any given point in R maps into the origin. As a matter of fact, this establishes the existence of the solution of the Dirichlet problem for R.

In Sec. 5.4, we introduced the concept of a Green's function for the solution of the Dirichlet problem. We now can obtain a simple expression for the

[1] For a proof of this important theorem see J. W. Dettman, "Applied Complex Variables," The Macmillan Company, New York, 1965, pp. 256–259.

Green's function in terms of the conformal map of R onto the interior of the unit circle. Let $f(z)$ be the mapping function which takes the point ζ in R into the origin. Then the Green's function is

$$G(\xi,\eta;x,y) = -\frac{1}{2\pi} \log |f(z)|$$

where $\zeta = \xi + i\eta$. We show this as follows.

$$G = \operatorname{Re} \left\{ -\frac{1}{2\pi} \log f(z) \right\}$$

and therefore G is harmonic everywhere in R except at ζ, where $f(\zeta) = 0$. Also since $f(z)$ is analytic at ζ

$$f(z) = f(\zeta) + a_1(z - \zeta) + a_2(z - \zeta)^2 + \cdots$$
$$= (z - \zeta)g(z)$$

where $g(\zeta) = f'(\zeta) \neq 0$, because $f(z)$ is conformal at ζ. Therefore,

$$-\frac{1}{2\pi} \log |f(z)| = -\frac{1}{2\pi} \log |z - \zeta| - \frac{1}{2\pi} \log |g(z)|$$

$$= -\frac{1}{2\pi} \log \sqrt{(x - \xi)^2 + (y - \eta)^2} + H(\xi,\eta;x,y)$$

where H is harmonic at ζ. This shows that G has the right kind of singularity at ζ. Finally, since $|f(z)| = 1$ when z is on the boundary of R, we have

$$G = -\frac{1}{2\pi} \log |f(z)| = -\frac{1}{2\pi} \log 1 = 0$$

on the boundary of R.

The second problem we shall consider here is the Neumann problem. This is the problem of finding a function ψ harmonic in a bounded region R, bounded by a finite number of closed smooth contours, such that $\nabla^2\psi = 0$ in R, ψ and its first partial derivatives are continuous in the closure of R, and the normal derivative of ψ takes on prescribed values on the boundary of R. We shall consider, for simplicity, only the case where R is simply connected and bounded by a closed smooth contour Γ. The fact that Γ is smooth assures us that it has a continuous tangent vector and therefore a continuous normal vector. Therefore, if ψ has continuous first partial derivatives in the closure of R we can define the normal derivative

$$\frac{d\psi}{dn} = \psi_x \cos \alpha + \psi_y \sin \alpha$$

where α is the angle between the positive x axis and the outward-pointing normal to Γ.

The solution to the Neumann problem, as stated above, is not unique because any constant could be added to a solution and we would still have a solution. However, if we specify that the average value of ψ on the boundary Γ is zero, then the solution, if it exists, is unique. To prove this, suppose there are two solutions ψ_1 and ψ_2 and let w be the difference; that is, $w = \psi_1 - \psi_2$. Then w is harmonic in R and has a vanishing normal derivative on Γ. Using Green's theorem we have

$$\iint\limits_{R} \left[\left(\frac{\partial w}{\partial x} \right)^2 + \left(\frac{\partial w}{\partial y} \right)^2 \right] dx\, dy = \iint\limits_{R} w \nabla^2 w \, dx\, dy + \int_{\Gamma_+} w \frac{dw}{dn}\, ds$$
$$= 0$$

Therefore, $w_x = w_y \equiv 0$ in R and $w = k$, a constant. But

$$0 = \int_{\Gamma} (\psi_1 - \psi_2)\, ds = \int_{\Gamma} w\, ds = k \int_{\Gamma} ds = kL$$

where L is the length of Γ. Therefore, $k = 0$ and $\psi_1 \equiv \psi_2$.

The actual solution of the Neumann problem can be effected by means of conformal mapping. The Riemann mapping theorem assures us that there is a conformal mapping which maps the simply connected region R onto the interior of a circle $C = \{w \mid |w| = \rho\}$ so that Γ maps onto C. As we have already seen, the values of the normal derivative map as follows:

$$\frac{d\psi}{dn_\Gamma} = |f'(z)| \frac{d\psi}{dn_C}$$

where $f(z)$ is the mapping function. Therefore, the problem is to solve the Neumann problem for the region bounded by the circle C. This we shall do using a Green's function, which we construct by analogy with Sec. 5.4.

Let $z = x + iy$ be the variable of the Green's function, and let $\zeta = \xi + i\eta$ be the parameter, the point in the region R where we wish to express the solution. The problem is two-dimensional and therefore we want the fundamental singularity to be of the form

$$-\frac{1}{2\pi} \log |z - \zeta| = -\frac{1}{2\pi} \log \sqrt{(x - \xi)^2 + (y - \eta)^2}$$

Elsewhere in the region we wish the Green's function to be harmonic and therefore we express it as follows:

$$G(\xi,\eta;x,y) = -\frac{1}{2\pi} \log |z - \zeta| + h(\xi,\eta;x,y)$$

where h is harmonic in the region. By analogy we might wish to require

$dG/dn = 0$ on the boundary, but $\int_\Gamma (dh/dn)\, ds = 0$ since h is harmonic in R, and

$$\frac{1}{2\pi} \int_{\Gamma_+} \frac{d}{dn} \log |z - \zeta|\, ds = 1$$

which implies that $\qquad \int_{\Gamma_+} \frac{dG}{dn}\, ds = -1$

If dG/dn is constant on Γ, then this constant must be $-1/L$, where L is the length of Γ. The conditions stated so far do not define G uniquely, since any constant could be added which would not effect any of these conditions. Therefore, we require in addition that $\int_{\Gamma_+} G\, ds = 0$. With this added requirement we can prove that the Green's function is unique. Suppose there are two functions

$$G_1 = -\frac{1}{2\pi} \log |z - \zeta| + h_1$$

$$G_2 = -\frac{1}{2\pi} \log |z - \zeta| + h_2$$

Then $w = G_1 - G_2 = h_1 - h_2$ is harmonic in R, has vanishing normal derivative on Γ, and has zero mean value on Γ. By the uniqueness of the solution of the Neumann problem $w \equiv 0$ and hence $h_1 \equiv h_2$ and $G_1 \equiv G_2$.

Before showing that we can express the solution of the Neumann problem in terms of the Green's function just described, we must put a condition on the given boundary data. Suppose $d\psi/dn = g(s)$ on Γ is the given data. Then $\int_{\Gamma_+} g(s)\, ds = 0$. This is required because, if ψ is harmonic in R, then $\int_{\Gamma_+} (d\psi/dn)\, ds = 0$.

Applying Green's theorem to the region $\bar{R}$ between Γ and a small circle of radius ρ centered on ζ, we have

$$0 = \iint_{\bar{R}} (\psi \nabla^2 G - G \nabla^2 \psi)\, dx\, dy$$

$$= \int_{\Gamma_+} \left(\psi \frac{dG}{dn} - G \frac{d\psi}{dn} \right) ds + \int_0^{2\pi} \left(G \frac{\partial \psi}{\partial r} \right)_{r=\rho} \rho\, d\theta - \int_0^{2\pi} \left(\psi \frac{\partial G}{\partial r} \right)_{r=\rho} \rho\, d\theta$$

where $r = |z - \zeta|$ and $\theta = \arg(z - \zeta)$. Now

$$\lim_{\rho \to 0} \int_0^{2\pi} \left(G \frac{\partial \psi}{\partial r} \right)_{r=\rho} \rho\, d\theta = 0 \qquad \lim_{\rho \to 0} \int_0^{2\pi} \left(\psi \frac{\partial G}{\partial r} \right)_{r=\rho} \rho\, d\theta = -\psi(\xi, \eta)$$

Therefore, $\qquad \psi(\xi, \eta) = \int_{\Gamma_+} gG\, ds + \frac{1}{L} \int_{\Gamma_+} g\, ds = \int_{\Gamma_+} gG\, ds$

Finally, we indicate the Green's function for the Neumann problem for the region bounded by the circle $C = \{z \mid |z| = \rho\}$. Let $r = |z|$, $\sigma = |\zeta|$, $r_1 = |z - \zeta|$, $r_2 = |z - (\rho^2/\sigma^2)\zeta|$. Then it is not difficult to show that

$$G(\xi,\eta\,;x,y) = -\frac{1}{2\pi} \log r_1 - \frac{1}{2\pi} \log r_2 + \frac{1}{2\pi} \log (\rho^3/\sigma)$$

The first term gives us the right singularity at ζ but is otherwise harmonic. The second term is harmonic inside C, since $\sigma < \rho$ and therefore $|(\rho^2/\sigma^2)\zeta| = \rho^2/\sigma > \rho$. The last term is constant, considered as a function of z. In polar coordinates $z = re^{i\theta}$, $\zeta = \sigma e^{i\phi}$, we have

$$G = -\frac{1}{4\pi} \log [r^2 + \sigma^2 - 2r\sigma \cos (\theta - \phi)]$$

$$-\frac{1}{4\pi} \log \left[\frac{1}{\rho^2} + \frac{r^2\sigma^2}{\rho^6} - \frac{2r\sigma}{\rho^4} \cos (\theta - \phi)\right]$$

Evaluating G on the boundary $r = \rho$

$$G = -\frac{1}{2\pi} \log \left[\frac{\rho^2 + \sigma^2 - 2\rho\sigma \cos (\theta - \phi)}{\rho^2}\right]$$

Therefore, the solution expressed in polar coordinates (σ,ϕ) and in terms of the given data $d\psi/dn = g(\theta)$ is

$$\psi(\sigma,\phi) = -\frac{\rho}{2\pi} \int_0^{2\pi} g(\theta) \log \left[\frac{\rho^2 + \sigma^2 - 2\rho\sigma \cos (\theta - \phi)}{\rho^2}\right] d\theta$$

for $\sigma < \rho$

$$\frac{\partial \psi}{\partial \sigma} = \frac{\rho}{2\pi\sigma} \int_0^{2\pi} g(\theta) \frac{\rho^2 - \sigma^2}{\rho^2 + \sigma^2 - 2\rho\sigma \cos (\theta - \phi)} d\theta - \frac{\rho}{2\pi\sigma} \int_0^{2\pi} g(\theta) \, d\theta$$

$$= \frac{\rho}{2\pi\sigma} \int_0^{2\pi} g(\theta) \frac{\rho^2 - \sigma^2}{\rho^2 + \sigma^2 - 2\rho\sigma \cos (\theta - \phi)} d\theta$$

Using arguments similar to those above for the Dirichlet problem we can show that

$$\lim_{\sigma \to \rho -} \frac{\partial \psi}{\partial \sigma} = g(\theta)$$

Exercises 7.10

1. Let $f(z)$ be analytic in a simply connected region R containing the circle $C = \{z \mid |z - z_0| = r\}$. Prove that $f(z_0) = (1/2\pi)\int_0^{2\pi} f(z_0 + re^{i\theta}) \, d\theta$. Infer that if $u(x,y)$ is harmonic in R, then $u(x_0,y_0) = (1/2\pi)\int_0^{2\pi} u(x_0 + r \cos \theta, y_0 + r \sin \theta) \, d\theta$.

2. Show that the Dirichlet problem for the upper half plane $y \geq 0$ has a unique solution, if it exists. The problem is to find a function $\psi(x,y)$ harmonic for $y > 0$, continuous for $y \geq 0$, such that $\psi(x,0) = g(x)$, a given function, and

$$\lim_{R \to \infty} \int_0^\pi [\psi(\partial\psi/\partial r)]_{r=R} R \, d\theta = 0$$

3. Referring to exercise 2, find a Green's function for the Dirichlet problem for the upper half plane, and express the solution in terms of the Green's function.

4. Let ψ be harmonic in a simply connected region R bounded by a smooth closed contour Γ. Let ψ and its first partial derivatives be continuous in the closure of R. Prove that $\int_\Gamma (d\psi/dn) \, ds = 0$.

5. Prove that $(1/2\pi) \int_{\Gamma+} d/dn \log |z - \zeta| \, ds = 1$, where Γ is any smooth closed contour with ζ in its interior, the integration is with respect to the variable z, and s is the arc length on Γ.

6. Prove that the Green's function for the Neumann problem for the circle $C = \{z \mid |z| = \rho\}$ satisfies all the required conditions.

7. If $w = f(z)$ is a conformal mapping which maps the simply connected region R onto the interior of the unit circle and ζ is a point in R, show that the Green's function for the Dirichlet problem in R is

$$G(\xi,\eta;x,y) = -\frac{1}{2\pi} \log \left| \frac{f(z) - f(\zeta)}{f^*(\zeta)f(z) - 1} \right|$$

where $\zeta = \xi + i\eta$.

References

Ahlfors, L. V.: "Complex Analysis," text ed., McGraw-Hill Book Company, New York, 1979.

Carrier, G. P., M. Krook, and C. E. Pearson: "Functions of a Complex Variable: Theory and Technique," McGraw-Hill Book Company, New York, 1966.

Churchill, R. V.: "Complex Variables and Applications," text ed., McGraw-Hill Book Company, New York, 1984.

Copson, E. T.: "An Introduction to the Theory of Functions of a Complex Variable," Oxford University Press, Fairlawn, N.J., 1935.

Dettman, J. W.: "Applied Complex Variables," Dover Publications, Inc., New York, 1984.

Hille, E.: "Analytic Function Theory," vols. I & II, 2d ed., Chelsea Publishers, Inc., New York, 1973.

Kellogg, O. D.: "Foundations of Potential Theory," Dover Publications, Inc., New York, 1953.

Nehari, Z.: "Conformal Mapping," Dover Publications, Inc., New York, 1975.

Whittaker, E. T., and G. N. Watson: "A Course of Modern Analysis," 4th ed., Cambridge University Press, London, 1952.

Chapter 8. Integral Transform Methods

8.1 Fourier Transforms

In Sec. 5.1, we considered the solution of certain nonhomogeneous boundary-value problems in which the unknown satisfies a nonhomogeneous partial differential equation of the type

$$\nabla^2 \phi = c \frac{\partial^2 \phi}{\partial t^2} - u(x,y,z)h(t)$$

and a homogeneous boundary condition $(d\phi/dn) + \alpha\phi = 0$. In the case where $h(t)$ is periodic, we expanded it in a Fourier series and effected a separation of variables, reducing the problem to solving a sequence of nonhomogeneous boundary-value problems of the type

$$\nabla^2 \psi + \lambda\psi = -u \quad \text{in } V$$

$$\frac{d\psi}{dn} + \alpha\psi = 0 \quad \text{on } S$$

where S is the surface of the bounded volume V. If $h(t)$ is not periodic, the separation-of-variables technique is not available. However, the problem may still be solvable by integral transform methods.

The representation of a function in a Fourier series for all values of the independent variable depends on the function's being periodic with a finite period. If the function is not periodic, it may still be possible to get a **Fourier integral representation** of the function. Roughly speaking, this can be thought of as the limiting form of the Fourier series when the period approaches infinity.[1] We shall approach the problem in a manner similar to that used in Sec. 2.1, where we attempted to obtain approximations to functions over a finite interval in terms of orthonormal sets of functions.

Let $f(x)$ be a complex-valued function of the real variable x defined

---

[1] See E. C. Titchmarsh, "The Theory of Functions," 2d ed., Oxford University Press, Fairlawn, N.J., 1937, pp. 432 and 433.

for $-\infty < x < \infty$, be piecewise continuous and have a piecewise continuous derivative in any finite interval, and be absolutely integrable, that is, $\int_{-\infty}^{\infty} |f(x)|\, dx < \infty$, and square integrable, that is, $\int_{-\infty}^{\infty} |f(x)|^2\, dx < \infty$. We attempt to find a function $g(t)$ which will minimize

$$\int_{-\infty}^{\infty} \left| f(x) - \frac{1}{\sqrt{2\pi}} \int_{-T}^{T} g(t) e^{ixt}\, dt \right|^2 dx$$

This is completely analogous to our approach to the generalized Fourier series of Sec. 2.1, where $g(t)$ plays the role of the Fourier coefficient, and instead of a summation from $-n$ to n we now have an integral from $-T$ to T.

To proceed further with the development of this section we shall need the results of the following lemmas, which we shall take time out to prove so that the following discussion can continue uninterrupted.

Lemma 1. If $f(x)$ is piecewise continuous for $a \le x \le b$, then

$$\lim_{R \to \infty} \int_{a}^{b} f(x) \sin Rx\, dx = 0$$

To prove this, we can assume without loss of generality that $f(x)$ is continuous, for the interval can always be broken up into a finite number of subintervals and proved for each of the subintervals where the function is continuous. We first make the change of variables $x = t + \pi/R$. Then

$$\int_{a}^{b} f(x) \sin Rx\, dx = -\int_{a-\frac{\pi}{R}}^{b-\frac{\pi}{R}} f\left(t + \frac{\pi}{R}\right) \sin Rt\, dt$$

$$2\int_{a}^{b} f(x) \sin Rx\, dx = \int_{a}^{b} f(t) \sin Rt\, dt - \int_{a-\frac{\pi}{R}}^{b-\frac{\pi}{R}} f\left(t + \frac{\pi}{R}\right) \sin Rt\, dt$$

$$= -\int_{a}^{b-\frac{\pi}{R}} \left[f\left(t + \frac{\pi}{R}\right) - f(t) \right] \sin Rt\, dt$$

$$+ \int_{b-\frac{\pi}{R}}^{b} f(t) \sin Rt\, dt - \int_{a-\frac{\pi}{R}}^{a} f\left(t + \frac{\pi}{R}\right) \sin Rt\, dt$$

Now if $f(x)$ is continuous for $a \le x \le b$, it is uniformly continuous, so we can pick R sufficiently large that $|f[t + (\pi/R)] - f(t)| < \epsilon/(b - a)$ for all t in the interval. Also we can pick R large enough that $\pi/R < \epsilon/2M$, where $|f(t)| < M$ in the interval. Hence,

$$2\left| \int_{a}^{b} f(x) \sin Rx\, dx \right| < \frac{\epsilon}{(b - a)}\,(b - a) + \frac{M\epsilon}{2M} + \frac{M\epsilon}{2M} = 2\epsilon$$

Since ϵ is arbitrary, this completes the proof.

Lemma 2. If $f(x)$ is piecewise continuous and has a piecewise continuous derivative in any finite interval and $\int_{-\infty}^{\infty} |f(x)|\, dx < \infty$, then

$$\lim_{R \to \infty} \int_{-T}^{T} f(x + t)\, \frac{\sin Rt}{t}\, dt = \frac{\pi}{2}[f(x+) + f(x-)]$$

where T may be finite or infinite.

We first prove it for T finite. Breaking up the integral into four parts, we have for $\delta > 0$

$$\int_{-T}^{T} f(x + t)\, \frac{\sin Rt}{t}\, dt = \int_{-T}^{-\delta} \frac{f(x + t)}{t}\, \sin Rt\, dt + \int_{-\delta}^{0} \frac{f(x + t)}{t}\, \sin Rt\, dt$$

$$+ \int_{0}^{\delta} \frac{f(x + t)}{t}\, \sin Rt\, dt + \int_{\delta}^{T} \frac{f(x + t)}{t}\, \sin Rt\, dt$$

The first and the last of these integrals approach zero as $R \to \infty$ by lemma 1. Now

$$\lim_{R \to \infty} \int_{0}^{\delta} \frac{\sin Rt}{t}\, dt = \lim_{R \to \infty} \int_{0}^{R\delta} \frac{\sin \tau}{\tau}\, d\tau = \int_{0}^{\infty} \frac{\sin \tau}{\tau}\, d\tau = \frac{\pi}{2}$$

Therefore, our first result follows if we can show that

$$\lim_{R \to \infty} \int_{0}^{\delta} \frac{f(x + t) - f(x+)}{t}\, \sin Rt\, dt = 0$$

$$\lim_{R \to \infty} \int_{0}^{\delta} \frac{f(x - t) - f(x-)}{t}\, \sin Rt\, dt = 0$$

But this follows from lemma 1, since

$$[f(x + t) - f(x+)]/t \qquad \text{and} \qquad [f(x - t) - f(x-)]/t$$

are piecewise continuous by the hypothesis that $f(x)$ has a piecewise continuous derivative. That the integration can be extended to infinity follows from the fact that

$$\int_{A}^{\infty} \frac{|f(t)|}{t}\, dt < \infty \qquad \text{and} \qquad \int_{-\infty}^{-B} \left| \frac{f(t)}{t} \right|\, dt < \infty$$

for $A > 0, B > 0$. Therefore for arbitrary $\epsilon > 0$, T can be chosen sufficiently large that

$$\int_{-\infty}^{-T} \left| \frac{f(t)}{t} \right|\, dt + \int_{T}^{\infty} \left| \frac{f(t)}{t} \right|\, dt < \epsilon$$

Getting back to the original problem, we have

$$\lim_{R \to \infty} \int_{-R}^{R} \left| f(x) - \frac{1}{\sqrt{2\pi}} \int_{-T}^{T} g(t)e^{ixt}\, dt \right|^2 dx$$

$$= \lim_{R \to \infty} \left[\int_{-R}^{R} |f|^2 d\,x - \frac{1}{\sqrt{2\pi}} \int_{-R}^{R} \int_{-T}^{T} f(x)g^*(t)e^{-ix\cdot}\, dt\, dx \right.$$

$$- \frac{1}{\sqrt{2\pi}} \int_{-R}^{R} \int_{-T}^{T} f^*(x)g(t)e^{ixt}\, dt\, dx$$

$$\left. + \frac{1}{2\pi} \int_{-R}^{R} \int_{-T}^{T} \int_{-T}^{T} g(t)g^*(\tau)e^{ix(t-\tau)}\, dt\, d\tau\, dx \right]$$

In the last integral we integrate first with respect to x.

$$\int_{-R}^{R} e^{ix(t-\tau)}\, dx = \int_{-R}^{R} \cos x(t-\tau)\, dx + i \int_{-R}^{R} \sin x(t-\tau)\, dx$$

$$= 2\frac{\sin R(t-\tau)}{t-\tau}$$

Next we integrate with respect to t.

$$\lim_{R \to \infty} 2 \int_{-T}^{T} g(t) \frac{\sin R(t-\tau)}{(t-\tau)}\, dt = \lim_{R \to \infty} 2 \int_{-T-\tau}^{T-\tau} g(\xi+\tau) \frac{\sin R\xi}{\xi}\, d\xi$$

$$= 2\pi g(\tau)$$

This follows from lemma 2, where of course we must assume that $g(t)$ satisfies the hypotheses of the lemma. Hence

$$\int_{-\infty}^{\infty} \left| f(x) - \frac{1}{\sqrt{2\pi}} \int_{-T}^{T} g(t)e^{ixt}\, dt \right|^2 dx$$

$$= \int_{-\infty}^{\infty} |f|^2\, dx + \int_{-T}^{T} |g|^2\, dt - \int_{-T}^{T} g(t)h^*(t)\, dt - \int_{-T}^{T} g^*(t)h(t)\, dt$$

$$= \int_{-\infty}^{\infty} |f|^2\, dx - \int_{-T}^{T} |h|^2\, dt + \int_{-T}^{T} |g(t) - h(t)|^2\, dt$$

where
$$h(t) = \frac{1}{\sqrt{2\pi}} \int_{-\infty}^{\infty} f(x)e^{-ixt}\, dx$$

The function $h(t)$ exists and is continuous because the integral is uniformly convergent in t. It is now obvious that to minimize the above expression we must choose $g(t) = h(t)$, and

$$\min \int_{-\infty}^{\infty} \left| f(x) - \frac{1}{\sqrt{2\pi}} \int_{-T}^{T} g(t)e^{ixt}\, dt \right|^2 dx = \int_{-\infty}^{\infty} |f(x)|^2\, dx - \int_{-T}^{T} |g(t)|^2\, dt \geq 0$$

This is true for all T.   Hence we have Bessel's inequality

$$\int_{-\infty}^{\infty} |g|^2 \, dt \le \int_{-\infty}^{\infty} |f|^2 \, dx < \infty$$

from which it follows that g is square integrable.

The function

$$g(t) = \frac{1}{\sqrt{2\pi}} \int_{-\infty}^{\infty} f(x) e^{-ixt} \, dx$$

is known as the **Fourier transform** of $f(x)$.   It exists if $f(x)$ is absolutely integrable, and it is square integrable if $f(x)$ is square integrable by Bessel's inequality.   The function

$$f_T(x) = \frac{1}{\sqrt{2\pi}} \int_{-T}^{T} g(t) e^{ixt} \, dt$$

is the best approximation in the least-mean-square sense to $f(x)$.   If we could show that

$$\int_{-\infty}^{\infty} |g(t)|^2 \, dt = \int_{-\infty}^{\infty} |f(x)|^2 \, dx$$

then

$$\lim_{T \to \infty} \int_{-\infty}^{\infty} |f(x) - f_T(x)|^2 \, dx = 0$$

or, in other words, $f_T(x)$ converges in mean to $f(x)$.   This is the analogue of completeness for sets of orthonormal functions.   In this case we again have convergence in mean.   This result is known as **Plancherel's Theorem,**[1] although we shall not attempt to prove it here.

Actually, under our present hypotheses, we can show directly that

$$\frac{1}{\sqrt{2\pi}} \int_{-\infty}^{\infty} g(t) e^{ixt} \, dt = \tfrac{1}{2}[f(x+) + f(x-)]$$

without using the theory of mean convergence or, for that matter, the square integrability of $f(x)$.   We have

$$\lim_{T \to \infty} \frac{1}{2\pi} \int_{-T}^{T} \int_{-\infty}^{\infty} f(\tau) e^{it(x-\tau)} \, d\tau \, dt$$

$$= \lim_{T \to \infty} \frac{1}{2\pi} \int_{-\infty}^{\infty} f(\tau) \frac{\sin T(x - \tau)}{(x - \tau)} \, d\tau$$

$$= \lim_{T \to \infty} \frac{1}{2\pi} \int_{-\infty}^{\infty} f(x + u) \frac{\sin Tu}{u} \, du$$

$$= \tfrac{1}{2}[f(x+) + f(x-)]$$

by lemma 2.

[1] *Ibid.*, pp. 436 and 437.

As an example, let us consider the transform of the function $f(x) = e^{-|x|}$. Then

$$g(t) = \frac{1}{\sqrt{2\pi}} \int_{-\infty}^{\infty} e^{-|x|} e^{-ixt}\, dx$$

$$= \frac{1}{\sqrt{2\pi}} \int_{-\infty}^{0} e^{x(1-it)}\, dx + \frac{1}{\sqrt{2\pi}} \int_{0}^{\infty} e^{-x(1+it)}\, dx$$

$$= \frac{1}{\sqrt{2\pi}} \left(\frac{1}{1-it} + \frac{1}{1+it} \right)$$

$$= \frac{1}{\sqrt{2\pi}} \frac{2}{1+t^2}$$

We can invert the transform in this case by complex contour integration. First consider $x > 0$. Then

$$\frac{1}{2\pi} \int_{-\infty}^{\infty} \frac{2}{1+t^2} e^{ixt}\, dt = \lim_{T \to \infty} \frac{1}{\pi} \int_{C} \frac{1}{1+z^2} e^{ixz}\, dz$$

where C is the contour shown in Fig. 18.

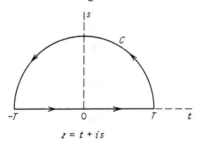

$$z = t + is$$

FIGURE 18

There is a simple pole at $z = i$ and by the residue theorem

$$\frac{1}{2\pi} \int_{-\infty}^{\infty} \frac{2}{1+t^2} e^{ixt}\, dt = 2\pi i \frac{e^{-x}}{2\pi i} = e^{-x}$$

provided we can show that the contribution to the contour integral on the semicircular arc goes to zero as $T \to \infty$. Letting $z = Te^{i\theta}$, we have

$$\left| \frac{1}{\pi} \int_{0}^{\pi} \frac{1}{1+T^2 e^{2i\theta}} e^{ixT(\cos\theta + i\sin\theta)} iTe^{i\theta}\, d\theta \right| \leq \frac{4T}{|T^2 - 1|} \to 0 \qquad \text{as } T \to \infty$$

For $x < 0$, we must close the contour with a semicircular arc below the real axis. Then the simple pole at $z = -i$ contributes e^{x}. Of course if $x = 0$, the integration can be performed directly; that is,

$$\frac{1}{2\pi} \int_{-\infty}^{\infty} \frac{2}{1+t^2}\, dt = \frac{1}{\pi} \tan^{-1} t \Big]_{-\infty}^{\infty} = 1$$

The Fourier transform is obviously a linear transformation, for letting

$$T[f] = \frac{1}{\sqrt{2\pi}} \int_{-\infty}^{\infty} f(x) e^{-ixt} \, dx$$

we have $\quad T[af + bg] = \dfrac{1}{\sqrt{2\pi}} \displaystyle\int_{-\infty}^{\infty} [af(x) + bg(x)] e^{-ixt} \, dx$

$$= \frac{a}{\sqrt{2\pi}} \int_{-\infty}^{\infty} f(x) e^{-ixt} \, dx + \frac{b}{\sqrt{2\pi}} \int_{-\infty}^{\infty} g(x) e^{-ixt} \, dx$$

$$= aT[f] + bT[g]$$

For square integrable functions we have already seen that the transform preserves norms; that is,

$$\|f\| = \left(\int_{-\infty}^{\infty} |f(x)|^2 \, dx \right)^{\frac{1}{2}} = \left(\int_{-\infty}^{\infty} |T[f]|^2 \, dt \right)^{\frac{1}{2}}$$

$$= \|T[f]\|$$

We also have Parseval's relation; that is,

$$\int_{-\infty}^{\infty} f^*(x) g(x) \, dx = \int_{-\infty}^{\infty} T^*[f] T[g] \, dt$$

This is proved as follows.[1]

$$\left| \int_{-\infty}^{\infty} f^*(x) g(x) \, dx - \int_{-R}^{R} T^*[f] T[g] \, dt \right|$$

$$= \left| \int_{-\infty}^{\infty} f^*(x) \left(g(x) - \frac{1}{\sqrt{2\pi}} \int_{-R}^{R} T[g] e^{ixt} \, dt \right) dx \right|$$

$$\leq \left(\int_{-\infty}^{\infty} |f|^2 \, dx \right)^{\frac{1}{2}} \left(\int_{-\infty}^{\infty} \left| g - \frac{1}{\sqrt{2\pi}} \int_{-R}^{R} T[g] e^{ixt} \, dt \right|^2 dx \right)^{\frac{1}{2}}$$

approaches zero as $R \to \infty$ by Plancherel's theorem.

Let us now turn to some of the other properties of the Fourier transform. First consider the effect of a translation of the independent variable, that is,

$$T[f(x - a)] = \frac{1}{\sqrt{2\pi}} \int_{-\infty}^{\infty} f(x - a) e^{-ixt} \, dx$$

$$= \frac{1}{\sqrt{2\pi}} \int_{-\infty}^{\infty} f(\xi) e^{-i(\xi + a)t} \, d\xi$$

$$= e^{-iat} \frac{1}{\sqrt{2\pi}} \int_{-\infty}^{\infty} f(\xi) e^{-i\xi t} \, d\xi$$

$$= e^{-iat} T[f(x)]$$

[1] Compare with Sec. 2.1.

A change of scale has the following effect.

$$T[f(ax)] = \frac{1}{\sqrt{2\pi}} \int_{-\infty}^{\infty} f(ax)e^{-ixt}\,dx$$

$$= \frac{1}{|a|}\frac{1}{\sqrt{2\pi}} \int_{-\infty}^{\infty} f(\xi)e^{-i\xi(t/a)}\,d\xi$$

$$= \frac{1}{|a|}\, T[f(x)]_{t\to\frac{t}{a}}$$

Since we are going to be using the Fourier transform to solve differential equations, we must know how it transforms the derivative of a function. Let $f(x)$ be a function which is continuous in any finite interval and is absolutely integrable. Assume that its first derivative is piecewise continuous and absolutely integrable. Then

$$T[f'(x)] = \frac{1}{\sqrt{2\pi}} \int_{-\infty}^{\infty} f'(x)e^{-ixt}\,dx$$

$$= \frac{1}{\sqrt{2\pi}}[f(x)e^{-ixt}]_{-\infty}^{\infty} + \frac{it}{\sqrt{2\pi}} \int_{-\infty}^{\infty} f(x)e^{-ixt}\,dx$$

$$= itT[f]$$

since $f(x) \to 0$ as $x \to \pm\infty$. Generalizing, we can say that if $f(x)$ and its first $k-1$ derivatives are all continuous and absolutely integrable and its kth derivative is piecewise continuous and absolutely integrable, then

$$T[f^{(k)}(x)] = (it)^k T[f]$$

For the purpose of solving certain integral equations we must determine how the integral of a function is transformed. Let $f(x)$ be a function which is piecewise continuous and absolutely integrable. Then

$$h(x) = \int_a^x f(\xi)\,d\xi$$

is continuous. Assume that $h(x)$ is absolutely integrable. Then

$$T[h(x)] = \frac{1}{\sqrt{2\pi}} \int_{-\infty}^{\infty} h(x)e^{-ixt}\,dx$$

$$= \frac{1}{\sqrt{2\pi}}\left[\frac{h(x)e^{-ixt}}{-it}\right]_{-\infty}^{\infty} + \frac{1}{it\sqrt{2\pi}} \int_{-\infty}^{\infty} f(x)e^{-ixt}\,dx$$

$$= \frac{1}{it}\, T[f]$$

We see then that integration has the effect of dividing the transform by it, whereas differentiation has the effect of multiplying the transform by it.

For the applications it is very important that we should know the transform of a certain integral, namely,

$$h(x) = \frac{1}{\sqrt{2\pi}} \int_{-\infty}^{\infty} f(x - \xi)g(\xi)\, d\xi$$

$$= \frac{1}{\sqrt{2\pi}} \int_{-\infty}^{\infty} g(x - \xi)f(\xi)\, d\xi$$

This is called the **convolution integral.** Taking the transform of $h(x)$, we have

$$T[h] = \frac{1}{2\pi} \int_{-\infty}^{\infty} g(\xi) \int_{-\infty}^{\infty} f(x - \xi)e^{-ixt}\, dx\, d\xi$$

$$= \frac{1}{2\pi} \int_{-\infty}^{\infty} g(\xi) \int_{-\infty}^{\infty} f(\eta)e^{-i(\xi+\eta)t}\, d\eta\, d\xi$$

$$= \left(\frac{1}{\sqrt{2\pi}} \int_{-\infty}^{\infty} g(\xi)e^{-i\xi t}\, d\xi\right) \left(\frac{1}{\sqrt{2\pi}} \int_{-\infty}^{\infty} f(\eta)e^{-i\eta t}\, d\eta\right)$$

$$= T[f]T[g]$$

Finally, before turning to the applications of Fourier transforms, let us consider the transform as a function of the complex variable $\zeta = t + is$. Suppose $f(x) = O(e^{\alpha x})$ as $x \to \infty$. If $\alpha \geq 0$, then $f(x)$ will not have a Fourier transform in the usual sense. Let $g(x) = e^{sx}f(x)$ with s real. Now $g(x) = O[e^{(\alpha+s)x}]$ as $x \to \infty$, and it will have a Fourier transform if $s < -\alpha$, provided it behaves properly as $x \to -\infty$. The Fourier integral theorem states that

$$g(x) = \frac{1}{2\pi} \int_{-\infty}^{\infty} e^{ixt} \int_{-\infty}^{\infty} g(\xi)e^{-i\xi t}\, d\xi\, dt$$

$$e^{sx}f(x) = \frac{1}{2\pi} \int_{-\infty}^{\infty} e^{ixt} \int_{-\infty}^{\infty} f(\xi)e^{-i\xi(t+is)}\, d\xi\, dt$$

$$f(x) = \frac{1}{2\pi} \int_{-\infty}^{\infty} e^{ix(t+is)} \int_{-\infty}^{\infty} f(\xi)e^{-i\xi(t+is)}\, d\xi\, dt$$

$$f(x) = \frac{1}{2\pi} \int_{-\infty+is}^{\infty+is} e^{ix\zeta} \int_{-\infty}^{\infty} f(\xi)e^{-i\xi\zeta}\, d\xi\, d\zeta$$

where the integral in the complex ζ plane is taken along the line $\zeta = t + is$, $-\infty < t < \infty$. **The complex Fourier transform** of $f(x)$ is defined as follows:

$$\phi(\zeta) = \frac{1}{\sqrt{2\pi}} \int_{-\infty}^{\infty} f(x)e^{-ix\zeta}\, dx$$

and the inverse transform is

$$f(x) = \frac{1}{\sqrt{2\pi}} \int_{-\infty+is}^{\infty+is} \phi(\zeta) e^{ix\zeta}\, d\zeta$$

Let us investigate more carefully the conditions under which the complex Fourier transform exists. Suppose $f(x)$ is continuous and has a piecewise continuous first derivative in any finite interval. Let $f(x) = O(e^{\alpha x})$ as $x \to \infty$ and $f(x) = O(e^{\beta x})$ as $x \to -\infty$; that is, there exist numbers M, N, and R such that $|f(x)| \le M e^{\alpha x}$ for $x \ge R$ and $|f(x)| \le N e^{\beta x}$ for $x \le -R$. Then

$$\phi(\zeta) = \frac{1}{\sqrt{2\pi}} \int_{-\infty}^{-R} f(x) e^{-i\zeta x}\, dx + \frac{1}{\sqrt{2\pi}} \int_{-R}^{R} f(x) e^{-i\zeta x}\, dx + \frac{1}{\sqrt{2\pi}} \int_{R}^{\infty} f(x) e^{-i\zeta x}\, dx$$

and

$$\left| \int_{-\infty}^{-R} f(x) e^{-i\zeta x}\, dx \right| \le N \int_{-\infty}^{-R} e^{x(\beta+s)}\, dx = \frac{N e^{-(\beta+s)R}}{\beta+s}$$

provided $s > -\beta$. Similarly,

$$\left| \int_{R}^{\infty} f(x) e^{-i\zeta x}\, dx \right| \le M \int_{R}^{\infty} e^{x(\alpha+s)}\, dx = \frac{M e^{(\alpha+s)R}}{s+\alpha}$$

provided $s < -\alpha$. Therefore, the integral exists for $-\beta < \operatorname{Im}(\zeta) < -\alpha$. In fact, the integrals converge uniformly in the strip

$$-\beta < \rho_1 \le \operatorname{Im}(\zeta) \le \rho_2 < -\alpha$$

where ρ_1 and ρ_2 are arbitrary numbers satisfying the inequality. Therefore, $\phi(\zeta)$ is an analytic function of ζ in the strip $-\beta < \operatorname{Im}(\zeta) < -\alpha$. If the strip $-\beta < \operatorname{Im}(\zeta) < -\alpha$ contains the t axis of the ζ plane, then the Fourier transform will exist in the ordinary sense. Furthermore, by the theory of analytic continuation,

$$\phi(\zeta) = \frac{1}{\sqrt{2\pi}} \int_{-\infty}^{\infty} f(x) e^{-ix\zeta}\, dx$$

is the only function analytic in the strip which takes on the values

$$\phi(t) = \frac{1}{\sqrt{2\pi}} \int_{-\infty}^{\infty} f(x) e^{-ixt}\, dx$$

when ζ is real.

Under rather general conditions the inverse transform can be computed by integrating along any line parallel to the real axis lying in the strip of analyticity of the complex Fourier transform; that is,

$$\frac{1}{\sqrt{2\pi}} \int_{-\infty+i\gamma_1}^{\infty+i\gamma_1} \phi(\zeta) e^{ix\zeta}\, d\zeta = \frac{1}{\sqrt{2\pi}} \int_{-\infty+i\gamma_2}^{\infty+i\gamma_2} \phi(\zeta) e^{ix\zeta}\, d\zeta$$

where γ_1 and γ_2 are any real numbers between $-\beta$ and $-\alpha$. This can be shown by applying Cauchy's theorem to the rectangular contour shown and then letting $T \to \infty$, provided

$$|\phi(T + is)| \leq M(T) \to 0 \qquad \text{and} \qquad |\phi(-T + is)| \leq N(T) \to 0$$

as $T \to \infty$ uniformly in s. The integrals on the vertical sides approach zero as $T \to \infty$ since

$$\left| \frac{1}{\sqrt{2\pi}} \int_{T+i\gamma_1}^{T+i\gamma_2} \phi(\zeta) e^{ix\zeta} \right| \leq \frac{M}{\sqrt{2\pi}} \int_{\gamma_1}^{\gamma_2} e^{-xs} \, ds \to 0 \qquad \text{as } T \to \infty$$

$$\left| \frac{1}{\sqrt{2\pi}} \int_{-T+i\gamma_1}^{-T+i\gamma_2} \phi(\zeta) e^{ix\zeta} \right| \leq \frac{N}{\sqrt{2\pi}} \int_{\gamma_1}^{\gamma_2} e^{-xs} \, ds \to 0 \qquad \text{as } T \to \infty$$

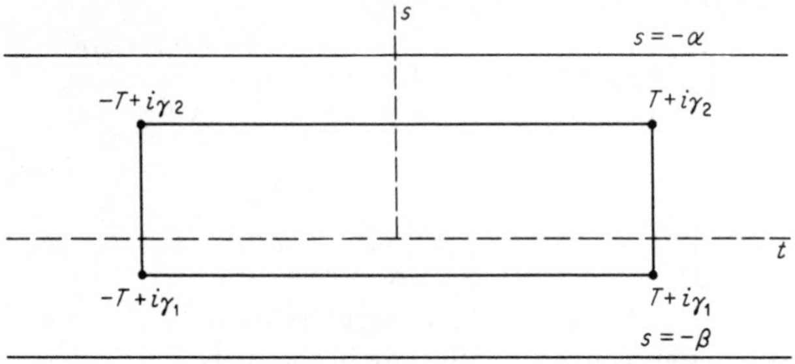

FIGURE 19

Therefore, since the closed contour lies entirely in the strip of analyticity, the result follows by Cauchy's theorem.

Exercises 8.1

1. Find the Fourier transform of $f(x) = e^{-ax}$, $x \geq 0$, $f(x) = 0$, $x < 0$, with $a > 0$. Verify the inverse transformation by direct integration. Note that the transform is not absolutely integrable, although it is square integrable.

*2. Show that $f(x) = e^{-x^2/2}$ is its own Fourier transform.

*3. If $f(x)$ is real and satisfies the conditions of Fourier's integral theorem, show that

$$\tfrac{1}{2}[f(x+) + f(x-)] = \frac{1}{\pi} \int_0^\infty \int_{-\infty}^\infty f(\xi) \cos t(x - \xi) \, d\xi \, dt$$

Furthermore, if $f(x)$ is even, show that

$$\tfrac{1}{2}[f(x+) + f(x-)] = \frac{2}{\pi} \int_0^\infty \cos xt \int_0^\infty f(\xi) \cos \xi t \, d\xi \, dt$$

and if $f(x)$ is odd,

$$\tfrac{1}{2}[f(x+) + f(x-)] = \frac{2}{\pi} \int_0^\infty \sin xt \int_0^\infty f(\xi) \sin \xi t \, d\xi \, dt$$

NOTE: $$\sqrt{\frac{2}{\pi}} \int_0^\infty f(x) \cos xt \, dx$$

is called the **Fourier cosine transform,** and

$$\sqrt{\frac{2}{\pi}} \int_0^\infty f(x) \sin xt \, dx$$

is called the **Fourier sine transform.**

4. Compute the complex Fourier transform of $f(x) = e^{-|x|}$ and show that it is analytic in the strip $-1 < \text{Im}(\zeta) < 1$.

*5. Let $f(x)$ be continuous and absolutely integrable for $-\infty < x < \infty$. Let $f'(x)$ be piecewise continuous and absolutely integrable. Show that if both $f(x)$ and $f'(x)$ are $O(e^{\alpha x})$ as $x \to \infty$ and $O(e^{\beta x})$ as $x \to -\infty$, $\phi(\zeta) = O(1/\zeta)$ in the strip $-\beta < \rho_1 \leq \text{Im}(\zeta) \leq \rho_2 < -\alpha$. Note that this is then a sufficient condition for the inversion integral to be independent of γ.

*6. Let $f(x)$ and $g(x)$ be absolutely integrable, $O(e^{\alpha x})$ as $x \to \infty$, and $O(e^{\beta x})$ as $x \to -\infty$. Show that the convolution

$$h(x) = \frac{1}{\sqrt{2\pi}} \int_{-\infty}^\infty f(x - \xi) g(\xi) \, d\xi$$

is $O(e^{\alpha_1 x})$ as $x \to \infty$ and $O(e^{\beta_1 x})$ as $x \to -\infty$, where $-\beta < -\beta_1 < -\alpha_1 < -\alpha$. Also show that

$$T[h] = T[f] T[g]$$

where these are the complex Fourier transforms and equality holds in the strip $-\beta < \text{Im}(\zeta) < -\alpha$, where all three transforms are analytic.

7. Let $f(x) = \sin \omega x$ for $x \geq 0$ and $f(x) = 0$ for $x < 0$. Find the complex Fourier transform of $f(x)$ and show that it is analytic for $\text{Im}(\zeta) < 0$. Verify the inversion integral directly, where $\gamma < 0$.

8.2 Applications of Fourier Transforms. Ordinary Differential Equations

We shall first apply the Fourier transform method to the solution of certain ordinary differential equations. To start with, consider the linear nth-order equation with constant coefficients

$$a_n \frac{d^n y}{dx^n} + a_{n-1} \frac{d^{n-1} y}{dx^{n-1}} + \cdots + a_1 \frac{dy}{dx} + a_0 y = f(x)$$

Denote the operator on the left by

$$L = a_n D^n + a_{n-1} D^{n-1} + \cdots + a_1 D + a_0$$

Then the equation can be written as

$$L(y) = f(x)$$

Assuming that we are interested in a solution for $-\infty < x < \infty$ where $f(x)$ is defined and has a Fourier transform, we can proceed as follows:

$$T[L(y)] = [a_n(it)^n + a_{n-1}(it)^{n-1} + \cdots + a_1(it) + a_0]T[y]$$
$$= P(it)T[y] = T[f] = F(t)$$

Then
$$T[y] = \frac{F(t)}{P(it)} = F(t)G(t)$$

where $G(t) = 1/P(it)$. Then, using the convolution theorem, we have

$$y(x) = T^{-1}[F(t)G(t)] = \frac{1}{\sqrt{2\pi}} \int_{-\infty}^{\infty} f(\xi)g(x - \xi)\,d\xi$$

where
$$g(x) = T^{-1}[G(t)] = T^{-1}\left[\frac{1}{P(it)}\right]$$

Proceeding formally we have the following interesting interpretation of the result. The Fourier transform of the Dirac delta function is[1]

$$T[\delta(x)] = \frac{1}{\sqrt{2\pi}} \int_{-\infty}^{\infty} \delta(x)e^{-ixt}\,dx = \frac{1}{\sqrt{2\pi}}$$

Then consider the following differential equation:

$$L(\gamma) = \delta(x)$$

Then
$$T[L(\gamma)] = P(it)T[\gamma] = \frac{1}{\sqrt{2\pi}}$$

$$\gamma(x) = T^{-1}\left[\frac{1}{\sqrt{2\pi}P(it)}\right] = \frac{1}{\sqrt{2\pi}}T^{-1}[G(t)] = \frac{1}{\sqrt{2\pi}}g(x)$$

so that
$$y(x) = \int_{-\infty}^{\infty} f(\xi)\gamma(x - \xi)\,d\xi$$

Therefore, γ behaves like a Green's function, that is, it is the response of the system to a delta function (unit impulse), and we have a superposition principle with $y(x)$, the response to $f(x)$, given by the above integral. The electrical engineer describes this phenomenon as follows. The function $f(x)$ is the **input,** while $y(x)$ is the **output.** The function $\gamma(x)$ is the **response to a unit impulse.** The transform of $\sqrt{2\pi}\gamma$ is the **admittance** $Y = 1/P(it)$. To get the response to a given input you find the transform of the input, multiply by the admittance, and then take the inverse transform of the product.

Let us illustrate these ideas by an elementary electrical circuit problem (see Fig. 20).

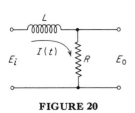

FIGURE 20

Here we have $E_0 = RI$ $L\dfrac{dI}{dt} + RI = E_i$

<hr>

[1] This can be made rigorous by treating the delta function as a generalized function.

Let $E_i = Ee^{-|t|}$. Then[1]

$$(is)LT[I] + RT[I] = \frac{2E}{\sqrt{2\pi}} \frac{1}{1 + s^2}$$

$$T[I] = \frac{2E}{\sqrt{2\pi}} \frac{1}{1 + s^2} \frac{1}{isL + R}$$

$$I(t) = \frac{E}{\pi} \int_{-\infty}^{\infty} \frac{e^{ist}\, ds}{(1 + s^2)(isL + R)}$$

The last integral can be evaluated by complex contour integration. For $t > 0$

$$I(t) = \frac{E}{\pi} 2\pi i \left(\text{residue at } s = i + \text{residue at } s = \frac{iR}{L} \right)$$

$$= 2Ei \left[\frac{e^{-t}}{2i(R - L)} + \frac{e^{-Rt/L}}{iL(1 - R^2/L^2)} \right]$$

$$= E \left(\frac{e^{-t}}{R - L} + \frac{2Le^{-Rt/L}}{L^2 - R^2} \right)$$

For $t < 0$

$$I(t) = \frac{-E}{\pi} 2\pi i (\text{residue at } s = -i) = \frac{Ee^{t}}{L + R}$$

At $t = 0$ the current is continuous; that is,

$$I(0) = \lim_{t \to 0} I(t) = \frac{E}{R + L}$$

If we are interested in a phenomenon which begins at a specific time, say $t = 0$, we can still use the Fourier transform technique if we exercise a bit of care in computing the transform. Suppose in the above example

$$E_i = Ee^{-at} \sin \omega t \qquad t \geq 0$$
$$E_i = 0 \qquad\qquad\quad t < 0$$

and we specify that $I = dI/dt = 0$ for $t < 0$, while $I(0+) = I_0$. Then

$$T[I'] = \frac{1}{\sqrt{2\pi}} \int_{0}^{\infty} \frac{dI}{dt} e^{-ist}\, dt$$

$$= \frac{-I_0}{\sqrt{2\pi}} + \frac{is}{\sqrt{2\pi}} \int_{0}^{\infty} I e^{-ist}\, dt$$

$$= \frac{-I_0}{\sqrt{2\pi}} + isT[I]$$

---

[1] Here we change to the variable s in the transform to avoid the time variable t in the differential equation.

Then we have

$$\frac{-I_0 L}{\sqrt{2\pi}} + (isL + R)T[I] = \frac{1}{\sqrt{2\pi}} \int_0^\infty E e^{-at} \sin \omega t \, e^{-ist} \, dt$$

$$= \frac{E}{2i\sqrt{2\pi}} \left(\frac{1}{a + is - i\omega} - \frac{1}{a + is + i\omega} \right)$$

Therefore,

$$T[I] = \frac{E}{2i\sqrt{2\pi}(isL + R)} \left(\frac{1}{a + is - i\omega} - \frac{1}{a + is + i\omega} \right) + \frac{I_0 L}{\sqrt{2\pi}(isL + R)}$$

Inverting the transform by complex contour integration, we see that for $t < 0$, $I = 0$, since in this case we close the contour below the real axis and there are no poles in the lower half plane. For $t > 0$ we close the contour above the real axis and evaluate the required residues. Then

$$I(t) = \frac{\omega E e^{-Rt/L}}{L[(a - R/L)^2 + \omega^2]} + I_0 e^{-Rt/L}$$

$$+ \frac{(R/L - a)E e^{-at} \sin \omega t}{L/[(a - R/L)^2 + \omega^2]} - \frac{\omega E e^{-at} \cos \omega t}{L/[(a - R/L)^2 + \omega^2]}$$

Notice that I is discontinuous at $t = 0$ but that $\lim_{t \to 0+} I(t) = I_0$ as specified.

If there is a capacitor in the circuit (see Fig. 21), then the basic equation is the integrodifferential equation

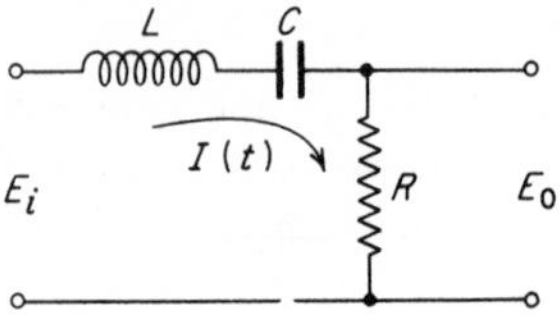

FIGURE 21

$$L \frac{dI}{dt} + RI + \frac{1}{C}\left(q_0 + \int_0^t I \, dt \right) = E_i(t)$$

where q_0 is the initial charge on the capacitor.

$$q = q_0 + \int_0^t I \, dt$$

is the charge, and obviously $dq/dt = I$. Let us assume that I, q, and E_i are all zero for $t < 0$, but that $I(0+) = I_0$ and $q(0+) = q_0$. Then

$$T[q] = \frac{1}{\sqrt{2\pi}} \int_0^\infty q(t) e^{-ist} \, dt$$

$$= \left[\frac{q(t)e^{-ist}}{-is\sqrt{2\pi}} \right]_0^\infty + \frac{1}{is\sqrt{2\pi}} \int_0^\infty I(t) e^{-ist} \, dt$$

$$= \frac{q_0}{\sqrt{2\pi}is} + \frac{T[I]}{is}$$

and we have

$$\frac{-I_0 L}{\sqrt{2\pi}} + \frac{q_0}{C\sqrt{2\pi} is} + \left(isL + R + \frac{1}{isC}\right)T[I] = T[E_i]$$

Then
$$T[I] = \frac{T[E_i] + I_0 L/\sqrt{2\pi} - q_0/(C\sqrt{2\pi} is)}{[isL + R + 1/(isC)]}$$

and the transform may be inverted using complex contour integration or by referring to a set of tables.[1]

Before leaving the discussion of ordinary differential equations, let us consider how the use of the complex Fourier transform facilitates the solution of some problems where the input does not have a Fourier transform in the usual sense. Consider the following input to the circuit of the last example:

$$E_i = 1 \qquad t > 0$$
$$E_i = 0 \qquad t < 0$$

This is the Heaviside function, or unit step function. This input does not have a Fourier transform in the ordinary sense, but it does have a complex transform; that is,

$$T[E_i] = \frac{1}{\sqrt{2\pi}} \int_0^\infty e^{-i\zeta t}\, dt$$

$$= \frac{1}{\sqrt{2\pi}\, i\zeta}$$

provided $\mathrm{Im}\,(\zeta) < 0$. The inversion integral yields E_i by complex contour integration. If $t > 0$

$$\frac{1}{2\pi} \int_{-\infty - i\gamma}^{\infty - i\gamma} \frac{e^{i\zeta t}}{i\zeta}\, d\zeta = 2\pi i \left(\text{residue of } \frac{e^{i\zeta t}}{2\pi i\zeta} \text{ at } \zeta = 0\right)$$
$$= 1$$

because here the contour is closed by a semicircular arc in an upper half plane and the pole at $\zeta = 0$ is inside the contour. If $t < 0$, the contour is closed below the real axis, there are no singularities inside the contour, and by Cauchy's theorem the integral is zero.

Let $I_0 = q_0 = 0$. Then we have, taking complex Fourier transforms,

$$\left[(i\zeta)L + R + \frac{1}{(i\zeta)C}\right]T[I] = \frac{1}{\sqrt{2\pi} i\zeta}$$

$$T[I] = \frac{1}{\sqrt{2\pi}[(i\zeta)^2 L + (i\zeta)R + 1/C]}$$

[1] See George A. Campbell and Ronald M. Foster, "Fourier Integrals for Practical Applications," D. Van Nostrand Company, Inc., Princeton, N.J., 1948.

Notice that the transform of the response to a unit step function is the same as the transform of the solution of the differential equation

$$L\frac{d^2\gamma}{dt^2} + R\frac{d\gamma}{dt} + \frac{1}{C}\gamma = \delta(t)$$

with the unit impulse as input.   This leads to the observation that the unit impulse is the derivative of the unit step function.[1]   Also, since

$$L\frac{d^2I}{dt^2} + R\frac{dI}{dt} + \frac{1}{C}I = \frac{d}{dt}E_i$$

for an arbitrary input E_i, we have the following result:

$$T[I] = \frac{1}{\sqrt{2\pi}[(i\zeta)^2 L + (i\zeta)R + 1/C]}T[E_i']$$

$$I = \int_{-\infty}^{\infty} E_i'(\tau)A(t-\tau)\,d\tau$$

where $A(t)$ is the **indicial admittance**, that is, the response to a unit step input.

$$A(t) = T^{-1}\left\{\frac{1}{2\pi[(i\zeta)^2 L + (i\zeta)R + 1/C]}\right\}$$

This is a very important result from the practical point of view, because it is much easier to simulate a unit step function than it is to simulate a unit impulse.   Hence the response to a unit step can be determined experimentally much more easily than the response to a unit impulse.

Exercises 8.2

1. Solve the differential equation $y'' + 3y' + 2y = e^{-t}$, $0 < t$, subject to $y(0+) = y_0$ and $y'(0+) = y_0'$, using the Fourier transform.

2. Solve the differential equation $y'' + \omega_0^2 y = \sin \omega t$, $0 < t$, $\omega \neq \omega_0$, subject to $y(0+) = 0$ and $y'(0+) = 0$, using the complex Fourier transform.   Also consider the case $\omega = \omega_0$.

3. Solve the integrodifferential equation $y' + \int_0^t y(\tau)\,d\tau = e^{-t}$, $0 < t$, subject to $y(0+) = y_0$, using the Fourier transform.

4. Solve the following pair of simultaneous differential equations for $t > 0$, using the Fourier transform.

$$2x' + y' - x + 2y = e^{-t}$$

$$x' + y' + x + y = e^{-2t}$$

subject to $x(0+) = x_0$, $y(0+) = y_0$.

---

[1] This statement is made rigorous in the theory of generalized functions. See Sec. 5.3.

8.3   Applications of Fourier Transforms.   Partial Differential Equations

Next we consider the Fourier transform technique for solving certain partial differential equations.   As a general class of problem consider the solution of the partial differential equation

$$\nabla^2\phi = a\phi + b\,\frac{\partial\phi}{\partial t} + c\,\frac{\partial^2\phi}{\partial t^2} - f(x,y,z,t)$$

to hold in a bounded volume V, bounded by the closed surface S.   On S we prescribe the boundary condition $d\phi/dn + \alpha\phi = 0$.   We also specify initially that $\phi(x,y,z,0) = u(x,y,z)$ and $(\partial\phi/\partial t)_{t=0} = v(x,y,z)$.   This problem was considered in Sec. 5.1, where the method of separation of variables was employed under certain special conditions on $f(x,y,z,t)$.   Let us assume that $\phi \equiv 0$ and that $f \equiv 0$ for $t < 0$.   We proceed by taking the Fourier transform of every term in the equation with respect to the variable t.   Let

$$T(x,y,z,s) = \frac{1}{\sqrt{2\pi}} \int_0^\infty \phi(x,y,z,t)e^{-ist}\,dt$$

$$F(x,y,z,s) = \frac{1}{\sqrt{2\pi}} \int_0^\infty f(x,y,z,t)e^{-ist}\,dt$$

Then we have

$$\nabla^2 T = aT - \frac{b}{\sqrt{2\pi}}\,u + isb\,T - \frac{c}{\sqrt{2\pi}}\,v - \frac{isc}{\sqrt{2\pi}}\,u - s^2 T - F(x,y,z,s)$$

or

$$\nabla^2 T + \lambda(s)T = -g(x,y,z,s)$$

where

$$\lambda(s) = -a + isb - s^2$$

$$g(x,y,z,s) = \frac{b}{\sqrt{2\pi}}\,u + \frac{cv}{\sqrt{2\pi}} + \frac{isc}{\sqrt{2\pi}}\,u + F(x,y,z,s)$$

Let us further assume that

$$\frac{dT}{dn} + \alpha T = \frac{1}{\sqrt{2\pi}} \int_0^\infty \left(\frac{d\phi}{dn} + \alpha\phi\right)e^{-ist}\,dt = 0$$

on S.   The transform technique has thus led us to the solution of a non-homogeneous boundary-value problem based on the Helmholtz equation of the type considered in Chap. 5.   Here s plays the role of a parameter, not a variable.   If the problem has a unique solution (which of course depends on the values that λ may attain), then we obtain T and attempt to find its inverse transform.

Of course we have made many gross assumptions, the most basic of which is that the solution of our problem has a Fourier transform. Even if it has, our procedure may not obtain it. To justify rigorously the technique at every step of the way would be impractical if not impossible. In this method one must take great liberties, but of course any possible solution so obtained must be checked against the stated conditions of the original problem. Surprisingly enough, sometimes a solution is obtained even though the intervening steps cannot be justified rigorously. The use of the complex Fourier transform will often help when some of the functions involved do not have transforms in the ordinary sense.

Let us illustrate the method in a few simple examples. Consider an elastic string of density ρ stretched between two fixed points ($x = 0$, $x = 1$) with a tension σ. Until time $t = 0$ the string is at rest. Then a force of linear density $F_0 \sin \pi x$ is applied to the string. What is the resulting motion? Let $\phi(x,t)$ be the displacement. The differential equation is

$$\frac{\partial^2 \phi}{\partial x^2} = \frac{\rho}{\sigma} \frac{\partial^2 \phi}{\partial t^2} - \frac{F_0}{\sigma} \sin \pi x \qquad 0 < x < 1; \ 0 < t$$

The boundary conditions are $\phi(0,t) = \phi(1,t) = 0$. The initial conditions are $\phi(x,0) = (\partial \phi/\partial t)_{t=0} = 0$. Taking the complex Fourier transform, we have

$$\frac{d^2 T}{dx^2} = -\zeta^2 \frac{\rho}{\sigma} T - \frac{F_0 \sin \pi x}{i\zeta\sigma\sqrt{2\pi}}$$

The solution of this equation is

$$T(x,\zeta) = A \sin \zeta \sqrt{\frac{\rho}{\sigma}} x + B \cos \zeta \sqrt{\frac{\rho}{\sigma}} x - \frac{F_0 \sin \pi x}{\sqrt{2\pi} i\zeta\sigma[\zeta^2(\rho/\sigma) - \pi^2]}$$

Using the derived conditions $T(0,\zeta) = T(1,\zeta) = 0$, we find $A = B = 0$. Hence,

$$T(x,\zeta) = - \frac{F_0 \sin \pi x}{\sqrt{2\pi} i\zeta\sigma[\zeta^2(\rho/\sigma) - \pi^2]}$$

By use of the complex inversion integral,

$$\phi(x,t) = \frac{-F_0 \sin \pi x}{2\pi i \sigma} \int_{-\infty+i\gamma}^{\infty+i\gamma} \frac{e^{i\zeta t} \, d\zeta}{\zeta[\zeta^2(\rho/\sigma) - \pi^2]}$$

where $\gamma < 0$. This can be evaluated by use of complex contour integration closing the contour with a semicircle in an upper half plane when $t > 0$ and in a lower half plane when $t < 0$. The result is

$$\phi(x,t) = \frac{F_0}{\pi^2 \sigma} \sin \pi x \left(1 - \cos \pi \sqrt{\frac{\sigma}{\rho}} t \right) \qquad t > 0$$

$$\phi(x,t) = 0 \qquad\qquad\qquad\qquad t < 0$$

We see that $\lim\limits_{t\to 0+} \phi(x,t) = 0$ and $\lim\limits_{t\to 0+} \partial\phi/\partial t = 0$, the prescribed initial conditions. $\phi(0,t) = \phi(1,t) = 0$, the prescribed boundary conditions. It is easily checked that $\phi(x,t)$ satisfies the differential equation and hence is the solution of the problem. We see that the result is a vibration about some deflected position produced by the constant distributed force.

The choice of force $F_0 \sin \pi x$, which vanishes at the ends of the string, made it easy to evaluate the constants A and B in the above example. This is not essential. For example, if the force were uniform, that is, $F = F_0$, $t > 0$, then we would have had

$$\frac{d^2 T}{dx^2} = -\zeta^2 \frac{\rho}{\sigma} T - \frac{F_0}{\sqrt{2\pi}i\zeta\sigma}$$

and
$$T = A \sin \zeta \sqrt{\frac{\rho}{\sigma}}\, x + B \cos \zeta \sqrt{\frac{\rho}{\sigma}}\, x - \frac{\rho F_0}{\sqrt{2\pi}i\zeta^3\sigma}$$

By use of the boundary conditions,

$$T(0,\zeta) = B - \frac{\rho F_0}{\sqrt{2\pi}i\zeta^3\sigma^2} = 0$$

$$T(1,\zeta) = A \sin \zeta \sqrt{\frac{\rho}{\sigma}} + B \cos \zeta \sqrt{\frac{\rho}{\sigma}} - \frac{\rho F_0}{\sqrt{2\pi}i\zeta^3\sigma^2} = 0$$

from which we get

$$B = \frac{\rho F_0}{\sqrt{2\pi}i\zeta^3\sigma^2}$$

$$A = \frac{\rho F_0}{\sqrt{2\pi}i\zeta^3\sigma^2}\, \frac{(1 - \cos \zeta\sqrt{\rho/\sigma})}{\sin \zeta\sqrt{\rho/\sigma}}$$

Then

$$T(x,\zeta) = \frac{\rho F_0}{\sqrt{2\pi}i\zeta^3\sigma^2}\left[\frac{\sin \zeta\sqrt{\rho/\sigma}\, x + \sin \zeta\sqrt{\rho/\sigma}(1 - x)}{\sin \zeta\sqrt{\rho/\sigma}} - 1\right]$$

$$= \frac{\rho F_0}{\sqrt{2\pi}i\zeta^3\sigma^2}\left[\frac{e^{i\zeta\sqrt{\rho/\sigma}x} - e^{-i\zeta\sqrt{\rho/\sigma}x} + e^{i\zeta\sqrt{\rho/\sigma}(1-x)} - e^{-i\zeta\sqrt{\rho/\sigma}(1-x)}}{e^{i\zeta\sqrt{\rho/\sigma}} - e^{-i\zeta\sqrt{\rho/\sigma}}} - 1\right]$$

$$= \frac{\rho F_0}{\sqrt{2\pi}i\zeta^3\sigma^2}\left[\sum_{n=0}^{\infty}(e^{i\zeta\sqrt{\rho/\sigma}(x-1-2n)} - e^{-i\zeta\sqrt{\rho/\sigma}(x+1+2n)})\right.$$

$$\left. + \sum_{n=0}^{\infty}(e^{-i\zeta\sqrt{\rho/\sigma}(x+2n)} - e^{-i\zeta\sqrt{\rho/\sigma}(-x+2+2n)}) - 1\right]$$

provided $|e^{-i2\zeta\sqrt{\rho/\sigma}}| = e^{2\eta\sqrt{\rho/\sigma}} < 1$; that is, $\eta = \text{Im}\,(\zeta) < 0$.

We shall assume that we can invert T term by term. To recognize the inverse transform consider the complex Fourier transform of the function

$$f(t) = (t - \tau)^2 \qquad t \geq \tau$$
$$f(t) = 0 \qquad t < \tau$$

Then

$$T[f] = \frac{1}{\sqrt{2\pi}} \int_\tau^\infty (t - \tau)^2 e^{-i\zeta t}\, dt$$

$$= \frac{e^{-i\zeta\tau}}{\sqrt{2\pi}} \int_0^\infty y^2 e^{-i\zeta y}\, dy$$

$$= \frac{2 e^{-i\zeta\tau}}{\sqrt{2\pi}(i\zeta)^3}$$

provided $\mathrm{Im}\,(\zeta) < 0$. Comparing this result with $T(x,\zeta)$, we have for $t > 0$

$$\phi(x,t)$$

$$= \frac{\rho F_0}{2\sigma^2} \Bigg\{ t^2 - \sum_{n=0}^\infty {}^* \left[t - \sqrt{\frac{\rho}{\sigma}}(2n + 1 - x) \right]^2 + \sum_{n=0}^\infty {}^* \left[t - \sqrt{\frac{\rho}{\sigma}}(2n + 1 + x) \right]^2$$

$$- \sum_{n=0}^\infty {}^* \left[t - \sqrt{\frac{\rho}{\sigma}}(2n + x) \right]^2 + \sum_{n=0}^\infty {}^* \left[t - \sqrt{\frac{\rho}{\sigma}}(2n + 2 - x) \right]^2 \Bigg\}$$

Here the asterisk denotes a convention that the summation shall contain the given term only when the quantity in the square bracket is positive. The convergence is assured by the fact that for any finite t and $0 \leq x \leq 1$ there are only a finite number of terms to be summed, since for n sufficiently large the quantities in all the square brackets can be made negative. It can be verified directly that this is a solution to the problem.

If the space variable ranges over an infinite interval, it may be convenient to transform the partial differential equation with respect to the space variable. For example, consider the infinite string as an approximation to a very long string. Suppose over an interval $a \leq x \leq b$, small compared with the length of the string, the string is given an initial displacement $\phi(x,0) = f(x)$ and an initial velocity $(\partial\phi/\partial t)_{t=0} = g(x)$. Then the boundary-value problem to be solved is the following:

$$\frac{\partial^2 \phi}{\partial x^2} = \frac{\rho}{\sigma} \frac{\partial^2 \phi}{\partial t^2} \qquad -\infty < x < \infty \qquad t > 0$$

$\phi \to 0$ as $x \to \pm\infty$. Assuming

$$T(\zeta,t) = \frac{1}{\sqrt{2\pi}} \int_{-\infty}^\infty \phi(x,t) e^{-i\zeta x}\, dx$$

we have

$$-\zeta^2 T(\zeta,t) = \frac{\rho}{\sigma} \frac{d^2 T}{dt^2}$$

Solving the transformed equation, we have

$$T(\zeta,t) = A \sin \zeta \sqrt{\frac{\sigma}{\rho}} t + B \cos \zeta \sqrt{\frac{\sigma}{\rho}} t$$

Now
$$T(\zeta,0) = \frac{1}{\sqrt{2\pi}} \int_{-\infty}^{\infty} \phi(x,0) e^{-i\zeta x} \, dx$$

$$= \frac{1}{\sqrt{2\pi}} \int_{-\infty}^{\infty} f(x) e^{-i\zeta x} \, dx = F(\zeta)$$

$$\left(\frac{dT}{dt}\right)_{t=0} = \frac{1}{\sqrt{2\pi}} \int_{-\infty}^{\infty} \left(\frac{\partial \phi}{\partial t}\right)_{t=0} e^{-i\zeta x} \, dx$$

$$= \frac{1}{\sqrt{2\pi}} \int_{-\infty}^{\infty} g(x) e^{-i\zeta x} \, dx = G(\zeta)$$

Using these derived conditions, we have

$$T(\zeta,0) = F(\zeta) = B$$

$$\dot{T}(\zeta,0) = G(\zeta) = \zeta \sqrt{\frac{\sigma}{\rho}} A$$

Hence,
$$T(\zeta,t) = \frac{G(\zeta)}{\zeta} \sqrt{\frac{\rho}{\sigma}} \sin \zeta \sqrt{\frac{\sigma}{\rho}} t + F(\zeta) \cos \zeta \sqrt{\frac{\sigma}{\rho}} t$$

$$= \tfrac{1}{2}[F(\zeta) e^{i\zeta \sqrt{\sigma/\rho}\, t} + F(\zeta) e^{-i\zeta \sqrt{\sigma/\rho}\, t}]$$

$$+ \frac{\sqrt{\rho/\sigma}}{2}\left[\frac{G(\zeta)}{i\zeta} e^{i\zeta \sqrt{\sigma/\rho}\, t} - \frac{G(\zeta)}{i\zeta} e^{-i\zeta \sqrt{\sigma/\rho}\, t}\right]$$

Using the inversion integral, we have

$$\phi(x,t) = \frac{1}{2\sqrt{2\pi}} \int_{-\infty}^{\infty} \left[F(\zeta) e^{i\zeta(x + \sqrt{\sigma/\rho}\, t)} + F(\zeta) e^{i\zeta(x - \sqrt{\sigma/\rho}\, t)} \right] d\zeta$$

$$+ \frac{\sqrt{\rho/\sigma}}{2\sqrt{2\pi}} \int_{-\infty}^{\infty} \left[\frac{G(\zeta)}{i\zeta} e^{i\zeta(x + \sqrt{\sigma/\rho}\, t)} - \frac{G(\zeta)}{i\zeta} e^{i\zeta(x - \sqrt{\sigma/\rho}\, t)}\right] d\zeta$$

$$= \tfrac{1}{2}\left[f\left(x + \sqrt{\frac{\sigma}{\rho}} t\right) + f\left(x - \sqrt{\frac{\sigma}{\rho}} t\right)\right]$$

$$+ \frac{\sqrt{\rho/\sigma}}{2}\left[\int_{a}^{x + \sqrt{\sigma/\rho}\, t} g(\xi) \, d\xi - \int_{a}^{x - \sqrt{\sigma/\rho}\, t} g(\xi) \, d\xi\right]$$

The choice of a as the lower limit in both integrals is dictated by the condition that $\phi(x,0) = f(x)$. We may therefore write

$$\phi(x,t) = \tfrac{1}{2}\left[f\left(x + \sqrt{\frac{\sigma}{\rho}} t\right) + f\left(x - \sqrt{\frac{\rho}{\rho}} t\right)\right] + \frac{\sqrt{\rho/\sigma}}{2}\left[\int_{x - \sqrt{\sigma/\rho}\, t}^{x + \sqrt{\sigma/\rho}\, t} g(\xi) \, d\xi\right]$$

It may be verified directly that this is a solution of the problem. Compare this result with Sec. 3.5.

Next let us consider some problems in the study of the heat equation. Consider an infinite heat-conducting rod as an approximation to a very long rod. We assume that the lateral surface is insulated against the flow of heat so that heat flows along the axis of the rod and the problem is essentially one-dimensional. The rod has some initial temperature distribution $\phi(x,0) = f(x)$, $-\infty < x < \infty$. We also assume that the ends are maintained at zero temperature and that $f(x)$ falls off to zero rapidly enough at $\pm\infty$ that it has a Fourier transform. The problem then is to solve the one-dimensional heat equation

$$\frac{\partial^2 \phi}{\partial x^2} = \frac{1}{a^2}\frac{\partial \phi}{\partial t} \qquad -\infty < x < \infty \qquad 0 < t$$

subject to $\phi(x,0) = f(x)$ and $\phi(x,t) \to 0$ as $x \to \pm\infty$. Let

$$T(\zeta,t) = \frac{1}{\sqrt{2\pi}}\int_{-\infty}^{\infty} \phi(x,t)e^{-ix\zeta}\,dx$$

Then

$$\frac{dT}{dt} + a^2\zeta^2 T = 0$$

$$T(\zeta,t) = T_0 e^{-\zeta^2 a^2 t}$$

$$T_0 = T(\zeta,0) = F(\zeta)$$

where

$$F(\zeta) = \frac{1}{\sqrt{2\pi}}\int_{-\infty}^{\infty} f(x)e^{-ix\zeta}\,dx$$

The transform of $\phi(x,t)$ can be inverted by use of the convolution theorem, but first we must determine $T^{-1}[e^{-\zeta^2 a^2 t}]$. Recall that

$$\frac{1}{\sqrt{2\pi}}\int_{-\infty}^{\infty} e^{-x^2/2}e^{-iux}\,dx = e^{-u^2/2}$$

We let $u^2/2 = \zeta^2 a^2 t$. Then

$$T^{-1}[e^{-\zeta^2 a^2 t}] = \frac{1}{\sqrt{2\pi}}\int_{-\infty}^{\infty} e^{-\zeta^2 a^2 t}e^{i\zeta x}\,d\zeta$$

$$= \frac{1}{2a\sqrt{\pi t}}\int_{-\infty}^{\infty} e^{-u^2/2}e^{iu(x/a\sqrt{2t})}\,du$$

$$= \frac{1}{a\sqrt{2t}}e^{-x^2/4a^2 t}$$

Therefore, $$T^{-1}[F(\zeta)e^{-\zeta^2 a^2 t}] = \frac{1}{2a\sqrt{\pi t}}\int_{-\infty}^{\infty} f(\xi)e^{-(x-\xi)^2/4a^2 t}\,d\xi$$

A slight modification of this technique will allow us to handle the case of the semi-infinite heat-conducting rod. In this case we will solve

$$\frac{\partial^2 \phi}{\partial x^2} = \frac{1}{a^2} \frac{\partial \phi}{\partial t} \qquad 0 < x < \infty \qquad 0 < t < \infty$$

subject to $\phi(x,0) = f(x)$, $0 < x$, $\phi(x,t) \to 0$ as $x \to \infty$. To solve our problem we must specify a missing boundary condition at $x = 0$. We note from the above solution for the infinite rod that if $f(x)$ is an odd function, $\phi(0,t) = 0$. Furthermore, $\phi(x,t)$ is then odd. Therefore, if we specify the boundary condition $\phi(0,t) = 0$, we will obtain a solution of our problem from the infinite-rod case if we merely continue $f(x)$ as an odd function. This is admittedly a special case but we shall now generalize it.

Alternatively, let us seek a solution for the semi-infinite rod with $\phi(x,0) = 0$ but $\phi(0,t) = g(t)$. In other words, we have a constant initial temperature distribution, but we are heating the end so that the temperature there is determined. The function $\phi(x,0) = 0$ is trivially an odd function, so we can again seek an odd solution of the differential equation. We know that the inverse transform of an odd function will give us zero at $x = 0$. However, we shall specify the boundary condition in the sense of a right-hand limit; that is,

$$\lim_{x \to 0+} \phi(x,t) = g(t)$$

This will mean that $\lim_{x \to 0-} \phi(x,t) = -g(t)$, but we are not concerned with what happens to the left of the origin. Assuming that $\phi(x,t) = -\phi(-x,t)$, we have

$$T[\phi] = T(\zeta,t) = \frac{1}{\sqrt{2\pi}} \int_{-\infty}^{\infty} \phi(x,t) e^{-ix\zeta} \, dx$$

$$= -\sqrt{\frac{2}{\pi}} \, i \int_0^{\infty} \phi(x,t) \sin x\zeta \, dx$$

$$T\left[\frac{\partial^2 \phi}{\partial x^2}\right] = -\sqrt{\frac{2}{\pi}} \, i \int_0^{\infty} \frac{\partial^2 \phi}{\partial x^2} \sin x\zeta \, dx$$

$$= -\sqrt{\frac{2}{\pi}} \, i \left[\frac{\partial \phi}{\partial x} \sin x\zeta\right]_0^{\infty} + \sqrt{\frac{2}{\pi}} \, i\zeta \int_0^{\infty} \frac{\partial \phi}{\partial x} \cos x\zeta \, dx$$

$$= \sqrt{\frac{2}{\pi}} \, i\zeta [\phi \cos x\zeta]_0^{\infty} + \sqrt{\frac{2}{\pi}} \, i\zeta^2 \int_0^{\infty} \phi \sin x\zeta \, dx$$

$$= -\sqrt{\frac{2}{\pi}} \, i\zeta g(t) - \zeta^2 T(\zeta,t)$$

The differential equation then transforms to

$$\frac{dT}{dt} + a^2\zeta^2 T = -\sqrt{\frac{2}{\pi}}\, i\zeta a^2 g(t)$$

subject to $T(\zeta,0) = \sqrt{2/\pi i} \int_0^\infty \phi(x,0) \sin x\zeta\, dx = 0$. Solving this differential equation, we have

$$T(\zeta,t) = -\sqrt{\frac{2}{\pi}}\, i\zeta a^2 \int_0^t g(\xi) e^{-a^2\zeta^2(t-\xi)}\, d\xi$$

We already know that

$$T^{-1}[e^{-a^2\zeta^2(t-\xi)}] = \frac{1}{a\sqrt{2(t-\xi)}}\, e^{-x^2/[4a^2(t-\xi)]}$$

Therefore, $T^{-1}[i\zeta e^{-a^2\zeta^2(t-\xi)}] = \dfrac{\partial}{\partial x} \dfrac{1}{a\sqrt{2(t-\xi)}}\, e^{-x^2/[4a^2(t-\xi)]}$

$$= \frac{-x}{2\sqrt{2}a^3(t-\xi)^{\frac{3}{2}}}\, e^{-x^2/[4a^2(t-\xi)]}$$

Hence, $\phi(x,t) = \dfrac{x}{2\sqrt{\pi}a} \displaystyle\int_0^t \dfrac{g(\xi)}{(t-\xi)^{\frac{3}{2}}}\, e^{-x^2/[4a^2(t-\xi)]}\, d\xi$

In inverting the transform we have assumed that integration and inversion are interchangeable. This of course does not have to be justified if the final result can be shown to satisfy the conditions of the original problem. This can be done, although we shall not do it here.

Finally, the general problem for the semi-infinite rod, that is,

$$\frac{\partial^2 \phi}{\partial x^2} = \frac{1}{a^2}\frac{\partial \phi}{\partial t} \qquad 0 < x < \infty \qquad 0 < t < \infty$$

with $\phi(x,0) = f(x)$ and $\phi(0,t) = g(t)$, can be solved by adding the solutions of the last two problems.

Exercises 8.3

1. Solve the following boundary-value problem using the Fourier transform method:

$$\frac{\partial^2 \phi}{\partial x^2} = \frac{\rho}{\sigma}\frac{\partial^2 \phi}{\partial t^2} - \frac{F_0}{\sigma} x(1-x) \qquad 0 < x < 1 \qquad 0 < t$$

subject to $\phi(0,t) = \phi(1,t) = 0, \phi(x,0+) = 0, \phi_t(x,0+) = 0$. Check the solution.

2. Solve the following boundary-value problem for the semi-infinite string:

$$\frac{\partial^2 \phi}{\partial x^2} = \frac{\rho}{\sigma}\frac{\partial^2 \phi}{\partial t^2} \qquad 0 < x \qquad 0 < t$$

subject to $\phi(0,t) = f(t)$, $\phi \to 0$ as $x \to \infty$, $\phi(x,0+) = 0$, $\phi_t(x,0+) = 0$. Check and interpret the result.

3. Using Fourier transform methods solve for the steady-state temperature distribution in a semi-infinite heat-conducting slab, $0 < x < \infty$, $0 \leq y \leq 1$, if the edges at $y = 0$ and $y = 1$ are maintained at temperature zero while the end $x = 0$ is heated to maintain a temperature distribution $\phi(0,y) = f(y)$. HINT: Continue the solution as an odd function of x, that is, $\phi(-x,y) = -\phi(x,y)$, and use the Fourier sine transform.

8.4   Applications of Fourier Transforms.   Integral Equations

The right-hand side of the Wiener-Hopf integral equation

$$f(x) = \int_{-\infty}^{\infty} k(x - \xi)\phi(\xi)\, d\xi$$

is in the form of a convolution integral, and this fact allows us to solve it using the Fourier transform. Let $F(\zeta)$ be the Fourier transform of $(1/\sqrt{2\pi})\,f(x)$, and let $K(\zeta)$ and $T(\zeta)$ be the transforms of $k(x)$ and $\phi(x)$, respectively. Then

$$F(\zeta) = K(\zeta)T(\zeta)$$

$$T(\zeta) = \frac{F(\zeta)}{K(\zeta)}$$

and $\phi(x)$ is the inverse transform of $F(\zeta)/K(\zeta)$. The procedure must of necessity be formal, because at the outset we do not even know if $\phi(x)$ has a Fourier transform. Therefore, any solution which the method produces must be checked in the integral equation as a last step.

As an example of the method, consider the integral equation

$$\sqrt{2\pi}e^{-x^2/2} = \int_{-\infty}^{\infty} e^{-|x-\xi|}\phi(\xi)\, d\xi$$

The transformed equation is

$$e^{-\zeta^2/2} = \frac{2}{\sqrt{2\pi}}\frac{T(\zeta)}{1 + \zeta^2}$$

It follows that

$$T(\zeta) = \frac{\sqrt{2\pi}(1 + \zeta^2)e^{-\zeta^2/2}}{2} = \frac{\sqrt{2\pi}}{2}e^{-\zeta^2/2} - \frac{\sqrt{2\pi}}{2}(i\zeta)^2e^{-\zeta^2/2}$$

$$\phi(x) = \frac{\sqrt{2\pi}}{2}e^{-x^2/2} - \frac{\sqrt{2\pi}}{2}\frac{d^2}{dx^2}e^{-x^2/2}$$

$$= \sqrt{2\pi}\left(e^{-x^2/2} - \frac{x^2}{2}e^{-x^2/2}\right)$$

Unfortunately, in many applications where the Wiener-Hopf equation arises, the function $f(x)$ is not completely known. See, for example, the half-plane diffraction problem of Sec. 5.6. Suppose that

$$h(x) = f(x) = \int_{-\infty}^{\infty} k(x - \xi)\phi(\xi)\, d\xi \qquad 0 < x$$

$$h(x) = g(x) = \int_{-\infty}^{\infty} k(x - \xi)\phi(\xi)\, d\xi \qquad x < 0$$

where $f(x)$ is known, but $g(x)$ is not. In the half-plane diffraction problem, $\phi(x) = 0$ for $x < 0$. Hence,

$$T(\zeta) = \frac{1}{\sqrt{2\pi}} \int_{0}^{\infty} \phi(x) e^{-i\zeta x}\, dx$$

is analytic in a lower half plane. We shall outline the **Wiener-Hopf technique** for solving integral equations of the Wiener-Hopf type under conditions of sufficient generality to handle the half-plane problem. For a more general discussion of Wiener-Hopf techniques see Benjamin Noble, "The Wiener-Hopf Technique," Pergamon Press, Inc., New York, 1958.

Transforming the integral equation, we have

$$H(\zeta) = F_-(\zeta) + G_+(\zeta) = K(\zeta)T(\zeta)$$

where
$$F_-(\zeta) = \frac{1}{2\pi} \int_{0}^{\infty} f(x) e^{-i\zeta x}\, dx$$

and is analytic in a lower half plane $\text{Im}\,(\zeta) < \alpha$.

$$G_+(\zeta) = \frac{1}{2\pi} \int_{-\infty}^{0} g(x) e^{-i\zeta x}\, dx$$

and is analytic in an upper half plane $\text{Im}\,(\zeta) > \beta$. The transformed equation will in general hold in a strip $\beta < \text{Im}\,(\zeta) < \alpha$. Suppose $K(\zeta)$ can be decomposed as follows:

$$K(\zeta) = \frac{K_-(\zeta)}{K_+(\zeta)}$$

where $K_-(\zeta)$ is analytic for $\text{Im}\,(\zeta) < \alpha$, and $K_+(\zeta)$ is analytic for $\beta < \text{Im}\,(\zeta)$. Then

$$K_-(\zeta)T(\zeta) = K_+(\zeta)F_-(\zeta) + K_+(\zeta)G_+(\zeta)$$

Furthermore, let us assume that the known function $K_+(\zeta)F_-(\zeta)$ can be decomposed as follows:

$$K_+(\zeta)F_-(\zeta) = P_+(\zeta) + Q_-(\zeta)$$

where $P_+(\zeta)$ is analytic for $\beta < \text{Im}\,(\zeta)$, and $Q_-(\zeta)$ is analytic for $\text{Im}\,(\zeta) < \alpha$.

Then we can write

$$E(\zeta) = K_+(\zeta)G_+(\zeta) + P_+(\zeta) = K_-(\zeta)T(\zeta) - Q_-(\zeta)$$
$$= E_+(\zeta) = E_-(\zeta)$$

and the equation holds in the strip $\beta < \text{Im}(\zeta) < \alpha$. Now $E(\zeta)$ is defined by $E_+(\zeta)$ in an upper half plane and by $E_-(\zeta)$ in a lower half plane, and the two are equal in a common strip. Therefore, they provide the analytic continuation of one another to the whole plane. Hence, $E(\zeta)$ is an entire function, that is, is analytic in the whole plane. If it can be further shown that $|E(\zeta)| \to 0$ as $|z| \to \infty$, then by Liouville's theorem $E(\zeta) \equiv 0$, and we have

$$T(\zeta) = \frac{Q_-(\zeta)}{K_-(\zeta)}$$

and we have obtained the transform of our unknown function without having to deal with the unknown $G_+(\zeta)$. The final steps are to invert the transform by complex contour integration and check the formal solution to see if it meets the conditions of the original problem, which may be a boundary-value problem from which the integral equation was derived.

To illustrate this method we shall return to the half-plane diffraction problem of Sec. 5.6. Recall that we obtained the following integral equation:

$$-2e^{ikx\cos\alpha} = i\int_{-\infty}^{\infty} H_0^{(1)}(k\,|x - \xi|)I(\xi)\,d\xi \qquad x > 0$$

$$\psi(x,0-) - \psi(x,0+) - 2e^{ikx\cos\alpha} = i\int_{-\infty}^{\infty} H_0^{(1)}(k\,|x - \xi|)I(\xi)\,d\xi \qquad x < 0$$

We can identify the present case with the general procedure outlined above as follows:

$$T(\zeta) = \frac{1}{\sqrt{2\pi}}\int_0^{\infty} I(x)e^{-i\zeta x}\,dx$$

$$F_-(\zeta) = -\frac{1}{\pi}\int_0^{\infty} e^{ikx\cos\alpha}e^{-i\zeta x}\,dx$$

$$= \frac{i}{\pi}\frac{1}{\zeta - k\cos\alpha}$$

$$G_+(\zeta) = \frac{i}{2\pi}\int_{-\infty}^0 e^{-i\zeta x}\int_{-\infty}^{\infty} H_0^{(1)}(k\,|x - \xi|)I(\xi)\,d\xi\,dx$$

$$K(\zeta) = \frac{i}{\sqrt{2\pi}}\int_{-\infty}^{\infty} H_0^{(1)}(k\,|x|)e^{-i\zeta x}\,dx$$

The transform of $iH_0^{(1)}(k\,|x|)$ exists, since for large $|x|$ it behaves like $e^{ik|x|}/|x|^{\frac{1}{2}}$ and for small $|x|$ it behaves like $\log|x|$, and the integral $\int_0^{\rho} \log r\,dr$ exists. To evaluate it we note that $H^{(1)}(kx)$ satisfies the differential equation

$$x \frac{d^2}{dx^2} H_0^{(1)}(kx) + \frac{d}{dx} H_0^{(1)}(kx) + k^2 x H_0^{(1)}(kx) = 0$$

for $x > 0$. If we make the change of variable $t = -x$, we have

$$(-t) \frac{d^2}{dt^2} H_0^{(1)}[k(-t)] - \frac{d}{dt} H_0^{(1)}[k(-t)] - k^2 t H_0^{(1)}[k(-t)] = 0$$

or

$$t \frac{d^2}{dt^2} H_0^{(1)}(k\,|t|) + \frac{d}{dt} H_0^{(1)}(k\,|t|) + k^2 t H_0^{(1)}(k\,|t|) = 0$$

for $t < 0$. Hence, $H_0^{(1)}(k\,|x|)$ satisfies

$$x \frac{d^2}{dx^2} H_0^{(1)}(k\,|x|) + \frac{d}{dx} H_0^{(1)}(k\,|x|) + k^2 x H_0^{(1)}(k\,|x|) = 0$$

for $-\infty < x < \infty$ except at $x = 0$. Transforming the equation, we have

$$i \frac{d}{d\zeta} [(i\zeta)^2 K(\zeta)] + i\zeta K(\zeta) + ik^2 \frac{d}{d\zeta} K(\zeta) = 0$$

or

$$K' + \frac{\zeta K}{\zeta^2 - k^2} = 0$$

$$K = \frac{C}{\sqrt{k^2 - \zeta^2}}$$

where C is a constant of integration. It can be shown that $C = \sqrt{2/\pi}$. Hence

$$K(\zeta) = \sqrt{\frac{2}{\pi}} \frac{1}{\sqrt{k^2 - \zeta^2}}$$

In the present argument we need a common strip of analyticity for all the transforms involved. To this end we must take k complex with a small imaginary part, although k was considered real in our earlier discussions of the wave equation. Actually this is closer to physical reality than the case with real k. For r large, the scattered wave behaves as

$$\frac{e^{ikr}}{\sqrt{r}} = \frac{e^{i(k_1 + ik_2)r}}{\sqrt{r}} = \frac{e^{ik_1 r}}{\sqrt{r}} e^{-k_2 r}$$

The term $e^{-k_2 r}$ then represents a dissipation, and a more careful study of electromagnetic theory would reveal that this is the correct effect of a medium with small dissipation. For those less physically inclined, the addition of a small imaginary part can be regarded as a formal procedure introduced to obtain a solution which can be checked against the stated conditions of the problem.

Letting $k = k_1 + ik_2$, we see that $K(\zeta)$ has branch points at $\zeta = \pm(k_1 + ik_2)$. We select a branch of the square root which has branch cuts lying outside the strip $-k_2 < \operatorname{Im}(\zeta) < k_2$. We assume that $I(x) = O(e^{\beta x})$ as $x \to \infty$ and, therefore, that $T(\zeta)$ is analytic for $\operatorname{Im}(\zeta) < -\beta$. Since

$$H_0^{(1)}(k\,|x - \xi|) = O\left(\frac{e^{ik|x-\xi|}}{\sqrt{k\,|x - \xi|}}\right)$$

as $|x - \xi| \to \infty$, for x large and negative,

$$i\int_0^\infty H_0^{(1)}(k\,|x - \xi|)I(\xi)\,d\xi = O(e^{-ikx}) = O(e^{k_2 x})$$

provided $\displaystyle\int_0^\infty \frac{e^{ik\xi}I(\xi)}{\sqrt{\xi}}\,d\xi$ exists. This integral will exist provided $-k_2 < -\beta$. Consequently $G_+(\zeta)$ will be analytic in the half plane $-k_2 < \operatorname{Im}(\zeta)$, and the equation

$$F_-(\zeta) + G_+(\zeta) = K(\zeta)T(\zeta)$$

will hold in the strip $-k_2 < \operatorname{Im}(\zeta) < \min[-\beta, k_2\cos\alpha]$. Specifically,

$$\frac{i}{\pi(\zeta - k\cos\alpha)} + G_+(\zeta) = \sqrt{\frac{2}{\pi}}\,\frac{T(\zeta)}{\sqrt{k^2 - \zeta^2}}$$

We write $\qquad K(\zeta) = \sqrt{\dfrac{2}{\pi}}\,\dfrac{1}{\sqrt{k^2 - \zeta^2}} = \sqrt{\dfrac{2}{\pi}}\,\dfrac{1}{\sqrt{k - \zeta}\,\sqrt{k + \zeta}}$

and $\qquad \sqrt{\dfrac{2}{\pi}}\,\dfrac{T(\zeta)}{\sqrt{k - \zeta}} = \dfrac{i\sqrt{k + \zeta}}{\pi(\zeta - k\cos\alpha)} + \sqrt{k + \zeta}\,G_+(\zeta)$

Next we write

$$\frac{i\sqrt{k + \zeta}}{\pi(\zeta - \cos\alpha)} = \frac{i\sqrt{k + k\cos\alpha}}{\pi(\zeta - k\cos\alpha)} + \frac{i\sqrt{k + \zeta} - i\sqrt{k + k\cos\alpha}}{\pi(\zeta - k\cos\alpha)}$$

$$= Q_-(\zeta) + P_+(\zeta)$$

We define $P_+(\zeta)$ at $\zeta = k\cos\alpha$ by

$$\lim_{\zeta \to k\cos\alpha} P_+(\zeta) = \frac{i}{2\pi}\,\frac{1}{\sqrt{k + k\cos\alpha}}$$

Then

$$P'_+(k\cos\alpha)$$
$$= \lim_{\Delta\zeta \to 0} \frac{P_+(k\cos\alpha + \Delta\zeta) - P_+(k\cos\alpha)}{\Delta\zeta}$$
$$= \frac{i}{\pi}\lim_{\Delta\zeta \to 0} \frac{(\sqrt{k + k\cos\alpha + \Delta\zeta} - \sqrt{k + k\cos\alpha})/\Delta\zeta - 1/(2\sqrt{k + k\cos\alpha})}{\Delta\zeta}$$
$$= \frac{i}{8\pi}\,\frac{-1}{(k + k\cos\alpha)^{\frac{3}{2}}}$$

Therefore, $P_+(\zeta)$ is analytic at $\zeta = k \cos \alpha$. It has a branch point at $\zeta = -k$ but is nevertheless analytic in an upper half plane $-k_2 < \text{Im}(\zeta)$. Finally, we have

$$\sqrt{\frac{2}{\pi}} \frac{T(\zeta)}{\sqrt{k - \zeta}} - \frac{i}{\pi} \frac{\sqrt{k + k \cos \alpha}}{\zeta - k \cos \alpha} = P_+(\zeta) + \sqrt{k + \zeta} G_+(\zeta)$$

The left-hand side is analytic in a lower half plane $\text{Im}(\zeta) < \min[-\beta, k \cos \alpha]$, while the right-hand side is analytic in an upper half plane $-k_2 < \text{Im}(\zeta)$. The equality holds in the strip $-k_2 < \text{Im}(\zeta) < \min[-\beta, k_2 \cos \alpha]$. Therefore, one side is the analytic continuation of the other to the whole plane.

To complete the argument we have to show that the complete analytic function defined by the analytic continuation approaches zero as $|\zeta| \to \infty$. We have the following order conditions: $P_+(\zeta) = O(\zeta^{-\frac{1}{2}})$, $T(\zeta)/\sqrt{k - \zeta} = O(\zeta^{-\frac{1}{2}})$, $1/(\zeta - k \cos \alpha) = O(\zeta^{-1})$, $\sqrt{k + \zeta} G_+(\zeta) = O(\zeta^{\frac{1}{2}})$ in their respective half planes. Therefore, by Liouville's theorem, the complete analytic function is constant; but this constant must be zero, since the function goes to zero in the lower half plane. Finally,

$$T(\zeta) = \frac{i}{\sqrt{2\pi}} \frac{\sqrt{k - \zeta} \sqrt{k + k \cos \alpha}}{\zeta - k \cos \alpha}$$

The inversion of this transform is a fairly involved problem in complex contour integration, which we shall not go into here.[1]

The Wiener-Hopf technique can also be used to solve certain integral equations of the second kind which come up in some applications. Suppose

$$\phi(x) = f(x) + \frac{1}{\sqrt{2\pi}} \int_0^\infty k(x - \xi)\phi(\xi) \, d\xi \qquad 0 < x$$

$$0 = g(x) + \frac{1}{\sqrt{2\pi}} \int_0^\infty k(x - \xi)\phi(\xi) \, d\xi \qquad x < 0$$

Then taking transforms

$$T(\zeta) = F_-(\zeta) + G_+(\zeta) + K(\zeta)T(\zeta)$$

where

$$T(\zeta) = \frac{1}{\sqrt{2\pi}} \int_0^\infty \phi(x)e^{-i\zeta x} \, dx$$

$$F_-(\zeta) = \frac{1}{\sqrt{2\pi}} \int_0^\infty f(x)e^{-i\zeta x} \, dx$$

$$G_+(\zeta) = \frac{1}{\sqrt{2\pi}} \int_{-\infty}^0 g(x)e^{-i\zeta x} \, dx$$

$$K(\zeta) = \frac{1}{\sqrt{2\pi}} \int_{-\infty}^\infty k(x)e^{-i\zeta x} \, dx$$

[1] See John W. Dettman, "Applied Complex Variables," The Macmillan Company, New York, 1965, pp. 388–398.

This time we write

$$[1 - K(\zeta)]T(\zeta) = F_-(\zeta) + G_+(\zeta)$$

$$1 - K(\zeta) = \frac{K_-(\zeta)}{K_+(\zeta)}$$

and proceed as before.   The actual factorization may be fairly involved in practice, but it is not our purpose here to give all the details that one might get involved with in the problem.   If the reader is interested in pursuing the subject further, he should consult additional books on the subject.   A very extensive bibliography can be found in Benjamin Noble's "The Wiener-Hopf Technique," previously mentioned.

8.5 Laplace Transforms.   Applications

We have already seen in some of the applications where the function $f(x)$, whose Fourier transform we wished to compute, was zero for $x < 0$.   In this case the Fourier transform is

$$T[f] = \frac{1}{\sqrt{2\pi}} \int_0^\infty f(x)e^{-ix\zeta}\, dx = \phi(\zeta)$$

and

$$T^{-1}[\phi] = \frac{1}{\sqrt{2\pi}} \int_{-\infty+i\gamma}^{\infty+i\gamma} \phi(\zeta)e^{ix\zeta}\, d\zeta$$

The theory holds under suitable restrictions on $f(x)$.   In particular, $f(x)$ must be piecewise continuous, implying that

$$\lim_{x\to 0+} f(x) = f(0+)$$

exists.   If $f(x) = O(e^{\alpha x})$ as $x \to \infty$, that is, $f(x)$ is of **exponential order,** then the transform $T[f]$ exists and is an analytic function of ζ for Im $(\zeta) < -\alpha$.

It is sometimes convenient to carry out a rotation of the ζ plane through an angle of $\pi/2$; that is, we introduce the new variable

$$z = \sigma + i\tau = i\zeta = -s + it$$

We define the **Laplace transform** of $f(x)$ as follows:

$$\phi(z) = L[f] = \int_0^\infty f(x)e^{-i\zeta x}\, dx$$

$$= \int_0^\infty f(x)e^{(s-it)x}\, dx$$

$$= \int_0^\infty f(x)e^{-(\sigma+it)x}\, dx$$

$$= \int_0^\infty f(x)e^{-zx}\, dx$$

If $f(x)$ is piecewise continuous and of exponential order, that is, $f(x) = O(e^{\alpha x})$ as $x \to \infty$, the Laplace transform will exist and be an analytic function of z in a right half plane Re $(z) > \alpha$.

It follows directly from the theory of Fourier transforms that if $f(x)$ and $f'(x)$ are piecewise continuous in any finite interval and $f(x) = O(e^{\alpha x})$, then

$$\tfrac{1}{2}[f(x+) + f(x-)] = \frac{1}{2\pi} \int_{-\infty}^{\infty} \phi(-s_0 + it) e^{i(t+is_0)x} \, dt$$

$$= \frac{1}{2\pi i} \int_{-i\infty+\sigma_0}^{i\infty+\sigma_0} \phi(\sigma_0 + i\tau) e^{(\sigma_0+i\tau)x} \, d(\sigma_0 + i\tau)$$

$$= \frac{1}{2\pi i} \int_{\sigma_0-i\infty}^{\sigma_0+i\infty} \phi(z) e^{zx} \, dz$$

where the integration in the z plane is along the line $\sigma_0 + i\tau$, $-\infty < t < \infty$, $\sigma_0 > \alpha$, and the value of the integral is the Cauchy principal value. Of course, where $f(x)$ is continuous, $\tfrac{1}{2}[f(x+) + f(x-)] = f(x)$. For $x = 0$ the integral converges to $\tfrac{1}{2}f(0+)$. Furthermore, if $f(x)$ is continuous, $f'(x)$ is piecewise continuous, and both are of exponential order $O(e^{\alpha x})$, then

$$f(x) = \frac{1}{2\pi i} \int_{\gamma-i\infty}^{\gamma+i\infty} \phi(z) e^{zx} \, dz$$

and the integration can be taken along *any* line $z = \gamma + i\tau$, $-\infty < t < \infty$, $\gamma > \alpha$.†

Now let us list some of the more important properties of the Laplace transform, which either can be verified directly or will follow from corresponding properties of the Fourier transform. All functions involved are assumed to be zero for $t < 0$, defined by their right-hand limit at $t = 0$, and of exponential order as $x \to \infty$.

1. The Laplace transform is a linear transformation; that is,

$$L[af + bg] = aL[f] + bL[g]$$

2. If $f(x)$ is continuous for $0 \le t$, and $f'(x)$ is piecewise continuous in any finite interval,

$$L[f'] = zL[f] - f(0)$$

3. If $f(x)$ is continuous along with its first $n - 1$ derivatives and $f^{(n)}(x)$ is piecewise continuous in any finite interval,

$$L[f^{(n)}] = z^n L[f] - z^{n-1}f(0) - z^{n-2}f'(0)$$

$$- \cdots - zf^{(n-2)}(0) - f^{(n-1)}(0)$$

† See exercise 5, Sec. 8.1.

4. If $f(x)$ is piecewise continuous, then $\int_0^x f(\xi)\, d\xi = F(x)$ is continuous, $F(0) = 0$, and

$$L[F] = \frac{1}{z} L[f]$$

5. If $\phi(z) = L[f(x)]$, then

$$\phi(z - a) = L[e^{ax}f(x)]$$

6. Let $u(x)$ be the unit step function; that is,

$$u(x) = 0 \qquad x < 0$$
$$u(x) = 1 \qquad 0 \leq x$$

Then $$L[f(x - a)u(x - a)] = e^{-az}L[f(x)]$$

7. If $\phi(z) = L[f]$, then

$$\frac{d}{dz}\phi(z) = -L[xf(x)]$$

8. Let $h(x)$ be the convolution integral; that is,

$$h(x) = \int_0^x f(\xi)g(x - \xi)\, d\xi$$

Then $$L[h] = L[f]L[g]$$

The following are the Laplace transforms of some elementary functions:

a. $L[u(x)] = \displaystyle\int_0^\infty e^{-zx}\, dx = \frac{1}{z}$

b. $L[x^n] = n!\, z^{-(n+1)}$

c. $L[x^a] = \Gamma(a + 1)z^{-(a+1)} \qquad a \geq 0$

d. $L[e^{ax}] = \dfrac{1}{z - a}$

e. $L[\sin ax] = \dfrac{a}{z^2 + a^2}$

f. $L[\cos ax] = \dfrac{z}{z^2 + a^2}$

g. $L[e^{-bx}\sin ax] = \dfrac{a}{(z + b)^2 + a^2}$

h. $L[e^{-bx}\cos ax] = \dfrac{z + b}{(z + b)^2 + a^2}$

i. $L[\sinh ax] = \dfrac{a}{z^2 - a^2}$

j. $L[\cosh ax] = \dfrac{z}{z^2 - a^2}$

Formula b can be derived from a by use of property 3 plus the fact that $u(x) = \dfrac{1}{n!} \dfrac{d^n}{dx^n} f(x)$ where

$$f(x) = x^n \qquad x \geq 0$$

$$f(x) = 0 \qquad x < 0$$

Then

$$\frac{n!}{z} = L[f^{(n)}(z)] = z^n L[f]$$

$$L[f] = \frac{n!}{z^{n+1}}$$

It is also a special case of formula c, which can be established directly from the definition of the gamma function

$$\Gamma(z) = \int_0^\infty t^{z-1} e^{-t}\, dt \qquad \text{Re}\,(z) > 0$$

Formula e can be established as follows: Let $y = \sin ax$. Then $y'' + a^2 y = 0$, and $y(0) = 0$, $y'(0) = a$. Therefore,

$$z^2 L[y] - a + a^2 L[y] = 0 \qquad \text{and} \qquad L[y] = a/(z^2 + a^2)$$

Formula g is a direct application of property 5 and formula e.

As a further example, let us find the Laplace transform of the zeroth-order Bessel function $J_0(kx)$, starting from the differential equation

$$xJ_0''(kx) + J_0'(kx) + k^2 x J_0(kx) = 0$$

and the initial conditions $J_0(0) = 1$, $J_0'(0) = 0$. Transforming the equation, we have

$$-\frac{d}{dz}(z^2 \phi - z) + z\phi - 1 - k^2 \frac{d}{dz}\phi = 0$$

where $\phi(z) = L[J_0(kx)]$. Then

$$\phi' + \frac{z}{k^2 + z^2}\,\phi = 0$$

$$\phi(z) = \frac{C}{\sqrt{k^2 + z^2}}$$

where C is a constant of integration. To evaluate C we recall that

$$\frac{d}{dx} J_0(kx) = -kJ_1(kx)$$

and that $J_1(kx)$ is bounded for all x.   It therefore follows from property 2 that

$$L[-kJ_1(kx)] = z\phi(z) - J_0(0)$$

$$= \frac{Cz}{\sqrt{k^2 + z^2}} - 1$$

Now
$$|L[-kJ_1(kx)]| \le \int_0^\infty |kJ_1(kx)| \, e^{-\sigma x} \, dx$$

$$\le \frac{M}{\sigma}$$

Then
$$\lim_{\sigma \to \infty} L[-kJ_1(kx)]_{z=\sigma} = \lim_{\sigma \to \infty} \frac{C\sigma}{\sqrt{k^2 + \sigma^2}} - 1$$

$$= C - 1 = 0$$

Therefore, $C = 1$ and

$$L[J_0(kx)] = \frac{1}{\sqrt{k^2 + z^2}}$$

We have seen how we can use the differential equation plus the initial conditions satisfied by a given function to find its Laplace transform.   Now let us take the opposite point of view, that is, to find the solution of a given differential equation by finding its Laplace transform.   Consider the simple series electrical circuit shown in Fig. 22.

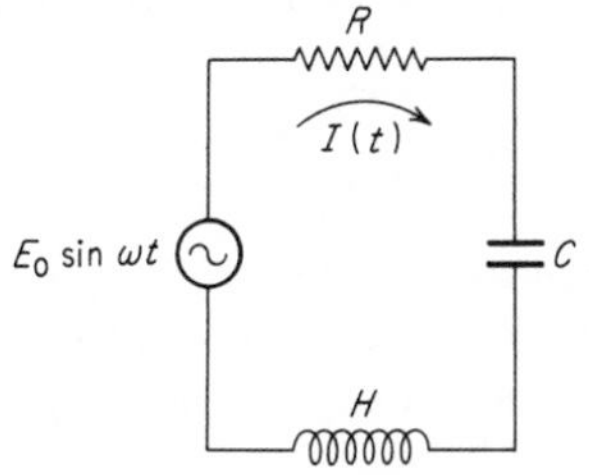

The current $I(t)$ is the solution of the following integrodifferential equation:[1]

$$H\frac{dI}{dt} + RI + \frac{1}{C}\left[q_0 + \int_0^t I(\tau)\,d\tau\right] = E_0 \sin \omega t$$

FIGURE 22

Let $I(0) = I_0$, and of course $q(0) = q_0$.   Let $\phi(z) = L[I]$.   Then

$$H(z\phi - I_0) + R\phi + \frac{1}{Cz}(q_0 + \phi) = \frac{E_0\omega}{z^2 + \omega^2}$$

Solving for ϕ, we have

$$\phi(z) = \frac{E_0\omega z}{(z^2 + \omega^2)(Hz^2 + Rz + 1/C)} + \frac{HI_0 z}{Hz^2 + Rz + 1/C} - \frac{q_0}{C(Hz^2 + Rz + 1/C)}$$

We can invert the transform by identifying the various parts with transforms

---

[1] We have used H for inductance to avoid the letter L designating the Laplace transform.

of known elementary functions. For example,

$$\frac{HI_0z}{Hz^2 + Rz + 1/C} = \frac{I_0z}{(z + R/2H)^2 + (1/HC - R^2/4H^2)}$$

$$= \frac{I_0(z + R/2H)}{(z + R/2H)^2 + \omega_0^2} - \frac{I_0R}{2H[(z + R/2H)^2 + \omega_0^2]}$$

where $\omega_0^2 = 1/HC - R^2/4H^2$. Now if $\omega_0^2 = 0$, we have the case of **critical damping** where

$$\frac{HI_0z}{Hz^2 + Rz + 1/C} = L\left[I_0e^{-(R/2H)t} - \frac{I_0R}{2H}te^{-(R/2H)t}\right]$$

If $\omega_0^2 < 0$, we have **overdamping** and

$$\frac{HI_0z}{Hz^2 + Rz + 1/C} = L\left[I_0e^{-(R/2H)t}\cosh\sqrt{\frac{R^2}{4H^2} - \frac{1}{HC}}\,t\right.$$

$$\left. - \frac{I_0R}{2H\sqrt{R^2/4H^2 - 1/HC}}e^{-(R/2L)t}\sinh\sqrt{\frac{R^2}{4H^2} - \frac{1}{HC}}\,t\right]$$

If $\omega_0^2 > 0$, we have **underdamping** and

$$\frac{HI_0z}{Hz^2 + Rz + 1/C} = L\left[I_0e^{-(R/2H)t}\cos\omega_0t - \frac{I_0R}{2H\omega_0}e^{-(R/2H)t}\sin\omega_0t\right]$$

We see then that the nature of the solution depends on the relative values of R, H, and C. This will also be true of the other two terms in the transform. For simplicity, therefore, we shall proceed with the underdamped case only. In this case,

$$\frac{q_0}{C(Hz^2 + Rz + 1/C)} = L\left[\frac{q_0}{HC\omega_0}e^{-(R/2H)t}\sin\omega_0t\right]$$

To identify the third term we must first perform a partial-fraction expansion.[1]

Let $$\frac{E_0\omega z}{(z^2 + \omega^2)(Hz^2 + Rz + 1/C)} = \frac{az + b}{z^2 + \omega^2} + \frac{cz + d}{Hz^2 + Rz + 1/C}$$

Then $$E_0\omega z = (az + b)\left(Hz^2 + Rz + \frac{1}{C}\right) + (cz + d)(z^2 + \omega^2)$$

---

[1] The method of partial-fraction expansion is important in the art of inverting Laplace transforms. The procedure can be formalized through a set of theorems which can be found in Ruel V. Churchill, "Operational Mathematics," 2d ed., McGraw-Hill Book Company, New York, 1958, pp. 57–64.

implying
$$\frac{b}{C} + d\omega^2 = 0$$

$$\frac{a}{C} + bR + c\omega^2 = E_0\omega$$

$$aR + bH + d = 0$$

$$aH + c = 0$$

Solving these equations, we obtain

$$a = \frac{-E_0 X}{R^2 + X^2}$$

$$b = \frac{\omega E_0 R}{R^2 + X^2}$$

$$c = \frac{E_0 X H}{R^2 + X^2}$$

$$d = \frac{-E_0 R}{\omega C(R^2 + X^2)}$$

where $X = \omega H - 1/\omega C$ is a quantity called the **reactance**. Let $Z = R + iX$ be the **complex impedance** and $\theta = \arg Z = \tan^{-1}(X/R)$ be the **phase.** Then

$$a = \frac{-E_0 \sin\theta}{|Z|}$$

$$b = \frac{\omega E_0 \cos\theta}{|Z|}$$

$$c = \frac{E_0 H \sin\theta}{|Z|}$$

$$d = \frac{-E_0 \cos\theta}{\omega C |Z|}$$

$$\frac{E_0\omega z}{(z^2 + \omega^2)(Hz^2 + Rz + 1/C)} = \frac{\omega E_0 \cos\theta}{|Z|(z^2 + \omega^2)} - \frac{z E_0 \sin\theta}{|Z|(z^2 + \omega^2)}$$

$$+ \frac{z E_0 H \sin\theta}{|Z|(Hz^2 + Rz + 1/C)} - \frac{E_0 \cos\theta}{\omega C |Z|(Hz^2 + Rz + 1/C)}$$

$$\frac{E_0\omega z}{(z^2 + \omega^2)(Hz^2 + Rz + 1/C)} = L\left[\frac{E_0}{|Z|}\sin(\omega t - \theta) + \frac{E_0}{|Z|}\sin\theta\, e^{-(R/2H)t}\cos\omega_0 t\right.$$

$$\left. - \left(\frac{E_0 R \sin\theta}{2H\omega_0 |Z|} + \frac{E_0 \cos\theta}{\omega\omega_0 CH |Z|}\right) e^{-(R/2H)t}\sin\omega_0 t\right]$$

Therefore the solution of the differential equation satisfying the initial conditions is

$$I(t) = \frac{E_0}{|Z|} \sin(\omega t - \theta) + \left(I_0 + \frac{E_0}{|Z|}\sin\theta\right)e^{-(R/2H)t}\cos\omega_0 t$$

$$+ \left(\frac{q_0}{HC\omega_0} - \frac{I_0 R}{2H\omega_0} - \frac{E_0 R \sin\theta}{2H\omega_0 |Z|} - \frac{E_0 \cos\theta}{\omega\omega_0 CH |Z|}\right)e^{-(R/2H)t}\sin\omega_0 t$$

This looks very complicated, but basically it is just the sum of the **steady-state solution** $(E_0/|Z|)\sin(\omega t - \theta)$ plus the **transient solution**

$$e^{-(R/2H)t}(A\cos\omega_0 t + B\sin\omega_0 t)$$

which dies out exponentially with time. The constants A and B are determined in order that the particular solution satisfy the initial conditions. One of the advantages of the Laplace transformation method is that it gives the values of A and B directly. In this particular example the work involved is such that there seems to be no advantage of the method over the conventional methods of elementary differential equations. However, the next example deals with a problem which cannot as readily be handled by the elementary methods.

Suppose the input to the circuit is a pulse of duration τ. This can be expressed as $E_0 u(t) - E_0 u(t - \tau)$ where $u(t)$ is the unit step function. The integrodifferential equation is

$$H\frac{dI}{dt} + RI + \frac{1}{C}\int_0^t I(\xi)\,d\xi = E_0 u(t) - E_0 u(t - \tau)$$

For simplicity we are assuming that $I_0 = q_0 = 0$. Transforming the equation, we have

$$zH\phi + R\phi + \frac{\phi}{Cz} = \frac{E_0}{z} - \frac{E_0 e^{-\tau z}}{z}$$

Solving for ϕ, we have

$$\phi(z) = \frac{E_0}{Hz^2 + Rz + 1/C} - \frac{E_0 e^{-\tau z}}{Hz^2 + Rz + 1/C}$$

Again we shall treat only the underdamped case. In this case,

$$\frac{E_0}{Hz^2 + Rz + 1/C} = \frac{E_0 \omega_0}{\omega_0 H[(z + R/2H)^2 + \omega_0^2]}$$

$$= L\left[\frac{E_0}{\omega_0 H}e^{-(R/2H)t}\sin\omega_0 t\right]$$

It therefore follows from property 6 that

$$\frac{E_0 e^{-\tau z}}{Hz^2 + Rz + 1/C} = L\left[\frac{E_0}{\omega_0 H}e^{-(R/2H)(t-\tau)}u(t - \tau)\sin\omega_0(t - \tau)\right]$$

and our solution is

$$I(t) = \frac{E_0}{\omega_0 H} \left[e^{-(R/2H)t} \sin \omega_0 t - e^{-(R/2H)(t-\tau)} u(t-\tau) \sin \omega_0 (t-\tau) \right]$$

Laplace transform methods can also be used to advantage in solving boundary-value problems in partial differential equations. We shall consider two cases, both based on the one-dimensional wave equation. Consider the semi-infinite string $x \geq 0$ with no initial displacement or initial velocity. Starting at $t = 0$, the end at $x = 0$ is displaced according to

$$\phi(0,t) = f(t)$$

We have then to solve the following problem:

$$\frac{\partial^2 \phi}{\partial x^2} = \frac{1}{a^2} \frac{\partial^2 \phi}{\partial t^2} \qquad x > 0 \qquad t > 0$$

$$\phi(x,0) = 0$$

$$\phi_t(x,0) = 0$$

$$\phi(0,t) = f(t)$$

Let $T(x,z) = \int_0^\infty \phi(x,t)e^{-tz} \, dt$. Then

$$\frac{d^2 T}{dx^2} - \frac{z^2}{a^2} T = 0$$

$$T(0,z) = \int_0^\infty f(t)e^{-tz} \, dt = F(z)$$

Hence $$T = Ae^{-(z/a)x} + Be^{(z/a)x}$$

Any displacement of the string must propagate with a finite velocity. Therefore, $\phi(x,t) \to 0$ as $x \to \infty$ for all t, and we assume consequently that

$$\lim_{x \to \infty} T(x,z) = 0$$

for Re $(z) > 0$. This requires that $B = 0$. Finally, $T(0,z) = A = F(z)$. Therefore,

$$T(x,z) = F(z)e^{-z(x/a)}$$

and $$\phi(x,t) = f\left(t - \frac{x}{a}\right) u\left(t - \frac{x}{a}\right)$$

It can easily be shown that this is a solution of the problem provided $f(t)$ has a continuous second derivative.

As a second case let $\phi(0,t) = 0$, $\phi(x,0) = g(x)$, $\phi_t(x,0) = h(x)$. The solution to a more general problem can be obtained by superimposing the solutions to this and the last problem. In the present case, the transformed equation becomes

$$\frac{d^2T}{dx^2} - \frac{z^2}{a^2} T = -\frac{1}{a^2} [zg(x) + h(x)]$$

subject to $T(0,z) = 0$ and $T(x,z) \to 0$ as $x \to \infty$ for Re $(z) > 0$. To solve the transformed problem we must develop the Green's function for the equation; that is, we find $G(y;x)$ such that

$$\frac{d^2G}{dx^2} - \frac{z^2}{a^2} G = 0 \qquad 0 \le x < y \qquad y < x < \infty$$

$$G(y;0) = \lim_{x \to \infty} G = 0$$

$$G(y;y-) = G(y;y+)$$

$$G'(y;y+) - G'(y;y-) = -1$$

We therefore have

$$G(y;x) = Ae^{-(z/a)x} + Be^{(z/a)x} \qquad 0 \le x < y$$

$$G(y;x) = Ce^{-(z/a)x} + De^{(z/a)x} \qquad y < x < \infty$$

$G(y;x) \to 0$ as $x \to \infty$ implies that $D = 0$. Also

$$G(y;0) = A + B = 0$$

$$G(y;y-) = Ae^{-(z/a)y} + Be^{(z/a)y} = G(y;y+) = Ce^{-(z/a)y}$$

$$G'(y;y+) - G'(y;y-) = -\frac{z}{a} Ce^{-(z/a)y} + \frac{z}{a} Ae^{-(z/a)y} - \frac{z}{a} Be^{(z/a)y} = -1$$

Solving for A, B, and C, we have

$$A = -\frac{a}{2z} e^{-(z/a)y}$$

$$B = \frac{a}{2z} e^{-(z/a)y}$$

$$C = -\frac{a}{2z} \left(e^{-(z/a)y} - e^{(z/a)y} \right)$$

so that $\quad G(y;x) = -\frac{a}{2z} \left(e^{-(z/a)(x+y)} - e^{(z/a)(x-y)} \right) \qquad 0 \le x < y$

$$G(y;x) = -\frac{a}{2z} \left(e^{-(z/a)(x+y)} - e^{(z/a)(y-x)} \right) \qquad y < x < \infty$$

In terms of the Green's function,

$$T(x,z) = \frac{1}{2a} \int_0^x g(y)(e^{-(z/a)(x-y)} - e^{-(z/a)(x+y)}) \, dy$$

$$+ \frac{1}{2a} \int_x^\infty g(y)(e^{-(z/a)(y-x)} - e^{-(z/a)(x+y)}) \, dy$$

$$+ \frac{1}{2az} \int_0^x h(y)(e^{-(z/a)(x-y)} - e^{-(z/a)(x+y)}) \, dy$$

$$+ \frac{1}{2az} \int_x^\infty h(y)(e^{-(z/a)(y-x)} - e^{-(z/a)(x+y)}) \, dy$$

$$= \frac{1}{2} \int_0^\infty g(x + a\xi)e^{-z\xi} \, d\xi + \frac{1}{2} \int_0^{x/a} g(x - a\xi)e^{-z\xi} \, d\xi$$

$$- \frac{1}{2} \int_{x/a}^\infty g(a\xi - x)e^{-z\xi} \, d\xi + \frac{1}{2z} \int_0^\infty h(x + a\xi)e^{-z\xi} \, d\xi$$

$$+ \frac{1}{2z} \int_0^{x/a} h(x - a\xi)e^{-z\xi} \, d\xi - \frac{1}{2z} \int_{x/a}^\infty h(a\xi - x)e^{-z\xi} \, d\xi$$

To identify the transform we define the functions

$$G(x) = g(x) \qquad x \geq 0$$
$$G(x) = -g(-x) \qquad x < 0$$
$$H(x) = h(x) \qquad x \geq 0$$
$$H(x) = -h(-x) \qquad x < 0$$

Then $G(-x) = -G(x)$ and $H(-x) = -H(x)$ for all x, and

$$T(x,z) = \frac{1}{2} \int_0^\infty G(x + a\xi)e^{-z\xi} \, d\xi + \frac{1}{2} \int_0^\infty G(x - a\xi)e^{-z\xi} \, d\xi$$

$$+ \frac{1}{2z} \int_0^\infty H(x + a\xi)e^{-z\xi} + \frac{1}{2z} \int_0^\infty H(x - a\xi)e^{-z\xi} \, d\xi$$

In this form we recognize the transform of

$$\phi(x,t) = \tfrac{1}{2}[G(x + at) + G(x - at)] + \frac{1}{2} \int_{-t}^t H(x + a\tau) \, d\tau$$

which can be verified as the solution of the problem.

Property 8, which involves taking the Laplace transform of the convolution integral, gives us a means for solving certain integral equations. Consider, for example, the integral equation of the first kind,

$$f(x) = \int_0^x k(x - \xi)\phi(\xi)\, d\xi$$

where $f(x)$ and $k(x)$ are known. $\phi(x)$ is the unknown to be found. The right-hand side is in the form of a convolution integral. Hence, we transform the equation and obtain

$$F(z) = K(z)T(z)$$

where $F(z)$, $K(z)$, and $T(z)$ are the transforms of $f(x)$, $k(x)$, and $\phi(x)$, respectively. Solving for $T(z)$, we have

$$T(z) = \frac{F(z)}{K(z)}$$

and we obtain $\phi(x)$ by inverting the transform. The procedure must of necessity be formal, because at the outset we do not even know if $\phi(x)$ has a transform. The solution thus obtained can be checked in the integral equation.

Abel's integral equation is an example of this type. The equation is

$$f(x) = \int_0^x \frac{\phi(\xi)}{\sqrt{x - \xi}}\, d\xi$$

Now $k(x) = x^{-\frac{1}{2}}$ is not piecewise continuous, but nevertheless it does have a Laplace transform

$$K(z) = \int_0^\infty x^{-\frac{1}{2}}\, e^{-zx}\, dx = \Gamma(\tfrac{1}{2})z^{-\frac{1}{2}} = \sqrt{\pi}z^{-\frac{1}{2}}$$

Hence,

$$T(z) = L[\phi] = \frac{F(z)}{K(z)} = z^{\frac{1}{2}} \frac{F(z)}{\sqrt{\pi}} = \frac{z}{\pi}[\sqrt{\pi}z^{-\frac{1}{2}}F(z)] = \frac{z}{\pi}K(z)F(z)$$

Therefore,

$$\phi(x) = \frac{1}{\pi}\frac{d}{dx}\int_0^x \frac{f(\xi)}{\sqrt{x - \xi}}\, d\xi$$

Certain Volterra integral equations can be solved by use of the Laplace transformation. Consider, for example, the equation

$$\phi(x) = f(x) + \int_0^x k(x - \xi)\phi(\xi)\, d\xi$$

where $f(x)$ and $k(x)$ are known and $\phi(x)$ is unknown. Let $F(z)$, $K(z)$, and $T(z)$ be the Laplace transforms of $f(x)$, $k(x)$, and $\phi(x)$, respectively. Then

$$T(z) = F(z) + K(z)T(z)$$

$$T(z) = \frac{F(z)}{1 - K(z)}$$

Inverting the transform, we obtain $\phi(x)$.   As an example of this, let us solve
exercise 6, Sec. 6.3, which has a well-known solution.   The equation is

$$\phi(x) = 1 - \int_0^x (x - \xi)\phi(\xi)\, d\xi$$

Then

$$T(z) = \frac{1}{z} - \frac{T(z)}{z^2}$$

$$T(z) = \frac{z}{z^2 + 1} = L[\cos x]$$

$$\phi(x) = \cos x$$

which can be verified as the solution of the equation.

Exercises 8.5

1. Prove that the Laplace transform of a piecewise continuous function of
exponential order approaches zero as Re (z) approaches infinity.
2. Let $f(x)$ be continuous for $0 < x < \infty$ and be of exponential order as $x \to \infty$
and of order x^p, $-1 < p$, as x approaches zero.   Prove that $f(x)$ has a Laplace
transform.
3. Show that $L[x^a] = \Gamma(a + 1)z^{-a+1}$, $-1 < a$.
4. Verify properties 4 to 7 for Laplace transforms.
5. Find the Laplace transforms of xe^{ax}, $x \sin ax$, $x^2 \cos ax$, and $e^{ax}J_0(kx)$.
6. Find the Laplace transform of $J_1(kx)$.
7. Find the inverse Laplace transforms of

$a.$ $\dfrac{1}{z(z^2 + a^2)}$

$b.$ $\dfrac{1}{z^2(z^2 - a^2)}$

$c.$ $\dfrac{1}{(z - a)^2(z - b)}$

$d.$ $\dfrac{e^{-ax}}{z^2}$

$e.$ $\dfrac{z}{(z^2 + a^2)^{\frac{3}{2}}}$

$f.$ $\dfrac{1}{az^2 + bz + c}$

8–11. Solve exercises 1 to 4 of Sec. 8.2, using Laplace transforms.
12. Using Laplace transform methods, find the displacement of a vibrating
string fixed at $x = 0$ and $x = L$ with initial displacement $\phi(x,0) = A \sin(\pi x/L)$
and initial velocity zero.
13. A uniform elastic bar of length L is fixed at $x = 0$ and at rest.   At time
$t = 0$ a force $F(t) = F_0t^2$ is applied longitudinally at the end $x = L$.   Find the

longitudinal displacement of the general cross section as a function of position and time as long as the material behaves elastically.

14. Solve the generalized Abel equation $f(x) = \int_0^x \phi(\xi)(x - \xi)^{-\alpha} d\xi, 0 < \alpha < 1.$

15. Solve the integral equation $\phi(x) = 1 + \int_0^x \phi(\xi) d\xi.$

8.6 Other Transform Techniques

Before we leave the general subject of integral transform methods, it is appropriate to mention some techniques based on other transforms. The reader should be aware of the fact that, although the Fourier and Laplace transforms are perhaps the most widely applied, there are other transforms which may be of great utility in specific applications. Some of these are the following: the **finite Fourier sine transform**, the **finite Fourier cosine transform**, the **Hankel transform**, and the **Mellin transform**. We shall discuss each of these briefly and give some examples of how they may be applied.

The **finite Fourier sine transform** of a function $f(x)$ is defined by

$$S_n[f] = \sqrt{\frac{2}{\pi}} \int_0^\pi f(x) \sin nx \, dx \qquad n = 1, 2, 3, \ldots$$

A sufficient condition for its existence is that $f(x)$ be piecewise continuous on the interval $0 \le x \le \pi$. We know from our previous work on Fourier series in Sec. 2.2 that

$$\tfrac{1}{2}[f(x+) + f(x-)] = \sqrt{\frac{2}{\pi}} \sum_{n=1}^\infty S_n \sin nx \qquad 0 \le x \le \pi$$

provided $f(x)$ and $f'(x)$ are both piecewise continuous. At a point where $f(x)$ is continuous the series converges to the function.

Let $f(x)$ and $f'(x)$ be continuous on the interval $0 \le x \le \pi$, while $f''(x)$ is piecewise continuous. Then

$$S_n[f''] = \sqrt{\frac{2}{\pi}} \int_0^\pi f''(x) \sin nx \, dx$$

$$= \sqrt{\frac{2}{\pi}} [f'(x) \sin nx]_0^\pi - n \sqrt{\frac{2}{\pi}} \int_0^\pi f'(x) \cos nx \, dx$$

$$= -n \sqrt{\frac{2}{\pi}} [f(x) \cos nx]_0^\pi - n^2 \sqrt{\frac{2}{\pi}} \int_0^\pi f(x) \sin nx \, dx$$

$$= n \sqrt{\frac{2}{\pi}} f(0) + (-1)^{n+1} n \sqrt{\frac{2}{\pi}} f(\pi) - n^2 S_n[f]$$

The **finite Fourier cosine transform** of a function $f(x)$ is defined by

$$C_n[f] = \sqrt{\frac{2}{\pi}} \int_0^\pi f(x) \cos nx \, dx \qquad n = 0, 1, 2, 3, \ldots$$

A sufficient condition for its existence is that $f(x)$ be piecewise continuous on the interval $0 \le x \le \pi$. If $f(x)$ and $f'(x)$ are both piecewise continuous, then

$$\tfrac{1}{2}[f(x+) + f(x-)] = \frac{1}{\sqrt{2\pi}} C_0 + \sqrt{\frac{2}{\pi}} \sum_{n=1}^\infty C_n \cos nx \qquad 0 \le x \le \pi$$

Where $f(x)$ is continuous, the series converges to the function.

If $f(x)$ and $f'(x)$ are continuous while $f''(x)$ is piecewise continuous, then

$$\begin{aligned}
C_n[f''] &= \sqrt{\frac{2}{\pi}} \int_0^\pi f''(x) \cos nx \, dx \\
&= \sqrt{\frac{2}{\pi}} [f'(x) \cos nx]_0^\pi + n \sqrt{\frac{2}{\pi}} \int_0^\pi f'(x) \sin nx \, dx \\
&= \sqrt{\frac{2}{\pi}} (-1)^n f'(\pi) - \sqrt{\frac{2}{\pi}} f'(0) - n^2 \sqrt{\frac{2}{\pi}} \int_0^\pi f(x) \cos nx \, dx \\
&= \sqrt{\frac{2}{\pi}} (-1)^n f'(\pi) - \sqrt{\frac{2}{\pi}} f'(0) - n^2 C_n[f]
\end{aligned}$$

Let $f(x)$ be continuous and $f'(x)$ be piecewise continuous on the interval $0 \le x \le \pi$. Then

$$\begin{aligned}
S_n[f'] &= \sqrt{\frac{2}{\pi}} \int_0^\pi f'(x) \sin nx \, dx \\
&= \sqrt{\frac{2}{\pi}} [f(x) \sin nx]_0^\pi - n \sqrt{\frac{2}{\pi}} \int_0^\pi f(x) \cos nx \, dx \\
&= -n C_n[f] \\
C_n[f'] &= \sqrt{\frac{2}{\pi}} \int_0^\pi f'(x) \cos nx \, dx \\
&= \sqrt{\frac{2}{\pi}} [f(x) \cos nx]_0^\pi + n \sqrt{\frac{2}{\pi}} \int_0^\pi f(x) \sin nx \, dx \\
&= \sqrt{\frac{2}{\pi}} (-1)^n f(\pi) - \sqrt{\frac{2}{\pi}} f(0) + n S_n[f]
\end{aligned}$$

The finite Fourier sine transform is ideally suited for solving for the vibrations of a finite string.   Consider the following boundary-value problem:

$$\frac{\partial^2 \phi}{\partial x^2} = \frac{1}{a^2} \frac{\partial^2 \phi}{\partial t^2} \qquad 0 \le x \le \pi \qquad 0 < t$$

$$\phi(0,t) = \phi(\pi,t) = 0$$

$$\phi(x,0) = f(x)$$

$$\phi_t(x,0) = g(x)$$

Let

$$S_n(t) = \sqrt{\frac{2}{\pi}} \int_0^\pi \phi(x,t) \sin nx \, dx$$

Then

$$-n^2 S_n(t) = \frac{1}{a^2} \ddot{S}_n(t)$$

$$S_n(t) = A_n \cos ant + B_n \sin ant$$

$$S_n(0) = F_n = A_n$$

$$\dot{S}_n(0) = G_n = anB_n$$

where

$$F_n = \sqrt{\frac{2}{\pi}} \int_0^\pi f(x) \sin nx \, dx \qquad G_n = \sqrt{\frac{2}{\pi}} \int_0^\pi g(x) \sin nx \, dx$$

Inverting, we have

$$\phi(x,t) = \sqrt{\frac{2}{\pi}} \sum_{n=1}^\infty \left(F_n \cos ant + \frac{G_n}{an} \sin ant \right) \sin nx$$

Next consider the following ordinary differential equation with constant coefficients:

$$ay''(x) + by'(x) + cy(x) = f(x)$$

Let $f(x)$ be periodic with period 2π; that is, $f(x + 2\pi) = f(x)$.   We look for a periodic solution of the differential equation.   This will not in general be the complete solution, but the general solution of the homogeneous equation can be added to it, giving the complete solution.   First we write

$$y(x) = \tfrac{1}{2}[y(x) + y(-x)] + \tfrac{1}{2}[y(x) - y(-x)]$$

$$= y_e(x) + y_o(x)$$

where $y_e(x)$ is even and $y_o(x)$ is odd.   We shall assume that $y(x)$ is continuous. This implies that $y_o(0) = y_o(\pi) = 0$.   The differential equation becomes

$$ay_e''(x) + ay_o''(x) + by_e'(x) + by_o'(x) + cy_e(x) + cy_o(x) = f_e(x) + f_o(x)$$

where we have similarly decomposed $f(x)$.   We first take the sine transform of the equation, and then the cosine transform.

$$-an^2S_n[y_o] - nbC_n[y_e] + cS_n[y_o] = S_n[f_o]$$
$$-an^2C_n[y_e] + nbS_n[y_o] + cC_n[y_e] = C_n[f_e]$$

Solving for $S_n[y_o]$ and $C_n[y_e]$, we have

$$S_n[y_o] = \frac{(c - an^2)S_n[f_o] + nbC_n[f_e]}{(c - an^2)^2 + n^2b^2}$$

$$C_n[y_e] = \frac{(c - an^2)C_n[f_e] - nbS_n[f_o]}{(c - an^2)^2 + n^2b^2}$$

Then

$$y(x) = y_e(x) + y_o(x) = \frac{1}{\sqrt{2\pi}} C_0 + \sqrt{\frac{2}{\pi}} \sum_{n=1}^{\infty} C_n \cos nx + \sqrt{\frac{2}{\pi}} \sum_{n=1}^{\infty} S_n \sin nx$$

The nature of the convergence of the series and the existence of the first and second derivatives can be determined when $f(x)$ is given.

The **Hankel transform** of a function $f(x)$ is defined by

$$H[f] = \int_0^\infty xf(x)J_\nu(tx)\,dx \qquad \nu \ge -\tfrac{1}{2}$$

It exists if $f(x)$ is piecewise continuous in any finite interval and absolutely integrable. If, in addition, $f'(x)$ is piecewise continuous in any finite interval we have the inversion integral[1]

$$\tfrac{1}{2}[f(x+) + f(x-)] = \int_0^\infty tH(t)J_\nu(tx)\,dt$$

Consider the following ordinary differential equation:

$$x\frac{d^2y}{dx^2} + \frac{dy}{dx} - \frac{\nu^2}{x}y = f(x)$$

Assume that $f(x)$ is absolutely integrable on $0 \le x < \infty$, that $y(0)$ and $y'(0)$ are finite, and that $y(x) = O(1/x)$ as $x \to \infty$. We obtain a formal solution by the use of the Hankel transform.

$$F(t) = \int_0^\infty f(x)J_\nu(tx)\,dx = \int_0^\infty J_\nu(tx)\left[\frac{d}{dx}\left(x\frac{dy}{dx}\right) - \frac{\nu^2}{x}y\right]dx$$

$$= [xJ_\nu(tx)y']_0^\infty - \int_0^\infty \left(xy'J_\nu' + \frac{\nu^2}{x}yJ_\nu\right)dx$$

$$= -[xyJ_\nu'(tx)]_0^\infty + \int_0^\infty y\left[(xJ_\nu')' - \frac{\nu^2}{x}J_\nu\right]dx$$

$$= -t^2\int_0^\infty xy(x)J_\nu(tx)\,dx = -t^2H[y]$$

[1] Ian N. Sneddon, "Fourier Transforms," McGraw-Hill Book Company, New York, 1951.

Hence, $H[y] = -F(t)/t^2$ and

$$y(x) = -\int_0^\infty \frac{F(t)J_\nu(xt)}{t}\,dt$$

The **Mellin transform** of a function $f(x)$ is defined by

$$M[f] = \int_0^\infty x^{z-1}f(x)\,dx = \phi(z)$$

Suppose $f(x)$ is piecewise continuous in any finite interval and

$$\int_0^\infty x^{\xi-1}\,|f(x)|\,dx$$

exists, where $\xi = \mathrm{Re}\,(z)$. This will be true in general in some strip

$$0 < \mathrm{Re}\,(z) < \beta$$

Then $\phi(z)$ will be an analytic function of z in this strip. Consider

$$\frac{1}{2\pi i}\int_{\gamma-i\infty}^{\gamma+i\infty} x^{-z}\phi(z)\,dz$$

with $0 < \gamma < \beta$. Let $z = \xi + i\eta$ and $x = e^t$. Then

$$\frac{1}{2\pi i}\int_{\gamma-i\infty}^{\gamma+i\infty} x^{-z}\phi(z)\,dz = \frac{1}{2\pi}\int_{-\infty}^\infty e^{-t(\gamma+i\eta)}\int_{-\infty}^\infty e^{\tau(\gamma+i\eta)}f(e^\tau)\,d\tau\,d\eta$$

$$= \frac{e^{-t\gamma}}{2\pi}\int_{-\infty}^\infty e^{-it\eta}\int_{-\infty}^\infty e^{i\tau\eta}e^{\gamma\tau}f(e^\tau)\,d\tau\,d\eta$$

$$= e^{-t\gamma}e^{t\gamma}\frac{f(e^{t+}) + f(e^{t-})}{2} = \tfrac{1}{2}[f(x+) + f(x-)]$$

provided $e^{\gamma t}f(e^t)$ satisfies the conditions of the Fourier integral theorem; that is,

$$\int_{-\infty}^\infty e^{\gamma t}|f(e^t)|\,dt = \int_0^\infty x^{\gamma-1}|f(x)|\,dx$$

exists and $f(x)$ and $f'(x)$ are piecewise continuous in any finite interval. The convergence of the integral is guaranteed, since $0 < \gamma < \beta$. We have thus obtained the inversion integral for the Mellin transform.

As an application of the Mellin transform, consider the following boundary-value problem:

$$\nabla^2\phi = 0 \qquad z = 0 \qquad -\infty < x < \infty \qquad 0 < y$$
$$\phi = f(|x|) \qquad y = 0$$
$$\phi = O\!\left(\frac{1}{\sqrt{r}}\right) \qquad \text{as } r \to \infty$$

In polar coordinates this problem becomes

$$\frac{\partial^2 \phi}{\partial r^2} + \frac{1}{r}\frac{\partial \phi}{\partial r} + \frac{1}{r^2}\frac{\partial^2 \phi}{\partial \theta^2} = 0 \qquad 0 < \theta < \pi \qquad 0 < r$$

$$\phi(r,0) = \phi(r,\pi) = f(r)$$

$$\phi = O\left(\frac{1}{\sqrt{r}}\right) \qquad \text{as } r \to \infty$$

We take the Mellin transform with respect to the variable r.   Let

$$M(z,\theta) = \int_0^\infty r^{z-1}\phi(r,\theta)\,dr$$

$$\frac{\partial^2 M}{\partial \theta^2} = \int_0^\infty r^{z-1}\frac{\partial^2 \phi}{\partial \theta^2}\,dr$$

$$= -\int_0^\infty r^z \frac{\partial}{\partial r}\left(r\frac{\partial \phi}{\partial r}\right)dr$$

$$= -\left[r^{z+1}\frac{\partial \phi}{\partial r}\right]_0^\infty + z\int_0^\infty r^z \frac{\partial \phi}{\partial r}\,dr$$

$$= [zr^z\phi]_0^\infty - z^2\int_0^\infty r^{z-1}\phi\,dr$$

$$= -z^2 M$$

Let $$\qquad F(z) = \int_0^\infty r^{z-1}f(r)\,dr = M(z,0) = M(z,\pi)$$

Solving and using the boundary conditions, we have

$$M(z,\theta) = A\cos z\theta + B\sin z\theta$$

$$F(z) = A$$

$$F(z) = A\cos \pi z + B\sin \pi z$$

$$F(z)(1 - \cos \pi z) = B\sin \pi z$$

$$2\sin^2 \frac{\pi z}{2}\,F(z) = 2B\cos \frac{\pi z}{2}\sin \frac{\pi z}{2}$$

$$B = F(z)\tan \frac{\pi z}{2}$$

$$M(z,\theta) = F(z)\frac{\cos z(\theta - \pi/2)}{\cos (\pi z/2)}$$

Inverting, we have

$$\phi(r,\theta) = \frac{1}{2\pi i}\int_{\gamma-i\infty}^{\gamma+i\infty} r^{-z}\frac{F(z)\cos z(\theta - \pi/2)}{\cos (\pi z/2)}\,dz$$

where $0 < \gamma < \frac{1}{2}$.

We could give other examples of integral transform techniques, but there seems to be little point in doing so.   By now the reader should be familiar with the general aspects of integral transform methods.   For a variety of other applications he may wish to refer to the book by Sneddon already cited.

Exercises 8.6

1. Solve the following boundary-value problem using the finite Fourier cosine transform:

$$\frac{\partial^2 \phi}{\partial x^2} = \frac{1}{a^2} \frac{\partial^2 \phi}{\partial t^2} \qquad 0 < x < \pi \qquad 0 < t$$

$$\phi_x(0,t) = \phi_x(\pi,t) = 0$$

$$\phi(x,0) = f(x)$$

$$\phi_t(x,0) = 0$$

What transform could be used if the interval is changed to $0 < x < L$?

2. Define the **finite Fourier transform** as

$$F_n[f] = \frac{1}{\sqrt{2\pi}} \int_{-\pi}^{\pi} f(x)e^{-inx}\,dx$$

State an inversion theorem giving sufficient conditions on $f(x)$ for its validity.

3. Solve the following boundary-value problem using the Hankel transform and cylindrical coordinates:

$$\nabla^2 \phi = 0 \qquad 0 < r \qquad 0 \le \theta \le 2\pi \qquad 0 < z$$

$$\phi(r,\theta,0) = f(r)$$

$$\phi = O\left(\frac{1}{r}\right) \qquad \text{as } r \to \infty$$

References

Campbell, George A., and Ronald M. Foster: "Fourier Integrals for Practical Applications," D. Van Nostrand Company, Inc., Princeton, N.J., 1948.

Churchill, Ruel V.: "Operational Mathematics," 2nd ed., McGraw-Hill Book Company, New York, 1958.

Dettman, John W.: "Applied Complex Variables," Dover Publications, Inc., New York, 1984.

Noble, Benjamin: "The Wiener-Hopf Technique," Pergamon Press, Inc., New York, 1958.

Sneddon, Ian N.: "Fourier Transforms," McGraw-Hill Book Company, New York, 1951.

Titchmarsh, E. C.: "The Theory of Functions," 2d ed., Oxford University Press, London, 1939.

Widder, David V.: "The Laplace Transform," Princeton University Press, Princeton, N.J., 1941.

Wiener, Norbert: "The Fourier Integral," Cambridge University Press, London, 1933.

Wolf, B. W.: "Integral Transforms in Science and Engineering," Plenum Press, New York, 1979.

Index

Index

A CATALOG OF SELECTED

DOVER BOOKS
IN SCIENCE AND MATHEMATICS

A CATALOG OF SELECTED
DOVER BOOKS
IN SCIENCE AND MATHEMATICS

QUALITATIVE THEORY OF DIFFERENTIAL EQUATIONS, V.V. Nemytskii and V.V. Stepanov. Classic graduate-level text by two prominent Soviet mathematicians covers classical differential equations as well as topological dynamics and ergodic theory. Bibliographies. 523pp. 5⅜ × 8½. 65954-2 Pa. $10.95

MATRICES AND LINEAR ALGEBRA, Hans Schneider and George Phillip Barker. Basic textbook covers theory of matrices and its applications to systems of linear equations and related topics such as determinants, eigenvalues and differential equations. Numerous exercises. 432pp. 5⅜ × 8½. 66014-1 Pa. $10.95

QUANTUM THEORY, David Bohm. This advanced undergraduate-level text presents the quantum theory in terms of qualitative and imaginative concepts, followed by specific applications worked out in mathematical detail. Preface. Index. 655pp. 5⅜ × 8½. 65969-0 Pa. $13.95

ATOMIC PHYSICS (8th edition), Max Born. Nobel laureate's lucid treatment of kinetic theory of gases, elementary particles, nuclear atom, wave-corpuscles, atomic structure and spectral lines, much more. Over 40 appendices, bibliography. 495pp. 5⅜ × 8½. 65984-4 Pa. $12.95

ELECTRONIC STRUCTURE AND THE PROPERTIES OF SOLIDS: The Physics of the Chemical Bond, Walter A. Harrison. Innovative text offers basic understanding of the electronic structure of covalent and ionic solids, simple metals, transition metals and their compounds. Problems. 1980 edition. 582pp. 6⅛ × 9¼. 66021-4 Pa. $15.95

BOUNDARY VALUE PROBLEMS OF HEAT CONDUCTION, M. Necati Özisik. Systematic, comprehensive treatment of modern mathematical methods of solving problems in heat conduction and diffusion. Numerous examples and problems. Selected references. Appendices. 505pp. 5⅜ × 8½. 65990-9 Pa. $12.95

A SHORT HISTORY OF CHEMISTRY (3rd edition), J.R. Partington. Classic exposition explores origins of chemistry, alchemy, early medical chemistry, nature of atmosphere, theory of valency, laws and structure of atomic theory, much more. 428pp. 5⅜ × 8½. (Available in U.S. only) 65977-1 Pa. $10.95

A HISTORY OF ASTRONOMY, A. Pannekoek. Well-balanced, carefully reasoned study covers such topics as Ptolemaic theory, work of Copernicus, Kepler, Newton, Eddington's work on stars, much more. Illustrated. References. 521pp. 5⅜ × 8½. 65994-1 Pa. $12.95

PRINCIPLES OF METEOROLOGICAL ANALYSIS, Walter J. Saucier. Highly respected, abundantly illustrated classic reviews atmospheric variables, hydrostatics, static stability, various analyses (scalar, cross-section, isobaric, isentropic, more). For intermediate meteorology students. 454pp. 6⅛ × 9¼. 65979-8 Pa. $14.95

THE FOUR-COLOR PROBLEM: Assaults and Conquest, Thomas L. Saaty and Paul G. Kainen. Engrossing, comprehensive account of the century-old combinatorial topological problem, its history and solution. Bibliographies. Index. 110 figures. 228pp. 5⅜ × 8½. 65092-8 Pa. $6.95

CATALYSIS IN CHEMISTRY AND ENZYMOLOGY, William P. Jencks. Exceptionally clear coverage of mechanisms for catalysis, forces in aqueous solution, carbonyl- and acyl-group reactions, practical kinetics, more. 864pp. 5⅜ × 8½. 65460-5 Pa. $19.95

PROBABILITY: An Introduction, Samuel Goldberg. Excellent basic text covers set theory, probability theory for finite sample spaces, binomial theorem, much more. 360 problems. Bibliographies. 322pp. 5⅜ × 8½. 65252-1 Pa. $8.95

LIGHTNING, Martin A. Uman. Revised, updated edition of classic work on the physics of lightning. Phenomena, terminology, measurement, photography, spectroscopy, thunder, more. Reviews recent research. Bibliography. Indices. 320pp. 5⅜ × 8¼. 64575-4 Pa. $8.95

PROBABILITY THEORY: A Concise Course, Y.A. Rozanov. Highly readable, self-contained introduction covers combination of events, dependent events, Bernoulli trials, etc. Translation by Richard Silverman. 148pp. 5⅜ × 8¼.
63544-9 Pa. $5.95

AN INTRODUCTION TO HAMILTONIAN OPTICS, H. A. Buchdahl. Detailed account of the Hamiltonian treatment of aberration theory in geometrical optics. Many classes of optical systems defined in terms of the symmetries they possess. Problems with detailed solutions. 1970 edition. xv + 360pp. 5⅜ × 8½.
67597-1 Pa. $10.95

STATISTICS MANUAL, Edwin L. Crow, et al. Comprehensive, practical collection of classical and modern methods prepared by U.S. Naval Ordnance Test Station. Stress on use. Basics of statistics assumed. 288pp. 5⅜ × 8½.
60599-X Pa. $6.95

DICTIONARY/OUTLINE OF BASIC STATISTICS, John E. Freund and Frank J. Williams. A clear concise dictionary of over 1,000 statistical terms and an outline of statistical formulas covering probability, nonparametric tests, much more. 208pp. 5⅜ × 8½. 66796-0 Pa. $6.95

STATISTICAL METHOD FROM THE VIEWPOINT OF QUALITY CONTROL, Walter A. Shewhart. Important text explains regulation of variables, uses of statistical control to achieve quality control in industry, agriculture, other areas. 192pp. 5⅜ × 8½. 65232-7 Pa. $7.95

THE INTERPRETATION OF GEOLOGICAL PHASE DIAGRAMS, Ernest G. Ehlers. Clear, concise text emphasizes diagrams of systems under fluid or containing pressure; also coverage of complex binary systems, hydrothermal melting, more. 288pp. 6½ × 9¼. 65389-7 Pa. $10.95

STATISTICAL ADJUSTMENT OF DATA, W. Edwards Deming. Introduction to basic concepts of statistics, curve fitting, least squares solution, conditions without parameter, conditions containing parameters. 26 exercises worked out. 271pp. 5⅜ × 8½. 64685-8 Pa. $8.95

TENSOR CALCULUS, J.L. Synge and A. Schild. Widely used introductory text covers spaces and tensors, basic operations in Riemannian space, non-Riemannian spaces, etc. 324pp. 5⅜ × 8¼.   63612-7 Pa. $8.95

A CONCISE HISTORY OF MATHEMATICS, Dirk J. Struik. The best brief history of mathematics. Stresses origins and covers every major figure from ancient Near East to 19th century. 41 illustrations. 195pp. 5⅜ × 8½.   60255-9 Pa. $7.95

A SHORT ACCOUNT OF THE HISTORY OF MATHEMATICS, W.W. Rouse Ball. One of clearest, most authoritative surveys from the Egyptians and Phoenicians through 19th-century figures such as Grassman, Galois, Riemann. Fourth edition. 522pp. 5⅜ × 8½.   20630-0 Pa. $10.95

HISTORY OF MATHEMATICS, David E. Smith. Nontechnical survey from ancient Greece and Orient to late 19th century; evolution of arithmetic, geometry, trigonometry, calculating devices, algebra, the calculus. 362 illustrations. 1,355pp. 5⅜ × 8½.   20429-4, 20430-8 Pa., Two-vol. set $23.90

THE GEOMETRY OF RENÉ DESCARTES, René Descartes. The great work founded analytical geometry. Original French text, Descartes' own diagrams, together with definitive Smith-Latham translation. 244pp. 5⅜ × 8½.
                                           60068-8 Pa. $7.95

THE ORIGINS OF THE INFINITESIMAL CALCULUS, Margaret E. Baron. Only fully detailed and documented account of crucial discipline: origins; development by Galileo, Kepler, Cavalieri; contributions of Newton, Leibniz, more. 304pp. 5⅜ × 8½. (Available in U.S. and Canada only)   65371-4 Pa. $9.95

THE HISTORY OF THE CALCULUS AND ITS CONCEPTUAL DEVELOPMENT, Carl B. Boyer. Origins in antiquity, medieval contributions, work of Newton, Leibniz, rigorous formulation. Treatment is verbal. 346pp. 5⅜ × 8½.
                                         60509-4 Pa. $8.95

THE THIRTEEN BOOKS OF EUCLID'S ELEMENTS, translated with introduction and commentary by Sir Thomas L. Heath. Definitive edition. Textual and linguistic notes, mathematical analysis. 2,500 years of critical commentary. Not abridged. 1,414pp. 5⅜ × 8½.   60088-2, 60089-0, 60090-4 Pa., Three-vol. set $29.85

GAMES AND DECISIONS: Introduction and Critical Survey, R. Duncan Luce and Howard Raiffa. Superb nontechnical introduction to game theory, primarily applied to social sciences. Utility theory, zero-sum games, n-person games, decision-making, much more. Bibliography. 509pp. 5⅜ × 8½.   65943-7 Pa. $12.95

THE HISTORICAL ROOTS OF ELEMENTARY MATHEMATICS, Lucas N.H. Bunt, Phillip S. Jones, and Jack D. Bedient. Fundamental underpinnings of modern arithmetic, algebra, geometry and number systems derived from ancient civilizations. 320pp. 5⅜ × 8½.   25563-8 Pa. $8.95

CALCULUS REFRESHER FOR TECHNICAL PEOPLE, A. Albert Klaf. Covers important aspects of integral and differential calculus via 756 questions. 566 problems, most answered. 431pp. 5⅜ × 8½.   20370-0 Pa. $8.95

CATALOG OF DOVER BOOKS

CHALLENGING MATHEMATICAL PROBLEMS WITH ELEMENTARY SOLUTIONS, A.M. Yaglom and I.M. Yaglom. Over 170 challenging problems on probability theory, combinatorial analysis, points and lines, topology, convex polygons, many other topics. Solutions. Total of 445pp. 5⅜ × 8½. Two-vol. set.
Vol. I 65536-9 Pa $7.95
Vol. II 65537-7 Pa. $6.95

FIFTY CHALLENGING PROBLEMS IN PROBABILITY WITH SOLUTIONS, Frederick Mosteller. Remarkable puzzlers, graded in difficulty, illustrate elementary and advanced aspects of probability. Detailed solutions. 88pp. 5⅜ × 8½.
65355-2 Pa. $4.95

EXPERIMENTS IN TOPOLOGY, Stephen Barr. Classic, lively explanation of one of the byways of mathematics. Klein bottles, Moebius strips, projective planes, map coloring, problem of the Koenigsberg bridges, much more, described with clarity and wit. 43 figures. 210pp. 5⅜ × 8½.
25933-1 Pa. $5.95

RELATIVITY IN ILLUSTRATIONS, Jacob T. Schwartz. Clear nontechnical treatment makes relativity more accessible than ever before. Over 60 drawings illustrate concepts more clearly than text alone. Only high school geometry needed. Bibliography. 128pp. 6⅛ × 9¼.
25965-X Pa. $6.95

AN INTRODUCTION TO ORDINARY DIFFERENTIAL EQUATIONS, Earl A. Coddington. A thorough and systematic first course in elementary differential equations for undergraduates in mathematics and science, with many exercises and problems (with answers). Index. 304pp. 5⅜ × 8½.
65942-9 Pa. $8.95

FOURIER SERIES AND ORTHOGONAL FUNCTIONS, Harry F. Davis. An incisive text combining theory and practical example to introduce Fourier series, orthogonal functions and applications of the Fourier method to boundary-value problems. 570 exercises. Answers and notes. 416pp. 5⅜ × 8½.
65973-9 Pa. $9.95

THE THEORY OF BRANCHING PROCESSES, Theodore E. Harris. First systematic, comprehensive treatment of branching (i.e. multiplicative) processes and their applications. Galton-Watson model, Markov branching processes, electron-photon cascade, many other topics. Rigorous proofs. Bibliography. 240pp. 5⅜ × 8½.
65952-6 Pa. $6.95

AN INTRODUCTION TO ALGEBRAIC STRUCTURES, Joseph Landin. Superb self-contained text covers "abstract algebra": sets and numbers, theory of groups, theory of rings, much more. Numerous well-chosen examples, exercises. 247pp. 5⅜ × 8½.
65940-2 Pa. $7.95

Prices subject to change without notice.
Available at your book dealer or write for free Mathematics and Science Catalog to Dept. GI, Dover Publications, Inc., 31 East 2nd St., Mineola, N.Y. 11501. Dover publishes more than 175 books each year on science, elementary and advanced mathematics, biology, music, art, literature, history, social sciences and other areas.